The Mathematics of Movement

Luca Giuggioli • Philip K. Maini
Editors

The Mathematics of Movement

An Interdisciplinary Approach to Mutual Challenges in Animal Ecology and Cell Biology

Springer

Editors
Luca Giuggioli
School of Engineering Mathematics
and Technology
University of Bristol
Bristol, UK

Philip K. Maini
Mathematical Institute
University of Oxford
Oxford, UK

ISBN 978-3-032-00872-5 ISBN 978-3-032-00873-2 (eBook)
https://doi.org/10.1007/978-3-032-00873-2

This Springer imprint is published by the registered company Springer Nature Switzerland AG
The registered company address is: Gewerbestrasse 11, 6330 Cham, Switzerland

'E pur si muove'
(And yet it moves)
Galileo Galilei

Foreword of Simon Levin

Theoretical approaches in population biology and ecology generally are traced back to fundamental work more than a century ago by Alfred Lotka and Vito Volterra, who independently developed equations to describe the interactions between well-mixed populations of competitors, and of predators and prey; but nature is more complicated than that, and it is the spatial heterogeneity of environments, much of it endogenously generated, that allows for the rich diversity that we see in the world. Among the key issues that arise are the conditions under which spatial heterogeneity can facilitate coexistence that would be impossible in homogeneous environments, the potential for self-organized spatial pattern, the role of geometry in the maintenance of diversity and implications for reserve design, and the description of the temporal dynamics of spatial spread of novel or invading species or genotypes. To begin to address these challenges, ecology borrowed freely from other disciplines, which had developed methods to address these challenges, and this cross-fertilization is in part what inspired this rich volume.

The confluence of mathematical methods that are the foundation of the mathematics of movement reflect the unity of biology, and indeed of science more broadly, in terms of the questions and challenges that arise. Robert Brown's observations of the random motions of pollen grains led to early mathematical descriptions by Louis Bachelier, and ultimately to the elegant analyses of Albert Einstein and Marian Smoluchowski in developing diffusion approximations of random walks. In population genetics, Ronald Fisher used such descriptions to characterize the rates of advance of novel alleles into new environments, leading to the sophisticated mathematical analyses of Andrey Kolmogorov, Ivan Petrovsky and Nikolai Piskunov, the experimental tests of the theory by Theodosius Dobzhansky and Sewall Wright, and JBS Haldane's consideration of clines. Much later, Alan Turing, better known for his work in computer science and crypto-analysis, became fascinated by the self-organizing development of the embryo, in particular how spatially complex structures like you and me could arise from an initially homogeneous egg; and his description of pattern formation via diffusive instabilities has inspired work across many disciplines, including ecology.

The incorporation of these methods into ecology proper can be traced to the 1951 paper of JG Skellam, who was interested in the rates of spread of species following the demise of glaciation, and the problem of the critical patch size of species that could support persistent populations. Similar issues arose in oceanography, in accounting for the patchy distribution of microorganisms in the ocean. I first confronted these problems in my own research in the early 1970s, inspired by all of the examples described above, in trying to understand how species escaped from the limitations of the competitive exclusion principle. I profited greatly from my close collaborators like Lee Segel and Akira Okubo, who were pioneers, and whose work is also foundational; by the elegant writings of Jim Murray and others in studying a variety of issues, including animal coat patterns; and by the organizers of the Newton program that led to this book, two of whom spent their postdoctoral years in my research group. This wonderful volume brings together in one place the theoretical and empirical, the mechanistic and the phenomenological, old and new methods in this rapidly growing field. The volume is a major contribution to the field.

Princeton University, Princeton, NJ, USA Simon Levin
April 2025

Simon Levin is the James S. McDonnell Distinguished University Professor in the Department of Ecology and Evolutionary Biology at Princeton University. He is a Fellow of the AAAS, and the American Academy of Arts and Sciences, and a Member of the National Academy of Sciences and the American Philosophical Society, and the recipient, among various prizes and honors, of the Robert H. MacArthur Award, the Okubo Prize for Mathematical Biology, the Heineken Prize for Environmental Sciences, the Kyoto Prize for Basic Sciences, the Tyler Prize for Environmental Achievement, the Margalef Prize in Ecology, the BBVA Frontiers of Science Award for Ecology and Conservation Biology, and the National Medal of Science.

Foreword of Ralf Metzler

Ronald Ross, in his 1905 treatise on how malaria is being spread by mosquitoes, formulated what we now call a random walk. Building on his exchange with Ross, the statistician Karl Pearson coined the concept of the "drunkards walk". At the same time Einstein and Smoluchowski published their probabilistic descriptions of Brownian motion. Today random walks and diffusion are routinely used to describe the spreading dynamics of "tracers", ranging from fluorescently labelled single molecules in living biological cells to macroscopic animal motion in movement ecology.

The beauty of the concept of probability and diffusion is that scales do not matter. The diffusive spreading of a sugar molecule in an (unstirred) cup of tea or the random exploration of mosquitoes in the rain forest underlies the same mathematical principles as the walk of Pearson's drunkards in a city. In the simplest case, a diffusive process is characterised by a single parameter, the diffusion coefficient. It can be measured in squared micrometres per day in groundwater contexts, or in squared kilometres per hour for flying birds. This property has been central for the wide applicability of these approaches. Nowadays, with the possibility to measure longer trajectories with ever better resolution, deviations of the spreading dynamics from the laws of Brownian motion with its linear-in-time mean-squared displacement and Gaussian displacement distribution are being found. Instead, anomalous diffusion, with a power-law time dependence of the mean-squared displacement, has been invoked. Starting with earlier works on using generalised random walk concepts such as Lévy walks, today refined anomalous stochastic processes find wide application in movement studies—they represent important modelling approaches next to the famous non-linear dynamic equations such as the Fisher-KPP or Lotka-Volterra equations associated with population statistics.

New concepts such as individual-based modelling, across-species interactions, or spatio-temporally heterogeneous parameters are being developed and applied to the statistical description in biophysical systems or in movement studies. In parallel, new diffusion or reaction-diffusion models; theories for first-passage statistics or the number of distinct sites visited; stochastic resetting concepts; collective motion; social interactions; or the mortality of walks are being introduced. All of these aspects eventually are found together in a unified mathematical modelling framework, the "Mathematics of Movement", as coined by the editors. Collecting such new ideas is at the heart of this beautiful book, bridging the language of mathematicians, physicists, ecologists, and cell biologists. The well-written chapters will be useful in the work of experimental and theoretical scientists alike, but they will also be a good guide for students starting in this field.

As the field is developing rapidly, Giuggioli and Maini's book will contribute to the importance of understanding movement and its relation to the large number of complex processes in ecology and cell biology. It could not have arrived at a better time.

University of Potsdam, Potsdam, Germany Ralf Metzler
May 2025

Ralf Metzler is Chair Professor for Theoretical Physics at the University of Potsdam, Germany. He was a Feodor Lynen Fellow of the Alexander von Humboldt Foundation, an Amos de Shalit Fellow of the Minerva Foundation, and a DFG Emmy Noether Fellow. He was a Canada Research Chair in Biological Physics and a Finland Distinguished Professor (Academy of Finland). For his work he received the 2017 SigmaPhi Prize "for outstanding achievements in Statistical Physics" and was awarded an Alexander von Humboldt Polish Honorary Research Scholarship as well as a Heilbronn Distinguished Fellowship. He was also Distinguished Fellow at the Asian Pacific Centre for Theoretical Physics and Distinguished Visiting Professor at the Indian Institute of Technology Bombay.

Preface

Quantifying and predicting the responses of biological entities to disturbances is key to the understanding of a vast array of processes occurring at all scales, from molecules and bacteria inside a microorganism, cells in a tissue, to humans and animals within an ecosystem. A natural metric used to infer such responses to perturbations is movement, which modern technology allows us to capture in great detail within cells, organs, tissues, and the environment. As the new generation of experimental probes and sensors is providing us with movement data of increasing resolution over extended periods of time and across multiple individuals simultaneously, we face the challenge of developing and integrating conceptual frameworks and methodologies to analyse and model the spatio-temporal dynamics of moving entities in a range of biological disciplines and beyond.

Given the ever-increasing level of sophistication required to interpret empirical observations, such challenges are customarily met by a continuous refinement and specialisation of the technical expertise within a discipline. But such advances, while certainly essential, bring with them the risk of siloed thinking, limiting the potential broader impact of knowledge transfer across disciplines. The set of mathematical tools to study movement is a paradigmatic example where such broader impact may be achieved, whereby progress from one area can be imported to the benefit of another area in a seemingly totally disconnected field. The reason that such possibilities arise is because the mathematical tools used to quantify movement processes, and their causes and consequences, constitute a language which, by its abstraction and generality, allows individuals from very distant disciplines to converse with one another. As illustrations of the similarities between applications in animal ecology and cell biology, one can think of the lessons learned in identifying the rules governing the spread of an infectious disease in an animal population that have guided quantitative studies on the growth of cancerous cells within a tissue, or the principles underlying cell chemotaxis that have helped quantify how environmental features influence animal migration. But cross-fertilisation has also occurred between biology and the engineering disciplines, e.g. the phenomenological laws explaining the collective murmuration of starlings and other birds have shaped the development of some of the path planning algorithms for robot swarms.

This book aims to disseminate the use of such a language, striving to facilitate a proliferation of techniques to quantify movement and enhance the reciprocal influence that seemingly unrelated fields of research may exert on each other. We thought it appropriate to coin the name *Mathematics of Movement* for such a language. It comprises the set of concepts, tools, and techniques that have been used across scientific domains to study movement in all its shapes and forms. Stochastic differential equations, partial differential equations, integro-differential equations, network theory, lattice and Brownian random walks, Monte Carlo methods, Bayesian statistics, and topological data analysis are all examples of quantitative approaches that form the backbone of the mathematics routinely used in the analysis and modelling of the movement of artificial objects and biological organisms. This book presents examples of some of these approaches with a focus on their application to animal ecology and cell biology.

It is within this philosophy of bridging disciplinary gaps that the idea of the book emerged. It came about following the 6-month research programme entitled "Mathematics of movement: an interdisciplinary approach to mutual challenges in animal ecology and cell biology", held in the second half of 2023 at the Isaac Newton Institute for Mathematical Sciences (INI) in Cambridge. The programme was organised and run by the present book editors in conjunction with Dr Guillermo Abramson from the Centro Atómico Bariloche and Instituto Balseiro, Bariloche, Argentina; Prof. Helen Byrne from the Mathematical Institute, University of Oxford, UK; Prof. Paul Kulesa from the Department of Biological Sciences, University of Notre Dame, USA; and Prof. Ran Nathan from the Department of Ecology, Evolution and Behaviour at the Hebrew University of Jerusalem, Israel. The programme saw the participation of around 200 scientists at different career stages, the majority of them being theoreticians but also with some experimentalists, from 24 countries across 5 continents for periods ranging from a few days to the entire duration of the programme. Participants came to present their research and discuss with other like-minded individuals interested in learning about other topics. It was precisely this strong interdisciplinary attitude that made the cross-fertilisation successful as evidenced by a survey conducted on 50 participants who overall indicated, as an outcome of their engagement with the programme, that they had initiated 44 new collaborations with scientists outside their own field.

As the book is the result of the many interactions that occurred at the INI, we first would like to thank all the programme participants for being the main source of inspiration. Our gratitude obviously does not stop here. As is customary with large endeavours, there are many other individuals and entities to thank starting from all the contributing authors, the referees who have helped maintain the high scientific standard of each chapter, the INI programme co-organisers, the administrative team at INI, the various colleagues that have encouraged us to complete the project, and the funding sources that supported the INI programme, which include the Engineering and Physical Sciences Research Council, the Development Assistance Committee, the Leverhulme Trust, the London Mathematical Society, the Newton Gateway, the Tuner Kirk Trust, the Rothschild Foundation, the Heilbronn Institute for Mathematical Research, and the University of Cambridge.

Finally, we conclude with a wish. And that is that the many scientists who already model and analyse movement phenomena across all application areas interact frequently, keep integrating their knowledge, and ultimately build a thriving interdisciplinary community, inching ever closer in making the mathematics of movement, for all intents and purposes, an interdisciplinary field of its own, advancing mathematics and science in tandem.

In memory of Paul Kulesa. On 15th March, 2025, our dear friend, colleague, and program co-organiser, Paul Kulesa passed away. Paul was a doctoral student of Professor James D. Murray FRS at the University of Washington, Seattle, where he received his doctorate in mathematical biology. Wishing to see model predictions being tested in the laboratory, he decided to become an experimentalist and trained under Professor Scott Fraser at the California Institute of Technology. He moved on to The Stowers Medical Center in Kansas to run his own laboratory, then to the University of Notre Dame, and most recently, to the Children's Mercy Hospital, Kansas City. Paul's work on neural crest has led to significant new findings in the field of collective cell movement, from the microscale genetic level right up to the macroscale tissue level. He was an excellent scientist who was always coming up with exciting, original ideas. He was warm, friendly, kind, highly supportive, and passionate about his science and had a wonderful sense of humour. He will be sorely missed by all who knew him. His work will live long into the future.

Bristol, UK — Luca Giuggioli
Oxford, UK — Philip K. Maini
March 2025

Contents

Chapter 1
A Cross-Disciplinary Way Forward in the Study of Movement

Luca Giuggioli and Philip K. Maini

1.1 Newton's Legacy

Newton's legacy has inspired generations of mathematicians and physicists who have dedicated their lives to the study of motion in all its shapes and forms. His accomplishments, together with those of other giants of yore that came after, have provided the foundation to advance the study of movement in classical mechanics, kinetic theory, partial differential equations, probability theory and stochastic calculus, to name but a few. As movement processes are ubiquitous, these areas of mathematics and theoretical physics have advanced due to an influx of problems from across the entire scientific spectrum creating a set of concepts, tools and techniques that may be collectively termed the mathematics of movement. While the external influx of problems has been instrumental in realising a high level of formal sophistication, it has created cultural and technical barriers between movement modellers working in different disciplines. Following the revolution in sensor technologies during the last few decades [1], these barriers have grown even further as the unprecedented resolution and amount of movement data [2, 3] have often pushed researchers to focus their efforts on tackling questions that are system or species specific.

To make technological advances valuable beyond the scientific areas from where they originated, we advocate for a more holistic framework with a common language that helps address general questions and contributes to new insights across the different types of movement processes. A fruitful first step to do this is to classify or

L. Giuggioli (✉)
School of Engineering Mathematics and Technology, University of Bristol, Bristol, UK
e-mail: luca.giuggioli@bristol.ac.uk

P. K. Maini
Mathematical Institute, University of Oxford, Oxford, UK
e-mail: philip.maini@maths.ox.ac.uk

L. Giuggioli, P. K. Maini (eds.), *The Mathematics of Movement*,
https://doi.org/10.1007/978-3-032-00873-2_1

categorise the modelling choices available to study movement. These classifications, even if broad, may help investigators understand the limits of their findings or realise when it is more profitable to change the type of modelling approach than to collect more data [4]. An essential classification is the scale of description, which customarily reduces to a binary choice between 'microscopic', i.e. at the level of the individual, and 'macroscopic' or coarse-grained, i.e. at the level of a group or an entire population, although multi-scale approaches to model biological systems are becoming increasingly important [5, 6]. The choice of the formalism being stochastic or deterministic is another important decision in the modelling paradigm to follow [7], even if it is often constrained by the numbers of moving agents involved, with random effects being negligible when population densities are high. These and other more detailed modelling choices, e.g. whether predictions are on transient dynamics or long-term dependencies and whether space is represented via continuous or discrete variables, i.e. a lattice or a network, bring about a set of concepts and techniques that are universal, whose regime of validity is well known, and can be applied independently of the species or environment where movement and interactions occur.

While many would agree on statements about the universality of mathematical tools, the specificity of the observations and the hard-to-measure interactions between biological organisms may cause some to hesitate considering of practical value the existence of underlying quantitative commonalities to describe movement. But this is precisely the reason why the applied disciplines historically have taken centre stage, not only in inspiring and motivating novel modelling approaches but also in influencing the development of new mathematics to quantify movement. Ascribing to a long-held philosophical belief in science [8], we consider that progress is best achieved by a dynamic dialogue between the unearthing of new concepts and the development of novel tools or refinement of old ones. In our context, a vehicle to ensure that such a dialogue occurs is via knowledge exchange, systematisation and integration of the ideas and techniques employed to analyse movement and interactions in animal ecology and cell biology. It was this strong belief that first inspired us, and our co-organisers, to conceive and run the Mathematics of Movement programme at the Isaac Newton Institute for Mathematical Sciences held in Cambridge during the second half of 2023 (see ref. [9] for a popular article on the Mathematics of Movement and interviews from three of the six programme co-organisers).

1.2 The Value of a Shared Approach

In highlighting the importance of bridging gaps and sharing knowledge between animal ecology and cell biology, we are drawn first to the common trajectory that the two fields have witnessed recently, namely, a leap forward in quality, types and number of movement data. In the last decades, as a result of empirical advances in animal tracking technologies [10, 11] and tissue imaging tools [12–14], the analysis and quantification of movement of organisms, and their environments, has undergone a significance upsurge.

In terms of data quality in animal studies, the ratio between the average speed of an organism and its size is now often larger than the sampling frequency of tracking data, making movement observations highly resolved in space and time. Across research areas experimental apparatuses have steadily improved. While laboratory experiments are linking movement decisions to neurological circuitry with the help of immersive volumetric virtual reality techniques [15], the increasing miniaturisation of tracking devices continues to expand the species whose movement can be recorded in the field, lowering the intrinsic limits dictated by the device body weight relative to that of the animal. In addition, the plethora of sensors [16, 17] measuring an animal's state, from its heart rate and body temperature to its locomotion and energetics, as well as the environmental information ranging from vegetation to chemical contents and climatic conditions [18], is creating new ways of thinking, analysis and modelling.

A recent milestone in ecological field studies to boost the number and accuracy of movement data has been the deployment of the ATLAS system [19–21], a high-throughput wildlife tracking system based on reverse-GPS (time-of-arrival) principles, which is capable of tracking multiple free-ranging wild individuals automatically and simultaneously, at high temporal frequency and spatial accuracy and low costs yielding up to 2.5 million data points (localisations) per day. This type of system plays a key role in uncovering how animals interact, not only with the environment but also with other conspecific or heterospecific individuals. While highly resolved GPS localisations have been instrumental in enabling us to infer interaction rules in various species of mammals [22, 23], the low cost, high resolution and high throughput of the ATLAS system is poised to provide the field counterpart of video tracking data in laboratory experiments, the latter traditionally employed to study the behaviour of insects [24, 25] and fish [26].

In cell biology, given the relative higher population density when compared to animal groups, the need to employ low frame rate image sequence to capture time-lapses with high resolution on multi-points requires appropriate software to ensure the proper identification of cells between consecutive images [27]. Improvement in optical microscopy has thus always been linked to the refinement of algorithms for track segmentation. Testament of this fundamental link between hardware and software is the ongoing Cell Tracking Challenge (http://celltrackingchallenge.net), an initiative in existence for more than 10 years, whose goal is to promote the development and evaluation of automated cell tracking algorithms (see ref. [28] for its most recent publication).

The nature and sophistication of data type have also been expanding, with spatio-temporal trajectories now routinely accompanied by a plethora of auxiliary information, e.g. real-time changes in cell shape, cell cycle, phenotypes and molecular transcriptional activities when imaging cell displacements. Of late, the analysis of fine-scale cellular imaging data via high-resolution confocal laser scanning microscopy (see, e.g. [29]) and the use of AI tools such as deep learning software (see, e.g. [30]) have allowed rapid progress in detecting differences in functional, molecular and anatomical properties of cells. These experimental approaches, and the associated analyses of the data, provide parameter information for mathematical models that marry descriptive details with the underlying mechanisms.

While the higher resolution of empirical observations has been beneficial to tackle many unsolved problems, the finer nuances required to analyse these novel datasets have created a series of conceptual and semantic barriers between the two disciplines, making modellers in animal ecology and cell biology move along parallel paths, with minimal exchange of ideas, tools and techniques. Given the vast methodological knowledge to model movement and resolve the data analysis challenges that the tracking revolution has brought, it behooves any researcher in one of the two disciplines to seek to profit from insights gained in the other. Although animals and cells move and interact in very different ways, many spatio-temporal movement patterns are common to both. Looking at them through the mathematics of movement language would bring a quantitative understanding that may not be available within the constraints of disciplinary thinking. We make these assertions more precise by presenting a couple of examples.

During some processes in morphogenesis and cancer development, cells move collectively through what appear to be velocity alignment processes and the establishment of heterogeneities with so-called leaders and followers (see, e.g. [31, 32]). During collective migration of epithelial cells, identifying the mechanical forces [33, 34] and chemical gradients [35] that drive this alignment over long distances has been key to understanding how polarity and movement occurs in a cell cluster. These studies have revealed that the leaders are able to pull the remaining individuals through a sustained propagation of extracellular signal-regulated kinase activation and contractile force generation, although the latest findings also indicate that collective cell migration requires a global mechanical organisation within the cell cluster [36].

Similar examples of group movement through alignment exist in schools of fish, herds of mammals and flocks of birds [37]. In these higher organisms, there are also examples where individuals respond not only to who is in front but also to individuals behind [38]. Addressing who is "paying attention" to whom, or who is mechanically or chemically connected to whom, appears key for interpreting and explaining collective movement data both in cell populations and in higher organisms. An interplay between multiple scales of alignment, attraction and avoidance is at the heart of the patterns that emerge from collective movement, but identifying the role that each type of interaction plays is the challenge. Techniques to do that are well-developed in the animal ecology literature (see, e.g. [39]) and could be exploited in analysing cell tracking data.

The spatial spread of an infectious disease in an animal population depends locally on individual interactions, namely, the encounter and transmission probabilities (or rates) between infectious and susceptible animals, which in turn are influenced by the animal movement statistics [40]. Globally the number of individuals depends on the environmental conditions in the habitat, as the latter determines the average population density that a region of space is able to sustain, the so-called carrying capacity. This direct relation between number of individuals and the availability of resources creates, in certain instances, natural hotspots or reservoirs of the disease, also called infection refugia [41], where the infected population never dies out. A classical example where infection waves emerge from refugia is the Hantavirus epidemic in mice populations in the Southwest USA. The emerging waves have been predicted by modelling the spatio-temporal dynamics of infected and susceptible populations in a spatially heterogeneous environment comprising optimal and sub-optimal habitats of the animal host, *Peromyscus maniculatus* [42]. The predictions of advancing and retracting waves of infection, in the form of pushed-type travelling fronts following vegetation gradients [43], have been recently verified and measured to reach the average speed of the order of tens of meters per day [44].

Wave-like propagation resulting from changes in the surrounding environment is also present in cell biology, the prototypical example occurring during durotaxis [45, 46], when cells move by latching and pulling onto the extracellular matrix, the fibre-like proteins in human and animal tissues. Recent modelling approaches via the Cellular Potts model [47] are providing mechanistic explanations of the measurements observed in live-cell cultures of local strain stiffening in biopolymer networks [48]. These types of mechanical waves provide a long-range guidance that allows cells to navigate at relatively lower speed, of the order of microns per minute or per hour [49], when compared to the Hantavirus example.

While the latter two examples above arise due to the interaction between organisms and their environment, the vastly different magnitudes of their speed may lead one to conclude that the movement of certain cells in the extracellular matrix and that of mice moving through different vegetation types is fundamentally different. And yet, while cells degrade and reshape the underlying fibre network leaving guidance signals for the ones that follow, mice consume the resources in the landscape and leave cues by depositing scent marks over the terrain. These marks contain a great deal of information and may direct others away or towards other conspecifics, thus generating a form of animal-animal interaction mediated by the environment. This form of indirect interaction, named stigmergy [50], is well-known to exist in eusocial insects [51] and has also been shown to be present in scent-marking mammals [52]. This deeper connection may be further exploited to encourage the use of Cellular Potts models to study how mice, while tracking vegetation gradients to forage, also influence the spatial spread of other individuals.

Quantitative similarities are not only present at the microscopic scale. They may also appear when coarse-graining the individual agent dynamics to describe the emerging pattern. This is precisely the context of the mathematical models of territorial scent-marking animals. The mesoscale mobility of the scented territories as a function of the time for which scent marks remain active displays spatial characteristics akin to those of two-dimensional colloids that possess a liquid-hexatic-solid melting scenario [53, 54]. While this so far remains a theoretical link, it may be another fertile ground to share tools and study how cells arrange spatially within tissues.

1.3 Some Common Research Challenges

Up to now we have presented examples of where movement modelling in one specific area may help in the study of movement processes in another, but there exist other challenges that go beyond the use of ad-hoc techniques in a specific scenario or a system and that could be addressed from the collaboration of movement modellers across disciplines. One such example is the development of a general theory of organism interaction. Being able to characterise from observations the type of mechanisms by which animals interact among themselves and with the environment is a big open question and it is at the core of pressing scientific and societal problems. At the ecosystem scale, to tackle the loss of biodiversity there is an urgent need for more rigorous predictions of how animals respond to changes in the environment [55, 56]. At the cellular level, quantifying how a tumour cell responds to the presence of nearby cells or the adjacent structures is critical to help determine disease initiation, maintenance and progression [57]. While mathematical techniques to estimate interaction events between moving organisms [58] and between an organism and its physical surroundings [59, 60] are starting to become available, they are often based on basic forms of interaction, e.g. spatial exclusion [61], attraction [62] or exchange of a binary form of information, e.g. infection status [63]. While more complex types of interactions, such as alignment processes, can be modelled explicitly [64], there is a need to develop a general mathematical theory where the motion parameters are coupled with the internal variables of the individuals (see, e.g. ref. [65] for an example of a model of collective sociogenesis). Such general development would be extremely valuable to animal ecologists and cell biologists and it will undoubtedly require a concerted effort from both communities. Recent theoretical developments of data-driven approaches to disentangle information about deterministic forces from stochastic effects [66–68] will play an important role in this respect and more generally in tackling the inverse problem of inferring from movement data the equations that govern the interaction dynamics.

Another fundamental open problem, rather ubiquitous in the context of phenomenological models, is the robustness of the behaviour of population-level descriptions to different hypotheses and assumptions employed for individual-level interactions, often refereed to as the micro-to-macro connection. In ecology, to tackle this issue one approach is to find the critical spatial, temporal and organisational scales of the system [69] and build a model that describes the basic interactions between a small number of individuals so that insights gained from the study of such simplified systems can be translated to complex population-level phenomena. Some of these methodologies have already influenced interdisciplinary approaches to study, for example, allometric scaling [70, 71], animal foraging patterns [72] and collective animal movement [37, 73, 74]. Connecting different levels of description has also taken centre stage in cell biology [75, 76]. A prototypical example is the historical development of the modelling approach to gliomas, a type of primary brain tumour arising from the glial cells of the nervous system. The early proliferation-infiltration glioma models [77] and 'go versus grow' hypothesis [78], proposing the switching between proliferative and migratory phenotypes, were superseded by anisotropic diffusion models [79] to account for the different movement spread occurring in the grey matter compared to the one occurring in the bundles of nerve fibres that make up the white matter tracts in the brain. But following the observations that glioma cells employ microtubes as a means for invasion and proliferation in the brain [80], more recently focus has shifted to mathematical representations of invading tumours that incorporate contact guidance and chemotaxis [81, 82].

Accumulating evidence from the high resolution of tracking data is pushing movement modellers to abandon the Markov assumption, that is, the condition for which the dynamics of a stochastic system depends only on its current state and not on its history. Although the Markov property has been used as a workhorse in theoretical investigations across disciplines, many biological examples do not fulfil that requirement. While animals are known to exploit their cognitive memory for foraging and navigation purposes [83], recently also cells have been shown to utilise a molecular mechanism in receptor signalling akin to a working memory in order to migrate (chemotaxis) within irregular and locally varying gradient cues [84]. Despite the ubiquity of non-Markovian dynamics across the life sciences and the existence of a wide variety of computational approaches to represent history-dependent movement processes, we still lack general null models that may help identify signatures of memory dependence from movement data. While recently some first-principled non-Markov movement models have appeared [85, 86], understanding mathematically how simple modifications, e.g. the introduction of an absorbing, permeable or reflecting boundary, affect the dynamics of a history-dependent movement process remains a challenge [87]. Given the high degree of spatial heterogeneities that animals perceive in the landscape or that cells detect in the surrounding tissue, it is imperative to quantify rigorously the interplay between spatial disorder and non-Markovian movement statistics.

Movement ecology [88–90] has been influential in bringing mathematical and physical approaches to improve methodologies to cluster movement data into distinct movement modes, importing from statistical physics techniques that allow for the analysis of individual trajectories and extraction of anomalous diffusion exponents [91–93], and ageing behaviour (loss of ergodicity) [94, 95]. At the conceptual level, movement ecology has given rise to a general framework to identify the components that shape animal movement, between the external ones and the internal ones, namely, the motivating factors, the motion capacity and the navigation capacity. As such a framework has demonstrably helped bridge knowledge gaps across ecological communities, an analogous framework that spans animal ecology and cell biology would also be valuable. One may look for various parallels, for example, the level of satiation of an animal may be related to the level of saturation in cell receptors, certain environmental triggers for animals may be compared to those present in the extracellular matrix, and an animal behavioural plasticity may be similar to a cell phenotypic plasticity. Initial efforts along this line of thinking are underway and appear promising [96].

While we have discussed here only a handful of common open problems, the abstraction that underpins mathematics allows us to compare many other quantitative approaches across disciplines. In exploiting the vast knowledge accumulated across different areas, it is imperative to establish a platform where scientists can learn about commonalities and where mathematical and modelling ideas can be easily translated from one setting to another in distinct scientific fields. This book aims to help build such a platform and to initiate a closer dialogue between animal ecology and cell biology by presenting studies at different organisational levels, across spatio-temporal scales and with distinct organisms.

1.4 Topics Covered in This Book

As the chapters straddle multiple areas, it has not been possible to group them around distinct themes—they have been ordered alphabetically based on the last name of the first author. Some of the topics introduce methods of analysis of data, while others present abstract/theoretical work, with the latter in some instances developed to interpret empirical observations. The chapters that include data analysis present specific techniques to extract movement characteristics either from laboratory observations, e.g. ant, bumblebee, fish and white blood cells, or from field recordings, namely, African buffalo, grey seal, jaguar, loggerhead sea turtle, raven and sheep, but also general procedures to quantify spatial patterns, such as self-organisation, across disciplines. Some of the theoretically focussed chapters analyse individual stochastic movement processes, e.g. epidemic propagation, resetting dynamics and resource-limited exploration, but also population-level movement models, e.g. dynamics of pattern formation with competition and exclusion, prospection and dispersal in meta-populations. Other chapters are more formal and tackle general methodological questions, e.g. deterministic versus stochastic descriptions in reaction-diffusion processes, information theoretic path segmentation techniques and the number of distinct and common visited locations by a random walk. A brief summary of each chapter within their respective literature context is provided below.

A major challenge in the simulation of stochastic spatially extended systems is that the number of individuals, be they molecules, cells or animals, can vary from large to small across the spatio-temporal domain of interest (see, e.g. [97]). In the regions of large numbers, modelling each individual is computationally expensive, and, typically, a partial differential equation (PDE) formulation is used, as random effects are negligible. However, for small numbers, deterministic PDEs are no longer viable and a stochastic description becomes necessary. This leads to what are called hybrid approaches for the regimes of intermediate numbers, and these are the focus of Chap. 2 by Alarcón et al. This chapter starts with a review of different hybrid formulations before moving on to the authors' recent work in addressing and overcoming a number of limitations of previous schemes. One of the recent advances in this context is when more than one species is being modelled, as the dynamics of such multi-species systems is much richer than that of a single species.

The comparative analysis of movement data of closely related species in the wild provides a valuable resource to study the consequences and future threat of global change. This is the philosophy underpinning the work in Chap. 3 by Ashkenazi et al. The authors compare the movement dynamics of two species of ravens living across the Dead Sea valley and the Judean Desert plateau in Israel, of which one, the fan-tailed raven, has been declining in numbers compared to the other one, the brown-necked raven. The study showcases the importance of employing multiple metrics [98]—in this case maximum displacement, start to end distance, linearity index, mean speed and home range size—over different temporal scales to identify disparity in foraging behaviour as animals may display divergent spatio-temporal responses to resource variability. As location, abundance and nutritional content of food sources may vary, among other factors, due to human-induced environmental changes, competition and landscape composition, similar sympatric species may exhibit different searching strategies, as is the case for the two raven species explored here.

Searching for targets, be they food items, mate or invaders, is a fundamental task that most organisms need to execute to reproduce and survive [99], and has been a central focus in the analysis of movement of biological organisms. The literature on search strategies of biological organisms has a long history (see, e.g. [100] in the ecological context and [101] for cell biology) and not without controversy in regard to animal foraging strategies [102–105]. Among random search strategies stochastic resetting has recently generated a large body of work [106]. Introduced as a movement model more than 20 years ago for continuous variables (Brownian walks) [107] and lately formalised for discrete variables (lattice random walks) [108], it entails a searcher returning at random times to its initial position so as to minimise the time spent searching unsuccessfully in the same region of space. For this reason it has been proposed as a potential mechanism for reducing the time to find targets in an unbounded domain. In the cellular transport context the searcher may be regarded as a motor complex carrying vesicles to synaptic targets in the axon of a neuron or the tip of a filament that delivers a morphogen to a cell during development, as explained by Bressloff in Chap. 4. This chapter is dedicated to the continuous variable formalisms that describe stochastic resetting for a diffusive walker, both at the probabilistic and stochastic level. For a one-dimensional system it explores the first-passage time problem when resetting is instantaneous, and the effects of finite return times and refractory periods when resetting is non-instantaneous.

A great number of movement models have traditionally assumed, for simplicity, Markovian dynamics [109, 110]. Among them, hidden Markov models (see, e.g. [111]), represent a versatile framework with which to identify behavioural switching between movement modes or states from animal movement paths, e.g. resting, running, soaring, etc. Chapter 5, by Campioni et al., extends the application of these models to when the statistics of sojourn times in each behavioural state are gamma distributed, the so-called semi-Markov case, rather than exponentially distributed. By using a continuous time Ornstein-Uhlenbeck process for the velocity vector and allowing for irregular sampling of the movement, a Monte Carlo expectation-maximisation algorithm allows for the extraction of information about directional persistence, mean speed and its variability, and sojourn statistics. The usefulness of the proposed framework is tested with synthetic data as well as data on grey seal and Merino sheep as case studies.

Since its inception from the early theoretical work in the 1980s [112, 113], the study of collective movement has fostered a culture of interdisciplinarity spanning a large number of disciplines, from biology, neuroscience and physics to engineering, computer science and mathematics (see, e.g. [114–119]). The imaging and tracking revolution, in conjunction with the theoretical development of movement models, share part of the recent successes in advancing the understanding of how animals move cohesively in groups. Chapter 6, by Escobedo et al., gives a detailed account of one of the latest advances in methodology to reconstruct social interactions from highly resolved animal trajectories and is applied to laboratory experiments with schooling fish, namely, *Hemigrammus rhodostomus*. Without making assumptions about the form or the intensity of the interactions, the procedure accounts for each individual animal interacting with its environment, e.g. physical obstacles or other conspecifics, through variables such as the instantaneous distance and orientation with respect to the physical boundaries of the experimental set-up and with respect to other individuals. By measuring fluctuations in speed and heading angles, the extracted pairwise attraction and alignment interactions are made up of the product of three components, namely, a probability governed by the relative distance, a probability dependent on the relative angle and a probability as a function of the viewing angle. Two of the methodological advantages, compared to other approaches, are the minimal computational power required and the ability to generate analytical expressions for the interaction probabilities.

The manner in which animals occupy space, the so-called animal space use, is a complex and highly integrated characteristic of the underlying movement processes [120]. It depends on multiple factors, including intra- and inter-specific interactions, environmental features, spatial memory effects and energy considerations. The spatial distributions that emerge may vary from migratory [121, 122] and nomadic patterns [123] to the formation of home ranges [124] and territories [61] or features displaying site fidelity [125]. How to quantify and extract information on these and other types of spatial patterns from movement observations is one of the important challenges in movement ecology. Chapter 7, by Fleming et al., explores the mathematical properties and biological meaning of various spatial distributions, namely, occurrence, individual and population range, corridor, encounter and revisitation. With the help of African buffalo GPS collar data, as well as a GPS jaguar dataset, the differences between interpolative predictions, e.g. occurrence distribution, and extrapolative predictions, e.g. range distribution, are made clear. As a result of the high resolution of movement data available, these topics are destined to attract further attention from the statistical physics community, which has a long tradition of devising theoretical approaches to access multi-target first-passage statistics, e.g. that of an individual revisiting certain locations or of a pair of individuals being in close proximity framed, respectively, as first-return or first-encounter events [126].

With movement tracks now reaching temporal resolution of seconds, there is growing interest in developing a general framework that allows track segmentation to be placed on a firm theoretical ground. Part of the motivation comes from the increase in the computational burden for the traditional approaches such as hidden Markov models and behavioural change point analysis (see, e.g. [127, 128]) and their limitations in identifying rare events such as parturition, migration initiation, etc., when the behavioural shift is not known a priori to have occurred [129]. Chapter 8, by Getz, makes advances in this direction by relying on concepts from Shannon's information theory [130]. Building on a recent multi-modal step-selection kernel framework [131], it proposes to cluster relocation data into segments characterised by different symbols and these, in turn, into words to build a behavioural classifier that automatically assigns a classification error rate. While the proposed approach requires a fixed length to avoid ambiguity in the number of elements to build any track segment, it is relatively flexible. It does not limit the type of AI clustering methods that can be applied to extract statistical movement elements and can accommodate for data obtained from inertial sensors such as accelerometers and gyroscopes, to help compensate for the intrinsic positional errors.

Chapter 9, by Kenkre, showcases the theoretical application of various statistical mechanics methods to ecological phenomena. The first method involves the determination of the probability of transmission of an infectious pathogen as a function of time from an infected animal to a susceptible one when living in separate but neighbouring burrows or dens. By employing techniques developed in solid state physics, namely, the problem of excitons mutually annihilating when close to one another [132], this work uncovers an interesting non-monotonicity in the temporal dependence of the overall infection probability. By modelling animals as Brownian walkers tethered to a focal point—the burrow or den—the overall infection rate exhibits a non-monotonic dependence as a function of the tendency of the animals to return to their respective home range centre. The second application is related to the dynamics of an entire animal population subject to the constraints imposed by the resources in the environment, the so-called carrying capacity. The growth of the population is described by a set of coupled Fisher-Kolmogorov-Petrovsky-Piskunov PDE equations [133, 134] with a spatially dependent carrying capacity. When the environment separates regions of high carrying capacity from regions where population survival is not possible, there exists a critical length below which the population disappears. While the existence of such a transition for a single region was known in the past [135, 136], it is the case of multiple neighbouring regions of high carrying capacity and their critical size dependence that is discussed in detail in this chapter. A further extension of the problem of a critical size domain is the case when an animal population may spread an infectious disease, which is used here to analyse the formation of the so-called refugia of infection in the environment. The final application concerns the so-called effective medium theory, which enables the study of movement in spatially disordered space providing analytical prescriptions that link features of the spatial heterogeneities to time non-locality in the equations of motion.

Chapter 10, by Klages et al., is a historical journey covering the main mathematical formalisms that model the movement of individual organisms. The journey is across the theoretical literature in active matter, anomalous diffusion and movement ecology, as the theory of active particles has often been applied to study the movement of organisms at the microscale [137], while the theory of diffusing or anomalously diffusing particles has formed the basis of many models of animals moving at the macroscale [7, 138, 139]. Two important modelling frameworks, namely, the correlated random walk [140, 141] and active Brownian particle models [142], are shown to be generally different as the former entails a description in a co-moving reference frame, whereas the latter is in a laboratory or Cartesian reference frame. While the special case of an overdamped Brownian particle with the velocity driven by Gaussian changes in the orientation angle is equivalent to a correlated random walk model, the two formulations lead, in general, to different dynamics. By taking an egocentric perspective [143], generalised overdamped coupled Langevin equations for the turning angle and the modulus of the velocity are proposed. The added value of the modelling framework is that the specific choice of the noise terms and non-linearities of the Langevin equations are reconstructed directly from movement data. Application of the model to data is illustrated for observations of bumblebees foraging in a laboratory arena, while other active particle models are applied to the movement data of neutrophil cells and the off-shore foraging of loggerhead sea turtles.

The study of spatial patterning as an emergent property of a system dates back to the classical work of Turing [144], which showed that diffusion could drive patterning in a two species system that had certain interaction dynamics, the most common being the combination of short-range activation and long-range inhibition [145]. While this mechanism is quite general, other patterning mechanisms are known to exist in cells [146] and animals [147], and identifying which underlying mechanism is responsible for the emerging population-level structure can be arduous. In Chap. 11 López et al. study theoretically the patterns for two such mechanisms, involving competitive-only or repelling interactions. The theoretical framework is the Dean-Kawasaki prescription (see, e.g. [148]), which describes density fluctuations in a system of a large number of indistinguishable particles undergoing diffusion. The mathematical object of interest, when the noise intensity in the system is small, is the so-called influence function, that is, the kernel of the PDE that approximately describes the entire population, whose shape controls the spatially non-local non-linear terms. With certain types of interactions, knowledge of the shape of this function, more precisely its Fourier transform, is sufficient to determine whether spatial patterns will emerge and whether they are of Turing type or not (see, e.g. [149]). In the context of a birth-death system this chapter also analyses the link between the influence function and the movement statistics, e.g. when individuals move following a Lévy process rather than diffusing. This link is explored by modifying phenomenologically the transport operator from a diffusive to a fractional one [150].

Fundamental concepts, such as the union and the intersection of a given number of sets, take centre stage in Chap. 12, authored by Majumdar and Schehr. For moving organisms the union may represent an entire habitat covered by multiple animals of the same species [151], while the intersection may refer to a resource-rich area for animals in the environment such as a water body [152] or a region of inflammation in the body to which neutrophils, that is, phagocytic white blood cells, are drawn [153]. Formally, these two statistical quantities can be represented, respectively, by the number of distinct [154] and commonly visited sites [155], by independent random walkers. This chapter explores the dependence of the dynamics of the probability distribution of the two quantities and their averages as a function of the number of individuals. By employing a link to extreme value statistics [156], the temporal scaling of the entire distribution in one dimension is obtained, while for arbitrary dimensions the temporal scaling of the averages is calculated. By relating the sought-after quantities to the probability that a site on a lattice is visited by a walker within a certain time, all findings are derived analytically. The key insight allowing for this analysis is to realise that for estimating scaling relations it suffices to replace the lattice random walker dynamics by the dynamics of a Brownian motion in continuous space and continuous time. The regimes of growth of the average number of commonly visited sites display an interesting transition in the exponent as a function of the spatial dimensions and the number of walkers.

Individual behavioural heterogeneity, also referred to as animal personality, is a burgeoning area of research in animal ecology and has been documented across many taxa [157–159]. With regard to movement, different traits in the exploratory behaviour of individuals may have evolutionary advantages since they support the ability of a population to adapt to changes or shifts of resources in the environment [160]. This is the premise that underscores the joint empirical/theoretical study on the foraging efficiency of the ant species *Aphaenogaster senilis* in Chap. 13, by Marris et al. In this species of ants within a colony, there exist two types of foragers with distinct movement statistics, the scouts, who tend to explore further from the nest, and the recruits, who tend not to stray too far away from the nest. To explain observations of *A. senilis* search behaviour in a laboratory experiment with multiple food patches on a finite Y-maze arena, a model of diffusive dynamics on honeycomb lattices [161] is extended to account for correlation in the movement steps [162]. The variable of interest is the *first* first-passage time to either of the food items within the patches and is studied by constructing a general analytic theory of the first first-passage probability for multiple random walkers with distinct movement statistics ranging from diffusive to correlated. The model allows for a rigorous quantification of how scouts outperform recruits in both exploring space and finding food items.

Understanding animal decision-making processes is a vast and rich area of investigation at the interface of behavioural ecology, computer science, ethology and neuroscience [163–165]. In recent developments there is a growing interest in employing quantitative approaches from the cognitive sciences to help provide a mechanistic explanation of how animals implement decisions to achieve specific goals, e.g. homeostasis, survival or reproduction [166]. The advances in tracking technology are playing an important part in this development. The ability to track groups of animals for long periods of time makes it possible to test hypotheses that were considered empirically out of reach just a decade ago. One example is prospection, i.e. an animal's ability to envisage future scenarios and organise current actions accordingly [167]. As empirical evidence is accumulating [168–170], general theoretical frameworks on prospection are starting to be proposed [171]. Chapter 14, by Molas and Campos, is part of this theoretical push and proposes an extension to the classical Voter model (see, e.g. [172]) in which information gathered from other individuals does not result in instantaneous decision-making but, instead, is accumulated before being acted upon. The new model explores the role that individual personal preferences and this accumulated "crowd" information play in making decisions. The main interest of the model is to determine the time to reach consensus, i.e. the time for all individuals within the system to reach the same decision among multiple options, and the so-called extinction time, which is the time required for one of the options to disappear.

In Chap. 15, Régnier et al. are concerned with the so-called starving random walk problem, a simple model representation of a diffusing organism performing a resource-limited exploration [173], a challenge that any living entity faces in order to uptake energy to move. The quantity of interest that helps to quantify the organism's success is the time elapsed between the finding of one food item and the next. It is represented through the so-called k-th inter-visit time, i.e. the time it takes for a random walker to find a site which has never been visited before, given that k sites have already been visited [174, 175]. The statistical distribution of the inter-visit times is analysed and shown to differ in its extreme values from the classical type, e.g. the Gumbel distribution [176]. Two important variables are studied: the lifetime of the walker and the total amount of food collected at starvation. Analytic scaling expressions for these variables are derived for arbitrary spatial dimensions in terms of a single parameter characterising the recurrence or transience of the walk [177]. As a null model of movement in an environment with depleting resources, the findings in this chapter offer various insights to guide explorations of resource-limited motion in more realistic scenarios.

One of the most striking features of the natural world is that individuals, be they cells, birds, fish, etc., self-organise to form spatial patterns [178, 179], and a significant part of the field of mathematical biology has been, and continues to be, devoted to developing a mechanistic understanding of this phenomenon. As a result, there are now a very large number of models that can produce an astonishing range of spatial patterns. To take these models to the next level requires validation against experimental data, and a major frustration in the field over the past decades has been the lack of suitable spatio-temporal data that is fit for this purpose. However, this situation has changed thanks to technologies that are now making it possible to harvest data at many scales across disciplines. In light of these developments one of the major challenges is how to extract suitable metrics, and this is the key question reviewed in Chap. 16 by Volkening. This chapter surveys a number of the methods that are being used to transform qualitative spatial data into quantitative information that can then be used to parametrise models, validate and test predictions or make comparative analyses. These methods include the identification of order parameters, the generation of pair correlation functions as well as techniques from the field of topological data analysis that are well suited to investigate both cells moving through a developing embryo and animals moving collectively.

We conclude by reminding the reader that, despite the impressive progress over the last decades chronicled in part in the abovementioned chapters, many open questions and unsolved problems remain. We are optimistic that this book will stimulate further developments to tackle these challenges, building new bridges and collaborations across animal ecology and cell biology. We hope that the state-of-the-art techniques and methods explored, the large variety of topics presented, the cited literature and the clarity of exposition of all contributions will make this book a go-to reference to learn about the mathematics of movement and its practical value in tackling challenges in the life sciences.

References

1. Cooke, S.J., Hinch, S.G., Wikelski, M., Andrews, R.D., Kuchel, L.J., Wolcott, T.G., Butler, P.J.: Biotelemetry: a mechanistic approach to ecology. Trends Ecol. Evol. **19**(6), 334–343 (2004)
2. Nathan, R., Monk, C.T., Arlinghaus, R., Adam, T., Alós, J., Assaf, M., Baktoft, H., Beardsworth, C.E., Bertram, M.G., Bijleveld, A.I., Brodin, T., Brooks, J.L., Campos-Candela, A., Cooke, S.J., Gjelland, K.Ø., Gupte, P.R., Harel, R., Hellström, G., Jeltsch, F., Killen, S.S., Klefoth, T., Langrock, R., Lennox, R.J., Lourie, E., Madden, J.R., Orchan, Y., Pauwels, I.S., Říha, M., Roeleke, M., Schlägel, U.E., Shohami, D., Signer, J., Toledo, S., Vilk, O., Westrelin, S., Whiteside, M.A.: Big-data approaches lead to an increased understanding of the ecology of animal movement. Science **375**(6582), eabg1780 (2022)
3. Matley, J.K., Klinard, N.V., Barbosa Martins, A.P., Aarestrup, K., Aspillaga, E., Cooke, S.J., Cowley, P.D., Heupel, M.R., Lowe, C.G., Lowerre-Barbieri, S.K., Mitamura, H., Moore, J.-S., Simpfendorfer, C.A., Stokesbury, M.J.W., Taylor, M.D., Thorstad, E.B., Vandergoot, C.S., Fisk, A.T.: Global trends in aquatic animal tracking with acoustic telemetry. Trends Ecol. Evol. **37**(1), 79–94 (2022)
4. Getz, W.M., Marshall, C.R., Carlson, C.J., Giuggioli, L., Ryan, S.J., Romañach, S.S., Boettiger, C., Chamberlain, S.D., Larsen, L., D'Odorico, P., O'Sullivan, D.: Making ecological models adequate. Ecol. Lett. **21**(2), 153–166 (2018)
5. Adam, T., Griffiths, C.A., Leos-Barajas, V., Meese, E.N., Lowe, C.G., Blackwell, P.G., Righton, D., Langrock, R.: Joint modelling of multi-scale animal movement data using hierarchical hidden Markov models. Methods Ecol. Evol. **10**(9), 1536–1550 (2019)
6. Khataee, H., Czirok, A., Neufeld, Z.: Multiscale modelling of motility wave propagation in cell migration. Sci. Rep. **10**, 8128 (2020)
7. Kenkre, V.M., Giuggioli, L.: Theory of the Spread of Epidemics and Movement Ecology of Animals: An Interdisciplinary Approach Using Methodologies of Physics and Mathematics. Cambridge University Press, Cambridge (2021)
8. Dyson, F.J.: Is science mostly driven by ideas or by tools? Science **338**(6113), 1426–1427 (2012)
9. Freiberger, M.: Know how it works see how it moves: why we need a mathematics of movement? (2024). Plus.maths.org. https://plus.maths.org/content/know-how-it-works-see-how-it-moves
10. Kays, R., Crofoot, M.C., Jetz, W., Wikelski, M.: Terrestrial animal tracking as an eye on life and planet. Science **348**(6240), aaa2478 (2015)
11. Hussey, N.E., Kessel, S.T., Aarestrup, K., Cooke, S.J., Cowley, P.D., Fisk, A.T., Harcourt, R.G., Holland, K.N., Iverson, S.J., Kocik, J.F., Mills Flemming, J.E., Whoriskey,

F.G.: Aquatic animal telemetry: a panoramic window into the underwater world. Science **348**(6240), 1255642 (2015)
12. Royer, L.A., Lemon, W.C., Chhetri, R.K., Wan, Y., Coleman, M., Myers, E.W., Keller, P.J.: Adaptive light-sheet microscopy for long-term, high-resolution imaging in living organisms. Nat. Biotechnol. **34**(12), 1267–1278 (2016)
13. Cutrale, F., Trivedi, V., Trinh, L.A., Chiu, C.L., Choi, J.M., Artiga, M.S., Fraser, S.E.: Hyperspectral phasor analysis enables multiplexed 5D in vivo imaging. Nat. Methods **14**(2), 149–152 (2017)
14. Merenich, D., Van Manen-Brush, K.E., Janetopoulos, C., Myers, K.A.: Advanced microscopy techniques for the visualization and analysis of cell behaviors. In: Cell Movement in Health and Disease, pp. 303–321. Academic Press, London (2022)
15. Sridhar, V.H., Li, L., Gorbonos, D., Nagy, M., Schell, B.R., Sorochkin, T., Gov, N.S., Couzin, I.D.: The geometry of decision-making in individuals and collectives. Proc. Natl. Acad. Sci. USA **118**, e2102157118 (2021)
16. Wilmers, C.C., Nickel, B., Bryce, C.M., Smith, J.A., Wheat, R.E., Yovovich, V.: The golden age of bio-logging: how animal-borne sensors are advancing the Frontiers of ecology. Ecology **96**, 1741–1753 (2015)
17. McMahon, C.R., Roquet, F., Baudel, S., Belbeoch, M., Bestley, S., Blight, C., Boehme, L., Carse, F., Costa, D.P., Fedak, M.A., Guinet, C., Harcourt, R., Heslop, E., Hindell, M.A., Hoenner, X., Holland, K., Holland, M., Jaine, F.R.A., Jeanniard du Dot, T., Jonsen, I., Keates, T.R., Kovacs, K.M., Labrousse, S., Lovell, P., Lydersen, C., March, D., Mazloff, M., McKinzie, M.K., Muelbert, M.M.C., O'Brien, K., Phillips, L., Portela, E., Pye, J., Rintoul, S., Sato, K., Sequeira, A.M.M., Simmons, S.E., Tsontos, V.M., Turpin, V., van Wijk, E., Vo, D., Wege, M., Whoriskey, F.G., Wilson, K., Woodward, B.: Animal borne ocean sensors–AniBOS–An essential component of the global ocean observing system. Front. Mar. Sci. **8**, 751840 (2021)
18. Dodge, S., Bohrer, G., Weinzierl, R., Davidson, S.C., Kays, R., Douglas, D., Cruz, S., Han, J., Brandes, D., Wikelski, M.: The environmental-data automated track annotation (Env-DATA) system: linking animal tracks with environmental data. Move. Ecol. **1**, 3 (2013)
19. Weiser, A.W., Orchan, Y., Nathan, R., Charter, M., Weiss, A.J., Toledo, S.: Characterizing the accuracy of a self-synchronized reverse-GPS wildlife localization system. In: 15th ACM/IEEE International Conference on Information Processing in Sensor Networks (IPSN), pp. 1–12. IEEE, Vienna (2016)
20. Toledo, S., Kishon, O., Orchan, Y., Shohat, A., Nathan, R.: Lessons and experiences from the design, implementation, and deployment of a wildlife tracking system. In: IEEE International Conference on Software Science, Technology and Engineering (SWSTE), pp. 51–60. IEEE, Be'er Sheva (2016)
21. Toledo, S., Shohami, D., Schiffner, I., Lourie, E., Orchan, Y., Bartan, Y., Nathan, R.: Cognitive map—based navigation in wild bats revealed by a new high-throughput tracking system. Science **369**(6500), 188–193 (2020)
22. Farine, D.R., Strandburg-Peshkin, A., Couzin, I.D., Berger-Wolf, T.Y., Crofoot, M.C.: Individual variation in local interaction rules can explain emergent patterns of spatial organization in wild baboons. Proc. R. Soc. B **284**(1853), 20162243 (2017)
23. Gómez-Nava, L., Bon, R., Peruani, F.: Intermittent collective motion in sheep results from alternating the role of leader and follower. Nat. Phys. **18**(12), 1494–1501 (2022)
24. Ouellette, N.T.: A physics perspective on collective animal behavior. Phys. Biol. **19**(2), 021004 (2022)
25. Cavagna, A., Di Carlo, L., Giardina, I., Grigera, T.S., Melillo, S., Parisi, L., Pisegna, G., Scandolo, M.: Natural swarms in 3.99 dimensions. Nat. Phys. **19**(7), 1043–1049 (2023)
26. Amichay, G., Li, L., Nagy, M., Couzin, I.D.: Revealing the mechanism and function underlying pairwise temporal coupling in collective motion. Nat. Commun. **15**(1), 4356 (2024)

27. Gómez-de-Mariscal, E., Grobe, H., Pylvänäinen, J.W., Xénard, L., Henriques, R., Tinevez, J.Y., Jacquemet, G.: CellTracksColab is a platform that enables compilation, analysis, and exploration of cell tracking data. PLoS Biol. **22**(8), e3002740 (2024)
28. Maška, M., Ulman, V., Delgado-Rodriguez, P., Gómez-de-Mariscal, E., Nečasová, T., Guerrero Peña, F.A., Ren, T.I., Meyerowitz, E.M., Scherr, T., Löffler, K., Mikut, R., Guo, T., Wang, Y., Allebach, J.P., Bao, R., Al-Shakarji, N.M., Rahman, G., Toubal, I.E., Palaniappan, K., Lux, F., Matula, P., Sugawara, K., Magnusson, K.E.G., Aho, L., Cohen, A.R., Arbelle, A., Ben-Haim, T., Riklin Raviv, T., Isensee, F., Jäger, P.F., Maier-Hein, K.H., Zhu, Y., Ederra, C., Urbiola, A., Meijering, E., Cunha, A., Muñoz-Barrutia, A., Kozubek, M., Ortiz-de-Solórzano, C.: The cell tracking challenge: 10 years of objective benchmarking. Nat. Methods **20**(7), 1010–1020 (2023)
29. van den Berg, M.C.W, MacCarthy-Morrogh, L., Carter, D., Morris, J., Bravo, I.R., Feng, Y., Martin, P.: Proteolytic and opportunistic breaching of the basement membrane zone by immune cells during tumor initiation. Cell Rep. **27**(10), 2837–2846 (2019)
30. Beau, M.,Herzfeld, D.J., Naveros, F., Hemelt, M.E., D'Agostino, F., Oostland, M., Sánchez-López, A., Chung, Y.Y., Maibach, M., Kyranakis, S., Stabb, H.N., Martínez Lopera, M.G., Lajko, A., Zedler, M., Ohmae, S., Hall, N.J., Clark, B.A., Cohen, D., Lisberger, S.G., Kostadinov, D., Hull, C., Häusser, M., Medina, J.F.: A deep-learning strategy to identify cell types across species from high-density extracellular recordings. Cell **188**, 1–17 (2025)
31. McLennan, R., Schumacher, L.J., Morrison, J.A., Teddy, J.M., Ridenour, D.A., Box, A.C., Semerad, C.L., Li, H., McDowell, W., Kay, D., Maini, P.K., Baker, R.E., Kulesa, P.M.: Neural crest migration is driven by a few trailblazer cells with a unique molecular signature narrowly confined to the invasive front. Development **142**(11), 2014–2025 (2015)
32. Zhang, J., Goliwas, K.F., Wang, W., Taufalele, P.V., Bordeleau, F., Reinhart-King, C.A.: Energetic regulation of coordinated leader–follower dynamics during collective invasion of breast cancer cells. Proc. Natl. Acad. Sci. USA **116**(16), 7867–7872 (2019)
33. Trepat, X., Wasserman, M.R., Angelini, T.E., Millet, E., Weitz, D.A., Butler, J.P., Fredberg, J.J.: Physical forces during collective cell migration. Nat. Phys. **5**(6), 426–430 (2009)
34. Serra-Picamal, X., Conte, V., Vincent, R., Anon, E., Tambe, D.T., Bazellieres, E., Butler, J.P., Fredberg, J.J., Trepat, X.: Mechanical waves during tissue expansion. Nat. Phys. **8**(8), 628–634 (2012)
35. Hino, N., Rossetti, L., Marín-Llauradó, A., Aoki, K., Trepat, X., Matsuda, M., Hirashima, T.: ERK-mediated mechanochemical waves direct collective cell polarization. Dev. Cell **53**(6), 646–660 (2020)
36. Rossetti, L., Grosser, S., Abenza, J.F., Valon, L., Roca-Cusachs, P., Alert, R., Trepat, X.: Optogenetic generation of leader cells reveals a force–velocity relation for collective cell migration. Nat. Phys. **20**, 1659–1669 (2024)
37. Vicsek, T., Zafeiris, A.: Collective motion. Phys. Rep. **517**(3–4), 71–140 (2012)
38. Giuggioli, L., McKetterick, T.J., Holderied, M.: Delayed response and biosonar perception explain movement coordination in trawling bats. PLoS Comput. Biol. **11**(3), e1004089 (2015)
39. Calovi, D.S., Litchinko, A., Lecheval, V., Lopez, U., Pérez Escudero, A., Chaté, H., Sire, C., Theraulaz, G.: Disentangling and modeling interactions in fish with burst-and-coast swimming reveal distinct alignment and attraction behaviors. PLoS Comput. Biol. **14**(1), e1005933 (2018)
40. Vargas Soto, J.S., Kosiewska, J.R., Grove, D., Metts, D., Muller, L.I., Wilber, M.Q.: How do non-independent host movements affect spatio-temporal disease dynamics? Partitioning the contributions of spatial overlap and correlated movements to transmission risk. Move. Ecol. **13**, 11 (2025)
41. Keppel, G., Van Niel, K.P., Wardell-Johnson, G.W., Yates, C.J., Byrne, M., Mucina, L., Schut, A.G., Hopper, S.D., Franklin, S.E.: Refugia: identifying and understanding safe havens for biodiversity under climate change. Glob. Ecol. Biogeogr. **21**(4), 393–404 (2012)
42. Abramson, G., Kenkre, V.M.: Spatiotemporal patterns in the hantavirus infection. Phys. Rev. E **66**(1), 011912 (2002)

43. Abramson, G., Giuggioli, L., Parmenter, R.R., Kenkre, V.M.: Quasi-one-dimensional waves in rodent populations in heterogeneous habitats: a consequence of elevational gradients on spatio-temporal dynamics. J. Theor. Biol. **319**, 96–101 (2013)
44. Parmenter, R.R., Glass, G.E.: Hantavirus outbreaks in the American Southwest: Propagation and retraction of rodent and virus diffusion waves from sky-island refugia. Int. J. Mod. Phys. B **36**(7–8), 2140052 (2022)
45. Pallarès, M.E., Pi-Jaumà, I., Fortunato, I.C., Grazu, V., Gómez-González, M., Roca-Cusachs, P., de la Fuente, J.M., Alert, R., Sunyer, R., Casademunt, J., Trepat, X.: Stiffness-dependent active wetting enables optimal collective cell durotaxis. Nat. Phys. **20**(10), 1659–1669 (2022)
46. Espina, J.A., Marchant, C.L., Barriga, E.H.: Durotaxis: the mechanical control of directed cell migration. FEBS J. **289**(10), 2736–2754 (2022)
47. Tsingos, E., Bakker, B.H., Keijzer, K.A., Hupkes, H.J., Merks, R.M.: Hybrid cellular Potts and bead-spring modeling of cells in fibrous extracellular matrix. Biophys. J. **122**(13), 2609–2622 (2023)
48. Van Helvert, S., Friedl, P.: Strain stiffening of fibrillar collagen during individual and collective cell migration identified by AFM nanoindentation. ACS Appl. Mater. Interfaces **8**(34), 21946–21955 (2016)
49. Katsuta, H., Okuda, S., Nagayama, K., Machiyama, H., Kidoaki, S., Kato, M., Sokabe, M., Miyata, T., Hirata, H.: Actin crosslinking by α-actinin averts viscous dissipation of myosin force transmission in stress fibers. Iscience **26**(3), 106090 (2023)
50. Grassé, P.P.: La reconstruction du nid et les coordinations interindividuelles chez bellicositermes natalensis et cubitermes sp. la théorie de la stigmergie: Essai d'interprétation du comportement des termites constructeurs. Insect. Soc. **6**, 41–80 (1959)
51. Theraulaz, G., Bonabeau, E.: A brief history of stigmergy. Artif. Life **5**(2), 97–116 (1999)
52. Giuggioli, L., Potts, J.R., Rubenstein, D.I., Levin, S.A.: Stigmergy, collective actions, and animal social spacing. Proc. Natl. Acad. Sci. USA **110**(42), 16904–16909 (2013)
53. Giuggioli, L., Kenkre, V.M.: Consequences of animal interactions on their dynamics: emergence of home ranges and territoriality. Move. Ecol. **2**, 1–22 (2014)
54. Heiblum Robles, A., Giuggioli, L.: Phase transitions in stigmergic territorial systems. Phys. Rev. E **98**, 042115 (2018)
55. Johnston, A.S.A, Boyd, R.J., Watson, J.W., Paul, A., Evans, L.C., Gardner, E.L., Boult, V.L.: Predicting population responses to environmental change from individual-level mechanisms: towards a standardized mechanistic approach. Proc. R. Soc. B **286**(1913), 20191916 (2019)
56. Musters, C.J.M., DeAngelis, D.L., Harvey, J.A., Mooij, W.M., van Bodegom, P.M., de Snoo, G.R.: Enhancing the predictability of ecology in a changing world: A call for an organism-based approach. Front. Appl. Math. Stat. **9**, 1046185 (2023)
57. Raluca, E., Trucu, D. (eds.): Modelling and Computational Approaches for Multi-scale Phenomena in Cancer Research: From Cancer Evolution to Cancer Treatment. World Scientific, London (2024)
58. Giuggioli, L., Sarvaharman, S.: Spatio-temporal dynamics of random transmission events: from information sharing to epidemic spread. J. Phys. A: Math. Theor. **55**(37), 375005 (2022)
59. Kay, T., Giuggioli, L.: Diffusion through permeable interfaces: fundamental equations and their application to first-passage and local time statistics. Phys. Rev. Res. **4**, L032039 (2022)
60. Sarvaharman, S., Giuggioli, L.: Particle-environment interactions in arbitrary dimensions: a unifying analytic framework to model diffusion with inert spatial heterogeneities. Phys. Rev. Res. **5**(4), 043281 (2023)
61. Giuggioli, L., Potts, J.R., Harris, S.: Animal interactions and the emergence of territoriality. PLoS Comput. Biol. **7**(3), e1002008 (2011)
62. Schellinck, J., White, T.: A review of attraction and repulsion models of aggregation: methods, findings and a discussion of model validation. Ecol. Model. **222**(11), 1897–1911 (2011)
63. Giuggioli, L., Pérez-Becker, S., Sanders, D.P.: Encounter times in overlapping domains: application to epidemic spread in a population of territorial animals. Phys. Rev. Lett. **110**(5), 058103 (2013)

64. McKetterick, T.J., Giuggioli, L.: Exact dynamics of stochastic linear delayed systems: application to spatiotemporal coordination of comoving agents. Phys. Rev. E **90**(4), 042135 (2014)
65. Neu, Z., Giuggioli, L.: Sociogenesis in unbounded space: modelling self-organised cohesive collective motion. Phys. Biol. **20**, 035002 (2023)
66. Frishman, A., Ronceray, P.: Learning force fields from stochastic trajectories. Phys. Rev. X **10**(2), 021009 (2020)
67. Nabeel, A., Karichannavar, A., Palathingal, S., Jhawar, J., Brückner, D.B., Raj M, D., Guttal, V.: Discovering stochastic dynamical equations from ecological time series data. Am. Nat. **10**(1086), 734083 (2024)
68. Gerardos, A., Ronceray, P.: Principled model selection for stochastic dynamics (2025). preprint arXiv:2501.10339
69. Levin, S.A.: The problem of pattern and scale in ecology. Ecology **73**(6), 1943–1967 (1992)
70. West, G.B., Brown, J.H., Enquist, B.J.: A general model for the origin of allometric scaling laws in biology. Science **276**(5309), 122–126 (1997)
71. Banavar, J.R., Maritan, A., Rinaldo, A.: Size and form in efficient transportation networks. Nature **399**(6732), 130–132 (1999)
72. Viswanathan, G.M., Da Luz, M.G.E., Raposo, E.P., Stanley, H.E.: The Physics of Foraging: An Introduction to Random Searches and Biological Encounters. Cambridge University Press, Cambridge (2011)
73. Vicsek, T., Czirók, A., Ben-Jacob, E., Cohen, I., Shochet, O.: Novel type of phase transition in a system of self-driven particles. Phys. Rev. Lett. **75**(6), 1226 (1995)
74. Lopez, U., Gautrais, J., Couzin, I.D., Theraulaz, G.: From behavioural analyses to models of collective motion in fish schools. Interface Focus **2**(6), 693–707 (2012)
75. Penington, C.J., Hughes, B.D., Landman, K.A.: Building macroscale models from microscale probabilistic models: a general probabilistic approach for nonlinear diffusion and multispecies phenomena. Phys. Rev. E **84**(4), 041120 (2011)
76. Pillay, S., Byrne, H.M., Maini, P.K.: Modeling angiogenesis: a discrete to continuum description. Phys. Rev. E **95**(1), 012410 (2017)
77. Swanson, K.R., Bridge, C., Murray, J.D., Alvord Jr., E.C.: Virtual and real brain tumors: using mathematical modeling to quantify glioma growth and invasion. J. Neurol. Sci. **216**(1), 1–10 (2003)
78. Giese, A., Loo, M.A., Tran, N., Haskett, D., Coons, S.W., Berens, M.E.: Dichotomy of astrocytoma migration and proliferation. Int. J. Cancer **67**(2), 275–282 (1996)
79. Swan, A., Hillen, T., Bowman, J.C., Murtha, A.D.: A patient-specific anisotropic diffusion model for brain tumour spread. Bull. Math. Biol. **80**, 1259–1291 (2018)
80. Osswald, M., Jung, E., Sahm, F., Solecki, G., Venkataramani, V., Blaes, J., Weil, S., Horstmann, H., Wiestler, B., Syed, M., Huang, L., Ratliff, M., Karimian Jazi, K., Kurz, F.T., Schmenger, T., Lemke, D., Gömmel, M., Pauli, M., Liao, Y., Häring, P., Pusch, S., Herl, V., Steinhäuser, C., Krunic, D., Jarahian, M., Miletic, H., Berghoff, A.S., Griesbeck, O., Kalamakis, G., Garaschuk, O., Preusser, M., Weiss, S., Liu, H., Heiland, S., Platten, M., Huber, P.E., Kuner, T., von Deimling, A., Wick, W., Winkler, F.: Brain tumour cells interconnect to a functional and resistant network. Nature **528**(7580), 93–98 (2015)
81. Conte, M., Loy, N.: A non-local kinetic model for cell migration: a study of the interplay between contact guidance and steric hindrance. SIAM J. Appl. Math. **84**(3), S429–S451 (2023)
82. Hillen, T., Loy, N., Painter, K.J., Thiessen, R.: Modelling microtube driven invasion of glioma. J. Math. Biol. **88**(1), 4 (2024)
83. Fagan, W.F., Lewis, M.A., Auger-Méthé, M., Avgar, T., Benhamou, S., Breed, G., LaDage, L., Schlägel, U.E., Tang, W.W., Papastamatiou, Y.P., Forester, J., Mueller, T.: Spatial memory and animal movement. Ecol. Lett. **16**(10), 1316–1329 (2013)
84. Nandan, A., Das, A., Lott, R., Koseska, A.: Cells use molecular working memory to navigate in changing chemoattractant fields. eLife **11**, e76825 (2022)

85. Boyer, D., Solis-Salas, C.: Random walks with preferential relocations to places visited in the past and their application to biology. Phys. Rev. Lett. **112**(24), 240601 (2014)
86. Falcón-Cortés, A., Boyer, D., Giuggioli, L., Majumdar, S.N.: Localization transition induced by learning in random searches. Phys. Rev. Lett. **119**(14), 140603 (2017)
87. Guggenberger, T., Pagnini, G. Vojta, T., Metzler, R.: Fractional Brownian motion in a finite interval: correlations effect depletion or accretion zones of particles near boundaries. New J. Phys. **21**(2), 022002 (2019).
88. Nathan, R.: An emerging movement ecology paradigm. Proc. Natl. Acad. Sci. USA **105**(49), 19050–19051 (2008)
89. Nathan, R., Getz, W.M., Revilla, E., Holyoak, M., Kadmon, R., Saltz, D., Smouse, P.E.: A movement ecology paradigm for unifying organismal movement research. Proc. Natl. Acad. Sci. USA **105**(49), 19052–19059 (2008)
90. Nathan, R., Giuggioli, L.: A milestone for movement ecology research. Move. Ecol. **1**, 1 (2013)
91. Metzler, R., Jeon, J.H., Cherstvy, A.G., Barkai, E.: Anomalous diffusion models and their properties: non-stationarity, non-ergodicity, and ageing at the centenary of single particle tracking. Phys. Chem. Chem. Phys. **16**(44):24128–24164 (2014)
92. Muñoz-Gil, G., Volpe, G., Garcia-March, M.A., Aghion, E., Argun, A., Hong, C.B., Bland, T., Bo, S., Conejero, J.A., Firbas, N., Garibo i Orts, Ò., Gentili, A., Huang, Z., Jeon, J.-H., Kabbech, H., Kim, Y., Kowalek, P., Krapf, D., Loch-Olszewska, H., Lomholt, M.A., Masson, J.-B., Meyer, P.G., Park, S., Requena, B., Smal, I., Song, T., Szwabiński, J., Thapa, S., Verdier, H., Volpe, G., Widera, A., Lewenstein, M., Metzler, R., Manzo, C.: Objective comparison of methods to decode anomalous diffusion. Nature Comm., **12**(1):6253 (2021)
93. Seckler, H., Metzler, R.: Bayesian deep learning for error estimation in the analysis of anomalous diffusion. Nature Comm. **13**, 6717 (2022)
94. Vilk, O., Aghion, E., Nathan, R., Toledo, S., Metzler, R., Assaf, M.: Classification of anomalous diffusion in animal movement data using power spectral analysis. J. Phys. A: Math. Theor. **55**, 334004 (2022)
95. Vilk, O., Orchan, Y., Charter, M., Ganot, N., Toledo, S., Nathan, R., Assaf, M.: Ergodicity breaking in area-restricted search of avian predators. Phys. Rev. X **12**, 031005 (2022)
96. Giuggioli, L., Abramson, G., Byrne, H., Kulesa, P., Maini, P.K., Nathan, R.: Mathematics of movement: integrating conceptual and quantitative approaches in animal ecology and cell biology. Preprint
97. Bar-Sinai, Y., Hoyer, S., Hickey, J., Brenner, M.P.: Learning data-driven discretizations for partial differential equations. Proc. Natl. Acad. Sci. USA **116**(31), 15344–15349 (2019)
98. Kays, R., Hirsch, B., Caillaud, D., Mares, R., Alavi, S., W Havmøoller, R., Crofoot, M.: Multi-scale movement syndromes for comparative analyses of animal movement patterns. Move. Ecol. **11**, 61 (2023)
99. Hein, A.M., Carrara, F., Brumley, D.R., Stocker, R., Levin, S.A.: Natural search algorithms as a bridge between organisms, evolution, and ecology. Proc. Natl. Acad. Sci. USA **113**(34), 9413–9420 (2016)
100. Bartumeus, F., Catalan, J.: Optimal search behavior and classic foraging theory. J. Phys. A: Math. Theor. **42**, 434002 (2009)
101. Krummel, M.F., Bartumeus, F., Gérard, A.: T cell migration, search strategies and mechanisms. Nat. Rev. Immunol. **16**(3), 193–201 (2016)
102. Sims, D.W., Southall, E.J., Humphries, N.E., Hays, G.C., Bradshaw, C.J.A., Pitchford, J.W., James, A., Ahmed, M.Z., Brierley, A.S., Hindell, M.A., Morritt, D., Musyl, M.K., Righton, D., Shepard, E.L.C., Wearmouth, V.J., Wilson, R.P., Witt, M.J., Metcalfe, J.D.: Scaling laws of marine predator search behaviour. Nature **451**(7182), 1098–1102 (2008)
103. Benhamou, S.: How many animals really do the Levy walk? Ecology **88**, 1962–1969 (2007)

104. Pyke, G.H.: Understanding movements of organisms: it's time to abandon the Lévy foraging hypothesis. Methods Ecol. Evol. **6**(1), 1–16 (2015)
105. Reynolds, A.: Liberating Lévy walk research from the shackles of optimal foraging. Phys. Life Rev. **14**, 59–83 (2015)
106. Kundu, A., Reuveni, S.: Preface: stochastic resetting—theory and applications. J. Phys. A: Math. Theor. **57**(6), 060301 (2024)
107. Evans, M.R., Majumdar, S.N.: Diffusion with stochastic resetting. Phys. Rev. Lett. **106**, 160601 (2011)
108. Das, D., Giuggioli, L.: Discrete space-time resetting model: application to first-passage and transmission statistics J. Phys. A: Math. Theor. **55**, 424004 (2022)
109. Holmes, E.E.: Are diffusion models too simple? A comparison with telegraph models of invasion. Am. Nat. **142**(5), 779–795 (1993)
110. van Kampen, N.G.: Remarks on non-Markov processes. Braz. J. Phys. 28:90–96 (1998)
111. Conners, M.G., Michelot, T., Heywood, E.I., Orben, R.A., Phillips, R.A., Vyssotski, A.L., Shaffer, S.A., Thorne, L.H.: Hidden Markov models identify major movement modes in accelerometer and magnetometer data from four albatross species. Move. Ecol. **9**, 7 (2021)
112. Aoki, I.: A simulation study on the schooling mechanism in fish. Bull. Jpn. Soc. Sci. Fish. **48**(8), 1081–1088 (1982)
113. Reynolds, C.W.: Flocks, herds and schools: a distributed behavioral model. ACM SIGGRAPH Comp. Graph. **21**(4), 25–34 (1987)
114. Gautrais, J., Ginelli, F., Fournier, R., Blanco, S., Soria, M., Chaté, H., Theraulaz, G.: Deciphering interactions in moving animal groups. PLoS Comput. Biol. **8**(9), e1002678 (2012)
115. Toner, J., Tu, Y.: Flocks, herds, and schools: a quantitative theory of flocking. Phys. Rev. E **58**, 4828–4858 (1998)
116. Couzin, I.D., Krause, J., James, R., Ruxton, G.D., Franks, N.R.: Collective memory and spatial sorting in animal groups. J. Theor. Biol. **218**, 1–12 (2002)
117. Sumpter, D.J.: The principles of collective animal behaviour. Philos. Trans. R. Soc. B **361**, 5–22 (2006)
118. Jiang, L., Giuggioli, L., Perna, A., Escobedo, R., Lecheval, V., Sire, C., Han, Z., Theraulaz, G.: Identifying influential neighbors in animal flocking. PLoS Comput. Biol. **13**(11), e1005822 (2017)
119. Cavagna, A., Giardina, I., Grigera, T.S.: The physics of flocking: correlation as a compass from experiments to theory. Phys. Rep. **728**, 1–62 (2018)
120. Gautestad, A.O.: Animal Space Use: Memory Effects, Scaling Complexity, and Biophysical Model Coherence. Dog Ear Publishing, Indianapolis (2015)
121. Dingle, H., Drake, V.A.: What is migration? BioScience **57**(2), 113–121 (2007)
122. Cooke, S.J., Piczak, M.L., Singh, N.J., Åkesson, S., Ford, A.T., Chowdhury, S., Mitchell, G.W., Norris, D.R., Hardesty-Moore, M., McCauley, D., Hammerschlag, N., Tucker, M.A., Horns, J.J., Reisinger, R.R., Kubelka, V., Lennox, R.J.: Animal migration in the Anthropocene: threats and mitigation options. Biol. Rev. **99**, 1242–1260 (2024)
123. Teitelbaum, C.S., Mueller, T.: Beyond migration: causes and consequences of nomadic animal movements. Trends Ecol. Evol. **34**(6), 569–581 (2019)
124. Börger, L., Dalziel, B.D., Fryxell, J.M.: Are there general mechanisms of animal home range behaviour? A review and prospects for future research. Ecol. Lett. **11**(6), 637–650 (2008)
125. Picardi, S., Abrahms, B., Gelzer, E., Morrison, T.A., Verzuh, T., Merkle, J.A.: Defining null expectations for animal site fidelity. Ecol. Lett. **26**(1), 157–169 (2023)
126. Das, D., Kenkre, V.M., Nathan, R., Giuggioli, L.: Misconceptions about quantifying animal encounter and interaction processes. Front. Ecol. Evol. **11**, 1230890 (2023)
127. Edelhoff, H., Signer, J., Balkenhol, N.: Path segmentation for beginners: an overview of current methods for detecting changes in animal movement patterns. Move. Ecol. **4**(1), 21 (2016)

128. Gurarie, E., Bracis, C., Delgado, M., Meckley, T.D., Kojola, I., Wagner, C.M.: What is the animal doing? Tools for exploring behavioural structure in animal movements. J. Anim. Ecol. **85**(1), 69–84 (2016)
129. Gundermann, K.P., Diefenbach, D.R., Walter, W.D., Corondi, A.M., Banfield, J.E., Wallingford, B.D., Stainbrook, D.P., Rosenberry, C.S., Buderman, F.E.: Change-point models for identifying behavioral transitions in wild animals. Move. Ecol. **11**(1), 65 (2023)
130. Shannon, C.E.: A mathematical theory of communication. Bell Syst. Tech. J. **27**, 379–423 (1948); ibid., 623–656 (1948)
131. Getz, W.M., Salter, R., Sethi, V., Cain, S., Spiegel, O., Toledo, S.: The statistical building blocks of animal movement simulations. Move. Ecol. **12**, 67 (2024)
132. Kenkre, V.M.: Theory of exciton annihilation in molecular crystals. Phys. Rev. B **22**, 2089–2098 (1980)
133. Fisher, R.A.: The wave of advance of advantageous genes. Ann. Eugenic. **7**(4), 355–369 (1937)
134. Kolmogorov, A.N., Petrovsky, N.G., Piskunov, N.S.: Study of the diffusion equation with growth of the quantity of matter and its application to a biology problem. Bull. Moscow State Univ. Ser. A. Math. Mech. **1**(6), 1–26 (1937)
135. Skellam, J.G.: Random dispersal in theoretical populations. Biometrika **38**, 196–218 (1951)
136. Ludwig, D., Aronson, D.G., Weinberger, H.F.: Spatial patterning of the spruce budworm. J. Math. Biol. **8**(3), 217–258 (1979)
137. Ramaswamy, S.: The mechanics and statistics of active matter. Annu. Rev. Condens. Matter Phys. **1**, 323–345 (2010)
138. Méndez, V., Campos, D., Bartumeus, F.: Stochastic Foundations in Movement Ecology. Springer, Berlin (2016)
139. Giuggioli, L., Sevilla, F.J., Kenkre, V.M.: A generalized master equation approach to modelling anomalous transport in animal movement. J. Phys. A: Math. Theor. **42**, 434004 (2009)
140. Goldstein, J.: On diffusion by discontinuous movements, and on the telegraph equation. Quart. J. Mech. **4**, 129–156 (1951)
141. Gillis, J.: Correlated random walk. Proc. Camb. Philos. Soc. **51**, 639–651 (1955)
142. Ebeling, W., Schweitzer, F., Tilch, B.: Active Brownian particles with energy depots. BioSystems **49**, 17–29 (1999)
143. Johnsen, S.H.W., Rytter, H.M.: Dissociating spatial strategies in animal research: Critical methodological review with focus on egocentric navigation and the hippocampus. Neurosci. Biobehav. Rev. **126**, 57–78 (2021)
144. Turing, A.M.: The chemical basis of morphogenesis. Philos. Trans. R. Soc. B **237**, 37–72 (1953)
145. Gierer, A., Meinhardt, H.: A theory of biological pattern formation Kybernetik **12**, 30–39 (1972)
146. Salazar-Ciudad, I., Jernvall, J., Newman, S.A.: Mechanisms of pattern formation in development and evolution. Development **130**, 2027–2037 (2003)
147. Cordeiro, E.M.G., Campbell, J.F., Phillips, T.W., With, K.A.: Behavioral and social mechanisms behind pattern formation: an experimental study of animal movement. Landsc. Ecol. **33**, 1881–1894 (2018)
148. Cornalba, F., Fischer, J.: The Dean-Kawasaki equation and the structure of density fluctuations in systems of diffusing particles. Arch. Rational Mech. Anal. **247**, 76 (2023)
149. Fuentes, M.A., Kuperman, M.N., Kenkre, V.M.: Nonlocal interaction effects on pattern formation in population dynamics. Phys. Rev. Lett. **91**(15), 158104 (2003)
150. Metzler, R., Klafter, J.: The random walk's guide to anomalous diffusion: a fractional dynamics approach. Phys. Rep. **339**(1), 1–77 (2000)
151. Jensen, S.P., Gray, S.J., Hurst, J.L.: Excluding neighbours from territories: effects of habitat structure and resource distribution. Anim. Behav. **69**(4), 785–795 (2005)
152. Gereta, E., Wolanski, E.: Wildlife-water quality interactions in the Serengeti National Park, Tanzania. Afr. J. Ecol. **36**, 1–14 (1998)

153. Weavers, H., Liepe, J., Sim, A., Wood, W., Martin, P., Stumpf, M.P.: Systems analysis of the dynamic inflammatory response to tissue damage reveals spatiotemporal properties of the wound attractant gradient. Curr. Biol. **26**, 1975–1989 (2016)
154. Larralde, H., Trunfio, P., Havlin, S., Stanley, H.E., Weiss, G.H.: Territory covered by N diffusing particles. Nature **355**, 423–426 (1992)
155. Majumdar, S.N., Tamm, M.V.: Number of common sites visited by N random walkers. Phys. Rev. E **86**, 021135 (2012)
156. Touchette, H.: The large deviation approach to statistical mechanics. Phys. Rep. **478**(1–3), 1–69 (2009)
157. Spiegel, O., Leu, S.T., Bull, C.M., Sih, A.: What's your move? Movement as a link between personality and spatial dynamics in animal populations. Ecol. Lett. **20**, 3–18 (2017)
158. Shaw, A.K.: Causes and consequences of individual variation in animal movement. Move. Ecol. **8**, 12 (2020)
159. Hertel, A.G., Niemelä, P.T., Dingemanse, N.J., Mueller, T.: A guide for studying among-individual behavioral variation from movement data in the wild. Move. Ecol. **8**, 30 (2020)
160. Goy, N., Glaser, S.M., Grüter, C.: The adaptive value of tandem communication in ants: Insights from an agent-based model. J. Theor. Biol. **526**, 110762 (2021)
161. Marris, D., Sarvaharman, S., Giuggioli, L.: Exact spatiotemporal dynamics of lattice random walks in hexagonal and honeycomb domains. Phys. Rev. E **107**(5), 054139 (2023)
162. Marris, D., Giuggioli, L.: Persistent and anti-persistent motion in bounded and unbounded space: resolution of the first-passage problem. New J. Phys. **26**, 073020 (2024)
163. Bell, A.M.: The evolution of decision-making mechanisms under competing demands. Trends Ecol. Evol. **39**(2), 141–151 (2024)
164. Houston, A.I., Fromhage, L., McNamara, J.M.: A general framework for modelling trade-offs in adaptive behaviour. Biol. Rev. **99**, 56–69 (2024)
165. Valone, T.J.: Probabilistic inference and Bayesian-like estimation in animals: Empirical evidence. Ecol. Evol. **14**(7), e11495 (2024)
166. Budaev, S., Jørgensen, C., Mangel, M., Eliassen, S., Giske, J.: Decision-making from the animal perspective: bridging ecology and subjective cognition. Front. Ecol. Evol. **7**, 164 (2019)
167. Suddendorf, T., Bulley, A., Miloyan, B.: Prospection and natural selection. Curr. Opin. Behav. Sci. **24**, 26–31 (2018)
168. Cerecedo-Iglesias, C., Bartumeus, F., Cortés-Avizanda, A., Pretus, J.L., Hernández-Matías, A., Real, J.: Resource predictability modulates spatial-use networks in an endangered scavenger species. Move. Ecol. **11**, 22 (2023)
169. Kralj, J., Ponchon, A., Oro, D., Amadesi, B., Arizaga, J., Baccetti, N., Boulinier, T., Cecere, J.G., Corcoran, R.M., Corman, A.M., Enners, L., Fleishman, A., Garthe, S., Grémillet, D., Harding, A., Igual, J.M., Jurinović, L., Kubetzki, U., Lyons, D.E., Orben, R., Paredes, R., Pirrello, S., Recorbet, B., Shaffer, S., Schwemmer, P., Serra, L., Spelt, A., Tavecchia, G., Tengeres, J., Tome, D., Williamson, C., Windsor, S., Young, H., Zenatello, M., Fijn, R.: Active breeding seabirds prospect alternative breeding colonies. Oecologia **201**, 341–354 (2023)
170. Bakner, N.W., Ulrey, E.E., Collier, B.A., Chamberlain, M.J.: Prospecting during egg laying informs incubation recess movements of eastern wild turkeys. Move. Ecol. **12**, 4 (2024)
171. Ponchon, A., Scarpa, A., Bocedi, G., Palmer, S.C.F., Travis, J.M.J.: Prospecting and informed dispersal: Understanding and predicting their joint eco-evolutionary dynamics. Ecol. Evol. **11**, 15289–15302 (2021)
172. Liggett, T.M.: Voter models. In: Stochastic Interacting Systems: Contact, Voter and Exclusion Processes, pp. 139–208. Springer, Berlin (1999)
173. Reader, S.M.: Causes of individual differences in animal exploration and search. Top. Cogn. Sci. **7**, 451–468 (2015)
174. Bénichou, O., Redner, S.: Depletion-controlled starvation of a diffusing forager Phys. Rev. Lett. **113**, 238101 (2014)
175. Régnier, L., Dolgushev, M., Redner, S., Bénichou, O.: Complete visitation statistics of one-dimensional random walks. Phys. Rev. E **105**, 064104 (2022)

176. Gumbel, E.J.: Statistics of Extremes. Columbia University Press, New York (1958)
177. Pólya, G.: Über eine Aufgabe der Wahrscheinlichkeitsrechnung betreffend die Irrfahrt im Straßennetz. Math. Ann. **84**(1), 149–160 (1921)
178. Bonabeau, E., Dorigo, M., Theraulaz, G.: Swarm Intelligence: From Natural to Artificial Systems. Oxford University Press, New York (1999)
179. Camazine, S., Deneubourg, J.L., Franks, N.R., Sneyd, J., Theraulaz, G., Bonabeau, E.: Self-Organization in Biological Systems. Princeton University Press, Princeton (2001)

Chapter 2
Hybrid Methods in Reaction-Diffusion Equations

Tomás Alarcón, Natalia Briñas-Pascual, Juan Calvo, Pilar Guerrero, and Daria Stepanova

2.1 Introduction

Collective cell motion is ubiquitous in many biological phenomena, both in normal physiological processes and in pathological situations. For example, collective cell migration plays an essential role in vertebrate development. A system that is particularly amenable to both experimental and computational investigation is the neural crest. Neural crest cell populations are known to perform long-distance collective migration. If they do not reach the proper target or colonise an incorrect location, then improper cell differentiation or uncontrolled proliferation can ensue. This system has been studied extensively from a multidisciplinary perspective [1–3]. These investigations have shown that follow-the-leader behaviour, phenotype heterogeneity, and interactions with the extracellular microenvironment, specifically

T. Alarcón (✉)
Institució Catalana de Recerca i Estudis Avançats, Passeig Lluís Companys, Barcelona, Spain

Centre de Recerca Matemàtica, Cerdanyola del Vallès, Spain

Departament de Matemàtiques, Universitat Autònoma de Barcelona, Cerdanyola del Vallès, Spain

Barcelona Collaboratorium for Predictive and Theoretical Biology, Barcelona, Spain
e-mail: talarcon@crm.cat

N. Briñas-Pascual · P. Guerrero
Grupo Interdisciplinar de Sistemas Complejos, Departamento de Matemáticas, Universidad Carlos III de Madrid, Madrid, Spain

J. Calvo
Departamento de Matemática Aplicada and Research Unit "Modeling Nature" (MNat), Facultad de Ciencias, Universidad de Granada, Granada, Spain

D. Stepanova
Centre de Recerca Matemàtica, Barcelona, Spain
e-mail: dstepanova@crm.cat

L. Giuggioli, P. K. Maini (eds.), *The Mathematics of Movement*,
https://doi.org/10.1007/978-3-032-00873-2_2

the extracellular matrix, are essential to coordinate the movement of neural crest cells effectively. Angiogenesis, i.e. the growth of de novo vascular networks by sprouting from the existing vascular plexus, also requires coordinated cell migration. Similarly to cell migration in the neural crest, the formation of proper vascular patterns also requires phenotype heterogeneity and switching as well as interactions with the extracellular matrix [4].

Pathological situations such as the growth of solid tumours also depend upon coordinated cell movement during invasion. This phenomenon demonstrates the emergence of cooperation between heterogeneous phenotypes to facilitate invasion within a dysregulated microenvironment, which is reminiscent of the collective cell motion occurring in development or wound-healing.

Collective cell motion, specifically invasion, has been modelled using several approaches, namely, individual-based models, stochastic models, and continuum models [2]. In particular, to study invasion phenomena, a commonly used framework is that of reaction-diffusion systems (RD hereafter—both in the stochastic and deterministic versions). Within this framework, the collective behaviours associated with invasion are usually analysed in the context of travelling wave (TW) solutions. TWs are solutions where invasion takes place as a moving front that advances at constant velocity.

Reaction-diffusion models are used in fields as diverse as physics, biology, and chemistry, to mention just a few. The fundamental entities in such models are individuals (molecules, animals, cells, etc.), which move and react stochastically. For large systems, a purely individual-based approach becomes computationally inefficient. If the size of the system is large enough, one can use several techniques to derive macroscopic limits of the microscopic (individual-based) model. These limits often take the form of partial differential equation (PDE) models, in which the stochastic behaviour of individuals is replaced by an averaged version of the system, usually referred to as the mean-field limit.

However, in many practical situations, the number of individuals is large in certain spatial regions and small in others, and the mean-field limit fails to capture the behaviour of the system in some regions having just a few individuals, whereas the stochastic or individual-based model may be inefficient in other regions where the number of individuals is huge.

To overcome this problem, the so-called hybrid methodologies have been developed. Such methods combine the mean-field and stochastic descriptions of the system into a single scheme in such a way that one takes advantage of both the accuracy of the stochastic description and the efficiency of the mean-field limit. The basis of all such methods is to use the stochastic system in some parts of the domain and its mean-field limit, i.e. the corresponding discretised PDE model, in some other regions. Therefore, all versions of a hybrid methodology must address two key issues. First, the hybrid method must provide a criterion (or criteria) for, at each time, deciding which regions of the system will be treated as deterministic and which ones are considered stochastic. The second issue hybrid methods must tackle is the definition of the transition region (or interface) between deterministic and stochastic subdomains. A standard assumption is that the interface between the two domains is chosen so that both representations (mean-field and stochastic) describe the system with sufficient accuracy. Such a criterion ensures that the errors associated with the fluxes of individuals between the stochastic and the deterministic regions are small.

The paradigmatic example that has been used as a benchmark for most proposals of hybrid methods in RD systems is that of TWs in one spatial dimension (1D). TWs are particular solutions of RD systems in which a stable phase invades an unstable one. TWs take the form of fronts propagating at a constant speed (asymptotically). A specific example of an RD system that exhibits this type of solution is the Fisher-Kolmogorov equation (also known as the Fisher-Kolmogorov-Petrovsky-Piskunov or FKPP equation; see, e.g. [5] and references therein). The FKPP equation provides a suitable test case against which the performance of hybrid methods can be tested. The front of a given travelling wave represents a transition region between a domain at the back of the moving front, where the population described by the FKPP equation is large (close to its carrying capacity), and a region ahead of the front, where the population is much lower (in fact, close to zero, i.e. close to extinction). As a result, the latter region would be better suited for a stochastic description, whereas the former can be described using the mean-field version without significant loss of accuracy. The interface can be therefore located within the front region at a position that allows for both descriptions to be applicable. Of course, such an interface should advance at (roughly) the same speed as the front.

Thus, the basic hybrid scheme should account for three key aspects, namely, (i) a criterion to decide which regions are mean-field and which ones are stochastic, (ii) appropriate boundary conditions at the interface so no artefacts are created by it, and, finally, (iii) a prescription for where the interface should be collocated so that it moves at the same rate as the fronts within the system. Several such proposals are summarised in Sect. 2.1.1.

However, this basic scheme falls short in many problems of interest, namely, as soon as we consider more than one species. RD systems can have a much richer range of behaviours than TWs, such as Turing-like patterns, standing waves, scroll waves, and more. Hybrid methods should be formulated so that they can cope with several interacting species and therefore account for these rich dynamics. In particular, hybrid methodologies should be able to account for more than one interface at once.

Another limitation of the basic hybrid schemes is more fundamental. The reaction-diffusion master equation (RDME), frequently used to describe system dynamics in stochastic subdomains, is afflicted with a serious problem. Namely, it does not converge to its microscopic description in dimensions greater than one [6], even though a meaningful PDE exists in the limit of compartment size going to zero [7]. The reason for the breakdown of the RDME as the lattice size tends to zero is that the waiting time for bimolecular reactions to occur tends to infinity, thus leading to accuracy deterioration as a consequence of mesh refinement, whereby mesh size must be taken to be much greater than the reaction radius between molecules. Two strategies have been used to address this problem. One approach is to propose several *generalised reaction rates* to address the lower-bound issue [8, 9].

The second strategy is to formulate a convergent version of the master equation for reaction-diffusion systems. The convergent RDME (cRDME) is a general approach proposed by [10], whereby an approximation of the Doi model for the binary reactions, $A + B \rightarrow C$, is put forward. In this model, molecules within different voxels are allowed to interact through bimolecular reactions, which are allowed to occur between particles in voxels that lie within a predefined reaction radius, thus overcoming the artefact of vanishing reaction rates as voxel size tends to zero. Thus, to move the current schemes from one to a higher dimension, one needs to develop hybrid methods for the cRDME.

On top of all this, we should also keep in mind that population dynamics models are often multiscale models, where cells have intracellular dynamics that regulate behaviour as a function of external inputs or cues (e.g. proliferation regulated by nutrient availability). The intracellular state is sometimes accounted for by a structure variable. We should also formulate hybrid methods for stochastic multiscale models of cell populations that extend the remit of existing hybrid methods for reaction-diffusion systems, by extending current efforts to structured (e.g. age-structured) populations.

In this chapter, we review our recent efforts to try and fill these gaps. The rest of this Introduction is devoted to reviewing the landscape of hybrid methods of models for RD systems. Among these, the case of multiscale population models is specifically discussed in Sect. 2.1.3. We then move on to summarise a hybrid method that accounts for several species and several interfaces, in Sect. 2.2. Next, in Sect. 2.3, we formulate a hybrid method for the cRDME. Finally, in the Discussion, we present our conclusions and elaborate on future directions.

2.1.1 Hybrid Methods in a Nutshell

RD systems describe the behaviour of various chemical, physical, and biological phenomena, from the formation of spatial patterns in nonlinear chemical reactions to the dynamics of populations. These systems model the spatio-temporal evolution of the concentrations (densities) of interacting species over time and space.

Usually, when studying these systems, there exist different techniques which exploit the existence of different inherent scales. Scale separation can be with regard to time scales, spatial scales, or the size of different species, among others. There exist four different scales in modelling algorithms that are relevant to our discussion. The macroscopic scale is the coarsest scale, where the number of particles of the system is large enough that species abundance can be represented by densities. Within this regime, PDEs are used. Within the mesoscopic scale, particles are compartmentalised in different subdomains. Particle distribution within compartments is assumed to be homogeneous. The number of particles changes because of diffusion between compartments and reactions within compartments. We will assume that the mesoscopic dynamics is described by a continuous-time Markov process. The microscopic scale is a finer scale where particles are studied as individuals with continuous trajectories. An example of this dynamics is Brownian motion. In the case of cellular systems, there is an intracellular scale, where a high level of detail is used to describe sub-cellular processes, such as signalling pathways (e.g. the cell cycle). Combining any two of these scales and their respective modelling tools into a hybrid model allows for the integration of the strengths of both methodologies.

The nature of reactions occurring in a biological system varies significantly and is influenced by factors such as the number of particles, the types of species involved, and other specific conditions. Chemical reactions in biological systems can be categorised as either fast or slow, based on the species involved and the rate of each reaction. This approach can be applied to reaction-diffusion systems by modelling diffusion deterministically and the remaining reactions stochastically, like the method described in [11], where a hybrid tau-leaping method is employed to first simulate the stochastic reactions and then use a deterministic diffusion operator. Also, in [12], an adaptive partitioning strategy is implemented based on the fractional propensity functions for each reaction or diffusion jump event. High-frequency processes are simulated using deterministic equations and low-frequency processes by applying the Gillespie algorithm [6].

2.1.1.1 Spatially Coupled Hybrid Methods

Systems that can be considered spatially homogeneous and have no further internal structure are typically amenable to simple simulation approaches. Hybrid methodologies can still play a role in cases where the evolution of very abundant species is coupled to much less abundant species, such as piecewise deterministic Markov processes [13, 14]. In general, more benefits are derived from the use of hybrid methods when we are describing systems with additional independent variables besides time. The most usual situation is that of systems with spatial degrees of freedom.

For concreteness, consider the Fisher-Kolmogorov system and its travelling wave solution. In this regime, the number of particles ahead of the front is much smaller than the number of particles behind it, where they approach the carrying capacity. Therefore, while the deterministic, mean-field PDE describes the system accurately behind the front, fluctuations at and, more importantly, ahead of the front must be considered. This is achieved by dividing the domain into two subdomains: one where the model is treated stochastically due to the low number of particles and another where a higher number of particles allow for a deterministic model. Note that in this specific situation, both subdomains are connected intervals. This need not be the case in more elaborate scenarios.

This approach of partitioning the domain into stochastic and deterministic subdomains has been implemented following several schemes and methodologies, which we review below. In general, hybrid methods are based on the idea of connecting the subdomains. Within the first region, populations are large enough to justify the use of a PDE. In the other region, stochastic algorithms such as

Table 2.1 Different hybrid methods

Reference	Domain decomposition	Type of interface	Other characteristics
Ferm et al. [15]	PDE + SSA + τ-leaping	Dynamic	Adaptive time steps
Flegg et al. [16]	Compartment-based model + Brownian dynamics	Fixed, no overlapping	Interface balanced flux
Franz et al. [17]	PDE + Brownian dynamics	Overlapping/empty intersection	Accurate simulation of mean and variance in the BD subdomain
Spill et al. [18]	PDE + Gillespie	Dynamic, overlapping	Adaptation for modelling competition of species
Harrison et al. [19]	PDE + compartment-based stochastic model	Dynamic, adaptive	Flux-based dynamics, adaptation for multiple species
Smith et al. [20]	PDE + compartment-based stochastic model	Fixed, zero flux boundary condition	Deterministic domain growth

the Gillespie algorithm must be applied. The main difference between the various hybrid methods lies in how the interface between regions is defined and the so-called interface conditions. A fixed interface can be considered, so that the boundaries between subregions do not change over time. Conversely, a dynamic interface can be defined, so the new boundaries of each subdomain are computed at each time step. Another characteristic to be considered is the part of the domain shared by both regions at the same time. They can be independent from one another or they can overlap. In the latter case, a third region must be considered where the stochastic and the deterministic algorithms coexist. See Table 2.1 for an account of several of the possibilities that are found in the literature.

The hybrid method builds upon these stochastic models by providing a computationally efficient framework that combines deterministic and stochastic regions, allowing for accurate simulation of travelling waves in complex environments.

We provide a generic description of a hybrid algorithm for a RD model. It is a basic version that relies on the default implementation of Gillespie algorithm [6].

Generic Hybrid Algorithm

1. Set initial time and provide initial settings.
2. Given a current time t, calculate the time to next event, τ, via the Gillespie algorithm [6]. This time is computed as

$$\tau = -\frac{1}{\sum_k a_k} \log(r_1), \tag{2.1}$$

   where a_k denotes any of the non-zero propensity functions of the model and r_1 is a random number sampled from a uniform distribution in the unit interval, i.e $r_1 \sim \mathcal{U}(0, 1)$.

3. Determine which event or reaction is happening at $t + \tau$. This is done by taking an independent second random number $r_2 \sim \mathcal{U}(0, 1)$ and looking for the event tag l such that

$$\frac{\sum_{k=1}^{l-1} a_k}{\sum_k a_k} < r_2 \leq \frac{\sum_{k=1}^{l} a_k}{\sum_k a_k}. \tag{2.2}$$

4. Advance up to the new time $t+\tau$ in the deterministic region via the corresponding partial differential equation.
5. Update interface positions and overlapped regions.
6. Return to step 2 and iterate until a prescribed final time is reached.

Some points above need a more detailed explanation. Again, these are easily understood if we have in mind the situation of a single TW moving from left to right.

Interface Location

The determination of the voxel(s) that constitutes the interface region must be done on the basis that both the deterministic and the stochastic description should hold true at the same time. This is crucial, as interactions with neighbouring voxels will be both deterministic and stochastic. Therefore, we need to enforce large concentrations in those interface voxels. Their location is based on a trade-off between performance and accuracy, since the mean-field PDEs neglect fluctuations but are numerically cheaper. Thus, expanding the mean-field domain improves performance. For large enough populations, fluctuations usually scale with the square root of the number N of individuals of a given species in a compartment k, $\sqrt{N(k)}$. A threshold Θ is set to classify each compartment as stochastic if $N(k) < \Theta$, or deterministic otherwise.

In the simple situation of TWs, assuming that the TW moves to the right, one sweeps the spatial domain from left to right, and we place the interface at those voxels that are highly populated but to their right the concentration drops widely, as represented in Fig. 2.1.

Fluxes and Reactions at Interfaces

For simplicity, consider a single interface compartment, located at position k_I. Three different events contribute to the dynamics. Fluxes into and from the deterministic compartment $(k_I - 1)$ are modelled via the mean-field equations. Fluxes into and from the stochastic compartment $(k_I + 1)$ and local reactions at k_I are accounted for by transition rates in a stochastic way.

Moving the Interface

As the number of particles for the species we are monitoring changes over time, the boundaries between the stochastic and deterministic regions need to be adjusted too, so the interfaces between them must be updated accordingly. This is done using the rules of interface placement described above.

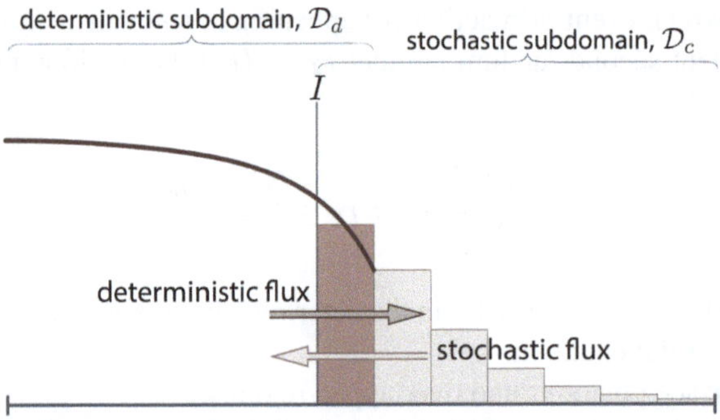

Fig. 2.1 Schematic of a hybrid model domain division. The domain is divided into a deterministic domain and a stochastic domain separated by an interface (dark brown region). The deterministic flux and the stochastic flux indicate the transfer of particles across the interface, ensuring mass conservation and accurate coupling between the deterministic and the stochastic model

Additionally, if the number of compartments in one region becomes too small, it may be more computationally efficient to merge this region with the other. Therefore, applying a minimum domain size condition helps to maintain efficiency by preventing very small regions.

Once the interface is properly allocated, we must ensure that all particle numbers in the stochastic domain are integer values and therefore density functions have to be updated too.

Renormalisation of Particle Distribution

When the interface changes its location, a compartment previously treated as deterministic can become part of the stochastic domain. Thus, after this is done we may get a particle number $N(k)$ which is not a natural number. We remedy this as follows. The fractional part

$$p = N(k) \mod (1) \tag{2.3}$$

is interpreted as the probability that an additional particle is in this compartment. By generating a random number $r \sim \mathcal{U}(0, 1)$ we can determine the allocation of the particle: if $r < p$ the particle is added to the box k; otherwise, it is placed in the mean-field domain. In the first case, the value $N(k)$ is increased by $(1 - p)$, thus becoming an integer; the compartment densities are already scaled accordingly. In the second case, $N(k)$ is decreased by p, and the densities are recomputed.

2.1.2 *The Hybrid Method in the Context of Pulled Front Theory*

Deterministic models of pulled fronts describe the propagation of waves in reaction-diffusion systems where the leading edge of the wave determines the speed. These models are essential for understanding the minimal wave speed required for the invasion of homogeneous media. The classical example is the FKPP equation, for which we can predict the speed of pulled fronts based on the reaction and diffusion coefficients. Namely, the wave speed v^* of a pulled front is given by:

$$v^* = 2\sqrt{Dr}, \tag{2.4}$$

where D is the diffusion coefficient and r is the growth rate. Recall that the FKPP equation was introduced to describe the spread of advantageous genes in a population. In this deterministic setting, this minimal wave speed is crucial for understanding how the advantageous trait invades the population.

In contrast, pushed fronts and waves in bistable media exhibit different propagation dynamics. Pushed fronts rely on the entire wave profile for propagation, while bistable waves require a threshold concentration for wave advancement. The distinction between pulled and pushed fronts is crucial for understanding wave dynamics in RD systems. Pulled fronts are driven by the leading edge, while pushed fronts depend on the bulk of the wave. Bistable waves, on the other hand, require a critical concentration threshold to propagate. Hybrid methods for RD systems need to be adapted specifically to handle these distinct scenarios.

In many instances of applying hybrid methods to RD systems, the focus is on the dynamic description of the propagation of fronts into unstable states. In the deterministic description, the state of the system in the region far ahead of the front is linearly unstable. We can therefore distinguish between the so-called pulled and pushed fronts in more technical terms as follows. For pulled fronts the asymptotic front speed (as time grows large) agrees with the linear spreading speed v^* of infinitesimal perturbations around the unstable state—the perturbation ahead of the front spreads at speed v^* and meanwhile pulls the rest of the front—whereas for asymptotic front speeds strictly greater than v^*, we talk about pushed fronts because it is the nonlinear growth in the region behind the leading edge that controls the overall dynamics, as a truly nonlinear effect. In the classical FKPP equation, pushed fronts correspond to travelling wave solutions with a speed greater than v^*. In this particular case, we can relate these notions to the stability properties of those travelling solutions, as it is the wave with the slowest admissible speed, v^*, the one that attracts the dynamics for initial conditions that do not fall off too slowly at infinity including, in particular, compactly supported initial conditions, which are the relevant ones for our discussion here (see [5]).

For more general FKPP equations of the form

$$u_t = u_{xx} + f(u), \tag{2.5}$$

where $f(0) = f(1) = 0$, $f'(0) = 1$, $f'(1) < 0$, there is a result ensuring that when $f(u)/u \leq f'(0)$ we will observe pulled fronts; this is to be expected, as under such conditions the nonlinearities will not be able to enhance growth and therefore we will not get pushed fronts. However, we lack a general mathematical framework which can predict whether, for a given RD equation, fronts will be pulled or pushed. The interested reader can have a look at [21, 22] and references therein for more information on these matters. To this point, we mention that in our discussion so far, the FKPP equation assumes a homogeneous medium, where the reaction and diffusion coefficients are constant; those restrictions can be lifted, and as a consequence, we can get even richer dynamics (see [23, 24]).

Stochastic and Heterogeneous Models of Pulled Fronts

Deterministic pulled fronts often fail to capture the effects of demographic noise and heterogeneity present in biological systems. Extensive research has been conducted on stochastic and heterogeneous models of pulled fronts. These models incorporate fluctuations and spatial variability, offering a more accurate depiction of wave propagation in natural systems. Key studies include work on large deviations and reaction-diffusion equations [25] and analysis of front speeds in heterogeneous media [24].

It seems therefore natural to bypass the aforementioned limitations with the development of hybrid methods, as they integrate stochastic and deterministic descriptions. In that regard, our focus is on stochastic descriptions given by discrete models, such as those reviewed in [21] (see also Section 7 in [22]). There are also stochastic descriptions based on continuous points of view, such as stochastic partial differential equations with multiplicative noise. Although these are not directly related to the family of hybrid models we review in this contribution, the reader can have a look at, for example, [21, 23, 26] and references therein.

Stochastic models introduce fluctuations that can significantly alter the dynamics of wave propagation. For instance, pulled fronts in noisy environments can exhibit subdiffusive behaviour, where the root mean square (rms) wandering of the front scales as $t^{1/4}$ rather than the $t^{1/2}$ scaling seen in deterministic or pushed fronts [23, 24]. Heterogeneous models further complicate this picture by introducing spatial variability in reaction rates and diffusion coefficients. Studies, such as those by [24], explore how deterministic heterogeneities influence front speeds using singular perturbation analysis and Hamilton-Jacobi dynamics. These models show that front speeds, profile, and location can vary significantly depending on the nature and scale of the heterogeneities, often requiring numerical simulations to capture the complex dynamics accurately [23, 24]. This is critical in various applications, from chemical kinetics to biological invasions, where spatial variations can significantly impact the spread of fronts.

Application to Pulled Fronts

Hybrid methods have primarily focused on pulled fronts due to their sensitivity to demographic noise at the leading edge. This is to be expected, as D. Panja notes [21]:

> It has to be noted that the majority of the studies on fluctuating fronts have been carried out on fluctuating "pulled" fronts (...). This is not very surprising. First of all, for the deterministic counterparts, it is the pulled fronts, for which there is a solid theoretical understanding. Secondly (and more importantly), it is the sensitivity of the pulled fronts to

the dynamics of the leading edge, which is severely affected by the discreteness of particles and the lattice, or by the presence of external noise terms.

Recall that for pushed fronts, their leading edges (which is the zone we would describe stochastically with a hybrid method) do not play any role in the front speed selection.

In the context of pulled fronts, hybrid methods have been particularly useful in studying the FKPP equation. The leading edge, characterised by low population densities, is treated stochastically, capturing the essential noise-driven dynamics. The bulk region, with higher densities, is modelled deterministically, reducing computational costs. This partitioning ensures that the front speed and shape are accurately simulated across the entire domain. However, applying these methods to bistable waves presents challenges, such as ensuring accurate interface placement and handling multiple stable states. Further research is needed to extend hybrid methodologies to these complex wave types.

Studies such as those by [26, 27] demonstrate the utility of hybrid methods in capturing the subdiffusive wandering of pulled fronts in noisy environments [24]. These methods bridge the gap between purely deterministic and purely stochastic models, providing a more complete and computationally feasible approach to studying complex wave propagation in reaction-diffusion systems.

2.1.3 Simulation Methods for Age-Structured Populations

In a nutshell, the dynamics of such complex biological systems, for example, those mentioned in Introduction, is an emergent property of all the layers involved. It is in this rich context where hybrid models for structured, spatially extended populations can be extremely helpful, as the computational burden to keep track of every detail of such systems can be overwhelming. There have been a number of contributions extending the remits of the hybrid methodology to structured RD systems, among which we mention the work done by [18, 28].

Including structure variables is particularly relevant when dealing with cell populations, as it allows for the incorporation of demographic noise through an age structure. Other applications are, of course, possible. Our main focus in this area has been multiscale models of tumour growth. Here, hybrid methods are particularly helpful as they can cope with the usual demographic noise present in RD systems with a reasonable computational cost. Thus, we can construct multiscale models that can account for fluctuations in both the number of cells and the intracellular dynamics. For the latter, a cell cycle model is adopted, where the age structure provides the coupling with higher levels of description.

The hybrid methodology put forward in [18, 28] has a number of distinctive features:

1. The decomposition into deterministic and stochastic domains is performed only with respect to the spatial variable.

2. A coarse-graining procedure is introduced, which integrates out the effects of the structure variable at those spatial regions where a deterministic description is used. The structure variable is fully taken into account in the stochastic regions, where the transition rates of the underlying birth-death-diffusion Markov process depend upon it.
3. The dynamic adaptation of the spatial interface involves a renormalisation procedure (explained earlier in this review) that might require the creation of new individuals. When this is necessary, new ages are sampled from the equilibrium age distribution, which can be theoretically computed.

The accuracy of this hybrid methodology is assessed by comparing it with full stochastic simulations, focusing on metrics such as the average position and speed of wavefronts [28]. Regardless of how severe these simplifications might appear, our hybrid methodology can retrieve the travelling wave speeds that are obtained with a fully stochastic model, at a much lower computational cost. In a nutshell, this methodology is versatile and can be easily adapted to other scenarios of interest. Future research in this area might focus on refining these procedures to run them in higher spatial dimensions.

2.2 Hybrid Method for Reaction-Diffusion Systems with Multiple Species

Reaction-diffusion (RD) models are a flexible framework that can be used to model the interactions between several (spatially extended) species in a variety of settings such as ecological populations, chemicals, and cellular lineages, among others. To proceed with the hybrid methodology in this case, the spatial domain is decomposed into regions for each species. However, since multi-species RD systems produce heterogeneous patterns (e.g. Turing patterns, spatio-temporal oscillations, scroll waves, etc.), the decomposition might differ between the species that are present, and the chance arises that at a certain spatial region, we find interacting species that are described using different mathematical frameworks (i.e. mean-field or stochastic).

Thus, the hybrid algorithm for multi-species RD systems involves multiple dynamic interfaces and key modifications to the standard setting to ensure efficient and accurate simulations. These adjustments are necessary to handle large and small stochastic time steps, deterministic interface reactions, and multiple interfaces for different species.

When the time to the next stochastic event, τ, is bigger than the maximum time step allowed by the numerical method used for solving the mean-field PDE, τ_{PDE}, several iterations of the finite difference scheme are required to evolve the PDE to the time τ. During these iterations, the interface is treated as a Neumann no-flux boundary to prevent the PDE solution from leaking into the stochastic domain. This ensures that the deterministic and stochastic regions remain accurately separated during the simulation.

Conversely, if the Gillespie time step is smaller than the PDE convergence time step, the algorithm is modified so that the mean-field domain is not iterated during every stochastic time step. Instead, it is iterated only after several stochastic steps, ensuring their cumulative time step is just under τ_{PDE}. This adjustment helps to maintain computational efficiency without compromising the accuracy of the simulation.

Reactions at the interface can also be described deterministically by modifying the governing equation of the interface compartment. This approach replaces the stochastic treatment of interface reactions with a deterministic update equation. This modification ensures that the transition rates describing stochastic reactions in the interface compartment are set to zero, providing a more stable and accurate representation of interface dynamics.

The complexity of managing multiple species in hybrid models necessitates the introduction of multiple dynamic interfaces. Different species may occupy distinct regions with high concentrations, requiring separate interfaces for each species. The interface condition must be applied separately for each species. In regions where some species are modelled stochastically and others deterministically, reactions involving both types of species are handled stochastically. This ensures accurate coupling between the stochastic and deterministic domains for different species.

The hybrid model for multiple species introduces several key differences from the single-species hybrid model. In the initialisation phase, multiple species are defined, each with its own set of parameters such as diffusion coefficients, reaction rates, and thresholds. Unlike the single-species model, where a single stochastic and deterministic domain is managed, the multi-species model requires separate management of interfaces for each species. This includes handling multiple dynamic interfaces that can move independently based on the local population of each species.

Moreover, the multi-species hybrid model must account for coupled reactions involving species in different domains, ensuring accurate interactions between stochastic and deterministic regions. This increases the computational complexity of the model, requiring more efficient handling of data structures and computational steps to manage the dynamics of multiple interacting species effectively.

The key differences with respect to single-species hybrid methods can be summarised as follows [18]:

- **Multiple Species Initialisation:** Unlike single-species models, multiple species are initialised, each with its own set of parameters (diffusion coefficients, reaction rates, thresholds).
- **Species-Specific Domains:** Each species may have its own stochastic and deterministic domains, requiring separate management of interfaces for each species.
- **Multiple Interfaces:** Separate and independently moving interfaces for each species, as opposed to a single interface in the single-species model.

- **Coupled Reactions:** Handling reactions that involve species in different domains, ensuring accurate coupling between stochastic and deterministic regions.
- **Computational Complexity:** Increased complexity due to multiple sets of dynamics and interactions, requiring more efficient handling of data structures and computational steps.

With all these considerations in mind, we can now give a generic description of a multi-species hybrid algorithm:

1. Initialise Model
    - **Define Spatial Domain**: Set up the spatial domain and discretise it into compartments.
    - **Initialise Species and Parameters**: Define multiple species, their initial populations, diffusion coefficients, reaction rates, and thresholds for switching between stochastic and deterministic descriptions.
    - **Set Up Interfaces**: Determine initial positions for interfaces between stochastic and deterministic domains for each species.
2. Main Loop
    - **Step 1: Generate Time to Next Stochastic Event**
        - Use Gillespie algorithm to determine the time step for the next event.
    - **Step 2: Determine Which Stochastic Event Occurs**
        - Identify the specific stochastic event that occurs based on the random number generated.
3. Update Deterministic Domain
    - **Finite Difference Scheme**: Apply the finite difference scheme to update the deterministic regions for each species.
    - For each species in the deterministic domain, update their concentrations using the PDEs.
4. Calculate Interface Conditions
    - **Stochastic to Deterministic Fluxes**: Calculate fluxes between stochastic and deterministic compartments for each species at the interfaces.
    - **Local Reactions at Interfaces**: Perform local reactions involving species from both stochastic and deterministic domains.
5. Move Interface if Necessary
    - **Check Thresholds**: For each species, check if the particle number in a compartment crosses the threshold, and adjust the interface position if necessary.
    - **Renormalise Particle Numbers**: Ensure the conservation of mass when adjusting the interfaces, and handle fractional particles appropriately.

6. Update Time
   - Increment the simulation time by the time step generated in step 1.
7. Repeat Until End Time
   - Continue looping through the steps until the specified end time is reached.

2.2.1 A Multiple Species Example: Hybrid Lotka-Volterra

The fine details concerning the algorithm steps we just outlined are better described by working out in detail an example with two species, including how to account for fluxes between regions, interface placement, and species interactions. The general case of more than two species, apart from combinatorial issues, does not bring in additional difficulties.

Therefore, the core of Sect. 2.2 will be constituted by the analysis of a hybrid model for a two species system, a classical example often described by the Lotka-Volterra equations; see Eq. (2.8). This system includes a stochastic component where individuals interact and move on a lattice and a deterministic component described by partial differential equations (PDEs). By integrating both methods, we aim to capture the detailed behaviour of individual organisms while also modelling the population-level dynamics, as will be seen below.

We explore a predator-prey model comprising predators (M) and prey (N). We denote by $M(k)$ and $N(k)$ the number of predators and prey, respectively, in voxel k. We shall write $M(t,k)$, $N(t,k)$ when we need to stress the temporal dependency.

The model is described in terms of its transition rates $W(B|A)$, which, to the first order in Δt, encode the probability of a system being in state A at time t to switch to state B during the time interval $[t, t+\Delta t)$, divided by Δt. For one-dimensional spatial systems, the state of each species is described via the state vector:

$$\mathcal{N} = (N(k_1), \ldots, N(k_{max})), \tag{2.6}$$

where $\{k_1, \ldots, k_{max}\}$ is the indexing set of the spatial lattice. Time dependency might be explicitly included if needed. The overall state of the system is described by the joint state vector of all the species. Hereafter we follow the convention that, when specifying transition rates, only the components of the state vector that change during the transition are explicitly written.

The stochastic model allows each species to migrate to neighbouring lattice sites. Prey reproduce at a rate a, while predators have a mortality rate c and reproduce by consuming prey at a rate b. The transition rates for these interactions are:

$$W\left(N(k)-1, N(k\pm 1)+1 | N(k), N(k\pm 1)\right) = \frac{D_N}{h^2} N(k),$$

$$
\begin{aligned}
W\left(M(k)-1, M(k\pm 1)+1|M(k), M(k\pm 1)\right) &= \frac{D_M}{h^2}M(k),\\
W\left(N(k)+1|N(k)\right) &= aN(k),\\
W\left(N(k)-1, M(k)+1|N(k), M(k)\right) &= bN(k)M(k),\\
W\left(M(k)-1|M(k)\right) &= cM(k),
\end{aligned}
\tag{2.7}
$$

where D_M and D_N denote the diffusion coefficients of $M(k)$ and $N(k)$, respectively, and h is the spatial discretisation size. For simplicity, we assume that each time a predator consumes a prey, a new predator is born. The mean-field and continuum limit lead to the classical spatial Lotka-Volterra equations:

$$
\begin{aligned}
\frac{\partial n}{\partial t} &= D_N\frac{\partial^2 n}{\partial x^2} + an - bnm,\\
\frac{\partial m}{\partial t} &= D_M\frac{\partial^2 m}{\partial x^2} + bnm - cm.
\end{aligned}
\tag{2.8}
$$

Here, $n = n(t, x)$ and $m = m(t, x)$ represent the prey and predator densities, respectively, analogous to $N(t, k)$ and $M(t, k)$ in the discrete model.

We solve Eqs. (2.8) using a finite difference approximation for spatial discretisation, using the same lattice as the stochastic model. For time integration, the fourth-order Runge-Kutta explicit method is used. All plots show normalised values, indicating the number of predators or prey in a box rather than their densities.

The model consists of four subdomains, determined by whether each species evolves deterministically or stochastically. For simplicity, we denote $N(k) = hn(k)$ and $M(k) = hm(k)$ where appropriate. We now specify which transition rates define the stochastic model and which PDEs correspond to the mean-field model solved in each subdomain. The interfaces between the subdomains are as described in Sect. 2.1.1.1:

1. Deterministic Predator and Prey System

   In this region, Eqs. (2.8) are solved, with the transition rates from Eqs. (2.7) set to zero, ensuring no stochastic reactions occur.
2. Deterministic Predator and Stochastic Prey

   The deterministic equations are:

$$
\frac{\partial n}{\partial t} = 0, \quad \frac{\partial m}{\partial t} = D_M\frac{\partial^2 m}{\partial x^2} - cm,
\tag{2.9}
$$

   and the transition rates are:

$$
\begin{aligned}
W\left(N(k)-1, N(k\pm 1)+1|N(k), N(k\pm 1)\right) &= \frac{D_N}{h^2}N(k),\\
W\left(M(k)-1, M(k\pm 1)+1|M(k), M(k\pm 1)\right) &= 0,
\end{aligned}
$$

$$
\begin{aligned}
W\left(N(k)+1 \mid N(k)\right) &= aN(k), \qquad (2.10)\\
W\left(N(k)-1, M(k)+1 \mid N(k), M(k)\right) &= bN(k)M(k),\\
W\left(M(k)-1 \mid M(k)\right) &= 0.
\end{aligned}
$$

3. Stochastic Predator and Deterministic Prey The deterministic equations are:

$$
\frac{\partial n}{\partial t} = D_N \frac{\partial^2 n}{\partial x^2} + an, \qquad \frac{\partial m}{\partial t} = 0, \qquad (2.11)
$$

and the transition rates are:

$$
\begin{aligned}
W\left(N(k)-1, N(k\pm 1)+1 \mid N(k), N(k\pm 1)\right) &= 0,\\
W\left(M(k)-1, M(k\pm 1)+1 \mid M(k), M(k\pm 1)\right) &= \frac{D_M}{h^2} M(k),\\
W\left(N(k)+1 \mid N(k)\right) &= 0, \qquad (2.12)\\
W\left(N(k)-1, M(k)+1 \mid N(k), M(k)\right) &= bN(k)M(k),\\
W\left(M(k)-1 \mid M(k)\right) &= cM(k).
\end{aligned}
$$

4. Stochastic Predator and Stochastic Prey
   In this scenario, both species evolve stochastically according to the transition rates defined by Eq. (2.7).

We now explore two scenarios within the spatial Lotka-Volterra system and compare the hybrid model to purely stochastic and deterministic models. Both scenarios involve solutions of the PDE that oscillate in space and time, but in one case, the oscillations bring the prey population so close to zero that extinction is possible in the stochastic model.

2.2.2 Oscillatory Behaviour Without Observable Extinction

In a domain of length $L = 20$ divided into $k_{max} = 101$ boxes ($h = 0.2$), the initial condition is a spatially homogeneous distribution of prey and predators, with $N(k, t = 0) = 50$ and $M(k, t = 0) = 5$. With Neumann boundary conditions, Eq. (2.8) maintain spatial homogeneity at all times, and both populations oscillate in time.

Figure 2.2 illustrates typical results, showing that peaks in prey numbers are followed by peaks in predator numbers. For the chosen parameter values, the minimum number of individuals of either species remains sufficiently large to make extinction in the stochastic model unlikely.

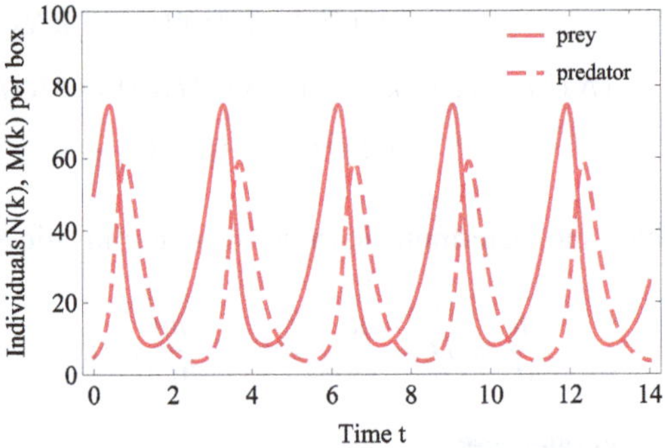

Fig. 2.2 The solution of the Lotka-Volterra Eq. (2.8) with parameters $D_N = D_M = 1$, $a = 2$, $b = 0.1$, $c = 3$ for spatially homogeneous initial conditions $N(t = 0, k) = 50$, $M(t = 0, k) = 5$. The plot shows the time evolution of the number of prey and predators in any given box k in the discretisation, for better comparison with the stochastic model. Since the diffusion terms do not contribute in the spatially homogeneous case, this solution is identical to the solution of the Lotka-Volterra ODEs

The results for the stochastic and hybrid models, using two different threshold values, are discussed in [18] observing differences in the result if the threshold is large. Figure 2.3 compares purely stochastic and hybrid methods with $\Theta = 10$. The left column illustrates the spatial distribution of predator and prey populations within a specific box at time $t = 4.1$, both for an individual simulation and the average of 256 simulations. When comparing the predator and prey populations, it is evident that the average values for both the stochastic and hybrid models are quite similar for the given threshold (refer to Fig. 2.3a, c). The individual simulations, however, show significant differences. As expected, the stochastic model exhibits considerable noise in predator and prey populations across the domain, whereas the hybrid model shows reduced noise once population sizes exceed the threshold. Figure 2.3c demonstrates that predator numbers are above the threshold, resulting in a smooth spatial distribution, although this distribution is not homogeneous, unlike the average profile.

The right column of Fig. 2.3 presents the temporal evolution of the spatial average number of prey, calculated as $\langle N \rangle_k = \frac{1}{k_{max}} \sum_{k=1}^{k_{max}} N(k)$. The mean values for the stochastic model (Fig. 2.3b) closely resemble those for the hybrid model, (Fig. 2.3d). In all cases, the system exhibits oscillatory behaviour around $n = 30$, the steady state of the corresponding PDE, with a diminishing amplitude over time. This contrasts with the PDE dynamics (Fig. 2.2), where oscillation amplitudes remain constant. Stochastic fluctuations cause the oscillations to desynchronise and average out, leading to a more pronounced damping effect in the mean of all realisations compared to individual ones, where amplitudes may vary over time. This indicates that the PDE does not accurately represent the mean over different realisations. Additionally, the spatial mean predator plots for the stochastic and hybrid models with different thresholds may look different; however, this variability is also seen

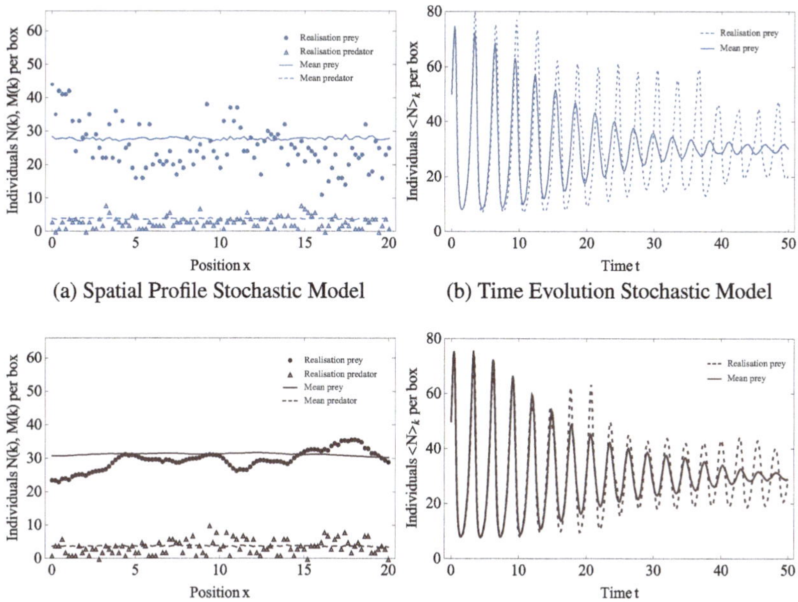

(a) Spatial Profile Stochastic Model (b) Time Evolution Stochastic Model

(c) Spatial Profile Hybrid $\Theta = 10$ (d) Time Evolution Hybrid $\Theta = 10$

Fig. 2.3 Simulations of the spatial Lotka-Volterra Model with parameters $D_N = D_M = 1$, $a = 2$, $b = 0.1$, $c = 3$, $k_{max} = 101$, $h = 0.2$ and initial values $N(t = 0, k) = 50$, $M(t = 0, k) = 5$ for $k = 1, \ldots, k_{max}$. We compare **(a)**–**(b)**, the stochastic model, to **(c)**–**(d)**, the hybrid model, with thresholds $\Theta = 10$. The figures on the left show the spatial profile at time $t = 4.1$ of the number of predators and prey of a single realisation as well as the mean of 256 different realisations, whereas the figures on the right show the time evolution of the spatial average of numbers of prey in a single realisation and the mean of realisations. The corresponding PDE solution is shown in Fig. 2.2

Table 2.2 Average prey population over space, time, and 256 realisations, along with predator numbers. The standard deviation, for prey and predators, are also displayed. Results for the corresponding mean-field model, where there are no different realisations, and thus no standard deviation, are included

	Stochastic	Hybrid $\Theta = 10$	Mean field
$\overline{\langle N\rangle}_{t,k}$	30.50 ± 0.05	30.30 ± 0.16	30.47
$\overline{\langle M\rangle}_{t,k}$	20.35 ± 0.04	20.22 ± 0.13	20.20
$\sqrt{\overline{\langle N\rangle^2_{t,k}} - \overline{\langle N\rangle}^2_{t,k}}$	16.0 ± 0.3	15.4 ± 0.4	
$\sqrt{\overline{\langle M\rangle^2_{t,k}} - \overline{\langle M\rangle}^2_{t,k}}$	12.8 ± 0.3	12.3 ± 0.4	

in different realisations of the same model. Therefore, quantitative measures are necessary for a proper comparison (see Table 2.2).

We calculate the spatial and temporal average of the mean number of prey across 256 different realisations,

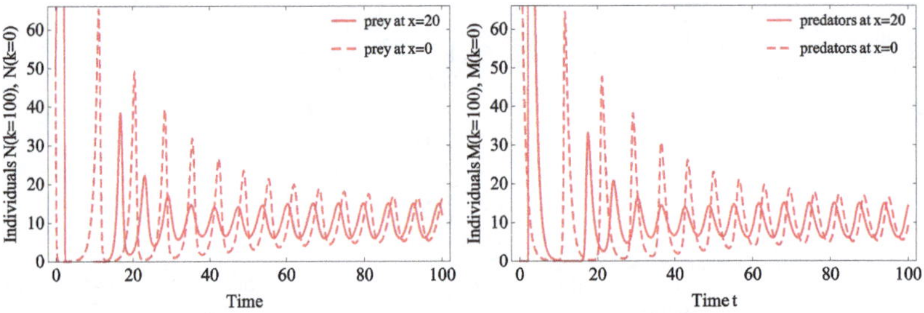

Fig. 2.4 Number of prey (left) and predators (right) at the boundaries ($x = 0$ and $x = 20$) over time, obtained from the deterministic model Eq. (2.8). The initial sharp rise in prey numbers at $x = 20$ is due to the lack of predators in that region and is followed by a predator peak. Early oscillations show population numbers close to zero, indicating possible extinction in the stochastic model. Later times show oscillations with minima sufficiently above zero, reducing the likelihood of extinction

$$\overline{\langle N \rangle}_{t,k} = \frac{1}{256 k_{max} t_{max}} \sum_{r=1}^{256} \sum_{k=1}^{k_{max}} \int_0^{t_{max}} N_r(t,k) dt,$$

over a simulation time of $t_{max} = 50$, and similarly for predators. This provides a single value with high statistical significance, facilitating easier comparison as individual time points may have oscillations that are out of synchrony. The standard deviation of the spatio-temporal averages, $\sqrt{\overline{\langle N \rangle^2_{t,k}} - \overline{\langle N \rangle}^2_{t,k}}$, for prey and predators shows close agreement between the stochastic model and the hybrid model with $\Theta = 10$. The hybrid model effectively reproduces the stochastic measures, such as standard deviation, with better accuracy for higher threshold values.

2.2.3 Extinction and Blow-Up

In this analysis, we set $k_{max} = 101$ and we consider an initial condition where prey are uniformly distributed, $N(t = 0, k) = 50$ for $k = 1, \ldots, k_{max}$, while predators are concentrated on the left side of the domain, with $M(k) = 100$ for $k = 1, \ldots, 9$, $M(10) = 98$, $M(11) = 50$, $M(12) = 2$, and $M(k) = 0$ for $k = 13, \ldots, k_{max}$. The boundary conditions are of Neumann type, and the parameters are set to $D_N = D_M = 1$, $a = 1$, $b = 0.1$, $c = 2$, with a domain length $L = 20$. This setup models a predator invasion into a prey population, causing spatial and temporal fluctuations even in a deterministic framework (Fig. 2.4).

Figure 2.4 illustrates the variations in predator and prey numbers at the boundaries ($x = 0$, $x = 20$) over time for the mean-field model. During the initial oscillations, both predator and prey populations dip close to zero. In the stochastic model, where population numbers are integers, values below one in the deterministic model suggest a high likelihood of extinction. Later on, regular oscillations with

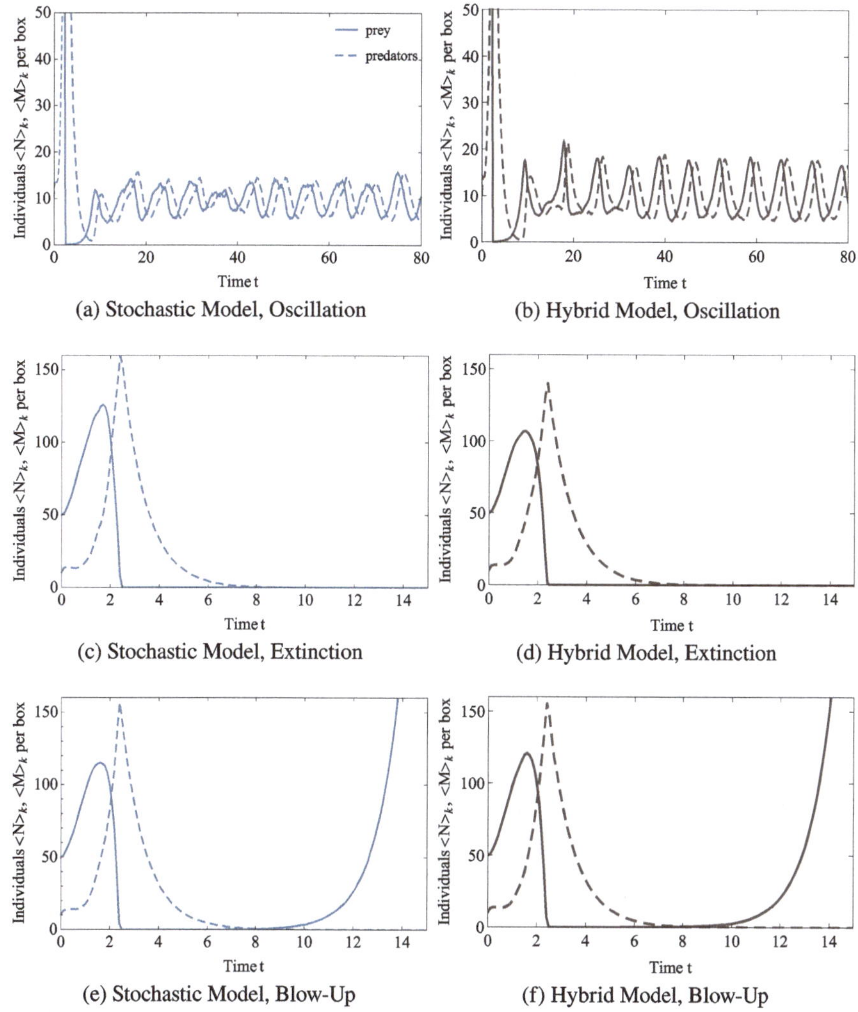

(a) Stochastic Model, Oscillation (b) Hybrid Model, Oscillation

(c) Stochastic Model, Extinction (d) Hybrid Model, Extinction

(e) Stochastic Model, Blow-Up (f) Hybrid Model, Blow-Up

Fig. 2.5 Simulations of the spatial Lotka-Volterra model with parameters $D_N = D_M = 1, a = 1, b = 0.1, c = 1, L = 20$, and $k_{max} = 101$. The spatial average $\langle N \rangle_k = \frac{1}{k_{max}} \sum_{k=1}^{k_{max}} N(k)$ of the number of predators and prey in a box is plotted over time for three realisations of the stochastic model (**a**, **c**, **e**), and three realisations of the hybrid model, (**b**, **d**, **f**),, with a threshold of $\Theta = 10$. These realisations show three different outcomes: oscillatory behaviour, extinction of both species, or predator extinction followed by prey population explosion

minima significantly above zero are observed, indicating that if extinction occurs, it is most probable during early times.

Figure 2.5 confirms these expectations. Figure 2.5a demonstrates that in the stochastic model, the spatial average of predator or prey numbers can exhibit oscillations similar to those in the deterministic model. However, extinction events

Table 2.3 Extinction probability for prey and predator populations in the Lotka-Volterra system for the described scenario. This probability is calculated by conducting 256 simulations for each model (stochastic and hybrid) with thresholds of 10, 25, 50, and 100 and recording at time $t = 80$ whether a population is extinct

	Stochastic	Hybrid 100	Hybrid 50	Hybrid 25	Hybrid 10
Prey	0.46	0.45	0.46	0.45	0.43
Predator	0.59	0.57	0.59	0.61	0.56

can also occur. Figure 2.5c shows that if prey die out first, predators will necessarily follow. Conversely, if predators go extinct first, prey numbers will rapidly increase (see Fig. 2.5e). Figure 2.5b, e, and f demonstrate that the hybrid model can replicate each of these three distinct scenarios. Since the mean and standard deviation statistics are not meaningful if the prey population explodes, we compare the extinction event frequencies in the stochastic and hybrid models.

The results in Table 2.3 indicate that the hybrid model with a threshold of $\Theta = 10$ or higher has a similar extinction probability as the stochastic model. Therefore, the hybrid model is a good approximation of the stochastic model in this case. Finally, we evaluate the performance gain achieved by using the hybrid model.

2.3 Hybrid Method for Convergent Reaction-Diffusion Master Equation

Stochastic reaction-diffusion systems are commonly modelled using the compartment-based reaction-diffusion master equation (RDME). This method divides the domain into non-overlapping compartments (voxels) to capture spatial inhomogeneities. In this mesoscopic description, each molecule's position is known up to the voxel size, with molecules within the same voxel treated as indistinguishable and well-mixed. Diffusion between compartments is modelled as a continuous-time random walk, and chemical reactions occur independently within each compartment. The RDME approach is based on the assumption of the existence of propensity functions (which is an alternative term to refer to transition rates in the context of chemical kinetics). For well-mixed systems, propensities are functions of the number of reactant molecules present in the system. However, when molecular species in compartments are present in low quantities, the well-mixed assumption ceases to be valid, and a straightforward generalisation of the master equation to reaction-diffusion systems may produce incorrect results. Specifically, there exists a lower bound for the size of each compartment below which the probability of bimolecular reactions vanishes [7, 29]. This results in waiting times for bimolecular reactions tending to infinity as the lattice size decreases, causing inaccuracies with finer meshes. Therefore, for the RDME approach to be valid, the mesh size must be significantly larger than the reaction radius between molecules.

Two strategies have been employed to address this issue. One method introduces generalised reaction rates (propensities) for bimolecular reversible reactions of the type $A + B \rightleftarrows C$, which are allowed to occur between neighbouring voxels [8, 9]. To distinguish between reactions occurring within the same voxel and those between neighbouring ones, an additional correcting parameter is introduced [8]. This parameter can be derived by matching the mean binding-rebinding times at micro- and mesoscales, resulting in the corresponding reaction rates for the Smoluchowski interaction model in 2D and 3D. This modified RDME is accurate down to a smaller lower mesh size bound, approximately equal to the interaction radius.

A different and more general approach was proposed by [10]. This work introduces the convergent RDME (cRDME) for bimolecular reactions of type $A + B \rightarrow C$ within the Doi interaction model. The cRDME formulation allows interaction between molecules from different voxels that lie within a predefined reaction radius. This addresses the issue of vanishing reaction rates as voxel size approaches zero. The corresponding reaction rates are obtained from a finite volume discretisation of the spatially continuous stochastic reaction-diffusion model. It was demonstrated that for large mesh sizes, the cRDME coincides with the classical RDME, thus extending the existing approach. The cRDME can be simulated using the traditional stochastic simulation algorithm (SSA) [6] or its more efficient formulations, such as the next subvolume (NSV) method [30].

While mesoscopic approaches offer computational advantages over microscopic molecular-based simulations, they often encounter practical limitations in real-world applications. As a result, alternative methods are frequently needed to simulate complex reaction-diffusion systems more effectively. In this section, we introduce one such approach, which consists in formulating a hybrid method for the cRDME. As an application, we choose the stochastic Fisher-Kolmogorov system, which is particularly well-suited to study the implementation of hybrid methods, as it exhibits travelling wave behaviour [18, 27, 28, 31]. We begin with a brief description of this system within the convergent framework in Sect. 2.3.1. In Sect. 2.3.2, we proceed with a description of our implementation of a hybrid method for cRDME. We finally compare its performance with the purely stochastic simulations via the NSV method in Sect. 2.3.3.

2.3.1 Convergent Formulation of the Stochastic Fisher-Kolmogorov Model

The stochastic Fisher-Kolmogorov (sFK) system for a single species is described by two processes, corresponding to branching (birth) and binary annihilation

$$A \xrightarrow{\bar{\lambda}} 2A,$$

$$2A \xrightarrow{\bar{\mu}} \emptyset,$$

where $\bar{\lambda}$ and $\bar{\mu}$ characterise mesoscopic birth and annihilation (death) rates, respectively. They can be obtained from the macroscopic reaction rates, λ and μ, respectively, by appropriate scalings, which we describe in more detail below. We denote by $\mathbf{N}(t)$ the total number of molecules present in the system. Starting from a positive initial condition, $\mathbf{N}_0 = \mathbf{N}(t = 0)$, this system equilibrates when the population reaches its carrying capacity, given by the ratio of the macroscopic reaction rates, $\mathcal{K} = \lambda/\mu$. If the spatial distribution of molecules plays a relevant role, this description must be complemented by diffusion dynamics.

The birth reaction, being first-order (unimolecular), does not present convergence issues. However, second-order annihilation reactions require special consideration. Whether using the RDME or its convergent version (cRDME), the mesoscopic rate for bimolecular annihilation, $\bar{\mu}$, is obtained by scaling the corresponding macroscopic rate with the lattice spacing and the system size, Ω [18]. The system size parameter has units of area (volume) and allows to perform conversions between the number of molecules and molecular concentrations, i.e. $n = \mathbf{N}/\Omega$. The implementation of the convergent formulation also requires that the molecules in the sFK system interact between different voxels within the reaction radius [10]. Diffusion of molecules will be modelled in the standard way as a continuous-time unbiased random walk between neighbouring compartments. We will consider reflecting zero-flux boundary conditions to represent a closed system. For simplicity, we will present our results in one dimension. Nonetheless, it is straightforward to extend the formulation of the convergent sFK system and the hybrid method described later to higher dimensions.

Consider a one-dimensional domain $\mathcal{D} = [-l, l]$. We partition $\mathcal{D}$ into $2K$ equal voxels or compartments of size $h = l/K$. Then a voxel k is given by an interval $v_k = [(k-1)h,\ kh]$ for $k = \{-K+1, \ldots, K-1, K\}$. We denote by $N(k)$ the number of particles in voxel k. Molecules in each voxel are assumed to be well-mixed and identical. The vector $\mathcal{N} = (N(-K+1) \ldots N(K))$ describes the current state of the system. Its dynamics are then described by a master equation of the form:

$$\frac{dP(t, \mathcal{N})}{dt} = \sum_{\widetilde{\mathcal{N}}} \left[W(\mathcal{N}|\widetilde{\mathcal{N}}) P(t, \widetilde{\mathcal{N}}) - W(\widetilde{\mathcal{N}}|\mathcal{N}) P(t, \mathcal{N}) \right]. \tag{2.13}$$

Here, $P(t, \mathcal{N})$ represents the probability distribution of system states at time t. The transition rates, W, must be properly specified according to the cRDME approach introduced in [10] to ensure its convergence.

2.3.1.1 Mesoscopic Reaction Rates

Diffusion We describe the random movement of molecules between voxels in the usual manner using the following rates:

$$W_d\left(N(k)-1, N(k\pm 1)+1 | N(k), N(k\pm 1)\right) = \overline{D}N(k). \tag{2.14}$$

Since these rates are linear with respect to $N(k)$, they are not affected by convergence issues and therefore require no modification. The scaled diffusion coefficient, $\overline{D}$, equals Dh^{-2}.

Birth We assume that birth events occur at a constant rate per molecule. The associated propensity is given as follows:

$$W_b\left(N(k)+1 | N(k)\right) = \overline{\lambda}N(k), \tag{2.15}$$

with $\overline{\lambda} = \lambda$. When a birth event occurs, the new molecule is placed in the same voxel k.

Annihilation In order to account for nonlinear (i.e. multi-molecular) reactions, we use the Doi model of partial absorption. The Doi model for molecular interaction assumes that annihilation can only occur if the distance between molecules is less than the prescribed reaction radius, R_a. Since molecule positions are known only up to the compartment size, particles from voxels k and m annihilate each other at a reaction rate determined by three factors: the rescaled death rate, μ; a geometric correction factor, $p(k,m)$; and a combinatorial factor associated with second-order kinetics [6].

The rescaled annihilation rate, $\overline{\mu}$, is given by $\overline{\mu} = \mu/(2h\Omega)$ for $m \neq k$. Here, μ is the macroscopic annihilation rate and Ω is the system size. The factor of $1/2$ appears due to the fact that annihilation reactions are symmetric. Thus, for a pair of voxels k and m, the total rate is equal to $2\overline{\mu}$, where this factor comes into play to cancel the additional $1/2$. For annihilation reactions within the same voxel, i.e. $m = k$, $\overline{\mu} = \mu/(h\Omega)$ [18].

The correction factor, $p(k,m)$, accounts for the statistical effects of partial coverage of compartments by the reaction region. Binary reactions occur if and only if $v_m \cap B_{R_a}(k) \neq \emptyset$, where $B_{R_a}(k)$ is a ball of radius R_a centred at the midpoint of voxel k. Thus, the so-called reaction ratio, $p(k,m) = |v_m \cap B_{R_a}(k)|/|v_m|$, represents the probability of the voxels k and m interacting when they are only partially covered by the reaction region. Here, $|\cdot|$ denotes the length of the one-dimensional interval. In the 1D case, the explicit formula for $p(k,m)$ reads as follows:

$$p(k,m) = \begin{cases} \min\left(\frac{2R_a}{h}, 1\right), & \text{if } m = k; \\ \max\left[0, \min\left(\frac{R_a}{h} - \mid m-k \mid + 0.5, 1\right)\right], & \text{otherwise.} \end{cases} \tag{2.16}$$

Similarly, the probability of selecting two voxels must be appropriately defined. Although there is no unique way to define this probability, any reasonable choice should result in similar dynamics. For simplicity, we choose a uniform distribution. Thus, the probability for an annihilation reaction between two voxels within the reaction radius to occur is equal to $h/(2R_a)$. The factor of 1/2 appears because reactions are possible on both sides of the voxel.

Finally, according to the law of mass action, the resulting reaction rate is proportional to the number of all possible combinations of molecules that participate in the reaction. In the case of distinct voxels, it is simply the product, $N(k)N(m)$, while within a single voxel, the number of distinct combinations is $1/2N(k)(N(k)-1)$. We note the factor of 1/2, which was previously absent in the rescaled reaction rate $\bar{\mu}$ for $k = m$, but now appears for the number of molecule combinations. In summary, the annihilation transition rate in the convergent formulation is given by the following expression:

$$\begin{aligned} &W_a\left(N(k)-1, N(m)-1|N(k), N(m)\right) \\ &= \frac{\mu}{4R_a\Omega}\, p(k,m) \begin{cases} N(k)N(m), & \text{if } m \neq k; \\ N(k)(N(k)-1), & \text{if } m = k. \end{cases} \end{aligned} \tag{2.17}$$

With the reaction rates given by Eqs. (2.14), (2.15), and (2.17), the cRDME for the sFK system reads:

$$\begin{aligned} &\frac{dP(\mathcal{N},t)}{dt} \\ &= \underbrace{\sum_{k,m=k\pm1} (\mathcal{E}_k^1\mathcal{E}_m^{-1}-1)W_d\left(N(k)-1, N(m)+1|N(k), N(m)\right)P(\mathcal{N},t)}_{\text{diffusion}} \\ &+ \underbrace{\sum_k (\mathcal{E}_k^{-1}-1)W_b\left(N(k)+1|N(k)\right)P(\mathcal{N},t)}_{\text{birth}} \\ &+ \underbrace{\sum_{k,m} (\mathcal{E}_k^1\mathcal{E}_m^1-1)W_a\left(N(k)-1, N(m)-1|N(k), N(m)\right)P(\mathcal{N},t)}_{\text{annihilation}}. \end{aligned} \tag{2.18}$$

Here, we used the shift operator, $\mathcal{E}_k^r$, which 'shifts' the number of molecules in voxel k by r units:

$$\begin{aligned} \mathcal{E}_k^r f(N(-K+1), \ldots, N(k), \ldots, N(K)) = f(N(-K+1), \ldots, N(k) \\ +r, \ldots, N(K)). \end{aligned}$$

Equation (2.18) with the transition rates specified above represents the cRDME for the sFK system. Its limit for $h \to 0$ is well-defined since bimolecular reactions do not vanish. Smaller lattice size simply implies molecular interactions between higher-order neighbouring compartments.

2.3.1.2 Macroscopic Limit

To derive the mean-field description for this system, we define the concentration of molecules in voxel k as $n_k = N(k)/(\Omega h)$. In the limit $h \to 0$, this discretisation yields a continuous concentration function, $n(x, t)$.

For the standard Fisher-Kolmogorov system [32], the mean-field limit equation is given as follows:

$$\frac{\partial n}{\partial t}(t, x) = D\frac{\partial^2 n}{\partial x^2}(t, x) + \lambda n(t, x) - \mu n^2(t, x). \tag{2.19}$$

In the convergent formulation corresponding to cRDME from Eq. (2.18), the non-locality of annihilation events results in the integral term for this reaction

$$\frac{\partial n}{\partial t}(t, x) = D\frac{\partial^2 n}{\partial x^2}(t, x) + \lambda n(t, x) - \mu n(t, x)\int_{\mathcal{D}} b(x - y)n(t, y)dy, \tag{2.20}$$

with $b(x, y) = b(x - y)$ being a symmetric kernel called the *influence function*. For the uniform probability distribution used in the annihilation transition rate, this kernel function has the following expression:

$$b(x - y) = \frac{1}{2R_a}\mathbb{1}_{\mathcal{R}(x)}(y)\,. \tag{2.21}$$

Here, the set $\mathcal{R}(x) = \{y : |x - y| < R_a\}$ and $\mathbb{1}_{\mathcal{R}(x)}$ denotes the indicator function for this set.

It can be proven that the space-homogeneous stationary solution of this non-local Fisher-Kolmogorov integro-partial differential equation (IPDE) coincides with the stationary state of the standard Fisher-Kolmogorov system, given by λ/μ.

We complement Eq. (2.20) with no-flux boundary conditions for $x \in \partial\mathcal{D} = \{-l, l\}$, along with an initial condition for the initial molecule distribution:

$$\frac{\partial n}{\partial x}(t, -l) = \frac{\partial n}{\partial x}(t, l) = 0\,, \tag{2.22}$$

$$n(t = 0, x) = \phi(x)\,. \tag{2.23}$$

2.3.2 Hybrid Method for cRDME

The key ingredient of a wide class of hybrid methods to simulate reaction-diffusion systems [18, 19, 27] consists in defining an interface region within which the macroscopic mean-field and stochastic descriptions coexist. The coupling between these descriptions is achieved by prescribing appropriate boundary conditions at the boundaries of the interface region. Although the exact formulation of these boundary conditions varies among different hybrid methods, their aim is to minimise the effects due to the presence of interfaces. Similarly, our primary goal will be to ensure that the interface region does not impact the system's dynamics. This will be achieved by balancing the fluxes of molecules transitioning between the subdomains.

The approach used here is based on the previous work, which introduced a hybrid method for systems exhibiting travelling wave behaviour [19]. In particular, we assume that the simulation domain, $\mathcal{D}$, is divided into two overlapping subdomains (see Fig. 2.6a): one modelled using a stochastic compartment-based approach and the other using a macroscopic mean-field description. In what follows, we assume that the compartment-based description is valid in the right part of $\mathcal{D}$ (see Fig. 2.6a). The method can be similarly formulated if the positions of the subdomains are reversed.

The stochastic subdomain, $\mathcal{D}_c = [I_c, l]$, is partitioned into $K + M$ voxels of uniform size $h_c = (l - I_c)/(K + M)$. Out of these $K + M$ compartments, M voxels lie within the interface region, $\mathcal{D}_I = [I_c, I_d]$. We denote by $N(k)$ the number of molecules in voxel k for $k \in [1, K + M]$, while the corresponding concentration

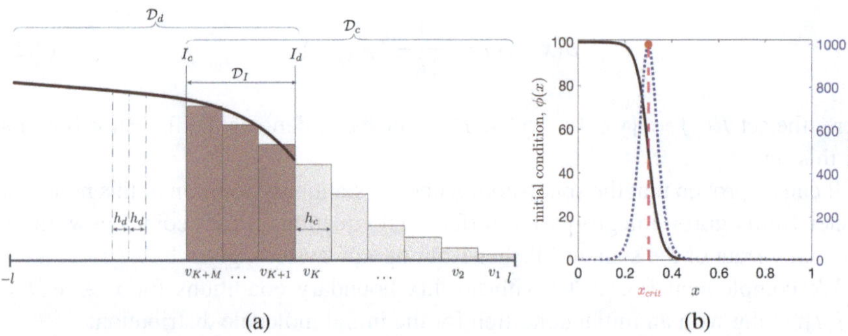

Fig. 2.6 (**a**) An illustration demonstrating the partitioning of the domain $\mathcal{D} = [-l, l]$ into the compartment-based subdomain, $\mathcal{D}_c$, simulated by the cRDME, and the deterministic subdomain, $\mathcal{D}_d$, described by the mean-field IPDE. The overlap region, $\mathcal{D}_I = [I_c, I_d] = \mathcal{D}_c \cap \mathcal{D}_d$, where both descriptions are valid, is determined by the positions of two interfaces, I_c and I_d. Here, h_c denotes the uniform voxel size, while h_d is the mesh size for the finite difference discretisation in the deterministic subdomain. (**b**) An illustration of the critical point position, x_{crit}, determining the centre of the interface region at the initial time point. Here, x_{crit} corresponds to the point at which the maximum of the absolute value of the gradient of the initial molecular distribution, $\phi(x)$, occurs

is given by $n_k = N(k)/(\Omega h_c)$. In the limit for the lattice size h_c approaching zero, a continuous description, $n_c(x, t)$, valid within the compartment-based subdomain, can be obtained.

The macroscopic subdomain, $\mathcal{D}_d = [-l, I_d]$, is covered by a uniform mesh of size, h_d, for finite difference discretisation of the mean-field description, $n_d(x, t)$. In general, it is not necessary for the mesh points of $\mathcal{D}_d$ to coincide with the boundaries of the voxels (within the overlap region). The mesh size, h_d, must be chosen to ensure that the numerical schemes used to solve the mean-field limit IPDE provide accurate approximations.

2.3.2.1 General Overview

We now provide a general overview of the algorithm for our hybrid method. More detailed descriptions of the initial placement of interfaces, their dynamic update, and coupling between the stochastic and mean-field regimes are discussed below:

1. Set parameter values. Set the initial time, $t = 0$, and the time counter for updating the mean-field solution, $t_d = 0$.
2. Set the initial condition, $n(x, t = 0) = \phi(x)$. We choose an initial condition such that the system evolves as a travelling wave.
3. Calculate the initial position of the interfaces (Sect. 2.3.2.2) and assign the corresponding subdomains for the compartment-based and mean-field regimes accordingly.
4. Set time step, τ, as the waiting time to the next stochastic event using one of the algorithms for simulating stochastic systems (e.g. SSA or NSV method).
5. Update the state of the system within the compartment-based domain, $\mathcal{D}_c$, according to the chosen stochastic algorithm.
6. Update the time variable, $t = t + \tau$.
7. If $(t - t_d < \tau_d)$, where τ_d is the time step for the mean-field discretisation, update the deterministic subdomain:

    a. Solve the mean-field IPDE in the time interval, $[t_d, t)$, within the deterministic subdomain, $\mathcal{D}_d$.
    b. Update the time counter, $t_d = t$.
    c. If necessary, update the position of the overlap region (Sect. 2.3.2.3).

8. Renormalise the number of molecules within the overlap region so that it is an integer. The renormalisation method used is explained in detail in [18].
9. Iterate steps 4–9 until some stopping condition is satisfied. Here, we used the stopping criterion of reaching the final simulation time, i.e. $t \geq t_{final}$.

2.3.2.2 Initial Placement of the Interfaces

Here, we assume that a single pair of interfaces suffices to capture the spatial inhomogeneities of the molecule distribution. However, the method can be easily extended to multiple overlap regions. For each interface region, we need to specify (i) its centre and (ii) its width.

To determine the initial position of the centre of the overlap region, we identify a critical point, $x_{crit} \in \mathcal{D}$, corresponding to the maximum absolute value of the gradient of $\phi(x)$, where $\phi(x)$ represents the initial distribution of molecules. Figure 2.6b illustrates the position of x_{crit} for a sigmoidal initial condition. The algorithm can be extended to identify multiple critical points, including local extrema or all points where the absolute value of the gradient of $\phi(x)$ exceeds a certain threshold. In such cases, an overlap region must be defined for each critical point.

The interface region must be non-empty to ensure an accurate approximation of the system's dynamics [19]. Additionally, the non-local nature of the bimolecular reactions in the convergent sFK system results in molecule fluxes between subdomains within the reaction radius, R_a. Therefore, the size of the overlap region must be at least R_a. In this study, we generally set the width of the interface region to $d = 2R_a$ or $d = 4R_a$.

Therefore, the positions of the two interfaces are calculated as follows:

$$I_c = \max\{-l, x_{crit} - 0.5d\} \text{ and } I_d = \min\{l, x_{crit} + 0.5d\}.$$

2.3.2.3 Dynamic Adaptation of the Interface Region

To update the interface region, we formulate mutually exclusive criteria for extending one of the subdomains. We use the local concentration of molecules at the centre of the overlap region, $\mathcal{D}_I$, to determine whether the subdomains should be updated. As described above, the critical point defining the centre of the interface region is situated approximately at the centre of the wavefront. We introduce algorithm parameters, $0 < r_1 < r_2 < 1$, to control the adaptation of the interfaces. Although both descriptions are valid in this region, the deterministic concentration of molecules is not subject to fluctuations. Thus, if the mean-field concentration of molecules $n_d(t, x_{crit})$ falls below $r_1\mathcal{K}$ (indicating the wavefront has moved towards I_c), we extend $\mathcal{D}_c$ by adding a new voxel, with the number of molecules approximated by the deterministic concentration at the new voxel's centre. At the same time, the mean-field domain is reduced by the voxel width at the interface, I_d. Conversely, if the concentration $n_d(t, x_{crit})$ exceeds $r_2\mathcal{K}$, we extend the domain $\mathcal{D}_d$ by replacing the voxel next to the interface I_d with its mean-field description. The stochastic subdomain is then reduced by one compartment at the interface, I_c. The two interfaces and the centre point between them must then be updated accordingly. If changes in the size of the compartment-based domain affect the stochastic algorithm (e.g. if the NSV method is used), it must also be updated to ensure its correct performance. Typically, in our simulations, we use $r_1 \in [0.4, 0.6]$ and $r_2 \in [0.8, 1)$.

Since we use the mean-field molecule density as the criterion for updating the interface region, domain adaptation is performed at most once every deterministic time step, τ_d, which is typically much larger than the time step for more frequent stochastic reactions, τ. This ensures that the interface update algorithm does not significantly increase the computational complexity of the method.

2.3.2.4 Coupling the Mean-Field and the Stochastic Descriptions at the Overlap Region

The coupling algorithm of our hybrid method must ensure consistent molecule concentration across the interfaces. The non-local nature of the convergent formulation of the sFK system (see Sect. 2.3.1) implies that stochastic and deterministic densities influence each other through molecular fluxes from reactions occurring within the reaction radius, R_a, beyond the subdomain borders. This direct coupling between the two descriptions must be accounted for in our hybrid method.

Following the approach proposed in [19], we balance molecular fluxes at the interfaces, I_c and I_d. This is achieved by calculating the additional flux of molecules entering the stochastic subdomain through I_c for voxels within the reaction radius, R_a, from this interface. For the mean-field density, we prescribe an additional annihilation flux coming from the stochastic subdomain and set an appropriate Dirichlet boundary condition at the interface, I_d. As before, we assume that the compartment-based domain is on the right.

Coupling for the Mean-Field Description Within the overlap region, the evolution of the deterministic mean-field molecular concentration, $n_d(x, t)$, corresponding to the convergent sFK system, is governed by the following IPDE:

$$\begin{aligned}\frac{\partial n_d}{\partial t}(t, x) &= D\frac{\partial^2 n_d}{\partial x^2}(t, x) + \lambda n_d(x, t) \\ &\quad - \mu n_d(t, x)\underbrace{\left[\int_{\mathcal{D}_d} b(x-y)n_d(t, y)dy + \mathcal{A}(t, x)\right]}_{\mathcal{A}_d(t,x)}, \end{aligned} \tag{2.24}$$

$$n_d(t, I_d) = \bar{n}, \qquad \frac{\partial n_d}{\partial x}(t, -l) = 0. \tag{2.25}$$

Here, we note the Dirichlet boundary condition, $\bar{n}$, at the interface, I_d, and the additional annihilation flux $\mathcal{A}(t, x)$ coming from the compartment-based subdomain (compare with Eq. (2.20)). The expression for the former was derived in [19] and reads:

$$\bar{n} := \frac{N(K) + N(K+1)}{2\Omega h_c} = \frac{n_K + n_{K+1}}{2}. \tag{2.26}$$

We recall that voxel K (respectively, $K + 1$) lies to the right (respectively left) of the interface, I_d (see Fig. 2.6a). It can be shown [19] that the boundary condition given by Eq. (2.26) ensures the consistency of the molecular density across I_d, i.e. $n_d(t, I_d) = n_c(t, I_d) + O(h_c^2) \approx n_c(t, I_d)$.

We further introduce the annihilation flux, $\mathcal{A}(t, x)$ (see Eq. (2.24)), which arises due to the non-local reaction kinetics to ensure convergence of the RDME. It is calculated as follows:

$$\mathcal{A}(t, x) = \begin{cases} \frac{1}{2R_a} \sum\limits_{k \in \mathcal{I}_d^r(x) \cap \mathcal{I}_d^r(I_d)} p(x, k) \frac{N(t, k)}{\Omega}, & \text{if } I_d - R_a < x \leq I_d, \\ 0, & \text{otherwise,} \end{cases} \tag{2.27}$$

where $p(x, k) = |v_k \cap B_{R_a}(x)|/|v_k|$ is the continuous version of $p(k, m)$ from Eq. (2.16) and the index set, $\mathcal{I}_d^r(\cdot)$, is as described in Fig. 2.7a.

It is straightforward to check that in the limit $h_c \to 0$, the annihilation fluxes through the interface I_d, $\mathcal{A}_d(t, x)$ and $\mathcal{A}_c(t, x)$, for stochastic and deterministic descriptions, respectively, are equal. Here, $\mathcal{A}_d(t, x)$ is outlined in Eq. (2.24) with $\mathcal{A}(t, x)$ given by Eq. (2.27), whereas the stochastic flux due to annihilation events is given by the following expression:

$$\mathcal{A}_c(t, x_k) = \frac{\mu}{2R_a} \frac{1}{\Omega h_c} N(t, k) \sum_{m \in \mathcal{I}_d^l(x_k) \cup \mathcal{I}_d^r(x_k)} p(k, m) \frac{N(t, m)}{\Omega}. \tag{2.28}$$

Here, the index set, $\mathcal{I}_d^r(x)$, is defined in Fig. 2.7a.

Coupling for the Compartment-Based Regime At the interface, I_c, compartments lying to the right within the reaction radius R_a are influenced by the molecular flux from the deterministic subdomain. Specifically, for all voxels v_k in $\mathcal{D}_c$ with $k \in \mathcal{I}_c^r(I_c)$ (see Fig. 2.7a), we must account for the additional flux, $\psi_k^a(t, \Delta t)$, due to annihilation reactions between molecules from different subdomains occurring at time t during the time step Δt. To achieve this, we introduce ghost voxels to the left of the interface I_c (assuming the stochastic subdomain is on the right). The number of ghost voxels corresponds to the number of indices in the set $\mathcal{I}_c^r(I_c)$. The number of molecules in the ghost voxels is determined by interpolating the molecular concentration at the voxel centre using the two nearest points from the $n_d(t, x)$ discretisation. Figure 2.7b illustrates this process. Consequently, the additional annihilation flux into voxel v_k for $k \in \mathcal{I}_c^r(I_c)$ is given by the following expression:

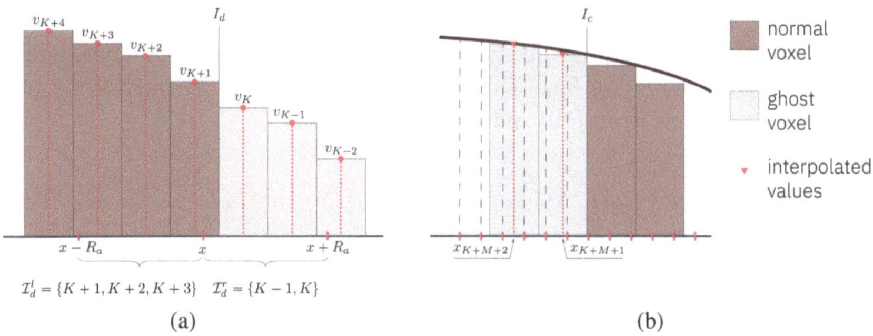

Fig. 2.7 (**a**) We introduce index sets, $\mathcal{I}_i^l(x)$ and $\mathcal{I}_i^r(x)$, of voxels which lie within the reaction radius, R_a, from point x to the left and right of the interface I_i, respectively. Here, $i \in \{c, d\}$ is the interface label. Specifically, left-lying sets of voxels are defined as $\mathcal{I}_i^l(x) := \{k \mid (l - (k - 0.5)h_c) \in [x - R_a, I_i)\}$, while the indices of the voxels to the right of the interface I_i are given by $\mathcal{I}_i^r(x) := \{k \mid (l - (k - 0.5)h_c) \in [I_i, x + R_a])\}$. We note that these index sets can include ghost voxels (see (**b**)). The illustration shows these index sets for the interface, I_d. (**b**) An illustration of linear interpolation for ghost voxels $(K + M + 1)$ and $(K + M + 2)$

$$\psi_k^a(t, \Delta t) = \Delta t \, \mu \frac{h_c}{2R_a} n_c(t, x_k) \sum_{m \in \mathcal{I}_c^l(x_k) \cap \mathcal{I}_c^l(I_c)} p(k, m) \, \widetilde{n}_m. \tag{2.29}$$

Here, $\widetilde{n}_m$ denotes the concentration of molecules in ghost voxel m, and x_k is the centre of the k-th voxel. The summation in Eq. (2.29) encompasses all voxels within the reaction radius from x_k that are not part of the stochastic subdomain (i.e. ghost voxels, $\widetilde{n}_m$). This equation, therefore, accounts solely for the molecular flux originating from the deterministic subdomain. The stochastic dynamics within compartments $v_k \subset \mathcal{D}_c$, including annihilation events, are defined as described previously in Sect. 2.3.1.

In the limit as the lattice size, h_c, approaches zero, it can be shown that the annihilation fluxes, $\mathcal{A}_c(t, x_k)$ and $\mathcal{A}_d(t, x_k)$, for $x_k \in [I_c, I_c + R_a]$ are identical:

$$\begin{aligned} &\mu \frac{h_c}{2R_a} n_c(t, x_k) \sum_{m \,\in\, \mathcal{I}_c^l(x_k) \,\cup\, \mathcal{I}_c^r(x_k)} p(k, m) \, n_c(t, x_m) \\ &\overset{h_c \to 0}{=} \mu \frac{1}{2R_a} n_d(t, x_k) \int_{x_k - R_a}^{x_k + R_a} n_d(t, y) dy. \end{aligned}$$

To verify this, we note that within the overlap region, $n_d(t, x)$ serves as a good approximation of $n_c(t, x)$. Therefore, for any $x \in [I_c, I_c + R_a]$, $n_d(t, x) \approx n_c(t, x)$. Additionally, to the left of the interface, I_c, we used the deterministic molecule concentration for the linear interpolation of $n_c(t, x)$ in ghost voxels. Thus, these fluxes are identical, and the presence of the interface, I_c, does not affect the molecular concentration.

For voxel $(K + M)$, adjacent to the interface I_c (see Fig. 2.6a), incoming molecules from the deterministic domain due to diffusion must be considered in addition to the annihilation flux. The diffusion flux through the interface, I_c, was calculated in previous work [19] and is given as follows:

$$\psi^{d}_{K+M}(t, \Delta t) = \Delta t \frac{D}{h_c^2} (\widetilde{n}_{K+M+1} - n_{K+M}) . \tag{2.30}$$

In summary, for voxel $(K + M)$ at the interface I_c, we must correct its molecular concentration by incorporating molecular fluxes from both annihilation events and diffusion (Eqs. (2.29)–(2.30)). For all other voxels $k \neq K + M$ influenced by the deterministic subdomain, we adjust their molecular concentrations solely based on the flux from annihilation using Eq. (2.29). This ensures the consistency of molecular concentration across the I_c interface.

2.3.3 Simulation Results

We illustrate our hybrid method for the cRDME with characteristic simulation results for the sFK model. The parameter values used in these simulations are listed in Table 2.4. Figure 2.8 displays four snapshots of the wavefront propagation over time. We note that, in these simulations, the noise parameter in the stochastic regime was set to a high value to demonstrate the robustness of our hybrid method (corresponding system size parameter, $\Omega = 100$). The small hump forming at the domain border is due to the no-flux boundary condition and the non-locality of annihilation reactions.

Table 2.4 Parameters used in numerical simulations of the convergent Fisher-Kolmogorov system

Param.	Value	Description	Param.	Value	Description
$h_c = h_d$	0.01	Lattice spacing	D	$4e-3$ (Fig. 2.8); $4e-5$ (Fig. 2.9)	Diffusion coefficient
λ	1.0	Birth rate	$\mathcal{K}$	100	Carrying capacity
R_a	0.05	Interaction radius for bimolecular reactions	t_{final}	7 (Fig. 2.8); 30 (Fig. 2.9)	Simulation time
r_1	0.5	Interface adaptation parameter	r_2	0.8	Interface adaptation parameter

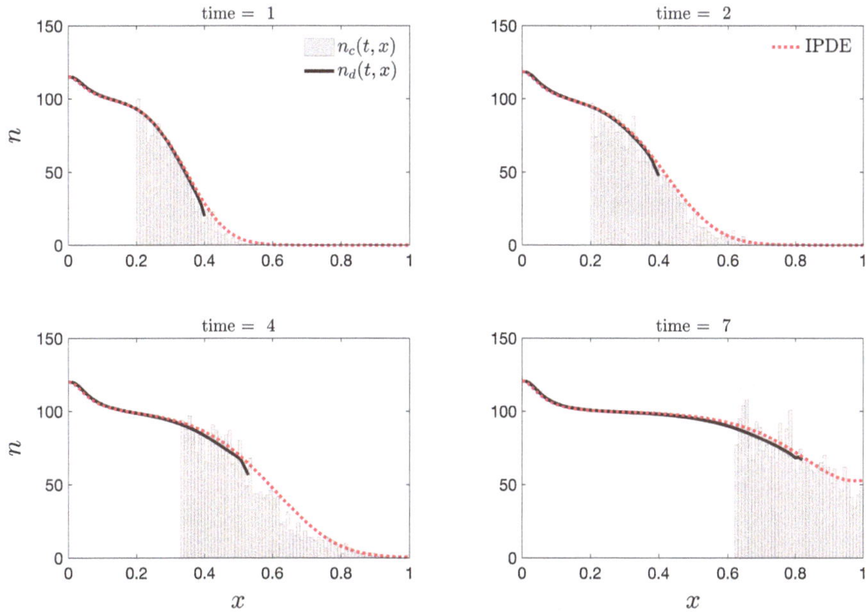

Fig. 2.8 Four snapshots of a numerical simulation of our hybrid method for the non-local Fisher-Kolmogorov system. Dashed lines indicate the solution of the deterministic IPDE given by Eq. (2.20) at the same times. The width of the interface region was set to $d = 4R_a$ and the system size to $\Omega = 100$. The remaining parameters are listed in Table 2.4. The initial molecular concentration used for this simulation is given by a smooth sigmoidal function, $\phi(x) = 0.5\mathcal{K}(1 - \tanh(20(x - 0.3)))$

2.3.3.1 Comparison

Although Fig. 2.8 showcases a single simulation, a slight delay in wave propagation speeds between the hybrid algorithm and the purely deterministic IPDE can already be observed. In the final snapshot at $t = 7$, the IPDE front is notably advancing faster. This discrepancy arises because the wavefront advancement in our hybrid method is governed by the stochastic regime, which typically exhibits slower wave propagation speeds that depend on the noise level and approach the deterministic mean-field estimate as the noise tends to zero (equivalently, system size in stochastic simulations tends to infinity). The mean-field estimation for the front advancement in the non-local Fisher-Kolmogorov system (Eq. (2.20)) is not affected by the integral term and is given by $c_d \approx 2\sqrt{D\lambda}$ for an infinite domain [33]. In order to evaluate the performance of our hybrid method, in this section, we compare wave propagation speeds in our hybrid method, purely stochastic simulations using the NSV method, and the mean-field limit estimate, c_d.

In order to obtain estimates of the wave propagation speed for stochastic and hybrid simulations, we employ the method introduced in [34]. This method relies on the linear temporal growth of the (averaged) total molecular mass in a system

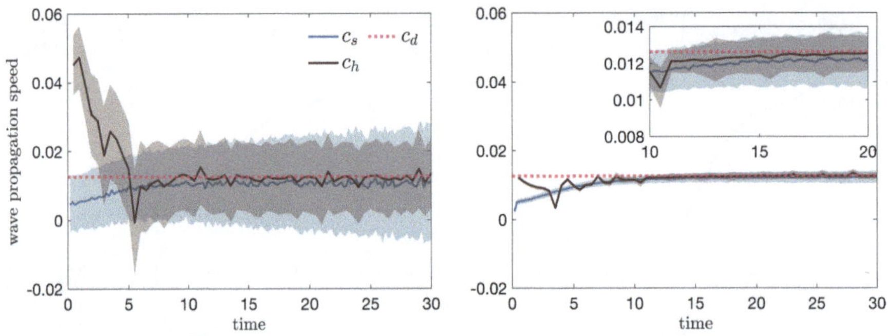

Fig. 2.9 Wave propagation speeds for the convergent Fisher-Kolmogorov system estimated from purely stochastic (c_s, in blue) and hybrid (c_h, in dark brown) simulations, with system sizes, $\Omega = 100$ (left panel) and $\Omega = 1000$ (right panel), and the theoretical mean-field estimate, c_d. Solid lines indicate the mean values averaged over 500 realisations, while shaded regions show their standard deviation. Here, the width of the interface region was set to $d = 2R_a$ and the initial condition, $\phi(x)$, is as in Fig. 2.8. The remaining parameters are listed in Table 2.4

with travelling wave behaviour, where the slope of its growth corresponds to the wavefront speed. We denote by $\mathcal{M}_s(t)$ and $\mathcal{M}_h(t)$ the total masses of the system at the time point t in purely stochastic and hybrid simulations, respectively. Then, the wave speeds can be approximated by the following expressions:

$$c_s \approx \frac{\langle \mathcal{M}_s(t_2) \rangle - \langle \mathcal{M}_s(t_1) \rangle}{t_2 - t_1}, \qquad c_h \approx \frac{\langle \mathcal{M}_h(t_2) \rangle - \langle \mathcal{M}_h(t_1) \rangle}{t_2 - t_1}, \tag{2.31}$$

where $\langle \cdot \rangle$ denotes the mean taken over multiple stochastic realisations, and the total molecular masses are given by:

$$\mathcal{M}_s(t) = \frac{1}{\Omega} \sum_{k=-K+1}^{K} N(k), \qquad \mathcal{M}_h(t) = \frac{1}{\Omega} \sum_{k=1}^{K+M} N(k) + \int_{\mathcal{D}_d \setminus \mathcal{D}_I} n_d(t, x) dx. \tag{2.32}$$

Figure 2.9 presents the results of the wavefront propagation estimation for two system sizes: $\Omega = 100$ (left panel) and $\Omega = 1000$ (right panel). After an initial transient phase ($t \lesssim 10$), the system stabilises into a travelling wave propagating at constant average speeds, c_s and c_h, for the stochastic and hybrid simulations, respectively. As the noise level decreases (right panel), the estimates converge more closely to the deterministic limiting speed, c_d. The average wave propagation speeds are $c_s \approx 0.010676$ and $c_h \approx 0.012132$ for $\Omega = 100$ and $c_s \approx 0.012131$ and $c_h \approx 0.012596$ for $\Omega = 1000$. The macroscopic limit estimate for the parameters used in these simulations is $c_d = 0.012649$. As expected, the purely stochastic simulations show the slowest wave propagation speed, while the hybrid method's estimate lies between the purely stochastic and mean-field theoretical estimates.

We also recorded the computational times for performing these simulations using stochastic and hybrid approaches. For an individual realisation, the mean time to run a stochastic simulation with the NSV implementation was 20.9306 ± 0.8807 seconds. This time was reduced by approximately 48-fold for hybrid simulations, which required around 0.43309 ± 0.0333 seconds per realisation. These estimates were obtained for a system size of $\Omega = 100$. All simulations were performed on a standard personal PC. While simulation times for the relatively simple sFK model are manageable even for the purely stochastic approach, the reduction in simulation time provided by the hybrid method can offer significant advantages for more complex models.

In summary, hybrid methods for convergent RDME require the overlap region width to be on the order of the interaction radius for multi-molecular reactions. While this adds to the computational complexity compared to the traditional RDME approach, hybrid methods offer significant reductions in simulation time compared to purely stochastic methods. This introductory section demonstrates the formulation of such a hybrid method for the convergent Fisher-Kolmogorov system. However, more complex chemical reaction networks can also be implemented within this framework.

2.4 Conclusion and Discussion

This chapter reviewed hybrid models of travelling waves in reaction-diffusion systems, emphasising their relevance to biological processes such as animal and cell movement. We discussed the broader context of deterministic pulled fronts and highlighted the differences between various wave types. Future work will focus on extending hybrid methods to bistable waves and other complex systems, enhancing our ability to model biological invasions accurately.

Hybrid methodologies have been extensively shown to be powerful tools for effective simulation of stochastic reaction-diffusion systems. In this chapter, we have reviewed the different methodologies and formulations proposed in the literature (see Sect. 2.1), as well as our recent work devoted to addressing several limitations in the previous implementations. In particular, in Sect. 2.1.3, we have reviewed our recent work on extending the hybrid methodology to multiscale models of tumour growth. Stochastic multiscale models account for fluctuations in both the number of cells (the usual demographic noise present in reaction-diffusion systems) and the intracellular cell dynamics. The situation analysed in Sect. 2.1.3 illustrates the importance of properly considering the latter.

In Sect. 2.2, we have summarised our extension of the hybrid method to account for multi-species reaction-diffusion systems (e.g. Lotka-Volterra systems). Since these systems are known to produce heterogeneous spatial patterns, one must tackle the emergence of several interfaces [18]. Finally, in Sect. 2.3, we proposed a hybrid method that allows us to simulate the convergent version of the reaction-diffusion master equation. This method involves developing an approach to address non-local reaction mechanisms.

Beyond the extensions discussed in this chapter, there is still plenty of room for further development and applications of hybrid methods. A recent example is the paper by [35], where they studied the stochastic effects of the interaction between viruses and particles lacking replication genes that can interfere with wild-type viruses and serve as potential therapeutics in a two-dimensional model of infection. Extension of the current algorithms to higher spatial dimensions is an essential step to extend the current remit of hybrid methods regarding their applications.

Fronts are observed in systems beyond the ones we have described in detail in this review, and it would be a subject of interest to extend hybrid methodologies to such systems, for example, in the contexts of waves and fronts in bistable and excitable systems.

Another avenue of potential extensions of the current hybrid schemes is growing and compliant domains [20]. The mechanical microenvironment has revealed itself to be a key factor in many biological and biomedical problems from developmental biology to tumour growth, and therefore, it should be accounted for in order to further extend the range of application of hybrid methods.

Acknowledgments D.S. and T.A. would like to thank the Isaac Newton Institute for Mathematical Sciences, Cambridge, for their support and hospitality during the programme *Mathematics of movement: an interdisciplinary approach to mutual challenges in animal ecology and cell biology*, supported by EPSRC grant EP/R014604/1. This work has been funded by the Spanish Research Agency (AEI), through the Severo Ochoa and Maria de Maeztu Programme for Centres and Units of Excellence in R&D (CEX2020-001084-M). D.S. and T.A. thank CERCA Program/Generalitat de Catalunya for their institutional support. D.S. and T.A. have been funded by grant PID2021-127896OB-I00 from MCIN/AEI/10.13039/501100011033 'ERDF A way of making Europe'. J.C. has been partially supported by Grant C-EXP-265-UGR23 funded by Consejería de Universidad, Investigación e Innovación and ERDF/EU Andalusia Program; by Grant PID2022-137228OB-I00 funded by the Spanish Ministerio de Ciencia, Innovación y Universidades, MICIU/AEI/10.13039/501100011033 and 'ERDF/EU A way of making Europe'; and by Grant QUAL21-011 (Modeling Nature) Consejería de Universidad, Investigacion e Innovacion of the Junta de Andalucía. N.B. and P.G have been supported by Grant PID2022-141802NB-I00 (BASIC), the research network RED2022-134573-T funded by MCIN/AEI/10.13039/501100011033, and by 'ERDF A way of making Europe'.

References

1. McLennan, R., Schumacher, L.J., Morrison, J.A., Teddy, J.M., Ridenour, D.A., Box, A.C., Semerad, C.L., Li, H., McDowell, W., Kay, D., Maini, P.K., Baker, R.E., Kulesa, P.M.: Neural crest migration is driven by a few trailblazer cells with a unique molecular signature narrowly confined to the invasive front. Development **142**(11), 2014–2025 (2015)
2. Giniūnaitė, R., Baker, R.E., Kulesa, P.M., Maini, P.K.: Modelling collective cell migration: neural crest as a model paradigm. J. Math. Biol. **80**(1), 481–504 (2020)
3. Martinson, W.D., McLennan, R., Teddy, J.M., McKinney, M.C., Davidson, L.A., Baker, R.E., Byrne, H.M., Kulesa, P.M., Maini, P.K.: Dynamic fibronectin assembly and remodeling by leader neural crest cells prevents jamming in collective cell migration. eLife **12**, e83792 (2023)
4. Stepanova, D., Byrne, H.M., Maini, P.K., Alarcón, T.: A multiscale model of complex endothelial cell dynamics in early angiogenesis. PLoS Comput. Biol. **17**(1), e1008055 (2021)

5. Aronson, D.G., Weinberger, H.F.: Multidimensional nonlinear diffusion arising in population genetics. Adv. Math. **30**(1), 33–76 (1978)
6. Gillespie, D.T.: A general method for numerically simulating the stochastic time evolution of coupled chemical reactions. J. Comput. Phys. **22**(4), 403–434 (1976)
7. Smith, S., Grima, R.: Breakdown of the reaction-diffusion master equation with nonelementary rates. Phys. Rev. E **93**(5), 052135 (2016)
8. Hellander, S., Petzold, L.: Reaction rates for a generalized reaction-diffusion master equation. Phys. Rev. E **93**(1), 013307 (2016)
9. Erban, R., Chapman, S.J.: Stochastic modelling of reaction–diffusion processes: algorithms for bimolecular reactions. Phys. Biol. **6**(4), 046001 (2009)
10. Isaacson, S.A.: A convergent reaction-diffusion master equation. J. Chem. Phys. **139**(5), 054101 (2013)
11. Rossinelli, D., Bayati, B., Koumoutsakos, P.: Accelerated stochastic and hybrid methods for spatial simulations of reaction–diffusion systems. Chem. Phys. Lett. **451**(1–3), 136–140 (2008)
12. Haseltine, E.L., Rawlings, J.B.: Approximate simulation of coupled fast and slow reactions for stochastic chemical kinetics. J. Chem. Phys. **117**(15), 6959–6969 (2002)
13. Davis, M.H.A.: Piecewise-deterministic Markov processes: a general class of non-diffusion stochastic models. J. R. Stat. Soc. **46**(3), 353–376 (1984)
14. Bressloff, P.C., Lawley, S.D.: Mean first passage times for piecewise deterministic Markov processes and the effects of critical points. J. Stat. Mech. Theor. Exp. **2017**(6), 063202 (2017)
15. Ferm, L., Hellander, A., Lötstedt, P.: An adaptive algorithm for simulation of stochastic reaction–diffusion processes. J. Comput. Phys. **229**(2), 343–360 (2010)
16. Flegg, M.B., Chapman, S.J., Erban, R.: The two-regime method for optimizing stochastic reaction–diffusion simulations. J. R. Soc. Interface **9**(70), 859–868 (2012)
17. Franz, B., Flegg, M.B., Chapman, S.J., Erban, R.: Multiscale reaction-diffusion algorithms: PDE-assisted Brownian dynamics. SIAM J. Appl. Math. **73**(3), 1224–1247 (2013)
18. Spill, F., Guerrero, P., Alarcón, T., Maini, P.K., Byrne, H.M.: Hybrid approaches for multiple-species stochastic reaction–diffusion models. J. Comput. Phys. **299**, 429–445 (2015)
19. Harrison, J.U., Yates, C.A.: A hybrid algorithm for coupling partial differential equation and compartment-based dynamics. J. R. Soc. Interface **13**(122), 20160335 (2016)
20. Smith, C.A., Yates, C.A.: Incorporating domain growth into hybrid methods for reaction–diffusion systems. J. R. Soc. Interface **18**(177), 20201047 (2021)
21. Panja, D.: Effects of fluctuations on propagating fronts. Phys. Rep. **393**(2), 87–174 (2004)
22. van Saarlos, W.: Front propagation into unstable states. Phys. Rep. **386**, 29–222 (2003)
23. Xin, J.: Front propagation in heterogeneous media. SIAM Rev. **42**(2), 161–230 (2000)
24. Méndez, V., Fort, J., Rotstein, H.G., Fedotov, S.: Speed of reaction-diffusion fronts in spatially heterogeneous media. Phys. Rev. E **68**(4), 041105 (2003)
25. Freidlin, M.: Limit theorems for large deviations and reaction-diffusion equations. Ann. Probab. **13**(3), 639–675 (1985)
26. Rocco, A., Ebert, U., Van Saarloos, W.: Subdiffusive fluctuations of "pulled" fronts with multiplicative noise. Phys. Rev. E **62**(1), R13 (2000)
27. Moro, E.: Hybrid method for simulating front propagation in reaction-diffusion systems. Phys. Rev. E **69**(6), 060101 (2004)
28. de la Cruz, R., Guerrero, P., Calvo, J., Alarcón, T.: Coarse-graining and hybrid methods for efficient simulation of stochastic multi-scale models of tumour growth. J. Comput. Phys. **350**, 974–991 (2017)
29. Gillespie, D.T., Petzold, L.R., Seitaridou, E.: Validity conditions for stochastic chemical kinetics in diffusion-limited systems. J. Chem. Phys. **140**(5), 02B604_1 (2014)
30. Elf, J., Ehrenberg, M.: Spontaneous separation of bi-stable biochemical systems into spatial domains of opposite phases. Syst. Biol. **1**(2), 230–236 (2004)
31. Yates, C.A., Flegg, M.B.: The pseudo-compartment method for coupling partial differential equation and compartment-based models of diffusion. J. R. Soc. Interface **12**(106), 20150141 (2015)

32. Breuer, H.-P., Huber, W., Petruccione, F.: Fluctuation effects on wave propagation in a reaction-diffusion process. Phys. D: Nonlinear Phenom. **73**(3), 259–273 (1994)
33. Gourley, S.A.: Travelling front solutions of a nonlocal fisher equation. J. Math. Biol. **41**(3), 272–284 (2000)
34. Robinson, M., Flegg, M.B., Erban, R.: Adaptive two-regime method: application to front propagation. J. Chem. Phys. **140**(12), 124109 (2014)
35. Liang, Q., Yang, J., Fan, W.-T. L., Lo, W.-C.: Patch formation driven by stochastic effects of interaction between viruses and defective interfering particles. PLoS Comput. Biol. **19**(10), e1011513 (2023)

Chapter 3
Classification of Similarities and Differences in Movement Between Two Sympatric Raven Species Across Multiple Spatiotemporal Scales

Bar Ashkenazi, Miguel de Guinea, Rachel Zuckerman, Ran Nathan, and Michael Assaf

3.1 Introduction

Over the last two centuries, human activities have expanded worldwide accelerating the modification of natural ecosystems for agriculture, hunting, recreation or tourism [1, 2]. While some animal species are able to prevail and even thrive in human-dominated landscapes, others fail to adjust resulting in species considered as "winners" and "losers" [3]. Since human-induced changes directly alter the distribution of threats (i.e. roads) and food resources in the landscape (i.e. food subsidies; [1, 4]), the capacity of animals to cope with environmental changes will be reflected in their movement patterns [5]. Yet, how animals adjust their

---

Bar Ashkenazi and Miguel de Guinea authors contributed equally to this work.

B. Ashkenazi
Racah Institute of Physics and Minerva Center for Movement Ecology, The Hebrew University of Jerusalem, Jerusalem, Israel

Movement Ecology Lab, Department of Ecology, Evolution and Behavior, Alexander Silberman Institute of Life Sciences, Faculty of Science, The Hebrew University of Jerusalem, Jerusalem, Israel

M. de Guinea · R. Zuckerman · R. Nathan
Minerva Center for Movement Ecology, The Hebrew University of Jerusalem, The Hebrew University of Jerusalem, Jerusalem, Israel

Movement Ecology Lab, Department of Ecology, Evolution and Behavior, Alexander Silberman Institute of Life Sciences, Faculty of Science, The Hebrew University of Jerusalem, Jerusalem, Israel

M. Assaf (✉)
Racah Institute of Physics, The Hebrew University of Jerusalem, Jerusalem, Israel
e-mail: michael.assaf@mail.huji.ac.il

L. Giuggioli, P. K. Maini (eds.), *The Mathematics of Movement*,
https://doi.org/10.1007/978-3-032-00873-2_3

movements largely depends on the nature of the disturbance [6]. For instance, while the construction of roads and human infrastructures may reduce animal movement by limiting habitat connectivity, increased landscape fragmentation could conversely increase movement to explore for, and exploit, more distant and less connected foraging opportunities [7, 8]. Likewise, the biological attributes of the species can determine the response of wild animals to human disturbances [5]. For instance, even though behavioural plasticity and diet diversity are generally favourable biological attributes for species to inhabit urban areas, other biological attributes such as litter size or body size may condition species capacity to dwell in urban areas [9]. Hence, examining the movement patterns of phylogenetically close species that share the same area undergoing human-induced rapid environmental changes is an ideal scenario to decipher the biological attributes that mediate animals' behavioural responses to human disturbances [10].

Comparing species' movement patterns requires not only selecting biologically meaningful metrics but also a temporal resolution that reflects an animals' behavioural response [11, 12]. Terrestrial mammals, for example, responded variably to COVID-19 lockdowns when examining displacements at 1-hour and 10-day scales. Here, their displacement at the 10-day scale increased during lockdowns, indicating increased landscape permeability due to human confinement to living spaces and reduced human mobility in green spaces [13]. Displacement and space use also differed among ungulate species depending on their foraging strategy but only when analysing movement at temporal scales above 40 minutes [14]. Thus, grazers travelled longer distances than browsers but such difference could not be detected at short temporal scales [14]. In contrast, birds in homogeneous landscapes such as deserts or grasslands travelled longer distances than birds inhabiting heterogeneous landscapes when considering both 1-hour and 10-day temporal scales [15]. Thus, the distribution of food resources is arguably a main driver of movement decisions across temporal scales in birds [15].

Recent studies have further emphasized the importance of multiscale analysis in understanding animal movement patterns. For instance, research on avian predators has revealed complex behaviours in area-restricted search and ergodicity breaking in localized movements, highlighting the need to consider movement at different scales [16, 17]. Additionally, investigations into the origins of anomalous diffusion across various biological systems, from molecular movements to migrating storks, have provided insights into the underlying principles governing biological movement patterns [18, 19]. Therefore, segmenting movement data across multiple time scales, rather than focusing on a single scale, is crucial for identifying the temporal threshold at which behavioural changes become detectable. If these changes vary depending on the resolution of the time scale, it may indicate distinct biological processes driving those behavioural shifts. In addition, incorporating other movement metrics than displacement and space use is essential to detect behavioural changes through movement analyses [20]. For instance, under food scarcity an increase in displacement at large temporal scales in combination with low linearity of movement paths will likely indicate increased searching behaviour while increased speed combined with high linearity might reflect goal-directed commuting movement towards predictable feeding sites [21].

Arid ecosystems such as deserts are particularly vulnerable to rapid environmental changes induced by human activities due to their limited resources [22]. The transition from resource-poor, low-predictability ecosystems to resource-rich, high-predictability ecosystems in human-modified landscape typically incites animals to adjust their spatial behaviour to rely on anthropogenic subsidies [23, 24]. While animals inhabiting deserts typically need to travel long distances over large areas to forage because resources are highly sparse, human-dominated deserts increase the abundance of resources at key sites, allowing animals to shorten their between-patch movement [25]. In addition, low predictability in deserts will promote searching behaviour in wild animals, characterized by low linearity and speed, while animals in human-dominated deserts will be able to anticipate the presence of resources at key sites, leading to straighter and faster between-patch movement [21, 26]. Thus, the transition from low resource, low predictability to high resource, high predictability in deserts is likely to reduce the range used by wild animals, shortening their displacements and resulting in straighter and faster between-patch movements compared with animals inhabiting natural deserts. Alternatively, regardless of the benefits associated with foraging at increasingly abundant and predictable anthropogenic feeding sites, wild animals might avoid these sites to minimize their exposure to potential threats (e.g. human presence or poisoned food items [27]). Avoiding anthropogenic feeding sites in deserts increasingly dominated by human activities will result in longer distances between patches and increasing searching behaviour characterized by low straightness and speed. Hence, the response of wild animals to inhabiting deserts transitioning from low resource, low predictability to high resource, high predictability depends on the balance between the perceived benefits of foraging in anthropogenic sites and the perceived risks associated with human presence and other threats [5].

Corvids (specifically genus Corvus, crows and ravens) are not only successful colonizers of urban areas but have also expanded across the world occupying almost all of Earth's biomes [28]. A combination of traits that include enhanced cognitive skills [29], high neophobia [30] and high species diversification [28] has been argued to improve their capacity to efficiently adjust their behaviour and persist under rapidly changing environmental conditions [31, 32]. However, there is considerable behavioural and cognitive variation among corvid species that is likely to condition their capacity to cope with rapid environmental changes and human presence [30, 31]. Here, we examine the movement patterns of two sympatric corvid species in a desert ecosystem that has undergone rapid environmental changes over the past five decades: the Judean Desert and Dead Sea coastline in Israel. The fan-tailed raven (hereafter "FT"; *Corvus rhipidurus*) inhabits the Dead Sea coastline, where tourism has substantially expanded, while the brown-necked raven (hereafter "BN"; *C. ruficollis*) also occurs in other parts of the Judean Desert, where tourism remains scarce. Human activities along the Dead Sea coastline have increased the availability of provisioned food sources, creating predictable resource hotspots within a relatively small area. In contrast, the larger, more western part of the Judean Desert is subject to low human impact, with resources remaining scarce and unpredictable. Drawing on these observations about ravens' space use and landscape changes, and assuming a general two-phase movement behaviour involving commuting flights between resource patches and area-restricted search (ARS) within a patch [26, 33], we formulate the following three hypotheses. First, during the commuting phase, FTs are hypothesized to fly fast and straight between known, anthropogenic resource hotspots located short distances apart within the smaller area they inhabit. In contrast, BNs are hypothesized to fly more slowly and to employ a mix of straight and tortuous flights (hence overall less linear movement) to search for unpredictable resources. Their commuting routes are expected to be longer and more variable, consistent with the larger, resource-scarce less predictable area they occupy. Second, regarding the local ARS phase, alternative scenarios of differential movement behaviour may exist but require more complex reasoning to justify. We therefore adopt a null hypothesis, positing no interspecific differences in movement characteristics at this scale. Third, combining the above, we hypothesize that interspecific differences will depend on the temporal scale of analysis. At longer temporal scales relevant to the commuting phase (longer than 1 hour), interspecific differences in movement behaviour are expected. At shorter temporal scales relevant to ARS phase, no significant differences are anticipated.

3.2 Methodology

3.2.1 Study Site and Species

The Dead Sea coastline and other parts of the Judean Desert are considered arid ecosystems characterised by high temperatures and extreme dry periods in between May and October (T_{max}, 34.6–45.5 °C; mean rainfall, 0.0–5.0 mm) and lower temperatures with occasional rain in between October and April (T_{max}, 27.8–36.5°C; mean rainfall, 0.0–39.8 mm; Israeli Meteorological Archive, 2010–2024). The Dead Sea coastline (*ca.* 240 km^2) has experienced rapid expansion in tourism infrastructure involving popular recreational sites and three national parks (i.e. Qumran National Park, En Gedi Nature Reserve and Masada National Park). The mean number of daily visitors in the area increased by 60% from 3000 prior 2015 to 4810 in 2023 (Wendt, 2016; INPA, 2023), and it is expected to continue increasing in the next years. In contrast, human presence in most of the Judean plateau (*ca.* 900 km^2) is limited to a few sparsely distributed Bedouin villages and other settlements as well as military bases; see Fig 3.1.

The wind dynamics in the Dead Sea valley are generally stable driven by the Mediterranean breeze and the sharp elevation differences between the valley floor and the surrounding eastern and western mountain ridges (*ca.* 1000-meter difference). We demonstrated consistency in wind dynamics across the study area by comparing measures of wind speed collected at 10-minute intervals simultaneously

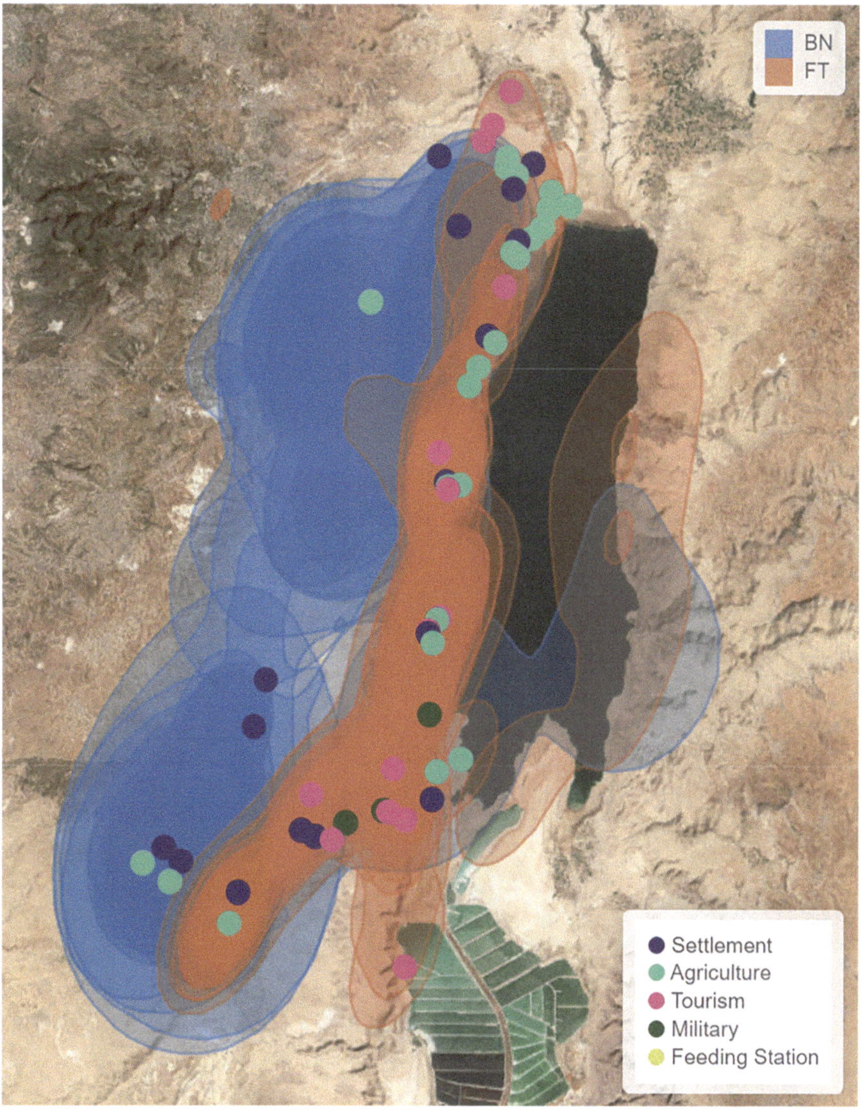

Fig. 3.1 Home ranges between June 2022 and August 2024 of fan-tailed (orange) and brown-necked (blue) ravens in the Dead Sea coastline, see text. Symbols represent relevant geographic locations in the area; see legend

across four meteorological stations. Wind speed was found to be strongly correlated among meteorological stations, with correlation between 0.28 and 0.37 (see Fig. 3.2), indicating that interspecific differences in movement patterns are unlikely to be caused by differences in wind.

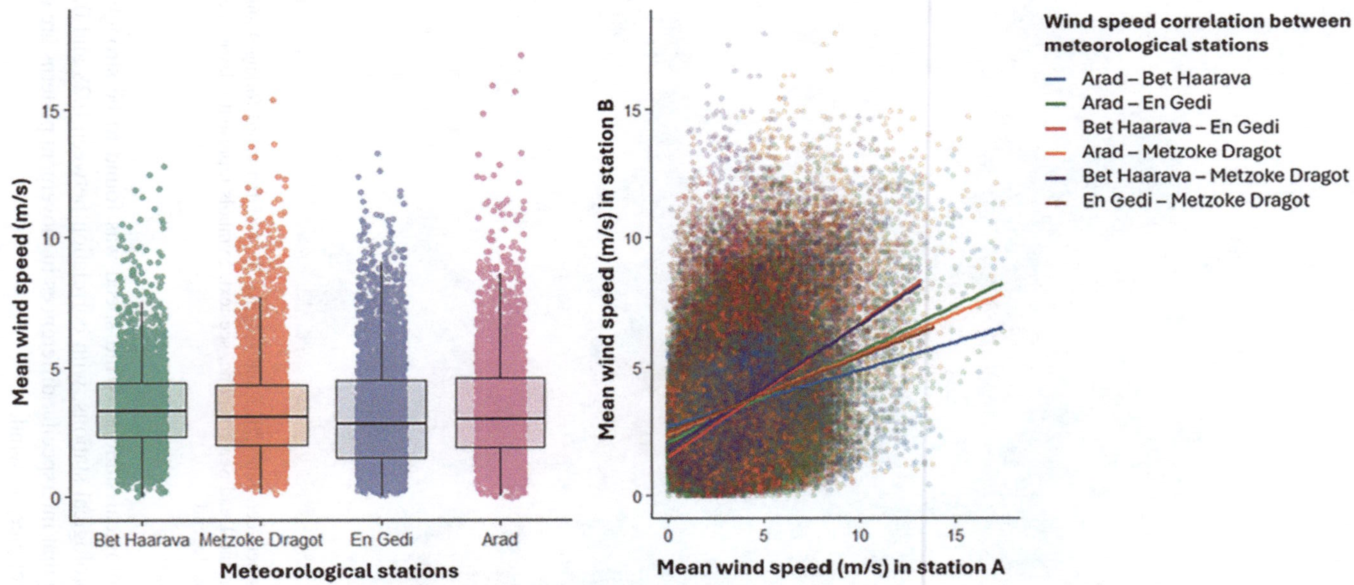

Fig. 3.2 Left panel shows the mean wind speed measured at 10-minute intervals in four meteorological stations present in the study area (i.e. Beit HaArava, Metzoke Dragot, En Gedi and Arad) between September and December of 2022 and 2023. Right panel shows the correlation between wind speed measurements collected simultaneously for each pair of meteorological stations present in the study area

The BNs are large Corvids (617.7 $\pm$ 62.44 g) with a distribution spanning North Africa, extending southward to Kenya and eastward to southern Iran. BNs prefer semi-desert and desert plains, mountain foothills including desert cultivation, desert settlements, rubbish dumps and army camps [34]. The FTs are smaller (478.27 $\pm$ 46.46 g), with a short, fan-shaped tail. FTs are distributed over sub-Saharan Africa, the South Middle East and the Arabian Peninsula, preferring high desert cliffs or open dry country [35]. A previous study [36] estimated the home range sizes for BN at 568 $\pm$ 385 km^2 and for FT at 212 $\pm$ 192 km^2 using auto-correlated kernel density estimation (AKDE) method at 95% isopleth. While the two species exhibited significant overlap in their ranges, their geographic distributions greatly differ; see Fig 3.1.

3.2.2 Data Collection

We trapped 43 FTs (21 males and 22 females) and 21 BNs (9 males and 12 females) across the Dead Sea coastline and the Judean Desert, respectively, between September 2021 and January 2024. We used Australian traps (4 $\times$ 4 m^2 and 2.5 m height) [37], located at En Gedi field school (31°28′12.0″N, 35°23′36.6″E), Hever vulture feeding station (31°24′15.2″N, 35o22′19.0″E) and Ras Mutreda, central Judean plateau (31°34′09.4″N35°20′53.0″E). We trapped from August to December to minimize the impact during breeding season (February to May). Subsequently, depending on the individuals' body mass, we deployed Ornitela 15 or 20 gram 4G GPS devices, which operate using Global Positioning System and tri-axial accelerometer. Notably, we selected GPS tags below 3% of the individuals' body mass to meet wildlife tagging convention [38]. The GPS tags were solar-charged which use cellular (GSM) networks for data transmission. The frequencies at which GPS fixes were collected depended on the battery charge level of the tag: (1) $>$ 90%, 1-second intervals; (2) 75–90%, 5-min ; (3) 50–75%, 10-min intervals; (4) 25–50%, 30-min intervals; and (5) $<$ 25%, 2-hours intervals.

3.2.3 Data Analyses

3.2.3.1 Data Processing

We processed GPS tracking data to remove outliers caused by satellite errors or device malfunction. Thus, we only kept GPS fixes that were collected: (i) within the Judean Desert area (i.e. only locations that fell within longitude [34.9, 35.7], latitude [30.9, 32.1]); (ii) by trilateration of at least six satellites; (iii) between September and December (to minimize the effects of breeding and heat waves during summer time); (iv) between the years 2022 and 2023 (to avoid human presence bias associated with the COVID-19 pandemic); (v) of individuals that have data for a minimum of

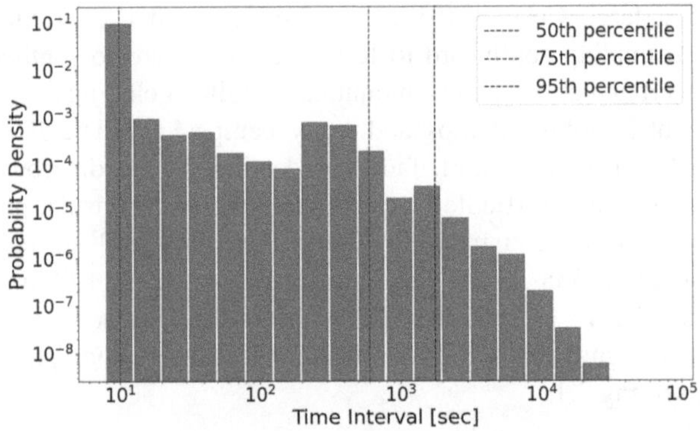

Fig. 3.3 Distribution of time intervals between records. Shown is the probability density of different time intervals between GPS records, after a temporal sub-sampling to a 10-second resolution. The first bin (< 15-second intervals) accounts for 56% of the records

three consecutive weeks; (vi) during active daylight hours (between 30 min prior to sunrise until 30 min after sunset); and (vii) with linear speed below 25 m/s between consecutive GPS records to remove outliers jumps. In addition, we sub-sampled GPS bursts (i.e. multiple GPS fixes collected at 1-second intervals) to a resolution of 10 seconds, by averaging all records within this time window, in order to reduce skewed sampling, caused by the rather asymmetric collection of GPS fixes between species. The distribution of temporal intervals between records is shown in Fig. 3.3. One can see that after the data resampling, the vast majority of temporal intervals falls within 1 minute and that approximately 95% of the temporal intervals are less than 30 min.

3.2.3.2 Segmentation

For all movement metrics, we segmented tracking data across four temporal scales: 20 minutes, 60 minutes, 1 day and 10 days. We selected these temporal scales for consistency with previous research in order to promote replicability [13–15]. For the 20- and 60-minute temporal scales, we identified data segments by ensuring that the time difference between the first and last record in each segment was within ±5% of the specified duration. We filtered out segments with less than 50 records (in 20-minute segments) and 150 records (in 60-minute segments). One-day segments were created using days with at least 5 hours. The longest day segment spanned 13.6 hours, resulting in significant variability in segment durations, and in the number of records per day; see Fig. 3.4 and its inset. To account for this variability, certain metrics were normalized to a full-day length of 13.5 hours; see below. This adjustment also accounted for the substantial variation in sunrise-to-sunset times over the 4 months of data collection. Finally, 10-day segments were defined as a range of 10 days in which at least 7 days contained valid day segments. As for the

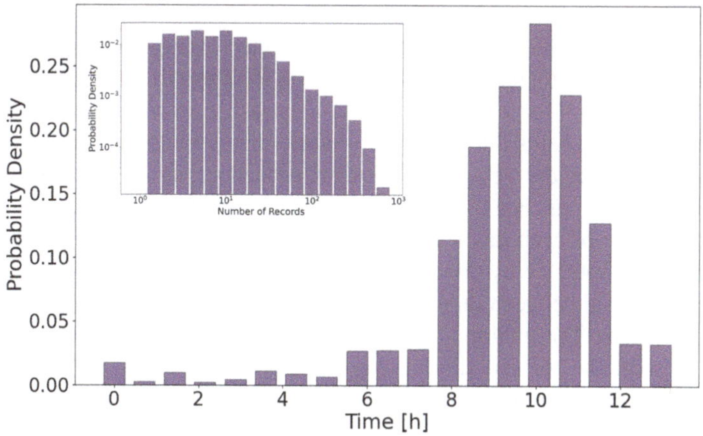

Fig. 3.4 Distribution of daily tracking duration. Shown is the probability distribution of total tracking time per day. Inset shows the probability distribution of the total number of records per day (after sub-sampling), for days with over 5 hours

1-day segments, for certain metrics the 10-day segments were then normalized by the total number of hours in the longest 10-day segment.

3.2.3.3 Movement Metrics

Using segments of varying temporal scales, we calculated a series of metrics derived from tracking data for each raven species. First, we calculated the *total distance travelled* (TDT) as the sum of the radial distances between every two consecutive records in a segment, representing the cumulative distance travelled within that segment. Due to the variability in the number of hours in these segments, normalization to a full segment length was applied to both 1-day and 10-day segments. Second, we calculated the *maximum displacement* (MD) as the maximum radial distance between any two records within a segment, providing an estimate of the largest extent of movement during that time. To complement the distance measurements, we additionally calculated the *start-end distance* as an estimate of the distance travelled by ravens between their roosting sites on consecutive nights. To ensure accuracy, we only accounted for days in which the first record was within half an hour before sunrise to 15 minutes after sunrise, and the last record was within 15 minutes before sunset to half an hour after sunset. Note that this complementary metric was calculated only for the 1-day scale. The third metric we calculated was the *straightness* (also known as linearity index [39]), which ranges from 0 to 1, with values close to 1 indicating movement along a straight path and values closer to 0 indicating more search-oriented, tortuous movement. In each segment, we identified the point farthest from the starting point and calculated the ratio of the radial distance between these two points to the TDT within this range. Similarly, we calculated the ratio between the farthest point and the endpoint of the segment. The straightness was then computed as the time-weighted average of these two values. This procedure was adopted because ravens tend to return to the same roost by the end of the day, which forces straightness to approach zero regardless of the actual straightness of their paths during the day. While this issue is most relevant for the 1-day and 10-day scales, we applied the same procedure across all scales to enable consistent comparisons. For 10-day segments, the straightness was calculated as the weighted average of each day's straightness. The fourth metric we computed was the *mean speed*, obtained by dividing the radial distance between consecutive rows by the time interval between them and taking the weighted average of all speeds in the same segment. The final metric we computed was the *home range size* (HRS), which represents the convex hull of a two-dimensional area formed by the polygon created by all points in a segment [40]. The latter metric turned out to be most significant in highlighting the interspecies differences; see Results section.

3.2.4 Statistical Analysis

For each metric and temporal scale we computed the mean and standard deviation of all data segments, removed outliers (farther than three standard deviations from the mean) and computed the various percentiles. To plot the histograms the data was binned and smoothed via a moving average of 3–5 bins, depending on the temporal scale. The histograms were then normalized to give an estimate of the probability density function.

3.3 Results

We first computed the TDT for the four temporal scales for the two different species; see Fig. 3.5. In this figure, we observe a relatively similar behaviour of the distributions for the 20-minute and 60-minute scales. For the 1-day temporal scale, the distributions start to deviate (especially the left tail), a tendency which becomes stronger for the 10-day scale. In both the 1-day and 10-day scales, the median of the TDT as well as the 5th and 95th percentiles are higher for the BNs compared to those of the FTs. The BN median is also higher for the shorter temporal scales. These results indicate that the total distance travelled by the BNs is higher (even if only by $\sim$10%) than that of the FTs.

Second, we computed the MD for the same temporal scales. In Fig. 3.6 we show the results for the 20-minute, 60-minute, 1-day and 10-day temporal scales. Importantly, in both Figs. 3.5 and 3.6 measuring the total and maximal distances, respectively, noticeable is the growing difference in the 5th percentile between the species as the temporal scale increases, which indicates a wider left tail of the FTs (which is especially evident for 60 minutes and above). While it does not affect the

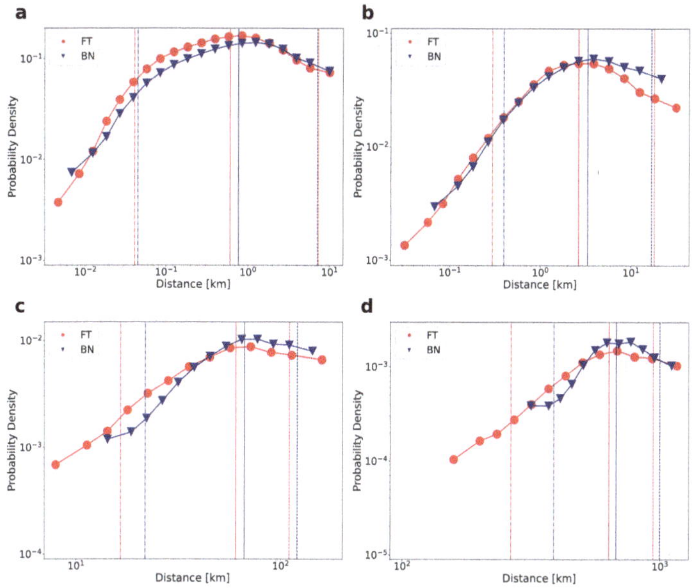

Fig. 3.5 Total distance travelled (TDT) at different temporal scales: 20 minutes (**a**), 60 minutes (**b**), 1 day (**c**) and 10 days (**d**). The solid, dash-dotted and dashed vertical lines represent the 50th (median), 5th and 95th percentiles, respectively

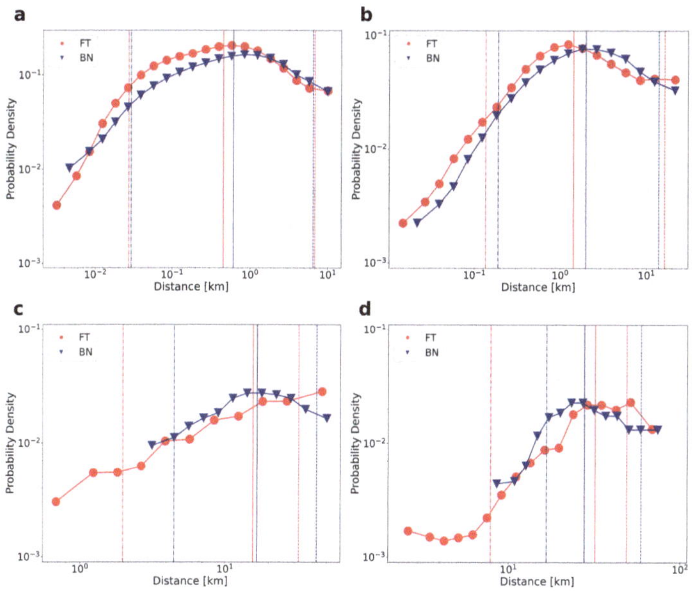

Fig. 3.6 Maximum displacement (MD) at different temporal scales: 20 minutes (**a**), 60 minutes (**b**), 1 day (**c**) and 10 days (**d**). The solid, dash-dotted and dashed vertical lines represent the 50th (median), 5th and 95th percentiles, respectively

medians, this difference indicates that among the FTs there are some individuals that occasionally move significantly shorter distances than the average (or median) in each temporal scale.

To complement the TDT and MD metrics, we measured the start-to-end distance averaged over days, as specified above; see Fig. 3.7. Here, the interspecies differences were significant. The median distance the BNs travel per day is over 10 times longer than that of the FT species, where a similar trend appears for the 5th and 95th percentiles. Moreover, the BNs display a very wide distribution, indicating that in some days, the roosting site of some individuals shifts by a distance on the order of tens of kilometres.

The third metric we have computed is the straightness index; see Fig. 3.8. While at the 20-minute and 60-minute temporal scales the distributions are statistically quite similar, for the 1-day and 10-day temporal scales the distribution is markedly different between the FT and BN species. Here, the FTs have a larger straightness index manifested by the median which is higher by ~30% compared to that of the BN. Moreover, the 95th percentile is much higher for the FTs, indicating that the tail spans much larger values of the straightness. The significant difference between the straightness of the FTs and BNs observed at the longer temporal scales strongly indicates that the tendency and likelihood of the FTs to follow a linear trajectory are much higher than those of the BNs.

Next, we computed the distribution of the mean speeds within temporal scales of 20 and 60 minutes, 1 day and 10 days. Here, while the distributions and percentiles were very similar for the 20-minute and 60-minute scales, for the 1-day and 10-day segments, we observed a slight tendency of the BNs to move faster; see Fig. 3.9. However, the differences between the medians and other percentiles were relatively small and insignificant.

Finally, we computed the home range size (HRS) by calculating the convex hull area of each segment. We did so for the two species, for the 20-minute, 60-minute, 1-day and 10-day temporal segments. While thus far, besides the straightness metric and daily start-end distance, the interspecies differences were mild, the HRS metric highlights perhaps the most noticeable difference between the FT and BN species; see Fig. 3.10. Here, as before, the FTs and BNs display similar behaviour over

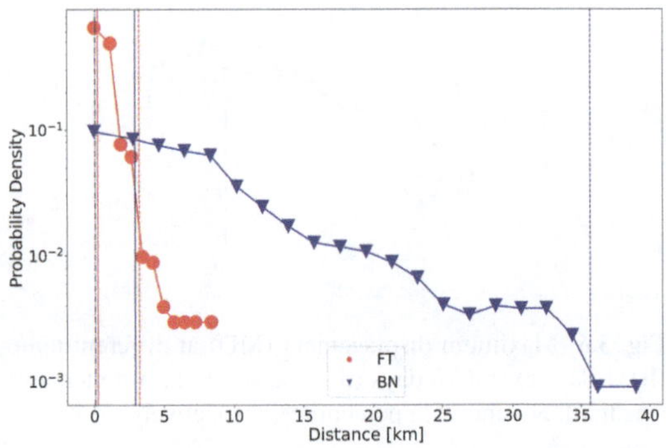

Fig. 3.7 Start-end distance. Shown is the distribution of the distances from sunrise until sunset for the FT and BN species. The solid, dash-dotted and dashed vertical lines represent the 50th (median), 5th and 95th percentiles, respectively

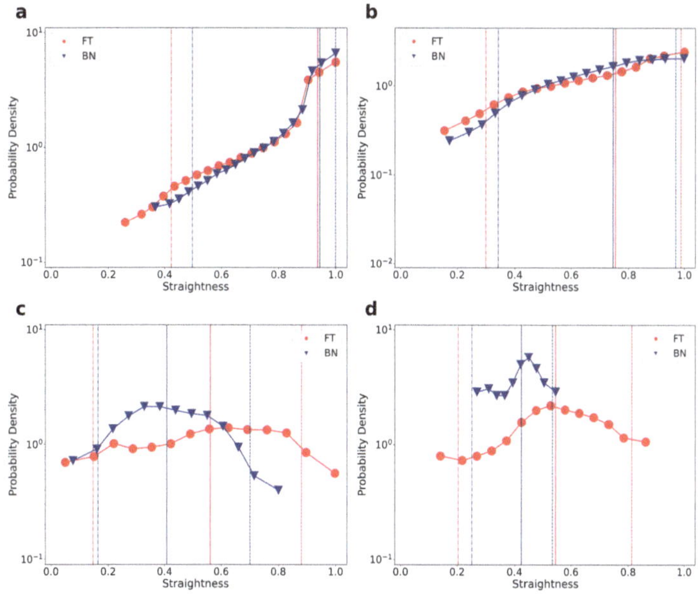

Fig. 3.8 Straightness at different temporal scales: 20 minutes (**a**), 60 minutes (**b**), 1 day (**c**) and 10 days (**d**). The solid, dash-dotted and dashed vertical lines represent the 50th (median), 5th and 95th percentiles, respectively

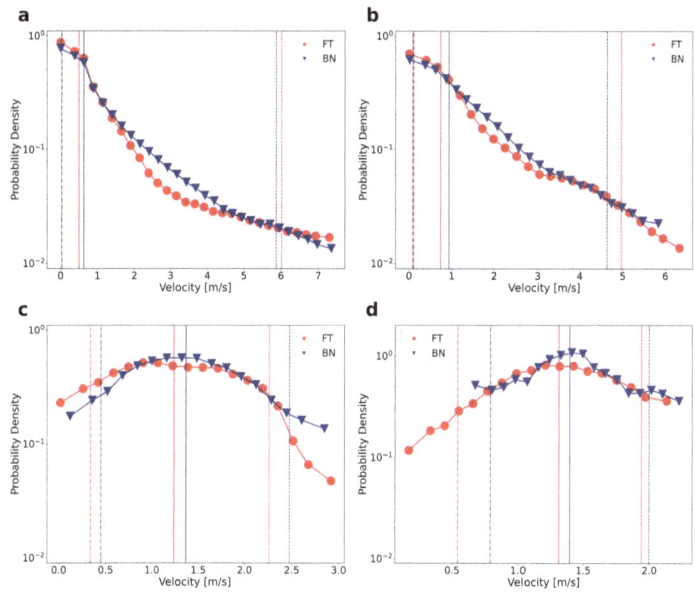

Fig. 3.9 Mean speed at different temporal scales: 20 minutes (**a**), 60 minutes (**b**), 1 day (**c**) and 10 days (**d**). The solid, dash-dotted and dashed vertical lines represent the 50th (median), 5th and 95th percentiles, respectively

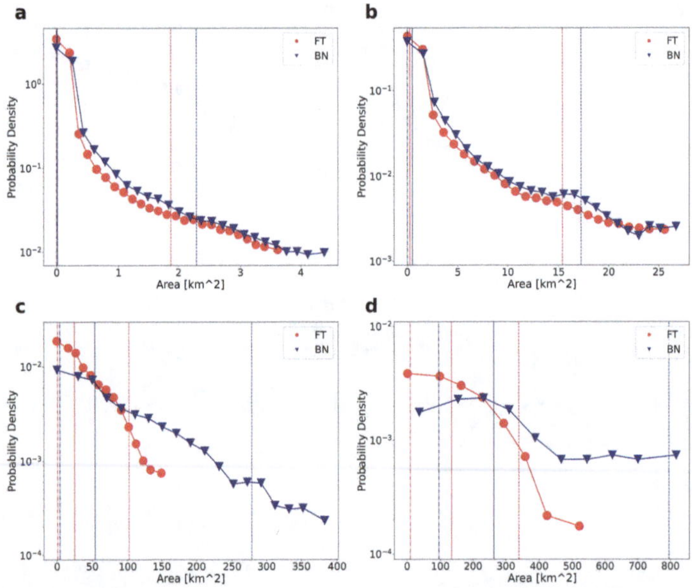

Fig. 3.10 The home range size (HRS) at different temporal scales: 20 minutes (**a**), 60 minutes (**b**), 1 day (**c**) and 10 days (**d**). The solid, dash-dotted and dashed vertical lines represent the 50th (median), 5th and 95th percentiles, respectively

shorter temporal scales (20 minutes and 60 minutes). In contrast, for the 1-day segments we see a dramatic difference in the median and 95th percentile, and a major difference in the distribution's right tail. For both the 1-day and 10-day temporal scales, the median of the BNs is over twice as large as the FTs median: $\sim$60 km^2 compared to $\sim$25 km^2 for the 1-day temporal scale and $\sim$260 km^2 compared to $\sim$130 km^2 for the 10-day temporal scale. A similar ratio between the BNs and FTs also appears in the 95th percentile for these temporal scales: $\sim$280 km^2 compared to $\sim$100 km^2 for the 1-day scale and $\sim$800 km^2 compared to $\sim$340 km^2 for the 10-day scale. These values are consistent with the area coverage shown in Fig. 3.1, which was 568 ± 385 km^2 for the BNs and 212 ± 192 km^2 for the FT, computed via the AKDE method at 95% isopleth. Notably, the 5th percentile displays an even larger difference between the BNs and FTs, even though it is not visible on the figure's scale.

3.4 Discussion

We have studied the differences between two raven species: the FT and BN, by measuring five different metrics over four different temporal scales and one complementary metric for the daily distance travelled. Our findings reveal clear interspecific differences in movement patterns between the two raven species, which vary according to the temporal scale at which movement metrics are segmented. First, we found interspecies differences at larger scales of 1 and 10 days, where FTs fly shorter distances and straighter than BNs during the commuting phase likely because they forage in anthropogenic resource hotspots instead of relying on unpredictable resources. Second, we did not find significant differences at shorter temporal scales of 20 and 60 minutes, which indicates similar interspecific behaviours in the local ARS phase. These results emphasize that different temporal scales may capture different behavioural modes, indicating that raven species may differ more markedly in behaviours detectable at broader temporal scales.

The fact that we did not observe interspecific differences in movement metrics at short temporal scales may also be an artefact of the resolution of GPS sampling and the sub-sampling we have performed. Indeed, previous studies comparing species movement metrics across temporal scales highlighted the importance of homogenizing sampling frequencies to reliably describe drivers of animal movement particularly at short temporal scales [13–15]. As a result, since here the GPS sampling frequency varied among individuals depending on device battery life, we had to use significant sub-sampling to homogenize the data, potentially hampering our ability to properly investigate shorter temporal scales on the order of minutes. This limitation stems from the trade-off between sampling frequency and tracking duration, which is common in many GPS-based studies. In contrast, high-throughput reverse-GPS tracking techniques like ATLAS routinely sample at higher frequencies (every 1 to 8 seconds), enabling analyses at finer temporal scales, such as 30 seconds or 1 minute [11]. However, GPS tracking was chosen for this study due to the extensive home range of ravens, which often exceeds the typical coverage area of most ATLAS systems (approximately 40 x 40 km). GPS tracking, therefore, enabled the detection of patterns at longer temporal scales of 1-day or 10-day segments. Here, we were able to clearly observe dramatic interspecific differences between the species, especially in the straightness, daily start-to-end distance and home range size.

Due to the complex nature of the data, our approach here was primarily metric based, attempting to find similarities and differences across the different temporal scales. Thus, our aim was not to model the movement in order to predict what the animal will do [41], but rather to understand what the animal is doing. In this aspect, a future direction would be to use numerical simulations that are mimicking the underlying dynamics, in order to increase our ability to validate our conceptual framework and hypotheses. This can be done, e.g. by implementing a multi-mode movement simulation based on the biased correlated random walk (BCRW) framework [42]. In these types of simulations, for each discrete movement step, the step lengths are sampled from a given distribution, while the turning angles are drawn from some other given distribution. Parameters are varied at various temporal scales as to reflect changes in behaviour, e.g. transition from within-patch search mode to goal-directed commute mode. By fitting the model parameters to empirical data, one can generate smoother and richer synthetic datasets consistent with empirical observations, which may increase our ability to perform quantitative analysis.

Importantly, different temporal scales may be capturing distinct behavioural modes [16]. At shorter temporal scales, movement data may reflect behaviours such as within-patch foraging, social interactions or brief resting periods in shaded areas. In contrast, at larger temporal scales, behaviours like between-patch foraging, long-distance commuting or extended resting periods may dominate the observed patterns. Consequently, FT and BN individuals may behave similarly at short temporal scales but differ at longer temporal scales, as these scales encompass behaviours more strongly influenced by ecological and functional adaptations specific to each species. Therefore, the distribution of food resources in the landscape occupied by each species likely determines their movement patterns. The rapid transition from resource-poor to resource-rich landscape due to the expansion of tourism along the Dead Sea coastline likely led FTs to maximize their foraging effort while reducing searching behaviour. In contrast, BNs may rely on opportunistic foraging in the resource-scarce desert, exhibiting movement patterns indicative of greater exploratory and between-patch foraging behaviour. This disparity underscores how resource distribution and landscape context drive interspecific differences in movement metrics, particularly at larger temporal scales.

Lastly, interspecific differences in movement patterns may not only be dictated by the composition of the landscape but also by other biologically relevant aspects of each species. The large body size of BNs enables them to dominate food sources in areas they co-shared with FTs further away from human presence, limiting FT access to food resources and space use patterns [36]. Hence, FTs are likely pressured to forage in tourist sites where food is more abundant but often nutritionally poor [27]. Reliance on low-quality food sources can negatively impact wildlife health, leading to nutritional deficiencies and increased susceptibility to diseases, and, in some cases, can even be lethal [43]. These pressures may be contributing to the documented decline in the population of FTs in Israel [de Guinea et al., *pers. comm.*], as reliance on human-provided food increases their vulnerability while reducing the resilience of their populations in natural habitats. Thus, our results possibly indicate competitive exclusion dynamics between BN and FT species, which may be driving the long term viability of the local population of FTs along the Dead Sea coastline. Future studies should examine differences in the internal state of individuals (i.e. glucocorticoids, microbiota diversity; [44]) as well as external conditions (i.e. cooling microhabitats, wind dynamics in wadis, human presence; [45]) influencing the movement patterns of each species to underscore the relationship between landscape composition, interspecies competition and population viability in the context of rapid human-induced environmental changes.

Acknowledgments This research was supported by the Ministry of Science of Israel (grant number 3-17405) and the Israeli Science Foundation Breakthrough Research Grant (grant number 2715/23). BA was partially supported by Israeli Science Foundation Research Grant (grant number 531/20). MdG was partially supported by the Lady Davis Postdoctoral Fellowship from the Hebrew University of Jerusalem.

References

1. Finn, C., Grattarola, F., Pincheira-Donoso, D.: More losers than winners: investigating anthropocene defaunation through the diversity of population trends. Biol. Rev. **98**(5), 1732–1748 (2023)
2. Lowry, H., Lill, A., Wong, B.B.: Behavioural responses of wildlife to urban environments. Biol. Rev. **88**(3), 537–549 (2013)
3. Sih, A., Mathot, K.J., Moiron, M., Montiglio, P.-O., Wolf, M., Dingemanse, N.J.: Animal personality and state–behaviour feedbacks: a review and guide for empiricists. Trends Ecol. Evol. **30**(1), 50–60 (2015)
4. Lambin, E.F., Geist, H.J., Lepers, E.: Dynamics of land-use and land-cover change in tropical regions. Annu. Rev. Environ. Resour. **28**(1), 205–241 (2003)
5. Sih, A.: Understanding variation in behavioural responses to human-induced rapid environmental change: a conceptual overview. Anim. Behav. **85**(5), 1077–1088 (2013)
6. Fahrig, L.: Non-optimal animal movement in human-altered landscapes. Funct. Ecol. **21**(6), 1003–1015 (2007)
7. Cosgrove, A.J., McWhorter, T.J., Maron, M.: Consequences of impediments to animal movements at different scales: a conceptual framework and review. Diversity Distrib. **24**(4), 448–459 (2018)
8. Doherty, T.S., Driscoll, D.A.: Coupling movement and landscape ecology for animal conservation in production landscapes. Proc. Roy. Soc. B: Biol. Sci. **285**(1870), 20172272 (2018)
9. Santini, L., González-Suárez, M., Russo, D., Gonzalez-Voyer, A., von Hardenberg, A., Ancillotto, L.: One strategy does not fit all: determinants of urban adaptation in mammals. Ecol. Lett. **22**(2), 365–376 (2019)
10. Doherty, T.S., Hays, G.C., Driscoll, D.A.: Human disturbance causes widespread disruption of animal movement. Nat. Ecol. Evol. **5**(4), 513–519 (2021)
11. Nathan, R., Monk, C.T., Arlinghaus, R., Adam, T., Alós, J., Assaf, M., Baktoft, H., Beardsworth, C.E., Bertram, M.G., Bijleveld, A.I., et al.: Big-data approaches lead to an increased understanding of the ecology of animal movement. Science **375**(6582), eabg1780 (2022)
12. Kays, R., Hirsch, B., Caillaud, D., Mares, R., Alavi, S., Havmøller, R.W., Crofoot, M.: Multi-scale movement syndromes for comparative analyses of animal movement patterns. Movement Ecol. **11**(1), 61 (2023)
13. Tucker, M.A., Schipper, A.M., Adams, T.S., Attias, N., Avgar, T., Babic, N.L., Barker, K.J., Bastille-Rousseau, G., Behr, D.M., Belant, J.L., et al.: Behavioral responses of terrestrial mammals to covid-19 lockdowns. Science **380**(6649), 1059–1064 (2023)
14. Tablado, Z., Revilla, E., Dubray, D., Saïd, S., Maillard, D., Loison, A.: From steps to home range formation: species-specific movement upscaling among sympatric ungulates. Funct. Ecol. **30**(8), 1384–1396 (2016)
15. Tucker, M.A., Alexandrou, O., Bierregaard Jr, R.O., Bildstein, K.L., Böhning-Gaese, K., Bracis, C., Brzorad, J.N., Buechley, E.R., Cabot, D., Calabrese, J.M., et al.: Large birds travel farther in homogeneous environments. Global Ecol. Biogeogr. **28**(5), 576–587 (2019)
16. Vilk, O., Campos, D., Méndez, V., Lourie, E., Nathan, R., Assaf, M.: Phase transition in a non-Markovian animal exploration model with preferential returns. Phys. Rev. Lett. **128**(14), 148301 (2022)
17. Vilk, O., Orchan, Y., Charter, M., Ganot, N., Toledo, S., Nathan, R., Assaf, M.: Ergodicity breaking in area-restricted search of avian predators. Phys. Rev. X **12**(3), 031005 (2022)
18. Vilk, O., Aghion, E., Nathan, R., Toledo, S., Metzler, R., Assaf, M.: Classification of anomalous diffusion in animal movement data using power spectral analysis. J. Phys. A: Math. Theor. **55**(33), 334004 (2022)
19. Vilk, O., Aghion, E., Avgar, T., Beta, C., Nagel, O., Sabri, A., Sarfati, R., Schwartz, D.K., Weiss, M., Krapf, D., Nathan, R., Metzler, R, Assaf, M.: Unravelling the origins of anomalous diffusion: from molecules to migrating storks. Phys. Rev. Res. **4**(3), 033055 (2022)

20. Reyna-Hurtado, R., Teichroeb, J.A., Bonnell, T.R., Hernández-Sarabia, R.U., Vickers, S.M., Serio-Silva, J.C., Sicotte, P., Chapman, C.A.: Primates adjust movement strategies due to changing food availability. Behav. Ecol. **29**(2), 368–376 (2018)
21. Janmaat, K.R., de Guinea, M., Collet, J., Byrne, R.W., Robira, B., van Loon, E., Jang, H., Biro, D., Ramos-Fernandez, G., Ross, C., et al.: Using natural travel paths to infer and compare primate cognition in the wild. Iscience **24**(4), 102343 (2021)
22. Oswald, K.N., Berger-Tal, O., Roll, U.: Small-scale land-use change effects on breeding success in a desert-living social bird. Behav. Ecol. **35**(3), arae023 (2024)
23. Restani, M., Marzluff, J.M., Yates, R.E.: Effects of anthropogenic food sources on movements, survivorship, and sociality of common ravens in the arctic. Condor **103**(2), 399–404 (2001)
24. Boarman, W.I., Kristan, III, W.B.: Trends in common raven populations in the Mojave and Sonoran deserts. (2006)
25. Zeller, K.A., Wattles, D.W., Conlee, L., DeStefano, S.: Black bears alter movements in response to anthropogenic features with time of day and season. Movement Ecol. **7**, 1–14 (2019)
26. Riotte-Lambert, L., Matthiopoulos, J.: Environmental predictability as a cause and consequence of animal movement. Trends Ecol. Evol. **35**(2), 163–174 (2020)
27. Oro, D., Genovart, M., Tavecchia, G., Fowler, M.S., Martínez-Abraín, A.: Ecological and evolutionary implications of food subsidies from humans. Ecol. Lett. **16**(12), 1501–1514 (2013)
28. Garcia-Porta, J., Sol, D., Pennell, M., Sayol, F., Kaliontzopoulou, A., Botero, C.A.: Niche expansion and adaptive divergence in the global radiation of crows and ravens. Nat. Commun. **13**(1), 2086 (2022)
29. Bugnyar, T.: Why are ravens smart? Exploring the social intelligence hypothesis. J. Ornithol. **165**(1), 15–26 (2024)
30. Miller, R., Lambert, M.L., Frohnwieser, A., Brecht, K.F., Bugnyar, T., Crampton, I., Garcia-Pelegrin, E., Gould, K., Greggor, A.L., Izawa, E.-I., et al.: Socio-ecological correlates of neophobia in corvids. Curr. Biol. **32**(1), 74–85 (2022)
31. Marzluff, J.M., Neatherlin, E.: Corvid response to human settlements and campgrounds: causes, consequences, and challenges for conservation. Biol. Conserv. **130**(2), 301–314 (2006)
32. Greggor, A.L., Berger-Tal, O., Blumstein, D.T., Angeloni, L., Bessa-Gomes, C., Blackwell, B.F., St Clair, C.C., Crooks, K., de Silva, S., Fernández-Juricic, E., et al.: Research priorities from animal behaviour for maximising conservation progress. Trends Ecol. Evol. **31**(12), 953–964 (2016)
33. Benhamou, S.: Of scales and stationarity in animal movements. Ecol. Lett. **17**(3), 261–272 (2014)
34. BirdLife International: Brown-necked raven (*corvus ruficollis*) - species factsheet (2024). Accessed 21 Nov 2024
35. BirdLife International: Fan-tailed raven (*corvus rhipidurus*) - species factsheet (2024). Accessed 21 Nov 2024
36. Zuckerman, R.: Interspecific competition between sympatric ravens in a human-modified desert ecosystem. Master's Thesis, The Hebrew University of Jerusalem, Jerusalem, Israel, 2024. Final work for a master's degree in natural sciences
37. Moran, S.: Control of hooded crows by modified Australian traps. Phytoparasitica **19**, 95–101 (1991)
38. Geen, G.R., Robinson, R.A., Baillie, S.R.: Effects of tracking devices on individual birds—a review of the evidence. J Avian Biol. **50**(2) (2019)
39. Benhamou, S.: How to reliably estimate the tortuosity of an animal's path: straightness, sinuosity, or fractal dimension? J. Theor. Biol. **229**(2), 209–220 (2004)
40. Getz, W.M., Wilmers, C.C.: A local nearest-neighbor convex-hull construction of home ranges and utilization distributions. Ecography **27**(4), 489–505 (2004)
41. Gurarie, E., Bracis, C., Delgado, M., Meckley, T.D., Kojola, I., Wagner, C.M.: What is the animal doing? Tools for exploring behavioural structure in animal movements. J. Anim. Ecol. **85**(1), 69–84 (2016)

42. McClintock, B.T., King, R., Thomas, L., Matthiopoulos, J., McConnell, B.J., Morales, J.M.: A general discrete-time modeling framework for animal movement using multistate random walks. Ecol. Monogr. **82**(3), 335–349 (2012)
43. Newsome, T.M., Dellinger, J.A., Pavey, C.R., Ripple, W.J., Shores, C.R., Wirsing, A.J., Dickman, C.R.: The ecological effects of providing resource subsidies to predators. Global Ecol. Biogeogr. **24**(1), 1–11 (2015)
44. Pannoni, S.B., Proffitt, K.M., Holben, W.E.: Non-invasive monitoring of multiple wildlife health factors by fecal microbiome analysis. Ecol. Evol. **12**(2), e8564 (2022)
45. Shaw, A.K.: Causes and consequences of individual variation in animal movement. Movement Ecol. **8**(1), 12 (2020)

Chapter 4
Probabilistic Formulations of Diffusive Search Processes with Stochastic Resetting

Paul C. Bressloff

4.1 Introduction

Stochastic resetting is a mechanism whereby a system is returned to its initial state at a random sequence of times that is typically generated by a Poisson process with constant rate r. The simplest example is a Brownian particle that instantaneously resets to its initial position $\mathbf{x}_0 \in \mathbb{R}^d$ [1–3]. There have been a wide range of generalisations at the single-particle level; see the review [4] and references therein. These include both modifications in the underlying stochastic dynamics in the absence of resetting and modifications in the resetting protocol itself (e.g. delays due to finite return times [5–10] and refractory periods [11, 12], randomly distributed reset points [2, 13, 14], non-Poissonian resetting [15], etc.) There is also an increasing interest in understanding the effects of stochastic resetting on interacting particle systems, as highlighted in the recent review [16]. One of the interesting features that naturally emerges in the analysis of multi-particle systems is the distinction between two modes of resetting protocol: (i) global resetting where all of the particles simultaneously reset and (ii) local resetting where each particle independently resets according to its own sequence of resetting times.

Within the specific application domains of animal movement and cell biology, stochastic resetting has been posited as a mechanism for reducing the expected time to find a hidden target $\mathcal{U} \subset \mathbb{R}^d$ within some large search domain. In the case of a foraging animal, the target represents a local region of resources such as food or shelter, and resetting mimics the observed tendency for an animal to return to its home base in order to rest or resupply [9]; see Fig. 4.1. Target detection is said to occur when the particle first reaches a point on the target boundary $\partial\mathcal{U}$, which can be implemented by taking $\partial\mathcal{U}$ to be a totally absorbing surface. The total search

P. C. Bressloff (✉)
Department of Mathematics, Imperial College London, London, UK
e-mail: p.bressloff@imperial.ac.uk

L. Giuggioli, P. K. Maini (eds.), *The Mathematics of Movement*,
https://doi.org/10.1007/978-3-032-00873-2_4

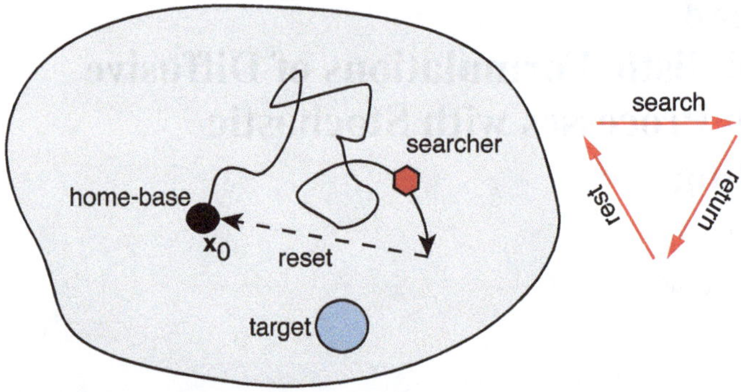

Fig. 4.1 A foraging animal diffusively searches for a hidden target $\mathcal{U}$ that contains resources such as food or shelter. Prior to finding the target, the forager may return to its home base at $\mathbf{x}_0$ in order to rest. The animal thus cycles through three phases until it finds the target [9]: a diffusive search phase, a return phase, and a refractory or rest phase. The duration of each search phase that resets before finding the target is usually assumed to be exponentially distributed with a constant resetting rate r. If $\mathbf{X}_n$ is the position of the particle after the nth reset event, then the time to return directly to home base at a constant speed V is $|\mathbf{X}_n - \mathbf{x}_0|/V$. Finally, the refractory period τ may be an additional random variable that is generated from a fixed waiting time density $\psi(\tau)$

time is given by the first passage time (FPT) $\widehat{T} = \inf\{t > 0, \mathbf{X}(t) \in \partial\mathcal{U}\}$, where $\mathbf{X}(t) \in \mathbb{R}^d$ is the position of the searcher at time t. If the searcher fails to find the target, then $\widehat{T} = \infty$. One of the limitations of a purely diffusive process as a stochastic search mechanism is that the mean first passage time (MFPT) $\langle\widehat{T}\rangle$ diverges as the size of the search domain goes to infinity. On the other hand, the introduction of a stochastic resetting protocol can support a finite MFPT that has a unique minimum as a function of the resetting rate [1–3]. Two examples of search processes with resetting at the cellular level are motor-based vesicular transport and cytoneme-based morphogenesis [10]. In the first case, the searcher represents a molecular motor complex transporting vesicles to synaptic targets in the axon or dendrites of a neuron, say, while resetting corresponds to the removal and return of the complex to the cell body. In the second case, the searcher represents the tip of a growing actin-rich filament (cytoneme) that delivers morphogen to target cells during embryonic development in vertebrates [10]. Resetting now takes into account the fact that a cytoneme can suddenly switch to a shrinkage phase and rapidly retract back to the source cell; see Fig. 4.2.

The stochastic processes shown in Figs. 4.1 and 4.2 cycle between a diffusive search phase, a ballistic return phase, and a refractory phase. If the delays associated with the last two phases are negligible, then the stochastic dynamics reduces to a search process with instantaneous stochastic resetting, as originally proposed in Refs. [1–3]. Most studies of diffusion with instantaneous resetting focus on the distribution of sample paths expressed as the solution to a forward or backward Fokker-Planck (FP) equation with additional source and sink terms. Recently, however, a complementary probabilistic approach has been developed in which

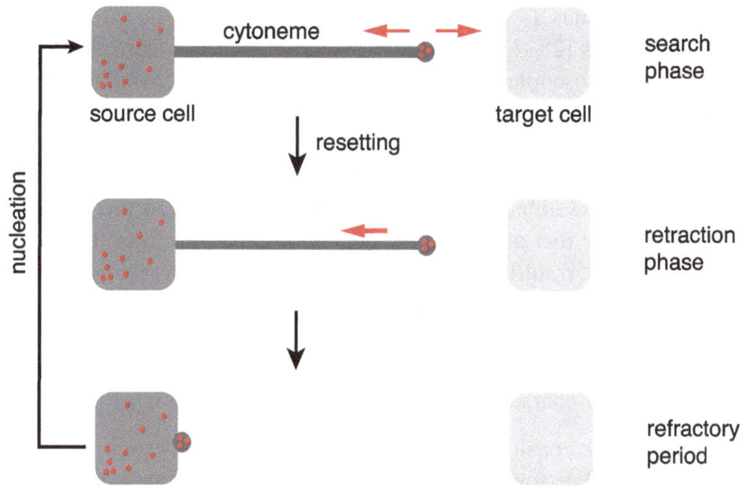

Fig. 4.2 Example of a subcellular diffusive search with resetting [10]. A cytoneme randomly grows and shrinks until it forms a contact with a target cell and delivers the morphogen at its tip. Prior to finding a target, the cytoneme may randomly switch to a retraction phase (resetting) and return to the origin at a constant speed V. After a refractory period (nucleation waiting time), a new cytoneme starts to grow. In the absence of resetting, the cytoneme may fail to find the target by growing in a direction that does not intersect the target (not shown)

individual sample paths are represented by solutions of a stochastic differential equation (SDE) for a jump-diffusion process [17, 18]. The latter probabilistic formulation has a number of useful features:

(i) There exists a generalised version of Itô's lemma for jump-diffusion processes that provides a direct link between sample paths and the underlying FP equation with resetting. This can be used to derive global density equations for populations of diffusing particles with local or global stochastic resetting [19].
(ii) The SDE for instantaneous resetting can be combined with a more general encounter-based formulation of diffusion-mediated surface absorption at a target boundary. The latter exploits the fact that diffusion in a domain with a totally reflecting surface can be implemented probabilistically in terms of so-called reflected Brownian motion, which involves the introduction of a Brownian functional known as the boundary local time [20–23]. The local time characterises the amount of time that a Brownian particle spends in the neighbourhood of a point on the boundary. In the encounter-based approach [24–27], the probability that the particle is absorbed at the boundary $\partial\mathcal{U}$ (finds the target) depends on the amount of surface contact time. This is implemented in terms of a stopping condition that requires the local time to cross some randomly generated threshold.

(iii) Stochastic calculus is playing an increasingly important role in stochastic thermodynamics [28–31]. The latter extends classical ideas of entropy, heat, and work to mesoscopic non-equilibrium systems. The generalised Itô's lemma for jump-diffusion processes has recently been used to determine the rate of stochastic entropy production along sample paths of diffusing particles with instantaneous resetting [18]. Averaging the stochastic entropy with respect to the ensemble of stochastic trajectories leads to the second law of thermodynamics that quantifies the degree of departure from thermodynamic equilibrium. The resulting analysis complements previous studies based on the Gibbs-Shannon entropy of the ensemble [32] or the ratio of the path probabilities of forward and time-reversed trajectories [33–36]. In the particular context of stochastic search processes, stochastic thermodynamics could provide a valuable tool for characterising the costs associated with resetting [37].

In this chapter, we consider probabilistic formulations of search processes with stochastic resetting. For simplicity, we focus on continuous one-dimensional (1D) diffusion processes. (For the analysis of discrete space-time stochastic resetting processes, see, e.g. Refs. [38].) We begin by showing how 1D diffusion with instantaneous resetting can be represented as a jump-diffusion process (Sect. 4.2). We construct the generalised Itô's lemma and use this to derive density equations for a single particle and for a population of non-interacting particles. In Sect. 4.3 we use the probabilistic formulation of instantaneous resetting to analyse the FPT problem for a single particle on the half-line with a partially absorbing target at the origin. Finally, in Sect. 4.4 we show how conditional expectations and renewal theory can be used to analyse the effects of finite return times and refractory periods on a diffusive search process with non-instantaneous resetting. Some background material on stochastic representations of Poisson processes is presented in the Appendix.

4.2 Instantaneous Resetting as a Jump-Diffusion Process

Let $X(t) \in \mathbb{R}$ denote the position of a single motile particle at time t that instantaneously resets to its initial position x_0 at the random times $\widehat{T}_n$ generated by a Poisson process $N(t)$ with rate r; see the Appendix. Let D and v denote the diffusion coefficient and velocity of the particle between resetting events. (In the case of a passive Brownian particle, v would be replaced by the term $F(x)/\gamma$, where $F(x)$ is a position-dependent force and γ is a friction coefficient satisfying the Einstein relation $D\gamma = k_B T$, where k_B is the Boltzmann constant and T is temperature.) The stochastic dynamics in the presence of resetting can be represented in terms of the jump-diffusion process [17, 18]

$$dX(t) = vdt + \sqrt{2D}dW(t) + (x_0 - X(t^-))dN(t), \tag{4.1}$$

where $W(t)$ is a Wiener process with

$$\langle W(t)\rangle = 0, \quad \langle W(t)W(s)\rangle = \min\{t, s\}. \tag{4.2}$$

The solution $X(t)$ is taken to be right-continuous so that

$$X(\widehat{T}_n) = \lim_{t\to\widehat{T}_n^-}(x_0 - X(t)) + \lim_{t\to\widehat{T}_n^-} X(t) = x_0, \tag{4.3}$$

which corresponds to perfect resetting. In standard formulations of stochastic resetting at the single-particle level [4], the SDE (4.1) is averaged over multiple realisations of the Poisson resetting process. Since $\mathbb{P}[\widehat{T}_n \in [t, t+dt]] = rdt$, Eq. (4.1) becomes

$$dX(t) = vdt + \sqrt{2D}dW \quad \text{with probability } 1 - rdt, \tag{4.4a}$$

$$dX(t) = x_0 - X(t) \text{ with probability } rdt. \tag{4.4b}$$

For a general introduction to jump-diffusion processes, see Ref. [39].

4.2.1 Generalised Itô's Lemma and the Fokker-Planck Equation

Let $f(x)$ be an arbitrary smooth bounded test function on $\mathbb{R}$. The infinitesimal $df(X(t))$ can be decomposed as

$$\begin{aligned} df(X(t)) &= f'(X(t))[vdt + \sqrt{2D}dW(t)] \\ &+ \frac{1}{2}f''(X(t)[vdt + \sqrt{2D}dW(t)]^2 \\ &+ \Big[f(X(t)) - f(X(t^-))\Big]dN(t) + o(dt). \end{aligned} \tag{4.5}$$

Using the identity $dW(t)^2 = dt$ and dropping all $o(dt)$ terms yield an extended version of Itô's lemma for the jump-diffusion process:

$$\begin{aligned} df(X(t)) &= [f'(X(t))v + Df''(X(t)]dt + \sqrt{2D}f'(X(t))dW(t) \\ &+ \Big[f(X(t)) - f(X(t^-))\Big]dN(t). \end{aligned} \tag{4.6}$$

The corresponding integral version of the extended Itô's lemma is

$$
\begin{aligned}
f(X(t) = f(x_0) + &\int_0^t \Big[v f'(X(s)) + D f''(X(s)) \Big] ds \\
&+ \sqrt{2D} \int_0^t f'(X(s)) dW(s) + \int_0^t \Big[f(X(s)) - f(X(s^-)) \Big] dN(s).
\end{aligned} \tag{4.7}
$$

Substituting for $dN(s)$ using (see the appendix)

$$
dN(s) := h(s)ds = \sum_{n\geq 1} \delta(t - \widehat{T}_n) ds \tag{4.8}
$$

implies that the final integral can be rewritten as

$$
\begin{aligned}
\mathcal{I}(t) &= \int_0^t \Big[\sum_{n\geq 1} \Big[f(X(s)) - f(X(s^-)) \Big] \delta(s - \widehat{T}_n) ds \\
&= \sum_{n\geq 1} \Big[f(X(\widehat{T}_n)) - f(X(\widehat{T}_n^-)) \Big] \Theta(t - \widehat{T}_n) \\
&= \sum_{n\geq 1} \Big[f(x_0) - f(X(\widehat{T}_n^-)) \Big] \Theta(t - \widehat{T}_n),
\end{aligned} \tag{4.9}
$$

where Θ is a Heaviside function. We have used the fact that $X(T_n) = x_0$.

Let $p_r(x,t)$ be the probability density for the particle to be at position x at time t with

$$
p_r(x,t) = \Big\langle \mathbb{E}\Big[\delta(x - X(t)) \Big] \Big\rangle. \tag{4.10}
$$

The subscript r means that p_r is the probability density in the presence of resetting at a rate r. Moreover, $\langle\cdot\rangle$ denotes expectation with respect to the white noise process and $\mathbb{E}$ denotes expectation with respect to the resetting process. We assume that the two sources of noise are uncorrelated. Consider the following equation for an arbitrary test function f:

$$
\Big[\int_{\mathbb{R}} f(x) \frac{\partial \rho(x,t)}{\partial t} dx \Big] dt = df(X(t)), \tag{4.11}
$$

where $\rho(x,t) = \delta(x - X(t))$. Applying Itô's lemma (4.6) and using Eq. (4.8) give

$$
\int_{\mathbb{R}} f(x) \frac{\partial \rho(x,t)}{\partial t} dx = \int_{\mathbb{R}} \rho(x,t) \Big[v f'(x) + D f''(x) + \sqrt{2D} f'(x) \xi(t) \Big] dx
$$

$$+ \sum_{n\geq 1} \delta(t-\widehat{T}_n) \int_{\mathbb{R}} [\delta(x-x_0) - \rho(x,t^-)] f(x) dx. \tag{4.12}$$

We have formally set $dW(t) = \xi(t)dt$ with $\xi(t)$, a white noise process, such that

$$\langle \xi(t) \rangle = 0, \quad \langle \xi(t)\xi(t') \rangle = \delta(t-t'). \tag{4.13}$$

Integrating by parts and using the arbitrariness of f lead to the stochastic partial differential equation (SPDE)

$$\begin{aligned} \frac{\partial \rho}{\partial t} &= D\frac{\partial^2 \rho(x,t)}{\partial x^2} - v\frac{\partial \rho(x,t)}{\partial x} + \sqrt{2D}\frac{\partial \rho(x,t)}{\partial x}\xi(t) \\ &\quad + \sum_{n\geq 1} \delta(t-\widehat{T}_n)[\delta(x-x_0) - \rho(x,t^-)]. \end{aligned} \tag{4.14}$$

Finally, taking expectations with respect to the white noise and resetting process recovers the standard FP equation for diffusion with resetting [4]:

$$\frac{\partial p_r(x,t)}{\partial t} = D\frac{\partial^2 p_r(x,t)}{\partial x^2} - v\frac{\partial p_r(x,t))}{\partial x} + r[\delta(x-x_0) - p_r(x,t)] \tag{4.15}$$

with $p_r(x,0) = \delta(x-x_0)$. We have used the identity $\mathbb{E}[\sum_{n\geq 1}\delta(t-\widehat{T}_n)dt] = \mathbb{E}[dN(t)] = rdt$; see the Appendix.

The above SDE formulation of stochastic resetting can be extended to the case where the resetting point is distributed according to some density $\sigma_0(x)$ along the lines considered in Refs. [2, 13, 14]. This means that at the nth reset time $\widehat{T}_n$ we have $X(\widehat{T}_n) = X_n$ with $\mathbb{P}[X_n \in [x, x+dx]] = \sigma_0(x)dx$. Formally speaking, Eq. (4.1) becomes

$$dX(t) = vdt + \sqrt{2D}dW(t) + \textstyle\sum_{n\geq 1}(X_n - X(t^-))\delta(t-\widehat{T}_n), \tag{4.16}$$

and Eq. (4.14) takes the form

$$\begin{aligned} \frac{\partial \rho(x,t)}{\partial t} &= D\frac{\partial^2 \rho(x,t)}{\partial x^2} - v\frac{\partial \rho(x,t)}{\partial x} + \sqrt{2D}\frac{\partial \rho(x,t)}{\partial x}\xi(t) \\ &\quad + \sum_{n\geq 1}[\delta(x-X_n) - \rho(x,t^-)]\delta(t-\widehat{T}_n). \end{aligned} \tag{4.17}$$

Finally, averaging Eq. (4.17) with respect to the white noise process and stochastic resetting gives

$$\frac{\partial p_r(x,t)}{\partial t} = D\frac{\partial^2 p_r(x,t)}{\partial x^2} - v\frac{\partial p_r(x,t)}{\partial x} + r[\sigma_0(x) - p_r(x,t)]. \tag{4.18}$$

4.2.2 Population of Non-interacting Searchers

As we mentioned in the Introduction, one of the applications of the probabilistic formulation of resetting is that it provides a mathematical framework for analysing populations of searchers. For simplicity, we assume that the searchers are non-interacting and have the same velocity and diffusivity. Let $X_j(t) \in \mathbb{R}$ denote the position of the jth searcher at time t, $j = 1, \ldots, \mathcal{N}$, where $\mathcal{N}$ is the number of searchers. Suppose that each searcher can reset to its initial position $x_{j,0}$ at a fixed Poisson rate r. Following Ref. [16], we distinguish between a global update rule, where all of the particles simultaneously reset, and a local resetting scheme in which the particles independently reset. Let $\widehat{T}_{j,n}$ denote the nth resetting time of the jth particle with $n \geq 0$. In the case of local resetting, the inter-reset intervals $\tau_{j,n} = \widehat{T}_{j,n+1} - \widehat{T}_{j,n}$ are independent and identically distributed random variables for all j, n with $\mathbb{P}[\tau_{j,n} \in [\tau, \tau + d\tau]] = re^{-r\tau}d\tau$. (For simplicity, we assume that all particles have the same resetting rate r.) On the other hand, in the case of global resetting we have $\widehat{T}_{j,n} = \widehat{T}_n$ for all $j = 1, \ldots, \mathcal{N}$ with $\tau_n = \widehat{T}_{n+1} - \widehat{T}_n$ exponentially distributed.

First, consider the effects of local resetting. The particle positions $X_j(t)$ evolve according to the jump-diffusion equations

$$dX_j(t) = vdt + \sqrt{2D}dW_j(t) + (x_{j,0} - X_j(t))dN_j(t), \tag{4.19}$$

where $W_j(t)$ is an independent Wiener process and $N_j(t)$ is an independent Poisson process. In order to derive a multi-particle analogue of the FP equation (4.14), we incorporate resetting into the 'hydrodynamic' formulation of Dean-Kawasaki [40, 41], following along analogous lines to Ref. [19]. Define the global density (or empirical measure)

$$\rho(x,t) = \frac{1}{\mathcal{N}}\sum_{j=1}^{\mathcal{N}} \delta(x - X_j(t)), \tag{4.20}$$

and introduce an arbitrary smooth test function $f(x)$ of compact support. Then

$$\left[\int_{\mathbb{R}} f(x)\frac{\partial \rho(x,t)}{\partial t}dx\right]dt = \frac{1}{\mathcal{N}}\sum_{j=1}^{\mathcal{N}} df(X_j(t)). \tag{4.21}$$

Using the generalised Itô's lemma (4.6), we find that

$$\int_{\mathbb{R}} f(x)\frac{\partial \rho(x,t)}{\partial t}dx = \int_{\mathbb{R}}\Bigg[\bigg(\frac{\sqrt{2D}}{\mathcal{N}}\sum_{j=1}^{\mathcal{N}}\delta(x-X_j(t))\xi_j(t) + +v\rho(x,t)\bigg)f'(x)$$
$$+Df''(x)\rho(x,t) + \frac{f(x)}{\mathcal{N}}\sum_{j=1}^{\mathcal{N}} h_j(t)[\delta(x-x_{j,0})$$
$$-\delta(x-X_j(t))]\Bigg]dx,$$

where $dW_j(t) = \xi_j(t)dt$ and

$$dN_j(\tau) := h_j(\tau)d\tau = \sum_{k=1}^{\infty}\delta(\tau - \widehat{T}_{j,k})d\tau. \tag{4.22}$$

Integrating by parts the various terms involving derivatives of f and using the fact that f is arbitrary yield the following SPDE for ρ in the weak sense:

$$\frac{\partial \rho(x,t)}{\partial t} = -\sqrt{\frac{2D}{\mathcal{N}^2}}\sum_{j=1}^{\mathcal{N}}\left[\xi_j(t)\frac{\partial \delta(x-X_j(t))}{\partial x}\right] + D\frac{\partial^2 \rho(x,t)}{\partial x^2} - v\frac{\partial \rho(x,t)}{\partial x}$$
$$+\frac{1}{\mathcal{N}}\sum_{j=1}^{\mathcal{N}}\left[\delta(x-x_{j,0}) - \delta(x-X_j(t))\right]h_j(t). \tag{4.23}$$

Setting $h_j = 0$ (no resetting) and performing a transformation of the Gaussian noise terms recover a version of the classical Dean-Kawasaki (DK) equation [40, 41]. On the other hand, taking expectations with respect to the independent Poisson processes shows that

$$\mathbb{E}\left[\frac{1}{\mathcal{N}}\sum_{j=1}^{\mathcal{N}}\Big(\delta(x-x_{j,0}) - \delta(x-X_j(t))\Big)h_j(t)\right] \tag{4.24}$$
$$= \frac{1}{\mathcal{N}}\sum_{j=1}^{\mathcal{N}}\mathbb{E}\left[\delta(x-x_{j,0}) - \delta(x-X_j(t))\right]\mathbb{E}[h_j(t)] = r[\rho_0(x) - \rho(x,t)],$$

with

$$\rho_0(x) = \frac{1}{\mathcal{N}}\sum_{j=1}^{\mathcal{N}}\delta(x-x_{j,0}). \tag{4.25}$$

Setting $p_r(x,t) = \langle \mathbb{E}[\rho(x,t)]\rangle$ and averaging Eq. (4.23) with respect to the Gaussian and Poisson noise processes result in a population version of Eq. (4.15):[1]

$$\frac{\partial p_r(x,t)}{\partial t} = D\frac{\partial^2 p_r(x,t)}{\partial x^2} - v\frac{\partial p_r(x,t)}{\partial x} + r\left[\rho_0(x) - p_r(x,t)\right]. \quad (4.26)$$

This is formally equivalent to the FP equation for a single-particle resetting to a position that is randomly selected from the distribution $\rho_0(x)$.

The situation is very different for a population of non-interacting searchers with global resetting, whose positions evolve according to

$$dX_j(t) = vdt + \sqrt{2D}dW_j(t) + (x_{j,0} - X_j(t))dN(t), \quad (4.27)$$

for a single global Poisson process $N(t)$. Equation (4.23) now reduces to the more compact form

$$\frac{\partial \rho(x,t)}{\partial t} = -\sqrt{\frac{2D}{\mathcal{N}^2}}\sum_{j=1}^{\mathcal{N}}\left[\xi_j(t)\frac{\partial \delta(x - X_j(t))}{\partial x}\right] + D\frac{\partial^2 \rho(x,t)}{\partial x^2} - v\frac{\partial \rho(x,t)}{\partial x}$$
$$+\left[\rho_0(x) - \rho(x,t)\right]h(t), \quad (4.28)$$

where we have set $dN(t) = h(t)dt$. In contrast to local resetting, averaging with respect to the white noise processes yields a closed stochastic equation for $\Phi(\mathbf{x},t) = \langle \rho(\mathbf{x},t)\rangle$:

$$\frac{\partial \Phi(x,t)}{\partial t} = D\frac{\partial^2 \Phi(x,t)}{\partial x^2} - v\frac{\partial \Phi(x,t)}{\partial x} + [\rho_0(x) - \Phi(x,t)]h(t). \quad (4.29)$$

This takes the form of a diffusion equation with stochastic resetting to an initial density $\rho_0(x)$. Taking expectations with respect to global resetting and defining

$$\overline{\Phi}(x,t) = \mathbb{E}[\Phi(x,t)], \quad C(x,y,t) = \mathbb{E}[\Phi(x,t)\Phi(y,t)], \quad (4.30)$$

we obtain the deterministic equations

$$\frac{\partial \overline{\Phi}(x,t)}{\partial t} = D\frac{\partial^2 \overline{\Phi}(x,t)}{\partial x^2} - v\frac{\partial \overline{\Phi}(x,t)}{\partial x} + r[\rho_0(x) - \overline{\Phi}(x,t)], \quad (4.31)$$

and

[1] The hydrodynamic formulation of Dean-Kawasaki can be extended to incorporate particle interactions [40, 41]. However, it is no longer possible to obtain a closed equation for the average one-particle density p_r, since it couples to the two-point correlation function, etc. One exception is to take the thermodynamic limit $\mathcal{N} \to \infty$ under a mean field ansatz [19].

$$\frac{\partial C(x,y,t)}{\partial t} = D\frac{\partial^2 C(x,y,t)}{\partial x^2} + D\frac{\partial^2 C(x,y,t)}{\partial y^2} - v\frac{\partial C(x,y,t)}{\partial x} - v\frac{\partial C(x,y,t)}{\partial y}$$
$$+r[C_0(x,y) - C(x,y,t)]. \tag{4.32}$$

It immediately follows from direct substitution that

$$C(x,y,t) \neq \mathbb{E}[\Phi(x,t)]\mathbb{E}[\Phi(y,t)]. \tag{4.33}$$

The existence of a non-zero two-point correlation function reflects the fact that global resetting induces statistical correlations between the particles, even though they are non-interacting.

4.3 Encounter-Based Models of Target Absorption

So far we have ignored the presence of a target. In this section we construct the FPT problem for a single particle that is restricted to the half-line $[0,\infty)$ with a partially reactive or absorbing target at $x = 0$ and instantaneous resetting to a point $x_0 > 0$. We show how to incorporate the encounter-based method for partially reactive targets [24–27] into the jump-diffusion formulation of instantaneous resetting.

4.3.1 *Boundary Local Time and Instantaneous Resetting*

First, suppose that the target is totally reflecting. Introduce the boundary local time [21–23]

$$L(t) = \lim_{a\to 0}\frac{\ell_a(t)}{a} = \lim_{a\to 0}\frac{D}{a}\int_0^t \Theta(a - X(\tau))d\tau = D\int_0^t \delta(X(\tau))d\tau, \tag{4.34}$$

where Θ is the Heaviside function. The function $\ell_a(t)$ measures the occupation time of the particle in a boundary layer of width a; see Fig. 4.3. It can be shown that $L(t)$ exists and is a non-decreasing stochastic process. In the presence of a reflecting boundary at $x = 0$, the jump-diffusion SDE (4.1) is modified according to

$$dX(t) = vdt + \sqrt{2D}dW(t) + (x_0 - X(t))dN(t) + dL(t). \tag{4.35}$$

Heuristically speaking, the differential of the local time generates a rightward impulsive kick whenever the particle hits the boundary.

Now suppose that the diffusion process is killed when the local time $L(t)$ at $x = 0$ exceeds a randomly generated threshold $\widehat{L}$ with

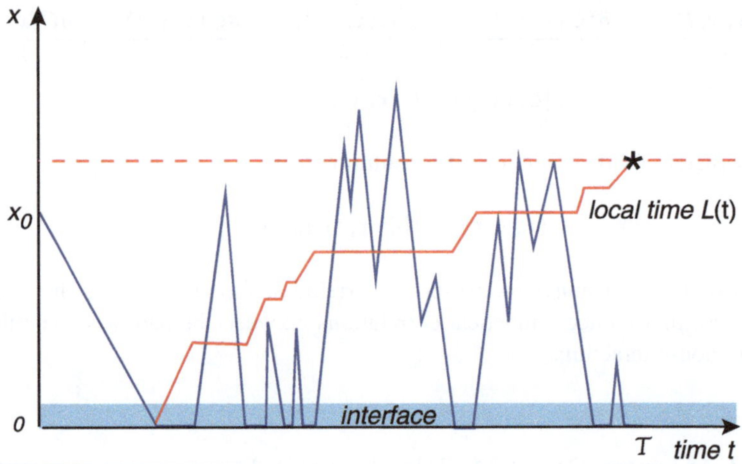

Fig. 4.3 Coarse-grained trajectory of Brownian motion on the half-line with a reflecting boundary at $x = 0$

$$P[\widehat{L} > l] \equiv \Psi(l). \tag{4.36}$$

That is, the particle is absorbed (finds the target) at the stopping time

$$\mathcal{T} = \inf\{t > 0 : \ L(t) > \widehat{L}\}. \tag{4.37}$$

Since $L(t)$ is a non-decreasing process, the condition $t < \mathcal{T}$ is equivalent to the condition $L(t) < \widehat{L}$. Hence, the corresponding SDE is

$$dX(t) = \left[vdt + \sqrt{2D}dW(t) + (x_0 - X(t))dN(t) + dL(t) \right] \Theta(\widehat{L} - L(t)). \tag{4.38}$$

4.3.2 Generalised FP Equation and the Local Time Propagator

The SDE (4.38) can be used to derive the generalised FP equation for instantaneous resetting on the half-line with a partially absorbing target at $x = 0$. First, we introduce the empirical measure

$$\rho(x, t) = \delta(x - X(t))\Theta(\widehat{L} - L(t)). \tag{4.39}$$

Let $f(x)$ be a bounded smooth test function and consider the identity

$$\left[\int_0^\infty f(x)\frac{\partial \rho(x,t)}{\partial t}dx\right]dt$$
$$= \left[df(X(t)) + f'(X(t))dL(t)\right]\Theta(\widehat{L} - L(t)) + f(X(t))d\Theta(\widehat{L} - L(t)), \tag{4.40}$$

with $df(X(t))$ given by Eq. (4.6). From the definition of the local time we also have $dL(t) = D\delta(X(t))dt$ and

$$f(X(t))d\Theta(\widehat{L} - L(t)) = -Df(0)\delta(X(t))\delta(L(t) - \widehat{L}). \tag{4.41}$$

Rewriting the right-hand side of Eq. (4.40) using $\rho(x,t)$, we obtain a modified version of Eq. (4.12):

$$\begin{aligned}
&\int_0^\infty f(x)\frac{\partial \rho(x,t)}{\partial t}dx \\
&= \int_0^\infty \rho(x,t)\left[vf'(x) + Df''(x) + D\delta(x)f'(0) + \sqrt{2D}f'(x)\xi(t)\right]dx \\
&\quad - Df(0)\delta(X(t))\delta(L(t) - \widehat{L}) \\
&\quad + \sum_{n\geq 1}\delta(t - \widehat{T}_n)\Theta(\widehat{L} - L(t))\int_0^\infty [\delta(x - x_0) - \delta(x - X(t^-))]f(x)dx.
\end{aligned} \tag{4.42}$$

Finally, integrating by parts and using the arbitrariness of f give

$$\begin{aligned}
\frac{\partial \rho}{\partial t} &= D\frac{\partial^2 \rho(x,t)}{\partial x^2} - v\frac{\partial \rho(x,t)}{\partial x} + \sqrt{2D}\frac{\partial \rho(x,t)}{\partial x}\xi(t) \\
&\quad + \sum_{n\geq 1}\delta(t - \widehat{T}_n)[\delta(x - x_0)\Theta(\widehat{L} - L(t)) - \rho(x,t^-)],
\end{aligned} \tag{4.43}$$

together with the boundary condition

$$v\rho(0,t) - D\frac{\partial \rho(0,t)}{\partial x} = -\sqrt{2D}\xi(t)\rho(0,t) - D\delta(X(t))\delta(L(t) - \widehat{L}). \tag{4.44}$$

The final step is to take expectations with respect to the Gaussian noise and the threshold $\widehat{L}$. We use the identities

$$\mathbb{E}\left[\Theta(\widehat{L} - L(t))\right] = \Psi(L(t)), \quad \mathbb{E}\left[\delta(L(t) - \widehat{L})\right] := \psi(L(t)) = -\Psi'(L(t)). \tag{4.45}$$

Introduce the pair of densities

$$p_r(x,t) = \mathbb{E}[\langle \rho(x,t)\rangle] = \langle \delta(x - X(t))\Psi(L(t))\rangle, \tag{4.46a}$$

$$\nu_r(x,t) = \mathbb{E}[\langle \delta(x - X(t))\delta(L(t) - \widehat{L})\rangle] = \langle \delta(x - X(t))\psi(L(t))\rangle, \tag{4.46b}$$

and define the marginal distribution

$$Q_r(t) = \int_0^\infty p_r(x,t)dx = \langle \Psi(L(t))\rangle, \tag{4.47}$$

which is the survival probability that the particle has not been absorbed at $x = 0$ in the time interval $[0, t]$. (We have suppressed the explicit dependence on the initial/reset point x_0.) We then obtain the extended FP equation

$$\frac{\partial p_r(x,t)}{\partial t} = -\frac{\partial J_r(x,t)}{\partial x} + r[\delta(x - x_0)Q_r(t) - p_r(x,t)], \tag{4.48a}$$

supplemented by the boundary condition

$$J_r(0,t) = -D\nu_r(0,t), \tag{4.48b}$$

with the probability flux

$$J_r(x,t) = vp_r(x,t) - D\frac{\partial p_r(x,t)}{\partial x}. \tag{4.49}$$

For a general choice of local time threshold distribution $\Psi(l)$, the system of Eq. (4.48) does not yield a closed equation for $p_r(x,t)$ due to the dependence of the boundary condition on $\nu_r(x,t)$. However, in the particular case of an exponential distribution,

$$\Psi(l) = \mathrm{e}^{-\kappa_0 l/D}, \quad \psi(l) = -\Psi'(l) = \kappa_0\Psi(l)/D, \tag{4.50}$$

we have $\nu(x,t) = \kappa_0 p(x,t)$. Equation (4.48b) then reduces to the Robin boundary condition

$$J_r(0,t) = -\kappa_0 p_r(0,t). \tag{4.51}$$

Note that in the limit $\kappa_0 \to 0$, the boundary at $x = 0$ becomes totally reflecting so that $Q_r = 1$ and we recover the standard forward equation for 1D diffusion with resetting [1, 2]. On the other hand, if $\kappa_0 \to \infty$ then the boundary is totally absorbing. In order to handle non-exponential distributions Ψ, we introduce the local time propagator

$$P_r(x,l,t) = \langle \delta(x - X(t))\delta(l - L(t))\rangle. \tag{4.52}$$

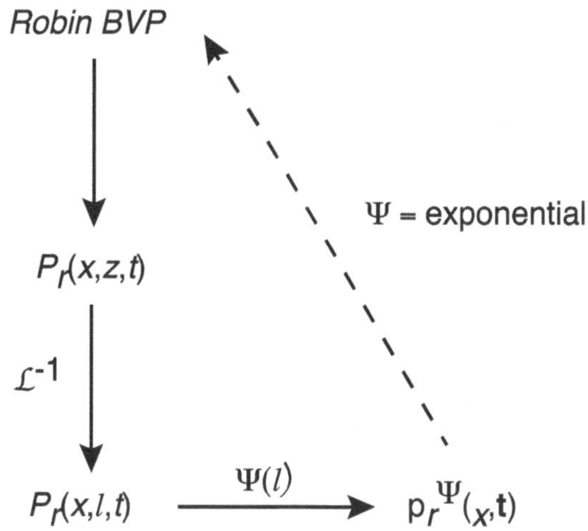

Fig. 4.4 Diagram illustrating the encounter-based framework for diffusion in a domain with a partially absorbing target. The solution of the BVP in the case of a constant reactivity κ_0 generates the Laplace transform $\widetilde{P}_r(x,z,t)$ of the local time propagator $P_r(x,l,t)$ with $z=\kappa_0/D$. The inverse Laplace transform determines the marginal probability density $p_r^{\Psi}(x,t)$ according to Eq. (4.54)

A crucial observation is that the solution to the Robin boundary value problem is the Laplace transform of the propagator with respect to l:

$$p_r(x,t)=\int_0^\infty \mathrm{e}^{-zl}P_r(x,l,t)dl=\widetilde{P}_r(x,z,t), \quad z=\frac{\kappa_0}{D}. \tag{4.53}$$

Hence, for a general distribution Ψ we have

$$p_r^{\Psi}(x,t)=\int_0^\infty \Psi(l)P_r(x,l,t)dl=\int_0^\infty \Psi(l)\mathcal{L}^{-1}[\widetilde{P}_r(x,z,t)]dl, \tag{4.54}$$

where $\mathcal{L}^{-1}$ denotes the inverse Laplace transform. The general construction is summarised in Fig. 4.4. Note that our derivation differs from previous studies of encounter-based models with resetting, which focused on the PDE for the full propagator [42, 43].

4.3.3 Solution of the Robin BVP (v = 0)

In order to illustrate the effects of resetting on the MFPT to find the target, we consider the particular case of zero drift ($v=0$) and a Robin boundary condition. Laplace-transforming equations (4.48) and (4.51) for $v=0$ give

$$D\frac{\partial^2\widetilde{p}_r(x,s)}{\partial x^2}-(r+s)p_r(x,s)=-[1+r\widetilde{Q}_r(s)]\delta(x-x_0), \quad x>0, \tag{4.55a}$$

$$D\frac{\partial \widetilde{p}_r(x,s)}{\partial x} = \kappa_0 \widetilde{p}_r(x,s), \quad x = 0. \tag{4.55b}$$

The general bounded solution is of the form

$$\widetilde{p}_r(x,s) = A_r(s)\mathrm{e}^{-\alpha x} + [1 + r\widetilde{Q}_r(s)]G(x,x_0), \tag{4.56}$$

where $\alpha = \sqrt{[r+s]/D}$. The first term on the right-hand side of Eq. (4.56) is the solution to the homogeneous version of Eq. (4.55) and G_r is the modified Helmholtz Green's function in the case of a totally absorbing boundary condition at $x = 0$:

$$D\frac{\partial^2 G_r}{\partial x^2} - (r+s)G = -\delta(x - x_0), \ 0 < x < \infty, \qquad G_r(0, \alpha|x_0) = 0, \tag{4.57}$$

with the solution

$$G_r(x,x_0) = \frac{1}{2D\alpha}\left[\mathrm{e}^{-\alpha|x-x_0|} - \mathrm{e}^{-\alpha|x+x_0|}\right]. \tag{4.58}$$

The unknown coefficient $A_r(s)$ is determined by imposing the Robin boundary condition (4.55b):

$$A_r(s) = \frac{1 + r\widetilde{Q}_r(s)}{\kappa_0 + \alpha D}\mathrm{e}^{-\alpha x_0}. \tag{4.59}$$

Hence, the full solution of the Laplace-transformed probability density with resetting can be decomposed as

$$\widetilde{p}_r(x,s) = [1 + r\widetilde{Q}_r(s)]\widetilde{p}_0(x, r+s), \quad 0 < x < \infty, \tag{4.60}$$

where $\widetilde{Q}_r(s) = \int_0^\infty \widetilde{p}_r(x,s)dx$ and $\widetilde{p}_0$ is the corresponding solution without resetting,

$$\widetilde{p}_0(x,s) = \frac{\mathrm{e}^{-\sqrt{s/D}(x+x_0)}}{\kappa_0 + \sqrt{sD}} + G_0(x,x_0). \tag{4.61}$$

Finally, integrating equation (4.60) with respect to x shows that

$$\widetilde{Q}_r(s) = [1 + r\widetilde{Q}_r(s)]\int_0^\infty \widetilde{p}_0(x, r+s)dx = [1 + r\widetilde{Q}_r(s)]\widetilde{Q}_0(r+s), \tag{4.62}$$

where $\widetilde{Q}_0$ is the Laplace transform of the survival probability without resetting:

$$\widetilde{Q}_0(s) = \frac{1 - e^{-\sqrt{s/D}x_0}}{s} + \frac{e^{-\sqrt{s/D}x_0}}{s + \kappa_0\sqrt{s/D}}. \tag{4.63}$$

Rearranging Eq. (4.62) thus determines the survival probability with resetting in terms of the corresponding probability without resetting:

$$\widetilde{Q}_r(s) = \frac{\widetilde{Q}_0(r+s)}{1 - r\widetilde{Q}_0(r+s)}. \tag{4.64}$$

The survival probability is related to the FPT density for absorption at $x = 0$ according to $f_r(t) = -\partial_t Q_r(t)$. It follows that the MFPT is

$$T_r := \int_0^\infty t f_r(t)dt = -\int_0^\infty t\partial_t Q_r(t)dt = \int_0^\infty Q_r(t)dt = \widetilde{Q}_r(0), \tag{4.65}$$

where we have used integration by parts and the result $Q_r(t) \to 0$ in the limit $t \to \infty$ when $\kappa_0 > 0$ and $r > 0$. Hence, setting $s = 0$ in Eq. (4.64) shows that

$$T_r = \frac{\widetilde{Q}_0(r)}{1 - r\widetilde{Q}_0(r)}. \tag{4.66}$$

In Fig. 4.5 we show sample plots of the MFPT T_r as a function of the resetting rate r for various reactivities κ_0. As expected, T_r is a unimodal function of r with

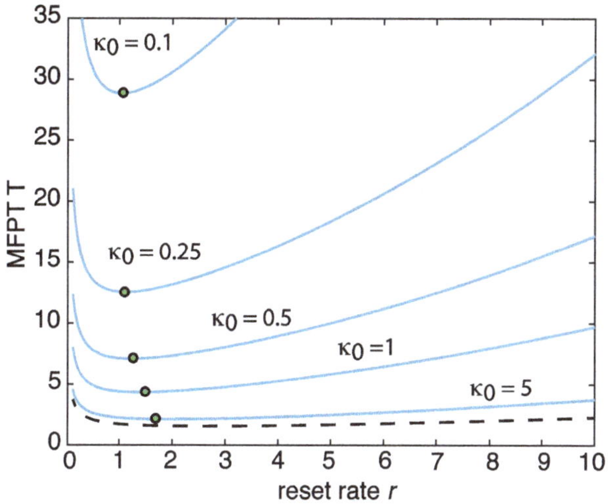

Fig. 4.5 Particle diffusing on the half-line $x \in [0, \infty)$ with a Robin boundary condition at $x = 0$. The particle resets to its initial position x_0 at a constant rate $r > 0$. Plot of the MFPT $T_r(x_0)$ given by Eqs. (4.63) and (4.66) as a function of the resetting rate r for various reactivities κ_0 with $D = 1$ and $x_0 = 1$. The dashed line indicates the case of a totally absorbing boundary ($\kappa_0 \to \infty$). Green dots indicate the optimal resetting rates

an optimal resetting rate $r_{\rm opt}$ that is a weakly increasing function of κ_0. Finally, note that the analysis can be extended to the case of a non-exponential local time threshold distribution $\Psi(l)$ as follows: First, substitute Eq. (4.64) into Eq. (4.60) with $\widetilde{Q}_0$ given by Eq. (4.63). Second, set $\widetilde{P}_r(x,z,s) = \widetilde{p}_r(x,s)$, $z = \kappa_0/D$, with $\widetilde{P}_r(x,z,s)$ the double Laplace transform of the local time propagator with respect to t and l. Third, calculate the inverse Laplace transform of $\widetilde{P}_r(x,z,s)$ with respect to z using a Bromwich integral. This then yields $\widetilde{p}^{\Psi}(x,s) = \int_0^\infty \Psi(l)\widetilde{P}_r(x,l,s)dl$, from which the MFPT can be extracted.

4.4 Non-instantaneous Resetting and Renewal Theory

In the case of search processes with non-instantaneous resetting (finite return times and refractory periods), it is no longer possible to write a compact SDE for the stochastic dynamics. However, probabilistic methods can be used to calculate the effects of delays on FPT problems [9, 44]. For concreteness, we follow the particular formulation of Ref. [44]. Let us return to the search process shown in Fig. 4.1. Suppose that following each reset event, the searcher switches to a ballistic state in which it returns to its home base $\mathbf{x}_0$ at a constant speed V. In addition, whenever the particle returns to $\mathbf{x}_0$, it is subject to a refractory period before the search begins again. The refractory period σ is a random variable with a corresponding waiting time density ϕ, which is taken to have a finite mean $\tau_{\rm ref}$. These features are illustrated in Fig. 4.6 for a random search process on the half-line with an absorbing target at the origin. In order to keep the analysis general, we will take resetting to occur at a random sequence of times generated by a probability density $\psi(\tau)$. It

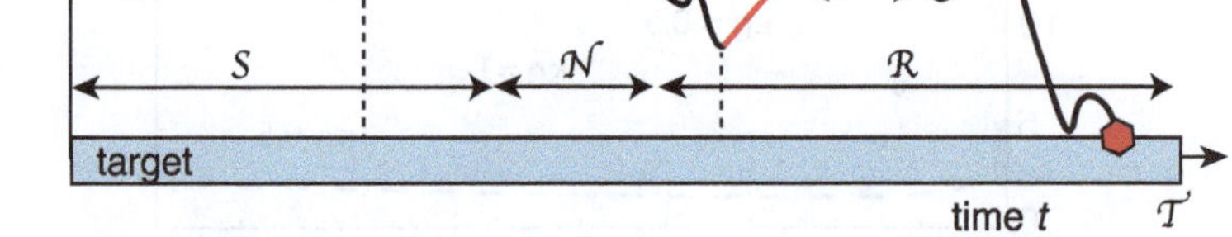

Fig. 4.6 Particle searching for a target at $x = 0$ on the half-line with an absorbing target at the origin, which is found after two resettings. After each resetting event, the particle returns to the point x_0 at a constant speed V, after which it remains at x_0 for a refractory period before re-entering the search phase. Also shown is the decomposition of the FPT, $\mathcal{T} = \mathcal{S} + \mathcal{N} + \mathcal{R}$

follows that $\Psi(\tau) = 1 - \int_0^\tau \psi(s)ds$ is the probability that no resetting has occurred up to time τ.

Let T denote the FPT to find the target $\mathcal{U}$, that is,

$$T = \inf\{t > 0,\ \mathbf{X}(t) \in \partial\mathcal{U}\}, \tag{4.67}$$

where $\partial\mathcal{U}$ denotes the target boundary. Finite return times and refractory periods modify the FPT as follows: First, since the particle returns to $\mathbf{x}_0$ at a constant speed V, there is an additional contribution to the FPT given by $|\mathbf{X}(\tau) - \mathbf{x}_0|/V$ where τ is either the first resetting time or the time since the last reset and $\mathbf{X}(\tau)$ is the position just prior to switching to the return phase. In general, $\mathbf{X}(\tau)$ will be a doubly stochastic variable, since both τ and $\mathbf{X}(t)$ are stochastic processes. Each resetting event also adds an additional contribution to the FPT due to refractoriness. Clearly both features increase the MFPT. We proceed by decomposing the various contributions to the FPT and conditioning on whether or not the particle resets at least once. This exploits the fact that once the particle has returned to $\mathbf{x}_0$ it has lost all memory of previous search phases. In other words, the stochastic process satisfies the strong Markov property.

Let $\mathcal{I}(t)$ denote the number of resetting events up to time t. Consider the following set of FPTs, analogous to the decompositions shown in Fig. 4.5:

$$\begin{aligned}
\mathcal{T} &= \inf\{t > 0; \mathbf{X}(t) \in \partial\mathcal{U},\ \mathcal{I}(t) \geq 0\},\\
\mathcal{S} &= \inf\{t > 0; \mathbf{X}(t) = \mathbf{x}_0,\ \mathcal{I}(t) = 1\},\\
\mathcal{R} &= \inf\{t > 0; \mathbf{X}(t + \mathcal{S} + \mathcal{N}) \in \partial\mathcal{U}_0,\ \mathcal{I}(t + \mathcal{S} + \mathcal{N}) \geq 1\}.
\end{aligned} \tag{4.68}$$

Here $\mathcal{T}$ is the FPT for finding the target $\mathcal{U}$ (equivalent to hitting a point on the boundary $\partial\mathcal{U}$ irrespective of the number of resettings), $\mathcal{S}$ is the FPT for the first resetting and return to the reset point $\mathbf{x}_0$ given that the particle is still free, $\mathcal{N}$ is the first refractory time, and $\mathcal{R}$ is the FPT for finding the target given that at least one resetting has occurred. Next, we introduce the sets $\Omega = \{\mathcal{T} < \infty\}$ and $\Gamma = \{\mathcal{S} < \mathcal{T} < \infty\} \subset \Omega$. Here, Ω is the set of all events for which the particle is eventually absorbed by the target (which has measure one), and Γ is the subset of events in Ω for which the particle resets at least once. It immediately follows that $\Omega\backslash\Gamma = \{\mathcal{T} < \mathcal{S} = \infty\}$, that is, $\Omega\backslash\Gamma$ is the set of all events for which the particle is captured by the target without any resetting. The Laplace transform of the FPT density with resetting is

$$\widetilde{f}_r(\mathbf{x}_0, s) = \mathbb{E}[\mathrm{e}^{-s\mathcal{T}} 1_\Omega]. \tag{4.69}$$

Following along the lines of Refs. [9, 10], consider the decomposition

$$\mathbb{E}[\mathrm{e}^{-s\mathcal{T}} 1_\Omega] = \mathbb{E}[\mathrm{e}^{-s\mathcal{T}} 1_{\Omega\backslash\Gamma}] + \mathbb{E}[\mathrm{e}^{-s\mathcal{T}} 1_\Gamma]. \tag{4.70}$$

The first expectation can be evaluated by noting that it is the FPT density for capture by the target without any resetting, and the probability density for such an event is $-\Psi(t)\partial_t Q_0(\mathbf{x}_0, t)d\tau$, where Q_0 is the survival probability. Hence,

$$\mathbb{E}[\mathrm{e}^{-s\mathcal{T}}1_{\Omega\backslash\Gamma}] = -\int_0^\infty \mathrm{e}^{-s\tau}\Psi(\tau)\frac{\partial Q_0(\mathbf{x}_0,\tau)}{\partial\tau}d\tau = 1 - \int_0^\infty \mathrm{e}^{-s\tau}\psi(\tau)Q_0(\mathbf{x}_0,\tau)d\tau$$
$$-s\int_0^\infty \mathrm{e}^{-s\tau}\Psi(\tau)Q_0(\mathbf{x}_0,\tau)d\tau. \tag{4.71}$$

The second expectation can be written as

$$\begin{aligned}\mathbb{E}[\mathrm{e}^{-s\mathcal{T}}1_\Gamma] &= \mathbb{E}[\mathrm{e}^{-s[\mathcal{S}+\mathcal{N}+\mathcal{R}]}1_\Gamma] \\ &= \int_0^\infty \psi(\tau_1)\left[\int_{\mathcal{U}\backslash\mathcal{U}_0} \mathrm{e}^{-s(\tau_1+|\mathbf{x}-\mathbf{x}_0|/V)}p(\mathbf{x},\tau_1|\mathbf{x}_0)d\mathbf{x}\right]d\tau_1 \\ &\quad\times\left[\int_0^\infty \mathrm{e}^{-s\tau_2}\phi(\tau_2)d\tau_2\right]\widetilde{f}_r(\mathbf{x}_0,s).\end{aligned} \tag{4.72}$$

We have used the fact that the probability that the first return is initiated in the interval $[\tau_1, \tau_1 + d\tau_1]$, given that $\mathbf{X}(\tau_1) = \mathbf{x}$ and the particle has not been captured by the target, is $\psi(\tau_1)p(\mathbf{x}, \tau_1|\mathbf{x}_0)d\tau_1$. The particle then takes an additional time $|\mathbf{x}-\mathbf{x}_0|/V$ to return to $\mathbf{x}_0$, after which it spends a time τ_2 in the refractory state with waiting time density $\phi(\tau_2)$. The remaining time to find the target has the same FPT density as $\mathcal{T}$.

Combining and rearranging Eqs. (4.70)–(4.72) yield the general result [9, 10]

$$\widetilde{f}_r(\mathbf{x}_0,s) = \frac{1-\int_0^\infty \mathrm{e}^{-s\tau}\psi(\tau)Q_0(\mathbf{x}_0,\tau)d\tau - s\int_0^\infty \mathrm{e}^{-s\tau}\Psi(\tau)Q_0(\mathbf{x}_0,\tau)d\tau}{1-\widetilde{\phi}(s)\int_0^\infty \psi(\tau_1)\left[\int_{\mathcal{U}\backslash\mathcal{U}_0}\mathrm{e}^{-s(\tau_1+|\mathbf{x}-\mathbf{x}_0|/V)}p(\mathbf{x},\tau_1|\mathbf{x}_0)d\mathbf{x}\right]d\tau_1}.$$

The Laplace transform of the FPT density is the moment generator of the FPT $\mathcal{T}$:

$$T_r^{(n)} = \mathbb{E}[\mathcal{T}^n 1_\Omega] = \left(-\frac{d}{ds}\right)^n \mathbb{E}[\mathrm{e}^{-s\mathcal{T}}1_\Omega]\bigg|_{s=0}. \tag{4.73}$$

For example, the MFPT $T_r = T_r^{(1)}$ is

$$T_r(\mathbf{x}_0) = \frac{\langle Q_0(\mathbf{x}_0,\tau)\rangle_\Psi - \langle\tau Q_0(\mathbf{x}_0,\tau)\rangle_\psi}{1-\langle Q_0(\mathbf{x}_0,\tau)\rangle_\psi}$$
$$+\frac{-\widetilde{\phi}'(0)\langle Q_0(\mathbf{x}_0,\tau)\rangle_\psi + \langle\tau Q_0(\mathbf{x}_0,\tau)\rangle_\psi + V^{-1}\langle F(\mathbf{x}_0,\tau)\rangle_\psi}{1-\langle Q_0(\mathbf{x}_0,\tau)\rangle_\psi}$$

$$= \frac{\langle Q_0(\mathbf{x}_0,\tau)\rangle_\Psi + \tau_{\text{ref}}\langle Q_0(\mathbf{x}_0,\tau)\rangle_\psi + V^{-1}\langle F(\mathbf{x}_0,\tau)\rangle_\psi}{1-\langle Q_0(\mathbf{x}_0,\tau)\rangle_\psi}, \tag{4.74}$$

where we have set

$$\langle Q_0(\mathbf{x}_0,\tau)\rangle_\psi = \int_0^\infty \psi(\tau)Q_0(\mathbf{x}_0,\tau)d\tau, \tag{4.75}$$

and similarly for $\langle Q_0(\mathbf{x}_0,\tau)\rangle_\Psi$ and $\langle F(\mathbf{x}_0,\tau)\rangle_\psi$

$$F(\mathbf{x}_0,t) = \int_{\mathcal{U}\backslash\mathcal{U}_0} |\mathbf{x}-\mathbf{x}_0| p(\mathbf{x},t|\mathbf{x}_0)d\mathbf{x}. \tag{4.76}$$

We have also used the result

$$-\widetilde{\phi}'(0) = \tau_{\text{ref}} = \int_0^\infty \tau\phi(\tau)d\tau.$$

In the case of exponential resetting,

$$\psi(t) = re^{-r\tau}, \quad \Psi(\tau) = \mathrm{e}^{-rt},$$

everything can be expressed in terms of Laplace transforms. For example, the MFPT reduces to

$$T_r(\mathbf{x}_0) = \frac{\widetilde{Q}_0(\mathbf{x}_0,r) + r\tau_{\text{ref}}\widetilde{Q}_0(\mathbf{x}_0,r) + r\widetilde{F}(\mathbf{x}_0,r)/V}{1-r\widetilde{Q}_0(\mathbf{x}_0,r)}, \tag{4.77}$$

where $\widetilde{F}(\mathbf{x}_0,r)$ is the Laplace transform of $F(\mathbf{x}_0,\tau)$. Similarly, the Laplace transform of the FPT density becomes

$$\widetilde{f}_r(\mathbf{x}_0,s) = \frac{1-(r+s)\widetilde{Q}_0(\mathbf{x}_0,r+s)}{1-r\widetilde{\phi}(s)\int_{\mathcal{U}\backslash\mathcal{U}_0}\mathrm{e}^{-s|\mathbf{x}-\mathbf{x}_0|/V}\widetilde{p}(\mathbf{x},r+s|\mathbf{x}_0)d\mathbf{x}}. \tag{4.78}$$

As an illustration of the above result, consider again a diffusing particle on the semi-infinite interval with a target at the origin; see Sect. 3.3. We now assume that the target is totally absorbing, which corresponds to the limit $\kappa_0 \to \infty$, and that resetting is non-instantaneous with the particle returning to x_0 at a constant speed V. It follows from Laplace-transforming equation (4.76) that $\widetilde{F}(x_0,r) = \int_0^\infty |x - x_0|\widetilde{G}(x,r|x_0)dx$, where

$$\widetilde{G}(x,r|x_0) = \frac{1}{2\sqrt{rD}}\left(\mathrm{e}^{-\sqrt{r/D}|x-x_0|} - \mathrm{e}^{-\sqrt{r/D}(x+x_0)}\right)$$

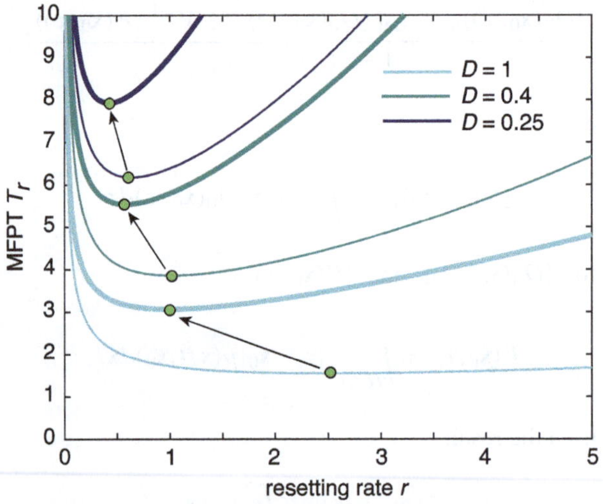

Fig. 4.7 Effect of finite return time on MFPT T_r for diffusion with resetting. Plot of T_r as a function of resetting rate r for various diffusivities D. Thick curves are for $V = 1$ and thin curves are for $V = \infty$. The reset point is $x_0 = 1$ and the target is at $x = 0$. The green dots indicate the optimal resetting rate for each curve, and the arrows indicate the shift due to a finite return time

is the Laplace transform of the Green's function for diffusion on the half-line with an absorbing boundary at $x = 0$. Evaluating the integral yields the result

$$\widetilde{F}(x_0, r) = \frac{1}{r}\sqrt{\frac{D}{r}}\left[1 - \sqrt{\frac{r}{D}}x_0 \mathrm{e}^{-\sqrt{r/D}x_0} - \mathrm{e}^{-2\sqrt{r/D}x_0}\right]. \tag{4.79}$$

Moreover, taking the limit $\kappa_0 \to \infty$ in Eq. (4.63) shows that $\widetilde{Q}_0(x_0, r) = 1 - \mathrm{e}^{-\sqrt{r/D}x_0}/r$. Substituting these expressions into (4.77) with $\tau_{\mathrm{ref}} = 0$ then determines the MFPT T_r. In Fig. 4.7 we plot T_r as a function of r for various diffusivities D and compare the two cases $V = 1$ and $V = \infty$ (instantaneous resetting). We see that there still exists a unique minimum of the MFPT curve when $V = 1$ (indicated by green dots in Fig. 4.7), but the minimum is shifted upwards and to the left. In particular, the value of the optimal resetting rate $r_{\min}$ is reduced in the case of a finite return time, with the degree of shift, a monotonically decreasing function of V. (A reduction in $r_{\min}$ also occurs for refractory periods.)

4.5 Discussion

In this chapter we considered various probabilistic formulations of diffusive search processes with stochastic resetting. In particular, we formulated diffusion with instantaneous resetting as a jump-diffusion process, analysed the FPT problem for a single particle on the half-line with a partially absorbing target at the origin, and used conditional expectations and renewal theory to analyse the effects of finite return times and refractory periods on a diffusive search process with non-instantaneous resetting. In order to develop the theory, we considered 1D models. However, all of the analysis can be extended to higher spatial dimensions.

There are a number of issues that warrant further consideration: first, extending the analysis of FPTs to multiple search-and-capture events. Again, there is a major difference between cellular transport processes and animal foraging. The former usually involves the delivery of cargo to a subcellular target such as a synapse, whereas the latter typically involves the extraction of some resource such as food or shelter. In the cases where cargo delivery triggers a downstream signalling pathway in a cell, the arrival time of the first searcher may be crucial, which can be analysed using, for example, extreme statistics. On the other hand, the first arrival time is not so relevant in the case of animal foraging, provided that there is a sufficient supply of resources. A complementary perspective on multiple rounds of search and capture is the accumulation of resources within a cellular target. As we have shown elsewhere [45], the latter can be mapped onto a queuing process as follows (see Fig. 4.8): individual searchers are analogous to customers, the delivery of a resource packet by a searcher corresponds to a customer arriving at a service station, and the degradation or utilisation of a resource packet is the analogue of a customer exiting the system after being serviced. If the packets are utilised independently

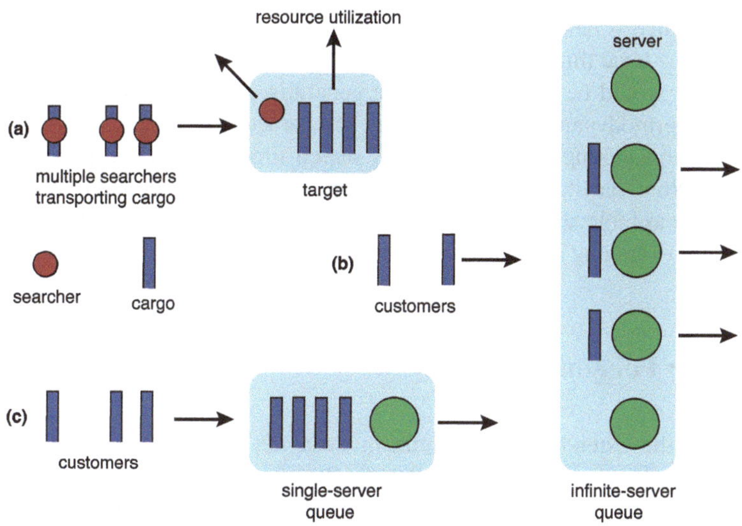

Fig. 4.8 Accumulation of resources in a target mapped onto a queuing process. (**a**) Multiple particles searching for a target according to a given FPT process. Each time a particle finds the target, it delivers a discrete packet of resources. Delivered resources are either utilised independently or sequentially (**b**) GM/∞ queue: customers arrive at a service station containing an infinite number of servers. (**c**) $G/M/1$ queue: customers arrive at a service station containing a single server

of each other, then the effective number of servers in the corresponding queuing model is infinite, that is, the presence of other customers does not affect the service time of an individual customer. It follows that the relevant queuing model is the $G/M/\infty$ system [46, 47]. The symbol G denotes a general customer inter-arrival time distribution (which is determined by the underlying FPT density of a single search process), the symbol M stands for a Markovian or exponential service-time distribution, and '∞' denotes infinite servers. On the other hand, if packets are utilised in the order that they are delivered, then we have a $G/M/1$ queue [48, 49]. Mapping onto a queuing problem means that the steady-state mean and variance of the number of resources in the target (or a set of targets) can be calculated using well-known probabilistic methods [45, 50].

The second issue is developing more detailed models of the target detection process itself in order to derive the appropriate target boundary condition. At the molecular level, one finds that various surface-based reactions are better modelled in terms of a reactivity that is a function of the local time [51, 52]. That is, the surface may need to be progressively activated by repeated encounters with a diffusing particle, or an initially highly reactive surface may become less active due to multiple interactions with the particle (passivation). In the case of a foraging animal, it is possible that reaching the target boundary is insufficient to extract the resource, which may require an exploration phase in the vicinity of the boundary. This leads to another possible modification of the search process, in which the particle has to enter the target domain $\mathcal{U}$ and search within the interior. In this case, the probability of finding a resource or delivering a cargo depends on the amount of time spent within $\mathcal{U}$, which is determined by the occupation time $A(t) = \int_{\mathcal{U}} \delta(x - X(t))dx$ rather than the local time on the boundary $\partial\mathcal{U}$ [26, 27].

Finally, it would be interesting to use generalised Itô calculus to explore how stochastic thermodynamics could be used to characterise the efficacy of a search process by considering the dual costs of search time and work expended during the search. Note that stochastic thermodynamics is typically applied to mesoscopic systems such as molecular motors or active colloids [30]. The analogous description at the level of foraging animals is less clear.

Appendix: Poisson Processes

In deriving the extended Itô's lemma in Sect. 4.2, we combined stochastic Itô calculus with a stochastic representation of a Poisson processes. In this appendix we provide some mathematical background regarding the latter. Consider a sequence $\tau_1, \tau_2, \ldots$ of independent exponential random variables with the same mean $1/\lambda$:

$$\mathbb{P}[\tau_n \in [\tau, \tau + d\tau]] = \lambda e^{-\lambda\tau} d\tau, \quad \ell \geq 1. \tag{A.1}$$

Introduce the jump times $\widehat{T}_n = \sum_{k=1}^{n} \tau_n$. The Poisson process $N(t)$ is a Markov process that counts the number of jumps occurring in the interval $[0, t]$. In other

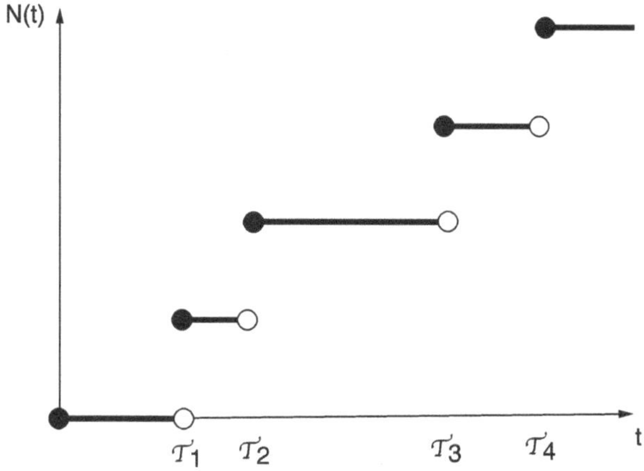

Fig. 4.9 One path of a Poisson process $N(t)$, illustrating that it is right-continuous. Jumps occur at the times $\widehat{\mathcal{T}}_\ell$, $\ell \geq 1$

words,

$$N(t) = n, \quad \widehat{\mathcal{T}}_n \leq t < \widehat{\mathcal{T}}_{n+1}. \tag{A.2}$$

Note that $N(t)$ is defined to be right-continuous, $\lim_{\epsilon \to 0^+} N(t+\epsilon) = N(t)$. In particular, $N(\widehat{\mathcal{T}}_n^-) = n-1$, whereas $N(\widehat{\mathcal{T}}_n) = n$. This is illustrated in Fig. 4.9. The probability distribution of the Poisson process is

$$\mathbb{P}[N(t) = n] = P_n(t) \equiv \frac{(\lambda t)^n \mathrm{e}^{-\lambda t}}{n!}. \tag{A.3}$$

Moreover, $N(t+s) - N(s)$, which is the number of jumps in the time interval $(s, t+s]$, has the same distribution as $N(t)$ so the Poisson process is time homogeneous. Introducing the moment-generating function

$$G(z,t) = \sum_{n \geq 0} z^n P_n(t) = \mathrm{e}^{(z-1)\lambda t}, \tag{A.4}$$

we deduce that

$$\mathbb{E}[N(t)] = \left. \frac{\partial G(z,t)}{\partial z} \right|_{z=1} = \lambda t, \tag{A.5}$$

and

$$\mathbb{E}[N(t)(N(t)-1)] = \left. \frac{\partial^2 G(z,t)}{\partial z^2} \right|_{z=1} = (\lambda t)^2. \tag{A.6}$$

In particular, $\mathrm{Var}[N(t)] = \lambda t$. Moreover, defining the infinitesimal $dN(t) = N(t + dt) - N(t)$ and using the fact that the Poisson process is time homogeneous, we have

$$\mathbb{E}[dN(t)] = \lambda dt. \tag{A.7}$$

Let f be an arbitrary bounded function on $\mathbb{Z}$. Suppose that $N(t) = n$. Using telescoping we have

$$\begin{aligned} f(N(t)) - f(0) &= \sum_{k=1}^{n} [f(k) - f(k-1)] = \sum_{k=1}^{n} [f(N(\widehat{T}_k)) - f(N(\widehat{T}_{k-1}))] \\ &= \sum_{k=1}^{n} [f(N(\widehat{T}_{k-1}) + 1) - f(N(\widehat{T}_{k-1}))] \\ &= \sum_{k=1}^{\infty} \int_0^t [f(N(s-) + 1) - f(N(s-))]\delta(s - \widehat{T}_k) ds, \end{aligned} \tag{A.8}$$

which is independent of the particular value n. Using the fact that

$$\int_s^t \left(\sum_{k=1}^{\infty} \delta(\tau - \widehat{T}_k) \right) d\tau = N(t) - N(s), \tag{A.9}$$

we have

$$f(N(t)) = f(0) + \int_0^t [f(N(s-) + 1) - f(N(s-))] dN(s), \tag{A.10}$$

with

$$dN(\tau) = \sum_{k=1}^{\infty} \delta(\tau - \widehat{T}_k) d\tau. \tag{A.11}$$

References

1. Evans, M.R., Majumdar, S.N.: Diffusion with stochastic resetting. Phys. Rev. Lett. **106**,160601 (2011)
2. Evans, M.R., Majumdar, S.N.: Diffusion with optimal resetting. J. Phys. A Math. Theor. **44**, 435001 (2011)
3. Evans, M.R., Majumdar, S.N.: Diffusion with resetting in arbitrary spatial dimension. J. Phys. A **47**, 285001 (2014)
4. Evans, M.R., Majumdar, S.N., Schehr, G.: Stochastic resetting and applications. J. Phys. A **53**, 193001 (2020)

5. Pal, A., Kusmierz, L., Reuveni, S.: Diffusion with stochastic resetting is invariant to return speed. Phys. Rev. E **100**, 040101 (2019)
6. Pal, A., Kusmierz, L., Reuveni, S.: Invariants of motion with stochastic resetting and spacetime coupled returns New J. Phys. **21**, 113024 (2019)
7. Maso-Puigdellosas, A., Campos, D., Mendez, V.: Transport properties of random walks under stochastic noninstantaneous resetting. Phys. Rev. E **100**, 042104 (2019)
8. Bodrova, A.S., Sokolov, I.M.: Resetting processes with non-instantaneous return. Phys. Rev. E **101**, 052130 (2020)
9. Pal, A., Kusmierz, L., Reuveni, S.: Home-range search provides advantage under high uncertainty. Phys. Rev. Res. **2**, 043174 (2020)
10. Bressloff, P.C.: Modeling active cellular transport as a directed search process with stochastic resetting and delays. J. Phys. A **53**, 355001 (2020)
11. Evans, M.R., Majumdar, S.N.: Effects of refractory period on stochastic resetting. J. Phys. A: Math. Theor. **52**, 01LT01 (2019)
12. Maso-Puigdellosas, A., Campos, D., Mendez, V.: Stochastic movement subject to a reset-and-residence mechanism: transport properties and first arrival statistics. J. Stat. Mech. **2019**, 033201 (2019)
13. Gonzalez, F.H., Riascos, A.P., Boyer, D.: Diffusive transport on networks with stochastic resetting to multiple nodes. Phys. Rev. E **103**, 062126 (2021)
14. Toledo-Marin, J.Q., Boyer, D.: First passage time and information of a one-dimensional Brownian particle with stochastic resetting to random positions. Physica A **625**, 129027 (2022)
15. Pal, A., Kundu, A., Evans, M.R.: Diffusion under time-dependent resetting. J. Phys. A: Math. Theor. **49** 225001 (2016)
16. Nagar, A., Gupta, S.: Stochastic resetting in interacting particle systems: a review. J. Phys. A: Math. Theor. **56**, 283001 (2023)
17. Magdziarz, M., Tázbierski, K.: Stochastic representation of processes with resetting. Phys. Rev. E **106**, 014147 (2022)
18. Bressloff, P.C.: Generalized Itô's lemma and the stochastic thermodynamics of diffusion with resetting. J. Phys. A. **57**, 445003 (2024)
19. Bressloff, P.C.: Global density equations for interacting particle systems with stochastic resetting: from overdamped Brownian motion to phase synchronization. Chaos **34**, 043101 (2024)
20. Lévy, P.: Sur certains processus stochastiques homogènes. Compos. Math. **7**, 283 (1939)
21. Itô, K., McKean, H.P.: Brownian motions on a half line Illinois J. Math. **7** 181–231 (1963)
22. McKean, H.P.: Brownian local time. Adv. Math. **15**, 91–111 (1975)
23. Grebenkov, D.S.: Partially reflected Brownian motion: a stochastic approach to transport phenomena. In: Velle, L.R. (ed.) Focus on Probability Theory, pp. 135–169. Nova Science Publishers, Hauppauge (2006)
24. Grebenkov, D.S.: Paradigm shift in diffusion-mediated surface phenomena. Phys. Rev. Lett. **125**, 078102 (2020)
25. Grebenkov, D.S.: An encounter-based approach for restricted diffusion with a gradient drift. J. Phys. A. **55**, 045203 (2022)
26. Bressloff, P.C.: Diffusion-mediated absorption by partially reactive targets: Brownian functionals and generalized propagators. J. Phys. A. **55**, 205001 (2022).
27. Bressloff, P.C.: Spectral theory of diffusion in partially absorbing media. Proc. R. Soc. A **478**, 20220319 (2022)
28. Sekimoto, K.: Stochastic Energetics. Lecture Notes in Physics. Springer, Berlin (2010)
29. Seifert, U.: Stochastic thermodynamics, fluctuation theorems and molecular machines. Rep. Prog. Phys. **75**, 126001 (2012)
30. Peliti, L., Pigolotti, S.: Stochastic Thermodynamics. Princeton University Press, Princeton (2021)
31. Róldan, E., Neri, I., Chetrite, R., Pigolotti, S., Jülicher, F., Sekimoto, K.: Martingales for physicists: a treatise on stochastic thermodynamics and beyond. Adv. Phys. **72**, 2317494 (2023)

32. Fuchs, J., Goldt, S., Seifert, U.: Stochastic thermodynamics of resetting. Europhys. Lett. **113**, 60009 (2016)
33. Pal, A., Rahav, S.: Integral fluctuation theorems for stochastic resetting systems. Phys. Rev. E **96**, 062135 2017
34. Busiello, D.M., Gupta, D., Maritan, A.: Entropy production in systems with unidirectional transitions. Phys. Rev. Res. **2**, 023011 (2020)
35. Pal, A., Reuven, S., Rahav, S.: Thermodynamic uncertainty relation for systems with unidirectional transitions. Phys. Rev. Res. **3**, 013273 (2021)
36. Mori, F., Olsen, K.S., Krishnamurthy, S.: Entropy production of resetting processes. Phys. Rev. Res. **5**, 123103 (2023)
37. Olsen, K.S., Gupta, D., Mori, F., Krishnamurthy, S.: Thermodynamic cost of finite-time stochastic resetting (2023). arXiv:2310.11267v2
38. Das, D., Giuggioli, L.: Discrete space-time resetting model: application to first-passage and transmission statistics. J. Phys. A **55**, 424004 (2022)
39. Bjork, T.: Point Processes and Jump Diffusions. Cambridge University Press, Cambridge (2021)
40. Dean, D.S.: Langevin equation for the density of a system of interacting Langevin processes. J. Phys. A **29**, L613–L617 (1996)
41. Kawasaki, K.: Microscopic analyses of the dynamical density functional equation of dense fluids. J. Stat. Phys. **93**, 527–546 (1998)
42. Bressloff, P.C.: Diffusion-mediated surface reactions and stochastic resetting. J. Phys. A **55**, 275002 (2022)
43. Benkhadaj, Z., Grebenkov, D.S.: Encounter-based approach to diffusion with resetting. Phys. Rev. E **106**, 044121 (2022)
44. Bressloff, P.C.: Search processes with stochastic resetting and multiple targets. Phys. Rev. E **102**, 022115 (2020)
45. Bressloff, P.C.: Queuing model of axonal transport. Brain Multiphysics **2**, 100042 (2021)
46. Takacs, L.: Introduction to the Theory of Queues. Oxford University Press, Oxford (1962)
47. Liu, L., Kashyap, B.R.K., Templeton, J.G.C.: On the GIX/G/infinity system. J. Appl Prob. **27**, 671–683 (1990)
48. Cooper, R.B.: Introduction to Queueing Theory, 2nd edn. North Holland, New York (1981)
49. Adan, I., Resing, J.: Queueing Systems. Lecture Notes (2015)
50. Giral Barajas, J., Bressloff, P.C.: Stochastic 1D search-and-capture as a G/M/c queueing model. J. Phys. A **58**, 355001 (2025)
51. Bartholomew, C.H.: Mechanisms of catalyst deactivation. Appl. Catal. A: Gen. **212**, 17–60 (2001)
52. Filoche, M., Grebenkov, D.S., Andrade Jr., J.S., Sapoval, B.: Sapoval, Passivation of irregular surfaces accessed by diffusion. Proc. Natl. Acad. Sci. **105**, 7636–7640 (2008)

Chapter 5
Scalable Non-Markovian State-Switching Models for Animal Movement

Nazareno Campioni, Dirk Husmeier, Jennifer Gaskell, Juan M. Morales, and Colin J. Torney

5.1 Introduction

Recent technological advances in the field of movement ecology have led to an increasing amount of high-resolution animal tracking data being collected [5, 20]. As a result, there has been an array of new statistical methodologies developed to tackle the challenges associated with the analysis of such large datasets. A common task in the study of animal movement is identifying changes in movement patterns that can be associated with different behavioural states. Here the canonical challenge is to segment the data into "states" based on quantitatively different statistical properties of segments of the full movement trajectory and to reconstruct the sequence of state switches. The term state is used as a proxy for a distinct animal behaviour; while movement data provides high-frequency locations, different behaviours are expected to show as a consequence of changes in the animals' activities over longer timescales, for example, a change from an exploratory behaviour to a resting behaviour.

In recent years, the task of identifying underlying behavioural states has been tackled by employing hidden Markov models (HMMs) [32], whereby the data are assumed to be generated by a movement process that is dependent on a latent (therefore hidden) behavioural process represented by a Markov chain. HMMs are formulated either in discrete time or in continuous time, and both formulations have their advantages and disadvantages. Discrete-time HMMs are currently the preferred model due to their mathematical simplicity and easily interpreted parameters. The

N. Campioni · D. Husmeier · J. Gaskell · C. J. Torney (✉)
School of Mathematics and Statistics, University of Glasgow, Glasgow, UK
e-mail: colin.torney@glasgow.ac.uk

J. M. Morales
School of Biodiversity, One Health & Veterinary Medicine, University of Glasgow, Glasgow, UK

L. Giuggioli, P. K. Maini (eds.), *The Mathematics of Movement*,
https://doi.org/10.1007/978-3-032-00873-2_5

standard discrete-time formulation requires the transformation of trajectories into a sequence of steps and turns in polar coordinates [17] and models the dynamics as an ordered series of step lengths and turn angles drawn from defined probability distributions. Extensive effort has been put in the development of user-friendly packages for the researcher designed to tackle switching problems for telemetry data analysis [13, 16]. However, discrete-time models are not ideal tools for the analysis of irregularly spaced data or in the presence of non-negligible measurement error [15, 21]. In particular, they require partitioning of the time domain into temporally regular time steps that must be specified a priori and that are assumed to match the scale at which behavioural decisions are made [12, 27].

At the expense of some mathematical simplicity, a natural extension of discrete-time HMMs are models where movement is defined using a continuous-time location process and the observation process is represented by adding measurement error to the true, noise-free locations. In [15], this approach was extended to a multi-state framework to address switching problems by introducing a hidden process representing the behavioural process, on which the location dynamics are dependent. The location process is described by a continuous-time correlated random walk (CTCRW) to account for autocorrelation in movement; specifically, the location is modelled through an integrated Ornstein-Uhlenbeck (OU) process [10, 15], whereas the behavioural process is described by a continuous-time Markov chain. Continuous-time models don't require specification of a timescale a priori and can therefore handle irregularly spaced data as well as different temporal scales [21]. This can overcome issues arising from missing data or limitations within the telemetry devices but also oversampling of movements.

While HMMs have seen an increase in popularity in the animal movement community and have been widely used in switching problems, the Markov property implies underlying assumptions on the behavioural process that are not realistic. In the absence of external variables that influence the switching behaviour, the residence (or sojourn, dwell) time in each behavioural state follows an exponential distribution (or geometric, in the discrete-time counterpart), for which short and frequent state changes are favoured. This implies that the behavioural process inherits the "memoryless" property of the exponential distribution, meaning the sojourn times are independent of the amount of time spent in a state. Thus, when using HMMs it is assumed that the amount of time an animal will remain in a behaviour of, say, resting does not depend on how long the animal has already rested for. Another drawback of using the exponential distribution lies in the application of HMMs in an unsupervised context, whereby the number of states is not defined a priori and is learnt during model fitting. If the distribution of the residence times is not exponential (or geometric), in other words if the underlying structure of the model is non-Markovian, as we would typically expect in most animal movement studies and has been shown empirically in the case of the distribution of foraging times of beaked whales [11], using model selection to define the number of states in HMMs is unreliable and expected to favour models having more behavioural states than there are in reality [22]. This can be avoided by including a non-Markovian structure into the model by using hidden semi-Markov models which allow for the informed specification of a dwell-time distribution, thus relaxing the assumption of exponentially distributed waiting times [24]. For a comprehensive examination of the pitfalls associated with HMMs, see [8].

It is also worth adding that current approaches are not scalable to large datasets due to the computational burden of standard schemes such as Markov chain Monte Carlo (MCMC) sampling for which evaluation of the full dataset is required at every step of the sampler. These methods therefore become prohibitively expensive to apply to long-term, high-frequency telemetry datasets.

Here we propose a novel method scalable to large datasets that can be employed in the context of switching problems and addresses the intrinsic assumptions of standard HMMs by introducing a memory in the process. In particular, similarly to [15] we employ a multi-state, CTCRW model where the location process is described by an integrated OU process; however, a continuous-time semi-Markov chain is employed to describe the behavioural process, thus allowing us to have control over the residence times. We employ a gamma distribution to model the residence times, nonetheless arbitrary distributions, or mixtures of distributions may be used within our framework. By employing the gamma distribution, we favour longer sojourn times and effectively introduce a memory in the switching process by ensuring that the more time spent in a state, the more likely to switch behaviour.

Our framework is intended to solve two inferential tasks: the first task aims at reconstructing the most probable hidden state sequence given observed trajectories, whereas the second task addresses the recovering of both the behavioural and the location process parameters. This is done by employing a Monte Carlo expectation-maximisation (MCEM) algorithm that performs the two tasks sequentially: the expectation step and the maximisation step. In the E-step, we employ a dynamic programming approach to generate samples from the posterior distribution of the latent states given the observations and the model parameters. In the M-step, the expectation is optimised with respect to the model parameters which are then updated and used in the subsequent iteration of the algorithm.

The remainder of the paper is structured as follows: in Sect. 5.2 we present the model formulation for both the movement and the switching processes, explain the inferential algorithm, and validate the method using synthetic data. In Sect. 5.3 we show the results of the method applied to data from a study of grey seal movement and from Merino sheep free roaming in Patagonia. We conclude with a discussion of our method.

5.2 Methods

5.2.1 State-Switching Movement Model

In this paper, we employ a three-layer state-space model where the first layer is the underlying, hidden switching state/behaviour process, the second layer is the location process which depends on the first layer, and the third and final layer is the observation process which accounts for measurement error. Here we focus on a continuous-time formulation of the switching process with a discrete state space.

In a standard implementation of an HMM, the switching process is assumed to be described by a Markov chain, for which state changes are dependent only on the state at the previous time point. As a consequence of the Markov property, the sojourn time of a state is a priori exponentially distributed. Our objective is to relax this assumption and present a more flexible behavioural process that can take into account biological factors that reflect on the duration of specific animals' activities.

Therefore, we make use of a continuous-time semi-Markov chain and we choose to model the sojourn time with a gamma distribution parametrised as follows:

$$\text{Gamma}\left(\alpha, \frac{\alpha}{m}\right), \tag{5.1}$$

where α is the shape parameter and m is the mean parameter, representing the mean residence time in a state.

The location process is modelled via an integrated OU process [10, 15], meaning that the velocity is modelled with an OU process and the position is derived by integration of the velocity model. By equipping the location process with an explicit state dependency, the two-dimensional location $\boldsymbol{\mu}(t)$ and velocity $\mathbf{v}(t)$ vectors are updated according to the following stochastic differential equations [15]:

$$\begin{aligned} d\boldsymbol{\mu}(t) &= \boldsymbol{v}(t)dt, \\ d\boldsymbol{v}(t) &= \beta_i(\boldsymbol{\gamma}_i - \boldsymbol{v}(t))dt + \sigma_i d\boldsymbol{w}(t), \end{aligned} \tag{5.2}$$

where the parameter β_i measures the persistence in the speed and direction of the movement, σ_i measures the variability in velocity, $\boldsymbol{w}(t)$ is a Wiener process, and $\boldsymbol{\gamma}_i$ is the mean velocity parameter which we take to be zero in all that follows. The subscript i represents the dependency on state i, meaning different movement characteristics are associated with different state values and the parameters β_i and σ_i are *state-specific* parameters. For example, lower values of β result into longer-term persistence in the direction and speed of the movement. Thus, we introduce the vector of model parameters for a model with n states

$$\boldsymbol{\theta} = [\boldsymbol{\beta}, \boldsymbol{\sigma}, \boldsymbol{\alpha}, \mathbf{m}, \omega], \tag{5.3}$$

where ω is the standard deviation of the observation noise and

$$\begin{aligned} \boldsymbol{\beta} &= [\beta_1, \ldots, \beta_n], \\ \boldsymbol{\sigma} &= [\sigma_1, \ldots, \sigma_n], \\ \boldsymbol{\alpha} &= [\alpha_1, \ldots, \alpha_n], \\ \mathbf{m} &= [m_1, \ldots, m_n], \end{aligned} \tag{5.4}$$

meaning that both the movement parameters and the behavioural switching parameters are state-dependent.

The last layer is represented by the observation process, i.e. the observed locations; this is obtained straightforwardly by augmenting the location process with noise:

$$\begin{aligned} \mathbf{y}(t) &= \boldsymbol{\mu}(t) + \boldsymbol{\chi}(t), \\ \chi_c &\sim N(0, \omega^2), \end{aligned} \tag{5.5}$$

where $c = x, y$ represents coordinate axes. Note that in our model we make an explicit assumption of isotropy in the $x - y$ plane. This means that the properties of the system are invariant under rotations in this plane. Consequently, the scalar values of β_i and σ_i are the same in both x and y directions, leading to rotational symmetry in movement. Equally, the observation noise is assumed to be isotropic and is incorporated as two independent univariate Gaussian distributions with the same variance ω^2.

5.2.2 Conditional Log-Likelihood

Let $\mathbf{Y} = \{\mathbf{y}_1, \ldots, \mathbf{y}_d\}$ be a trajectory, that is, a set of observed locations, where d is the dimension of the set. We refer to $\mathbf{Y}$ as the incomplete dataset. If knowledge of the hidden states $\mathbf{Z} = \{z_1, \ldots, z_d\}$ was available, with $z_t \in \{1, \ldots, n\}$, then we could access the complete dataset $\{\mathbf{Y}, \mathbf{Z}\}$. Given the complete dataset the likelihood of the movement parameters may be calculated using the Kalman filter since the integrated OU process is equivalent to a linear Gaussian state-space model. The transition, observation, and covariance matrices of this state-space model were initially described in [10] and then extended to the multi-state case by [15].

Since the complete dataset is unavailable due to the latent nature of the states $\mathbf{Z}$, we make use of the conditional log-likelihood

$$\ell_{cond} = \ln\ p(\mathbf{Y}|\mathbf{Z}, \boldsymbol{\theta}). \tag{5.6}$$

Unlike the incomplete data log-likelihood, the conditional likelihood is tractable. This will be exploited in our MCEM algorithm to approximately maximise the incomplete log-likelihood, as explained in the next section.

5.2.3 Monte Carlo EM Algorithm

We are faced with the task of reconstructing the history of the hidden behavioural states as well as recovering the state-dependent movement parameters of the model together with the measurement error parameters from the telemetry data. For this task, we employ a Monte Carlo expectation-maximisation algorithm [28] that enables us to combine a sampler with an efficient stochastic gradient descent [23] algorithm.

The expectation-maximisation (EM) algorithm is employed for finding maximum likelihood solutions for probabilistic models having latent variables [3, 7]. The algorithm is a two-step iterative algorithm: in the first step, the E-step, the expectation of the complete data log-likelihood is taken with respect to the posterior distribution of the hidden state sequences, given the current model parameters. In the second step, the M-step, this calculated expected value of the complete data log-likelihood is then optimised with respect to the model parameters. These optimised parameters are used to update the model, which is then employed in the subsequent E-step. This process is repeated until convergence.

In our settings, the goal is to find maximum likelihood estimates for the parameter set $\hat{\boldsymbol{\theta}}$ as well as the posterior distribution of the hidden state sequence given the trajectories and $\hat{\boldsymbol{\theta}}$, $p(\mathbf{Z}|\mathbf{Y}, \hat{\boldsymbol{\theta}})$. To obtain $\hat{\boldsymbol{\theta}}$, we need to maximise the incomplete data log-likelihood

$$\ln\, p(\mathbf{Y}|\boldsymbol{\theta}) = \ln\left(\sum_{\mathbf{Z}} p(\mathbf{Y}|\mathbf{Z}, \boldsymbol{\theta})\, p(\mathbf{Z}|\boldsymbol{\theta})\right), \tag{5.7}$$

which is generally intractable due to the fact that the number of possible state sequences increases exponentially with the trajectory length. To estimate $\ln\, p(\mathbf{Y}|\boldsymbol{\theta})$, we proceed as follows.

Given any proper probability distribution over the latent variables $q(\mathbf{Z})$, the following decomposition holds [3]:

$$\ln\, p(\mathbf{Y}|\boldsymbol{\theta}) = \mathcal{L}(q(\mathbf{Z}), \boldsymbol{\theta}) + KL(q(\mathbf{Z})||p(\mathbf{Z}|\mathbf{Y}, \boldsymbol{\theta})), \tag{5.8}$$

where the quantity

$$\mathcal{L}(q(\mathbf{Z}), \boldsymbol{\theta}) = \sum_{\mathbf{Z}} q(\mathbf{Z}) \ln\left(\frac{p(\mathbf{Y}, \mathbf{Z}|\boldsymbol{\theta})}{q(\mathbf{Z})}\right) \tag{5.9}$$

acts as a lower bound for the incomplete data log-likelihood and

$$KL(q(\mathbf{Z})||p(\mathbf{Z}|\mathbf{Y}, \boldsymbol{\theta})) = -\sum_{\mathbf{Z}} q(\mathbf{Z}) \ln\left(\frac{p(\mathbf{Z}|\mathbf{Y}, \boldsymbol{\theta})}{q(\mathbf{Z})}\right) \tag{5.10}$$

is the Kullback-Leibler divergence between $q(\mathbf{Z})$ and $p(\mathbf{Z}|\mathbf{Y}, \boldsymbol{\theta})$, where $KL \geq 0$ and $KL = 0$ if and only if $q(\mathbf{Z}) = p(\mathbf{Z}|\mathbf{Y}, \boldsymbol{\theta})$. Considering a general E-step and supposing the current parameter value is $\boldsymbol{\theta}^{-}$, the arbitrary $q(\mathbf{Z})$ distribution is set to the posterior distribution of the latent variables, giving [3]

$$\begin{aligned}\mathcal{L}(q(\mathbf{Z}), \boldsymbol{\theta}) = &\sum_{\mathbf{Z}} p(\mathbf{Z}|\mathbf{Y}, \boldsymbol{\theta}^{-}) \ln\,(p(\mathbf{Y}|\mathbf{Z}, \boldsymbol{\theta})\, p(\mathbf{Z}|\boldsymbol{\theta})) \\ &- \sum_{\mathbf{Z}} p(\mathbf{Z}|\mathbf{Y}, \boldsymbol{\theta}^{-}) \ln\, p(\mathbf{Z}|\mathbf{Y}, \boldsymbol{\theta}^{-}),\end{aligned} \tag{5.11}$$

that is, the expectation of the complete data log-likelihood with respect to the posterior distribution of the latent variables given the current parameter estimate $\boldsymbol{\theta}^-$. In the M-step, this quantity is maximised with respect to $\boldsymbol{\theta}$ to give a new parameter value $\boldsymbol{\theta}^+$ which is used in the next E-M cycle. Since the M-step maximises the lower bound (Eq. 5.9) and the E-step pushes the lower bound to the incomplete data log-likelihood because of Eqs. 5.8 and 5.10, every single cycle is guaranteed to increase the log-likelihood in Eq. 5.8 and it can be shown [7] that this converges to a zero-gradient point in the log-likelihood.

In our framework, we replace Eq. 5.11 with a numerical approximation using a Monte Carlo estimate

$$\mathcal{L}(q(\mathbf{Z}), \boldsymbol{\theta}) \approx \frac{1}{S} \sum_{i=1}^{S} \ln \left(p(\mathbf{Y}|\mathbf{Z}_i, \boldsymbol{\theta}) p(\mathbf{Z}_i|\boldsymbol{\theta}) \right) + const, \tag{5.12}$$

where the second term in Eq. (5.11) is constant with respect to $\boldsymbol{\theta}$, $\mathbf{Z}_i \sim p(\mathbf{Z}|\mathbf{Y}, \boldsymbol{\theta}^-)$ and S is the number of Monte Carlo samples. As this approximation uses samples from the posterior distribution of $\mathbf{Z}$, we are able to direct the Monte Carlo sum towards regions of higher likelihood; this is akin to importance sampling (see [3], Section 23.4) and results in substantially reduced computational cost. This wouldn't be feasible by using Eq. 5.7 directly, via the Monte Carlo approximation

$$\ln p(\mathbf{Y}|\boldsymbol{\theta}) \approx \ln \left(\frac{1}{S} \sum_{i=1}^{S} p(\mathbf{Y}|\mathbf{Z}_i, \boldsymbol{\theta}) \right), \tag{5.13}$$

with $\mathbf{Z}_i \sim p(\mathbf{Z}|\boldsymbol{\theta})$, as a finite sample from the prior distribution is very unlikely to include state sequences $\mathbf{Z}_i$ for which the argument of the Monte Carlo sum has significant weight, leading to slow convergence and high estimation variance.

To efficiently optimise the parameter set $\boldsymbol{\theta}$, we gather the observations into equally sized batches with each batch containing a pre-specified number of trajectories, where this number could represent the number of trajectories of different animals, the number of trajectories of the same animal over days, or a mixture of both: $\mathcal{B} = \{\mathbf{Y}_1, \ldots, \mathbf{Y}_b\}$, and we assume movement parameters are shared across individuals. Then, during the E-step we employ a dynamic programming approach to sample in parallel from the posterior distributions $p(\mathbf{Z}|\mathbf{Y}_j, \boldsymbol{\theta})$, for $j = 1, \ldots, b$, leading to $b \times S$ samples per E-step. Finally, in the M-step model parameters are optimised based on this set of samples and we run the optimisation routine until convergence. In the next sections, the E-step and the M-step will be described in detail.

5.2.3.1 Expectation Step

To optimise model parameters with respect to the expected latent state sequence, we employ a stochastic Monte Carlo approximation to the expectation as described in

Eq. 5.12. To generate these samples, we are required to draw from the distribution $p(\mathbf{Z}|\mathbf{Y}, \boldsymbol{\theta}^{-})$ which can be achieved using the forward filtering backward sampling algorithm in the case of a standard hidden Markov model, as described in [19]. For our model, there are two additional complications that must be accounted for. Firstly, we are assuming that state switches are non-Markovian and the probability of switching depends on the amount of time previously spent in the current state. Secondly, the movement process is modelled using a linear Gaussian state-space model, leading to dynamics which are commonly known as a switching Kalman filter [18]. We describe the details of the sampling scheme below, but in summary, we employ an augmented version of the forward-backward algorithm to address the first issue and employ the generalised pseudo-Bayesian (GPB) algorithm [2] of order 1 to address the second.

Since the underlying state transition process of our model is non-Markovian, a standard application of the forward-backward algorithm for HMMs is not possible. Instead, we employ a modified version of the algorithm that defines the forward-backward variables using the current state along with the remaining time until the next transition [29, 30]. The augmented state variable is therefore $\{z_t, \tau_t\}$ where z_t indicates the current state at time t and τ_t indicates the remaining time until the next switch. At each time step, one of two types of transitions may occur, either $\tau_t = 1$ and $z_{t+1} \neq z_t$, meaning the state occupied at time t has reached the end of its residence time, or $\tau_t > 1$ so that $z_{t+1} = z_t$ and $\tau_{t+1} = \tau_t - 1$, meaning there is no state transition and the time remaining in the current state decreases by 1. In two related works, [29, 30] present the forward-backward variables required to calculate the filtered and smoothed probabilities for the latent states given a set of emission probabilities, a distribution of residence times, and a maximum residence time for the model. For further details on the implementation of the forward-backward algorithm for hidden semi-Markov models, we refer the interested reader to these articles since we have directly implemented the algorithm described therein.

We do, however, make a significant alteration to accommodate the movement model described in Eq. 5.2 and associated Kalman filter. Since the state of the model is defined by the behavioural state as well as the latent position and velocity of an animal, we must include the filtered position and velocity probabilities within the forward-backward algorithm. The fact that the transition and covariance matrices of the Kalman filter depend on the behavioural state means that the probability of the latent position and velocity is a mixture of Gaussians, with the number of components in the mixture growing exponentially with time. For instance, given a model with two distinct behavioural states, the initial state $p(\boldsymbol{\mu}, \mathbf{v})$ is propagated forward to give a mixture of two Gaussians, with each component corresponding to the distribution associated with each of the possible states. At the next time step, the same process occurs giving rise to a Gaussian mixture with four components. Thus, the number of components of the mixture increases exponentially and inference becomes intractable. To address this issue, it is necessary to introduce an approximation that prevents the complexity of the algorithm increasing at each time step.

In this work, we adopt an approximation known as the generalised pseudo-Bayesian algorithm [2, 31] whereby the Gaussian mixture model at each time step is approximated by a single Gaussian distribution. This is referred to as the $O(1)$ GPB algorithm since the approximation uses a one-component distribution (higher-order algorithms are possible and the order of the algorithm corresponds to the number of components in the approximating mixture distribution). The approximation proceeds by "collapsing" the mixture into a single Gaussian by matching the first and second moments. Therefore, given a model with n behavioural states and an initial belief distribution defined by a Gaussian distribution with mean $\mathbf{X}_t$ and covariance $\mathbf{V}_t$, the updated distribution is a Gaussian mixture of n components with means $\mathbf{X}_{t+1}^j$ and covariances $\mathbf{V}_{t+1}^j$, corresponding to the updated probabilities conditioned on the behavioural state being state j. The weights w_{t+1}^j of the mixture distribution are defined as the filtered probability that state j is the true underlying state. To collapse the mixture into a single Gaussian, moment matching is used so that

$$\mathbf{X}_{t+1} = \sum_{j=1}^{n} w_{t+1}^j \mathbf{X}_{t+1}^j,$$

$$\mathbf{V}_{t+1} = \sum_{j=1}^{n} w_{t+1}^j \left(\mathbf{V}_{t+1}^j \left(\mathbf{X}_{t+1}^j - \mathbf{X}_{t+1} \right) \left(\mathbf{X}_{t+1}^j - \mathbf{X}_{t+1} \right)^T \right), \tag{5.14}$$

which provides the belief distribution for the next observation and in this way ensures that the number of components in the mixture distribution remains constant.

Using the approximation of the $O(1)$ GPB algorithm, the necessary forward and backward variables required for the method proposed in [30] may be calculated, meaning the filtered and smoothed probabilities of the behavioural states may be obtained and samples from the posterior distribution $p(\mathbf{Z}|\mathbf{Y}, \boldsymbol{\theta}^-)$ taken. In our implementation of the MCEM algorithm, these samples are then used in the maximisation step as we describe next.

5.2.3.2 Maximisation Step

In the E-step, the expectation of the complete data log-likelihood in Eq. 5.11 is approximated with a Monte Carlo estimate for b different trajectories

$$\zeta = \sum_{j=1}^{b} \sum_{i=1}^{S} \ln \left(p(\mathbf{Y}_j|\mathbf{Z}_i, \boldsymbol{\theta}) p(\mathbf{Z}_i|\boldsymbol{\theta}) \right), \tag{5.15}$$

where $\mathbf{Z}_i \sim p(\mathbf{Z}|\mathbf{Y}_j, \boldsymbol{\theta}^-)$.

In the M-step, maximisation of ζ with respect to $\boldsymbol{\theta}$ is carried out. To avoid mixture of strongly and weakly identifiable parameters, we reparametrise Eq. 5.2 so that

$$\tilde{\sigma} = \frac{\sigma}{\beta}, \quad \tau = \frac{1}{\beta}. \tag{5.16}$$

The parameter τ represents the autocorrelation time of the velocity process [9], whereas $\tilde{\sigma}^2$ is the long-term ($t >> \tau$) slope of the increase in the expected squared displacement over time.

We equip the optimiser with a convergence criterion to avoid expenditure of computational time with no return in improved model likelihood. We specify a tolerance parameter and compare the model loss before and after a single epoch of optimisation. If the change in the loss values is below the tolerance threshold for a specified number of iterations, then the optimisation is stopped and the maximisation step has completed. The algorithm iteratively performs the expectation and maximisation steps until parameter values have converged. Once the EM algorithm has converged, we then calculate the smoothed probabilities of the latent states to provide the posterior probability of occupying a given latent state at each time. The stochastic gradient descent optimiser and all other aspects of the algorithm are implemented within the machine learning framework TensorFlow [1]. This enables us to deploy the code on GPU-accelerated hardware which is highly efficient when the computational operations are parallelisable, such as is the case when using data from multiple individuals or when it is possible to split movement trajectories into batches.

5.2.4 Validation with Synthetic Data

In order to evaluate our framework, we generate synthetic data that matches the movement model used for inference. We simulate multiple individuals following an integrated OU process random walk with behavioural switching occurring at random intervals and the sojourn times in each behaviour following a gamma distribution. State transitions are equally likely for all states. For simulations of dynamics we do not numerically integrate equation 5.2 but use the exact solutions described in [15]:

$$\begin{aligned} v_c(t+\delta) &= e^{-\beta\delta} v_c(t) + \zeta_c(\delta), \\ \zeta_c(\delta) &\sim N\big(0, \sigma^2(1-e^{-2\beta\delta})/2\beta\big), \end{aligned} \tag{5.17}$$

and

Table 5.1 Simulation model ground truth parameter values

Symbol	True value	Description
τ	(60, 10, 10)	Autocorrelation time
$\tilde{\sigma}$	(10, 300, 50)	Long-term RMS displacement
α	(10, 10, 10)	Gamma shape
$\mathbf{m}$	(500, 500, 500)	Mean residence time
ω	0.1	Measurement error standard deviation

$$\begin{aligned}\mu_c(t+\delta) &= \mu_c(t) + v_c(t)\left(\frac{1-e^{-\beta\delta}}{\beta}\right) + \xi_c(\delta),\\ \xi_c(\delta) &\sim N\left(0, \frac{\sigma^2}{\beta^2}\left(\delta - \frac{2}{\beta}(1-e^{-\beta\delta}) + \frac{1}{2\beta}(1-e^{-2\beta\delta})\right)\right),\end{aligned} \tag{5.18}$$

where the subscript $c = x, y$ represents Easting and Northing and δ is the time interval. Observations were fixed at every 5 minutes and we used additive, independent, and identically distributed zero-mean isotropic Gaussian error for the measurement error.

In the simulation study, we generated 8 independent trajectories with 3 latent behavioural states consisting of 512 observations each with a fixed 5-minute time interval to simulate a moderately high-frequency GPS collar sampling rate. The model parameter vector therefore consisted of 13 parameters:

$$\boldsymbol{\theta} = \{\tau_1, \tau_2, \tau_3, \tilde{\sigma}_1, \tilde{\sigma}_2, \tilde{\sigma}_3, \alpha_1, \alpha_2, \alpha_3, m_1, m_2, m_3, \omega\}, \tag{5.19}$$

and the true values are displayed in Table 5.1.

We ran the EM algorithm until convergence, with each maximisation step optimising the model parameters until an early stopping criterion was reached and the algorithm was terminated once there was no improvement in the model during the optimisation. Each expectation step involved taking 512 samples from the posterior distribution over the latent states using the dynamic programming approach described above.

Once the parameter values had been optimised, we computed the smoothed probabilities of the latent states. The forward-backward algorithm took 3 minutes to compute these probabilities on an NVIDIA A10-24Q GPU, while the initial EM stage took 65 minutes. The probability of being in each state for each individual is reported in Fig. 5.1. In Fig. 5.2 we plot the optimised model parameters.

The sampler was able to reconstruct the underlying state sequences for all trajectories with some uncertainty occurring at the transition points. As for the optimiser, all parameters were optimised close to the true values except for the observation noise and the parameters of the gamma distributions. For the observation noise this is likely due to the frequency of observations, and far more frequent fixes would be required to accurately identify this parameter due to the indeterminacy

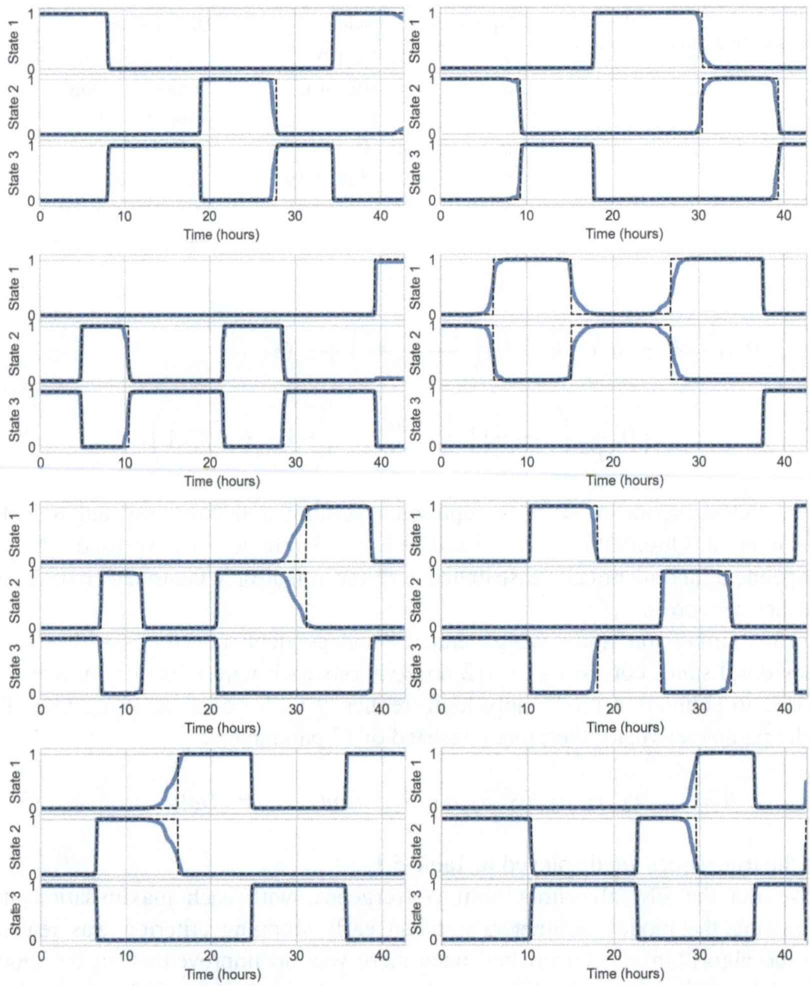

Fig. 5.1 Validation of the method using synthetic data. *Blue solid line:* Probability of being in a state. *Black dashed line:* Ground truth probability of being in a state

between movement and observation noise, whereas for the parameters of the gamma distributions this is likely due to the small effective data size in terms of the number of switches. Since under fairly mild regularity conditions the maximum likelihood estimator is asymptotically unbiased [6], we would expect that for a sufficiently large dataset, our parameter estimates will be approximately unbiased.

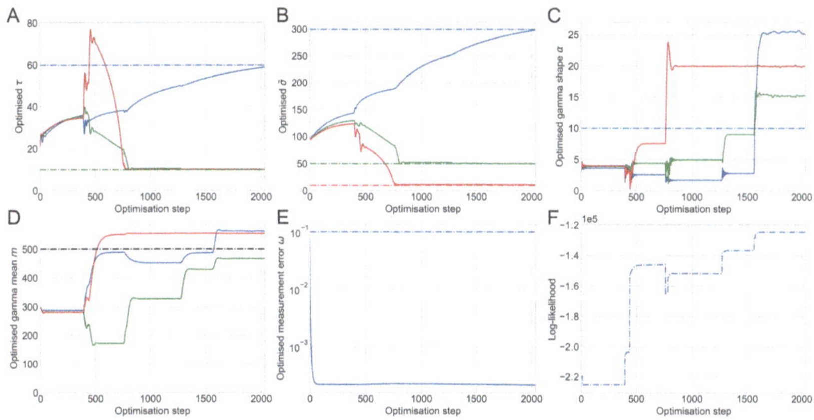

Fig. 5.2 Optimised model parameters and associated log-likelihood summed across individuals. (**a–b**) Optimised $\boldsymbol{\tau}$ and $\tilde{\boldsymbol{\sigma}}$ (solid lines) with corresponding true values (dashed dotted lines). (**c–d**) Solid lines, optimised gamma parameters α and m; dashed dotted lines, true values. (**e**) Solid line, optimised measurement error standard deviation; dashed dotted line, true value. (**f**) Total expected complete log-likelihood per optimisation step

5.3 Case Studies

5.3.1 Grey Seal Case Study

In order to provide a direct comparison to the method of [15], we apply our framework to the grey seal dataset first described in [25] and available in the Movebank database [14].

We considered a model with 3 behavioural states meaning there were 12 parameters to optimise relating to the movement and switching process,

$$\boldsymbol{\theta} = \{\tau_1, \tau_2, \tau_3, \tilde{\sigma}_1, \tilde{\sigma}_2, \tilde{\sigma}_3, \alpha_1, \alpha_2, \alpha_3, m_1, m_2, m_3\}. \tag{5.20}$$

We set the initial values for the movement parameters to be $\tau_1 = \tau_2 = \tau_3 = 2$ and $\tilde{\sigma}_{1,2,3} = \{1.0, 1.1, 0.1\}$. The gamma concentrations were initially set to 1.5 and all mean residence times were initialised as 5 hours. As the fixes are relatively infrequent, we assumed a negligible, fixed measurement error as in [15].

As in the simulation study, we ran the MCEM algorithm until convergence, alternating between the expectation and maximisation steps with the optimisation step lasting until we failed to see a significant improvement during an epoch of optimisation. Once the EM stage of the algorithm was complete, we ran the forward-backward algorithm to obtain the latent state probabilities.

We recovered optimised parameter values that were consistent with previous investigations of this dataset [15]; however, we found a three-state model that created a low motility state, a better fit to the data. The model therefore detected

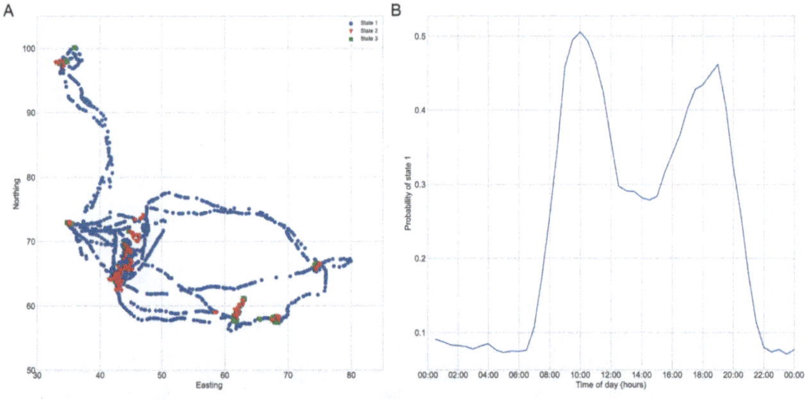

Fig. 5.3 (**a**) GPS location and behavioural states for the grey seal data. Points show a GPS location from the dataset, while the colour indicates the most likely state. Red locations correspond to the low autocorrelation movement, while blue locations indicate persistent movement, and green is the low motility state. (**b**) Posterior probability of behavioural state as a function of time of day for the Merino sheep dataset. Here, state 1 is the active mobile state, while state 2 corresponds to a low movement resting state. We observe clear peaks of activity around 10 a.m. and 7 p.m.

the following states, which we describe in terms of the optimised parameters and potential biological interpretation: State 1 displayed an average speed of 3.23 km/h and a velocity autocorrelation time of 4.76 hours. This state is interpreted as a transit state and it had a gamma mean parameter of 30 hours and a concentration parameter of 1, indicating no well-defined residence time in the state. State 2 had a similar average speed of 3.52 km/h but a much shorter autocorrelation time of 0.86 hour. The gamma mean parameter was 8 hours and the concentration parameter was 1.63, indicating that modelling this process with the gamma distribution provides a better approximation than the exponential distribution. We interpret this state as a foraging state since it consists of relatively fast movements, with low autocorrelation. The final state was characterised by a lower average speed (1.81 km/h) and low velocity autocorrelation (0.15 hour). As in state 1, the concentration parameter was 1, whereas the gamma mean parameter was 32 hours. This state corresponds to a period of resting or intensive and recurrent space use. A visualisation of the posterior distributions of the latent states is presented in Fig. 5.3a.

5.3.2 Merino Sheep Case Study

As a second example case study for our method, we analyse data collected from a long-term study of Merino sheep allowed to roam freely in a paddock of 700 hectares. Fieldwork was conducted at the “Campo Anexo Pilcaniyeu” from INTA (National Institute of Agricultural Technology) Bariloche, Patagonia, Argentina. The sheep were equipped with collars containing a GPS (CatLog-B, Perthold

Engineering www.perthold.de) that was programmed to record locations every 2 to 5 minutes. We analysed data collected from March to May 2019 from 34 sheep. Trajectories were split into segments of 1024 observations, and there were 581 of these segments resulting in 594,944 observations.

We considered a model with two behavioural states for a total of nine parameters to optimise:

$$\boldsymbol{\theta} = \{\tau_1, \tau_2, \tilde{\sigma}_1, \tilde{\sigma}_2, \alpha_1, \alpha_2, m_1, m_2, \omega\}. \tag{5.21}$$

We ran the MCEM algorithm with the same hyperparameters as for the previous case study and used the same process to assess when the algorithm had converged. Once the MCEM algorithm had finished, we then computed the posterior probability of the latent states. This took 5.7 hours on the same NVIDIA A10-24Q GPU used in the previous examples.

The optimised values for the two-state model correspond to a mobile state, state 1, with a mean speed of 0.74 km/h and an autocorrelation time of 0.1 hour, and a resting state, state 2, which had a mean speed of 0.28 km/h and an autocorrelation time of 0.35 hour. The mean residence time in state 1 was 1.1 hours and in state 2 it was 3.2 hours, while the gamma concentration parameter was 1.13 and 1, respectively, indicating an exponential sojourn time distribution would be a good approximation for state 2 (the resting state) but for state 1 (the mobile state) the mode of the residence time distribution was greater than zero.

From the smoothed probabilities of latent states, we calculated the expected state for the individuals as a function of time of day (converted to local time). This is shown in Fig. 5.3b which illustrates the two periods of high activity in the morning and evening, likely representing transit to and from grazing grounds. Our results show agreement with a previous analysis that revealed regular daily activity patterns in the sheep behaviour [26].

5.4 Discussion

We have presented a new framework that can infer latent behavioural states as well as recover the model parameters from a continuous-time, multi-state model of animal movement. The model is described by a latent continuous-time behavioural process and an observation process consisting of a location process augmented with measurement error. A novel degree of flexibility is introduced in the behavioural process by making use of a continuous-time hidden semi-Markov model so that the dwell times are not constrained to follow an exponential distribution, thus avoiding making unrealistic assumptions on the animals' state-switching behaviour. The specification of the residence time distribution is arbitrary, and we have used a gamma distribution in our work as its properties make up for more realistic assumptions on the animals' behaviour; thus, our method may be regarded as a generalisation of standard hidden Markov models. However, this specification can be changed on a case-by-case basis to accommodate the ecologist's prior assumptions.

Our inference scheme was developed to overcome the computational limitations of existing methods. This was achieved by employing the MCEM algorithm whereby the non-analytical E-step was approximated by a Monte Carlo sum and the non-analytical M-step by a stochastic gradient descent scheme. Existing state-of-the-art methods that use continuous-time models perform inference of model parameters via maximum likelihood estimation using the Kalman filter [10, 15] in combination with a hybrid MCMC scheme aimed at reconstructing the latent state sequence [4, 15]. The scheme is a Metropolis-within-Gibbs sampling scheme, and each iteration of the algorithm consists of an update for three groups: an update for the underlying state sequence, an update for the movement parameters, and an update for the state transition rates, respectively. Although such methods benefit from a proper uncertainty quantification in the parameter estimations, MCMC sampling schemes notoriously perform slowly with large datasets, and calculation of the likelihood requires utilisation of the whole dataset. Given the increasing availability of high-frequency data and the improvements in the machine learning field, we believe that putting effort in the creation of an animal movement model that can leverage highly optimised machine learning libraries to fully make use of large datasets should be prioritised.

These innovations expand the array of methodologies available to ecologists and we believe this method will positively impact new telemetry analyses as leveraging the information of big data is key to an increased understanding of animal movement.

Acknowledgments This work was supported by the Leverhulme Foundation (RPG-2018-398). JMM was supported by a Leverhulme Visiting Professorship (VP2-2018-0630). CJT is supported by a James S. McDonnell Foundation Studying Complex Systems Scholar Award. DH is supported by the Engineering and Physical Sciences Research Council (EPSRC), grant reference number EP/T017899/1.

References

1. Abadi, M., Barham, P., Chen, J., Chen, Z., Davis, A., Dean, J., et al.: TensorFlow: a system for large-scale machine learning. In: Proceedings of the 12th USENIX Symposium on Operating Systems Design and Implementation (OSDI '16), pp. 265–283 (2016)
2. Bar-Shalom, Y., Li, X.R., Kirubarajan, T.: Estimation with Applications to Tracking and Navigation: Theory Algorithms and Software. John Wiley & Sons, London (2004)
3. Bishop, C.M.: Pattern Recognition and Machine Learning. Springer, Berlin (2006)
4. Blackwell, P.G.: Bayesian inference for Markov processes with diffusion and discrete components. Biometrika **90**(3), 613–627 (2003)
5. Cagnacci, F., Boitani, L., Powell, R.A., Boyce, M.S.: Animal ecology meets GPS-based radiotelemetry: A perfect storm of opportunities and challenges. Philos. Trans. Roy. Soc. B **365**, 2157–2162 (2010)
6. Cramer, H.: Mathematical Methods of Statistics. Princeton University Press, Princeton (1946)

7. Dempster, A.P., Laird, N.M., Rubin, D.B.: Maximum likelihood from incomplete data via the EM algorithm. J. R. Stat. Soc. B **39**, 1–38 (1977)
8. Glennie, R., Adam, T., Leos-Barajas, V., Michelot, T., Photopoulou, T., McClintock, B.T.: Hidden Markov models: pitfalls and opportunities in ecology. Methods Ecol. Evol. **14**(1), 43–56 (2023)
9. Gurarie, E., Fleming, C.H., Fagan, W.F., Laidre, K.L., Hernandez-Pliego, J., Ovaskainen, O.: Correlated velocity models as a fundamental unit of animal movement: synthesis and applications. Move. Ecol. **5**, 13 (2017)
10. Johnson, D.S., London, J.M., Lea, M.A., Durban, J.W.: Continuous-time correlated random walk model for animal telemetry data. Ecology **89**(5), 1208–1215 (2008)
11. Langrock, R., Marques, T., Baird, R., Thomas, L.: Modeling the diving behavior of whales: a latent-variable approach with feedback and semi-Markovian components. J. Agric. Biol. Environ. Stat. **19**, 82–100 (2013)
12. McClintock, B., Johnson, D., Hooten, M., Ver Hoef, J., Morales, J.: When to be discrete: the importance of time formulation in understanding animal movement. Move. Ecol. **2**, 21 (2014)
13. McClintock, B.T., Michelot, T.: momentuHMM: R package for generalized hidden Markov models of animal movement. Methods Ecol. Evol. **9**(6), 1518–1530 (2018)
14. McConnell, B.J.: Data from: state-switching continuous-time correlated random walks. Movebank Data Repository (2019).
15. Michelot, T., Blackwell, P.G.: State-switching continuous-time correlated random walks. Methods Ecol. Evol. **10**, 637–649 (2019)
16. Michelot, T., Langrock, R., Patterson, T.A.: moveHMM: an R package for the statistical modelling of animal movement data using hidden Markov models. Methods Ecol. Evol. **7**(11), 1308–1315 (2016)
17. Morales, J.M., Haydon, D.T., Frair, J., Holsinger, K.E., Fryxell, J.M.: Extracting more out of relocation data: Building movement models as mixtures of random walks. Ecology **85**, 2436–2445 (2004)
18. Murphy, K.P.: Switching Kalman Filters. Compaq Cambridge Res Lab Tech Rep (1998)
19. Murphy, K.P.: Machine Learning: A Probabilistic Perspective. MIT Press, Cambridge (2012)
20. Nathan, R., Monk, C.T., Arlinghaus, R., Adam, T., Alós, J., Assaf, M., et al.: Big-data approaches lead to an increased understanding of the ecology of animal movement. Science **375**(6582) (2022)
21. Patterson, T.A., Parton, A., Langrock, R., Blackwell, P.G., Thomas, L., King, R.: Statistical modelling of individual animal movement: An overview of key methods and a discussion of practical challenges. ASTA-Adv Stat. Anal. **101**(4), 399–438 (2017)
22. Pohle, J., Langrock, R., van Beest, F., Schmidt, N.: Selecting the number of states in hidden Markov models: pitfalls, practical challenges and pragmatic solutions. J. Agric. Biol. Environ. Stat. (2017)
23. Robbins, H., Monro, S.: A stochastic approximation method. Ann. Math. Stat. **22**(3), 400–407 (1951)
24. Ruiz-Suarez, S., Leos-Barajas, V., Morales, J.M.: Hidden Markov and semi-Markov models when and why are these models useful for classifying states in time series data? J. Agric Biol. Environ. Stat. **27**(2), 339–363 (2022)
25. Russell, D.J., McClintock, B.T., Matthiopoulos, J., Thompson, P.M., Thompson, D., Hammond, P.S., et al.: Intrinsic and extrinsic drivers of activity budgets in sympatric grey and harbour seals. Oikos **124**(11), 1462–1472 (2015)
26. Torney, C., Morales, J., Husmeier, D.: A hierarchical machine learning framework for the analysis of large scale animal movement data. Move. Ecol. **9**, 1 (2021)
27. Turchin, P.: Quantitative Analysis of Movement: Measuring and Modeling Population Redistribution in Animals and Plants. Sinauer Associates (1998)
28. Wei, G.C.G., Tanner, M.A.: A Monte Carlo implementation of the EM algorithm and the poor man's data augmentation algorithms. J. Am. Stat. Assoc. **85**(411), 699–704 (1990)
29. Yu, S.Z., Kobayashi, H.: An efficient forward-backward algorithm for an explicit-duration hidden Markov model. IEEE Signal Proc. Lett. **10**(1), 11–14 (2003)

30. Yu, S.Z., Kobayashi, H.: Practical implementation of an efficient forward-backward algorithm for an explicit-duration hidden Markov model. IEEE Trans. Signal Process. **54**(5), 1947–1951 (2006)
31. Zeng, X., Ghanem, R.: Dynamics identification and forecasting of COVID-19 by switching Kalman filters. Comput. Mech. **66**, 1179–1193 (2020)
32. Zucchini, W., MacDonald, I., Langrock, R.: Hidden Markov Models for Time Series: An Introduction Using R. CRC Press, Boca Raton (2009)

Chapter 6
A Method for Reconstructing and Modeling Social Interactions in Moving Animal Groups from Trajectory Data

Ramón Escobedo, Clément Sire, and Guy Theraulaz

6.1 Introduction

Understanding the role played by social interactions in group-living species is one of the fundamental issues in the study of collective animal behavior [1, 2]. These interactions and their modulation by the abiotic parameters of the environment condition the way in which each individual adapts its movements to that of its conspecifics and also according to the environmental constraints it encounters [3–6]. The development of novel tracking methods in the laboratory [7–11] but also in the field [12–14] now provides high-resolution spatiotemporal data on moving animal groups opening the way for data-based modeling approaches that directly link individual trajectories to the emergence of group-level patterns. Several methods have also tried to infer the social interactions between individuals based on the analysis of their trajectories [15–18]. Thus, force maps have been used to visualize the impact of the interactions of a focal individual with a neighbor on its linear or radial speed [19–23]. However, due to the multifactorial dependence of interactions, force maps provide only an imperfect representation. One of their main limitations is the restrictions on the number of state variables they can handle—typically two, at most three. Additionally, these methods struggle to identify and disentangle intermediate mechanisms such as attraction, repulsion, and alignment, and they cannot distinguish the effects of variables (e.g., distance between fish)

R. Escobedo · G. Theraulaz (✉)
Centre de Recherches sur la Cognition Animale, Centre de Biologie Intégrative, CNRS, Université de Toulouse III – Paul Sabatier, Toulouse, France
e-mail: guy.theraulaz@utoulouse.fr

C. Sire
Laboratoire de Physique Théorique, CNRS, Université de Toulouse III – Paul Sabatier, Toulouse, France
e-mail: clement.sire@utoulouse.fr

L. Giuggioli, P. K. Maini (eds.), *The Mathematics of Movement*,
https://doi.org/10.1007/978-3-032-00873-2_6

from those of parameters (e.g., body length). Furthermore, it is challenging to derive analytical expressions that can be incorporated into a mathematical model. Above all, force maps do not provide a foundation for constructing quantitative models of behavior [24].

We have recently introduced a procedure for extracting analytical expressions describing both the social interactions between individuals and the effects of the environment on their movements [25]. This procedure consists in designing a model grounded on experimental observations, where the analytical expressions are obtained by imposing symmetry constraints that result exclusively from the fundamental principles of the physics governing the interactions. The procedure has already been successfully applied to characterize and compare the social interactions in different species of fish that use the same intermittent mode of locomotion [24], and to determine the response of one of these species to variations of light intensity [26].

Figure 6.1 provides a general overview of the procedure used to design the mathematical model and to reconstruct the social interactions. The procedure includes five consecutive steps, starting from experimental observations and ending with the numerical simulations of the model reproducing these observations. These steps can be repeated in a cycle, as the model predictions allow designing more targeted experiments, which in turn help to refine the model: the experiments feed the model, and the model guides the design of new experiments.

Here, we focus specifically on the reconstruction procedure that provides the analytical expressions describing the social interactions. We thus present only a short overview of the model derivation and put emphasis on both the fundamental hypotheses and the technical aspects involved in the derivation of such analytical expressions. We also show that the procedure is able to detect the absence of an explicit alignment interaction between fish as it has been shown to occur in some species [20, 27].

6.2 Method to Reconstruct and Model Social Interactions from Behavioral Data

The method consists in setting up a mathematical model that describes the relationship between the quantities that characterize the spontaneous behavior of an organism and its interactions with other organisms belonging to the same group or physical objects in the environment.

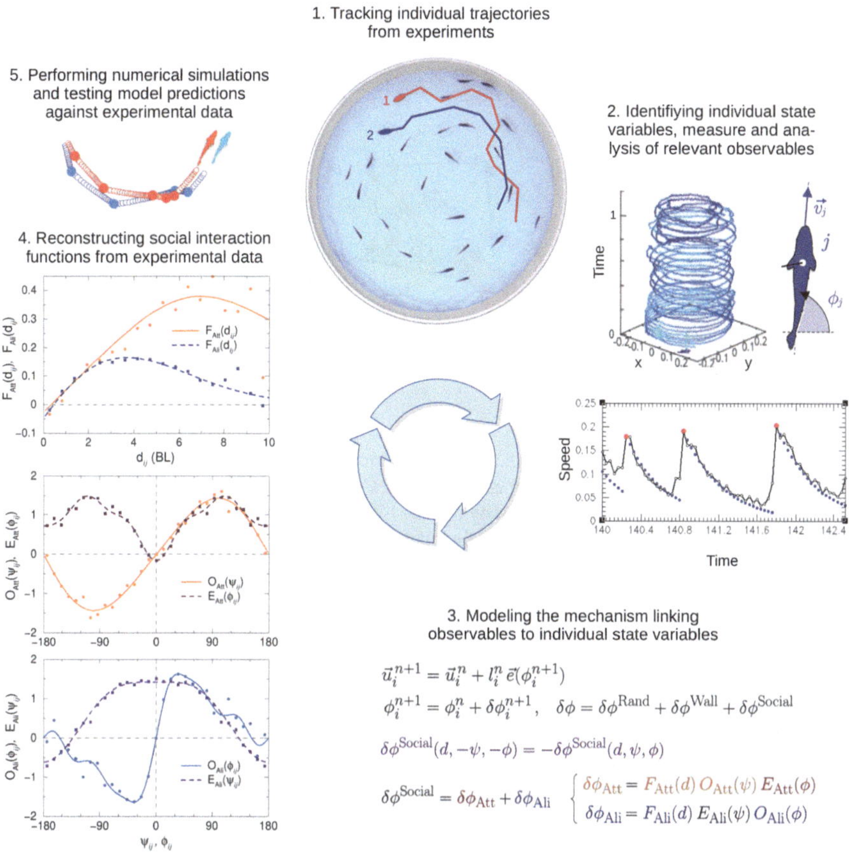

Fig. 6.1 General overview of the successive steps constituting our modeling process. Each step is illustrated by: Step 1: Fish trajectories (red and blue lines) in an experiment with a group of fish swimming in a circular tank. Step 2: Typical trajectory of a single fish swimming along the wall of the tank, individual variables describing the state of a fish (here velocity vector and heading angle), and time profile of speed showing the maxima (red dots) and the gliding phases (blue dots) of three kicks. Step 3: Model equations relating state variables, hypothesis of additive combination of the environmental effects acting on a fish, relation of symmetry, and decomposition of social interactions in the sum of attraction and alignment as products of single-variable functions. Step 4: Reconstructed interaction functions F_{Att}, F_{Ali}, E_{Att}, O_{Att}, O_{Ali}, and O_{Ali}, showing their dependence on the variables characterizing the state of fish i with respect to fish j: the distance between fish d_{ij}, the angle ψ_{ij} with which i perceives j, and the relative heading of j with respect to i, $\phi_{ij} = \phi_j - \phi_i$. See Fig. 6.2. Step 5: Trajectories of a pair of fish in numerical simulations of the model. The wide arrows in the center of the figure illustrate the possibility to start again, carrying out a new set of experiments or designing a new model, in order to check the model's predictions and to deepen the understanding of the underlying mechanisms

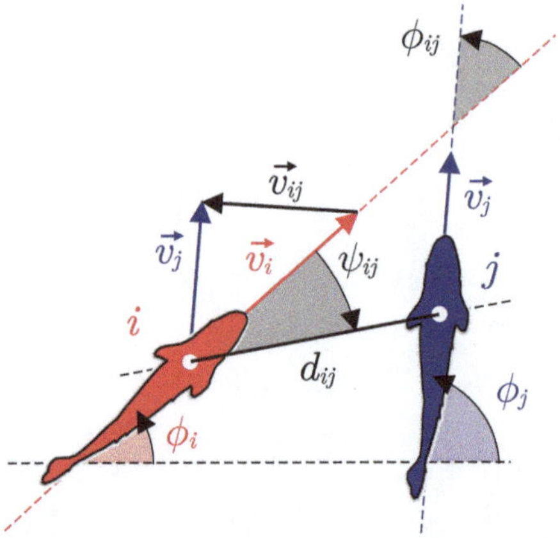

Fig. 6.2 State variables of a focal fish i (in red) with respect to its neighbor j (in blue). Fish position is determined by the center of mass of the fish. Collective measures (in black): d_{ij}, distance between fish; $\mathbf{v}_{ij} = \mathbf{v}_j - \mathbf{v}_i$, relative velocity of j with respect to i; ψ_{ij}, viewing angle with which i perceives j; and $\phi_{ij} = \phi_j - \phi_i$, relative heading of j with respect to i. Positive angles are measured in the counterclockwise direction and with respect to the horizontal line

6.2.1 *Overview of the Modeling Process*

The first step of the method consists in carrying out experiments motivated by either a fundamental question or by the results of the previous cycle of the ongoing modeling process. The data are individual trajectories obtained after tracking that have been conveniently purged from identification errors and other tracking issues (step 1 in Fig. 6.1), and stored in individual position vectors $\mathbf{u}_i(t)$, for $i = 1, \dots, N$, where N is the number of individuals. Velocity vectors $\mathbf{v}_i$ are calculated using backward finite differences, and heading angles ϕ_i representing the direction of motion of the individuals are extracted directly from the tracking data or, alternatively, taken as the orientation of the velocity vector.

The second step consists in identifying the state variables and the relevant observables that can be used to describe the phenomenon under study (step 2 in Fig. 6.1). In the case considered here, it is necessary to characterize both the individual state and the collective pairwise state of the fish. The individual state of a fish i is given by its position and velocity vectors. The relative state of a fish i with respect to another fish j is characterized by the distance between them d_{ij}, the relative speed v_{ij}, the viewing angle with which i perceives j, ψ_{ij} (which is not necessarily equal or opposite to ψ_{ji}), and their relative angle of orientation $\phi_{ij} = \phi_j - \phi_i$ (see Fig. 6.2).

Step 3 consists in defining a class of models that can provide the more suitable description of the observed phenomenon. This is by far the most important part of the method. It has been described in detail in [25] in the case of *H. rhodostomus* and is also applicable to other species using the same swimming mode, like *D. rerio*.

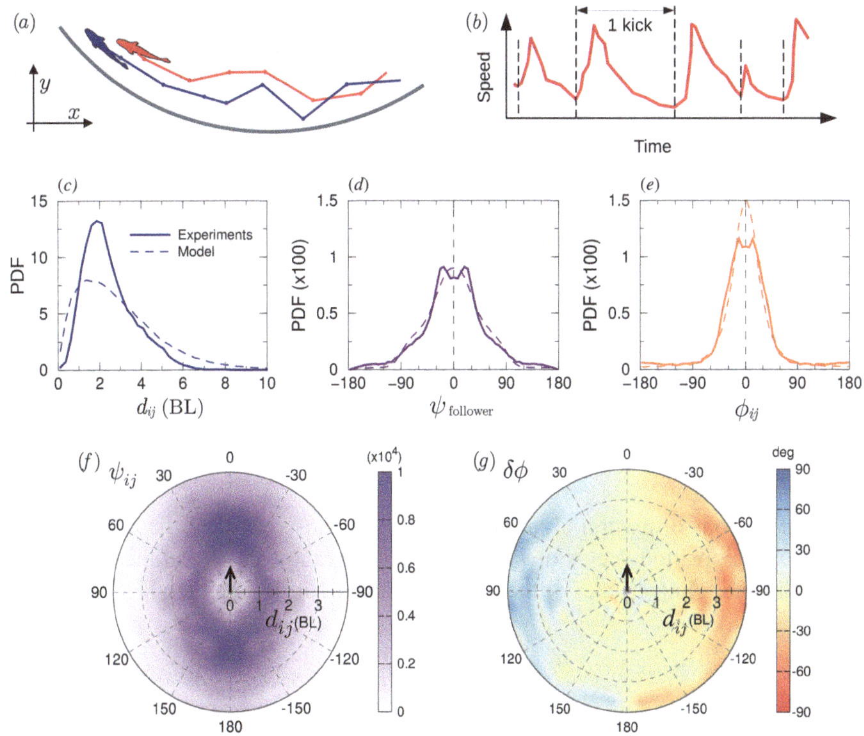

Fig. 6.3 Individual behavior and collective state in fish pairs. (**a**) Polygonal line segments describing fish trajectories in a circular tank. (**b**) Speed profile showing a kick formed by a short bursting phase and a long gliding phase. Probability density functions (PDF) of (**c**) the distance between fish d_{ij}, (**d**) the angle ψ_{follower} with which a fish i perceives j when $|\psi_{ij}| < |\psi_{ji}|$ (the "geometrical follower" being the fish which has to turn less to face the other fish, the latter being the "geometrical leader"), and (**e**) the difference of heading ϕ_{ij}. Solid lines, experimental data; dashed lines, numerical simulations of the model. One body length (1 BL) corresponds to about 3 cm in *H. rhodostomus*. (**f**) Density map of the location of a neighboring fish j in the system of reference centered on the focal fish i pointing north (black arrow), when fish i is further than 1 BL from the border of the tank, regardless of whether i is the leader or the follower. (**g**) Mean value of the angle change $\delta\phi$ depending on the relative location of the neighbor

Figure 6.3a, b shows the polygonal line segments that fish trajectories describe and the corresponding temporal speed profile. Fish swimming mode consists of successive straight segments of variable duration punctuated by abrupt accelerations called *kicks*. At each kicking instant, the fish adjusts its direction and maintains it unchanged while gliding with a decaying speed until the next kick is triggered [25, 28–30].

The gliding phase can be approximated by a decaying exponential with the same decay rate for every kick, as shown in the speed profile depicted in Fig. 6.1, step 2.

Figure 6.3c, d, e shows the probability density functions (PDF) of the collective state variables d_{ij}, ψ_{ij}, and ϕ_i corresponding to the experimental data of a pair of

H. rhodostomus. The two fish adopt a geometrical formation where one fish is in front and the other one follows it, quite close, at a distance of approximately 1 to 4 body lengths (1 BL $= 3$ cm), with the most frequent distance being around 2 BL, in a tank with a diameter of 16.6 cm. We define the *geometrical follower* (respectively, the *geometrical leader*) as the fish that needs to turn the least (respectively, the most) to have the other fish in front, i.e., the one with the smaller absolute viewing angle $|\psi_{ij}|$. In our experiments, fish alternate their geometrical position without preference. Most of the time, the geometrical leader is not only in front but directly ahead of the follower, with $|\psi_{ij}| < 30°$ (Fig. 6.3d), both fish remaining quite aligned most of the time, with $|\phi_{ij}| < 30°$ (Fig. 6.3e).

We assume that, when swimming in groups and in an environment with obstacles, the decision times at which fish integrate the effects of social interactions and obstacle avoidance are precisely the kicking instants. Despite this intermittent and asynchronous mode of swimming, fish groups manage to coordinate and self-organize, exhibiting sophisticated collective behavioral patterns such as schooling and milling.

The equations of motion of an individual fish i thus consist of the discrete-time update of the position vector $\mathbf{u}_i(t)$ and the heading angle $\phi_i(t)$ as functions of the individual and relative states of the fish at the instant of its n-th kick t_i^n,

$$\mathbf{u}_i^{n+1} = \mathbf{u}_i^n + l_i^n \, \mathbf{e}(\phi_i^{n+1}), \tag{6.1}$$

$$\phi_i^{n+1} = \phi_i^n + \delta\phi_i^{n+1}, \tag{6.2}$$

$$t^{n+1} = t_i^n + \tau_i^n, \tag{6.3}$$

where l_i^n and τ_i^n are the kick length and duration and $\mathbf{e}(\phi_i^{n+1})$ is the unitary vector in the direction marked by the angle ϕ_i^{n+1}. Positive angles are measured in the counterclockwise direction and with respect to the horizontal line. Kick length and duration are simply sampled from probability distributions fairly reproducing the experimental ones [25]. The heading angle change $\delta\phi_i^{n+1}$ is selected as follows.

The heading angle change $\delta\phi$ is responsible for incorporating the effects to which the individual fish is subjected. It can be decomposed as

$$\delta\phi(t) = \delta\phi^{\mathrm{R}}(t) + \delta\phi^{\mathrm{W}}(t) + \delta\phi^{\mathrm{S}}(t), \tag{6.4}$$

where superindices R, W, and S stand for random, wall, and social, respectively, accounting for the random angle change in spontaneous decisions of the fish, the angle change induced by the proximity of the wall of the tank, and the angle change due to the social interaction with other fish. Thus, a new position is calculated and the kick performed. If the new position of the fish $\mathbf{u}_i^{n+1}$ is outside the tank, we reject the kick and we sample new values of both the kick length l_i^n and the random angle change until the fish stays inside the tank.

The goal is now to obtain analytical expressions for these contributions in Eq. (6.4) (step 4 in Fig. 6.1). Once known, numerical simulations of the model are performed and the results are compared with the experimental observations. If the agreement is satisfactory, the modeling process is complete. Otherwise, the model has to be revised, additional observables have to be considered, or new experiments must be conducted.

To obtain analytical approximations of $\delta\phi^{\mathrm{R}}$, we perform experiments with a fish swimming alone and we focus on the data when it is around the center of the tank, where the effect of the wall can be neglected. We found no significant side preference in fish swimming, so a Gaussian random variable g with zero mean and unit variance is a suitable way to model the spontaneous heading angle change: $\delta\phi^{\mathrm{R}} = \gamma_{\mathrm{R}} g$, where γ_{R} is the amplitude of the angle changes.

In turn, obtaining analytical expressions of the wall effect $\delta\phi^{\mathrm{W}}$ and of the social interactions $\delta\phi^{\mathrm{S}}$ requires a more sophisticated procedure. We present it in the next section, with the focus on the reconstruction of the social interactions, the reconstruction of the interaction with the wall being a simplified application of the same procedure.

6.2.2 Pairwise Interaction Functions: Definition and Reconstruction Procedure

We first define the two kinds of pairwise interaction functions that can be identified with our procedure, then we describe the procedure, and finally, we show how the resulting analytical expressions can be used to measure the influence of each neighboring fish in a group, allowing us to extend the model to groups with more than two individuals.

Definition and Characterization of the Attraction and Alignment Interactions
We assume that the angle change of an individual fish due to the social interaction with other fish $\delta\phi_i^{\mathrm{S}}$ is the additive combination of the pairwise interactions with each of the other $N-1$ individuals in the group:

$$\delta\phi_i^{\mathrm{S}} = \sum_{j=1, j\neq i}^{N} \delta\phi_{ij}^{\mathrm{S}}, \tag{6.5}$$

where $\delta\phi_{ij}^{\mathrm{S}}$ is the pairwise interaction function (PIF) of fish i with fish j, which depends on the relative state of both fish, given by the quadruplet $(d_{ij}, \psi_{ij}, \phi_{ij}, \mathrm{v}_{ij})$.

To determine the pairwise effect of a specific neighbor, we use the experiments of two fish swimming in the tank, so that there is only one term in the sum in Eq. (6.5). Moreover, we focus on the data where fish are sufficiently far from the wall, so that the term $\delta\phi^{\mathrm{W}}$ in Eq. (6.4) can be neglected, and we assume that the random heading angle change has zero mean in each small neighborhood inside the tank, so, when averaging, the term $\delta\phi^{\mathrm{R}}$ can be neglected in Eq. (6.4).

The procedure is based on the fact that, at each instant of time and in every small neighborhood inside the tank, the heading angle change can be measured from the experimental data, allowing the calculation of its mean value in this neighborhood. This mean value can then be related to the state variables, which can also be calculated at each time instant. We thus discretize the four-dimensional space of the state variables d, ψ, ϕ, and v, in $\Omega = I \times J \times K \times M$ boxes, and calculate the mean value of $\delta\phi$ in each box. These multidimensional boxes are defined to ensure that the range of values for each state variable is fully covered, and that the number of bins for each variable is sufficient to capture the potential variations in the mean heading angle.

Then, Eq. (6.4) can be rewritten as a system of Ω equations and Ω unknowns,

$$\delta\phi^{\mathrm{S}}(d_i, \psi_j, \phi_k, \mathrm{v}_m) = \delta\phi_{ijkm}, \tag{6.6}$$

where $\delta\phi^{\mathrm{S}}$ is the unknown function, $(d_i, \psi_j, \phi_k, \mathrm{v}_m)$ is the representative node of the $ijkm$-box, and $\delta\phi_{ijkm}$ is the mean of the values of $\delta\phi$ that falls in the $ijkm$-box, that is, a *known* value calculated from the experimental data.

A crucial observation is that the intensity with which a fish reacts to a neighbor cannot depend on whether that neighbor is located to the right or left of the fish, nor on whether its misalignment is toward the left or the right. This observation imposes a symmetry constraint to the angle that the fish has to turn as a function of the relative state of a neighbor, specifically,

$$\delta\phi^{\mathrm{S}}(d, -\psi, -\phi, \mathrm{v}) = -\delta\phi^{\mathrm{S}}(d, \psi, \phi, \mathrm{v}). \tag{6.7}$$

Figure 6.4 illustrates this symmetry. The focal fish, shown in red, is expected to turn left in scenario (a) and to turn right in scenario (b), with the same angle magnitude but opposite direction. The vertical dashed line is the symmetry axis. Scenarios (c) and (d) show the same result if a horizontal axis of symmetry is considered.

Based on this observation, we hypothesize that the function $\delta\phi^{\mathrm{S}}$ can be expressed as the additive combination of products of functions of separate variables. There are only two ways for a product $g(\psi)h(\phi)$ to satisfy the symmetry condition (6.7): in one way, the first function is odd and the second is even, and in the other way, the first function is even and the second is odd. We thus decompose $\delta\phi^{\mathrm{S}}$ into the sum of exactly two functions, with no additional components, meaning that any social interaction is exclusively represented by one of these two forms:

$$\delta\phi^{\mathrm{S}}(d, \psi, \phi, \mathrm{v}) = \delta\phi_{\mathrm{Att}}(d, \psi, \phi, \mathrm{v}) + \delta\phi_{\mathrm{Ali}}(d, \psi, \phi, \mathrm{v}), \tag{6.8}$$

where

$$\delta\phi_{\mathrm{Att}}(d, \psi, \phi, \mathrm{v}) = f(d)\, g(\psi)\, h(\phi)\, l(\mathrm{v}), \tag{6.9}$$

$$\delta\phi_{\mathrm{Ali}}(d, \psi, \phi, \mathrm{v}) = \hat{f}(d)\, \hat{g}(\psi)\, \hat{h}(\phi)\, \hat{l}(\mathrm{v}). \tag{6.10}$$

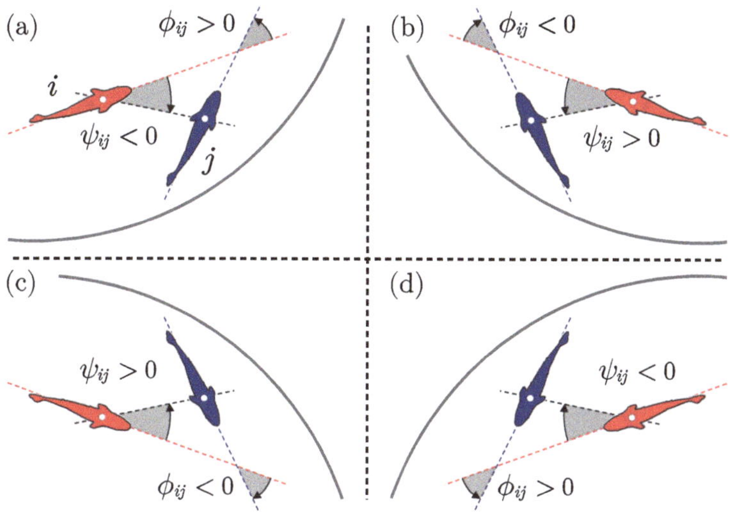

Fig. 6.4 Mirror symmetry property. Horizontal and vertical black dashed line are axes of symmetry. Scenarios (**b**) and (**c**) have the same probability of occurrence as scenarios (**a**) and (**d**). The fish turns the same absolute angle $|\delta\phi|$ in all scenarios, but the sign in (**a**) and (**d**) is opposite to the sign in (**b**) and (**c**): $\delta\phi^{(a)} = \delta\phi^{(d)} = -\delta\phi^{(b)} = -\delta\phi^{(c)}$. Note that (**d**) and (**c**) are respectively (**a**) and (**b**) rotated by an angle π

We thus *posit* that every interaction involving a product $g(\psi)h(\phi)$ that is odd in ψ and even in ϕ is an interaction of the *attraction* kind, and every interaction involving a product $\hat{g}(\psi)\hat{h}(\phi)$ that is even in ψ and odd in ϕ is an interaction of the *alignment* kind. This way, every information contained in the values of $\delta\phi_{ijkm}$ that verifies one of these two conditions is disentangled by our procedure and stored in the respective pairwise interaction function $\delta\phi_{\text{Att}}$ or $\delta\phi_{\text{Ali}}$.

Figure 6.5 shows the 2×2 possible states in which a pair of fish i and j can be found with respect to the two angles that define their relative state, ψ_{ij} and ϕ_{ij}, for the two types of interaction, attraction and alignment. The figure details the sign of the contribution of each angular function defining each interaction. Attraction contributes to making the fish turn right if the neighbor is at the right side of the fish. Thus, the contribution of $g(\psi_{ij})$ to the angle change $\delta\phi$ is negative when $\psi_{ij} < 0$. This is represented by a red wide arrow in Fig. 6.5a that points to the right (the decreasing or negative direction for angles). When the neighbor is on the left side, then $\psi_{ij} < 0$ and the red wide arrow points to the left in Fig. 6.5a'. In turn, the sign of the relative orientation ϕ_{ij} has no impact on the heading angle change, so the red arrow points in both directions in Fig. 6.5c, c'. A similar line of reasoning can be applied to alignment in Fig. 6.5d, d', b, b', respectively.

Most of the time the sign of the attraction is given by the sign of ψ_{ij}, and the sign of the alignment by the sign of ϕ_{ij}. This is because the even functions $\hat{g}(\phi_{ij})$ and $h(\psi_{ij})$ are in general positive. However, it can happen that one or two of these functions are negative in some specific interval of values of their respective variable.

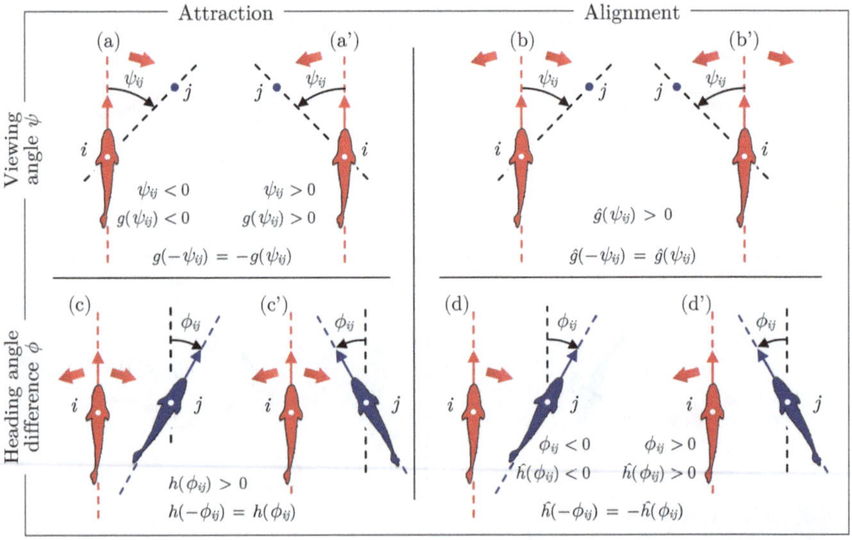

Fig. 6.5 Contributions and symmetry properties of the angular pairwise functions of the viewing angle ψ_{ij} and the heading angle difference ϕ_{ij} describing social interactions of the attraction and alignment kind. The focal fish i is shown in red. Blue circle in (**a, a', b, b'**) is the neighbor fish j, regardless of its orientation. The neighbor fish j is shown in blue in (**c, c', d, d'**) with a specific relative orientation with respect to the focal fish, regardless of its location. Wide red arrows in (**a, a', d, d'**) indicate the direction of the contribution of the interaction function to the heading angle change $\delta\phi$. Double wide arrows in (**b, b'**) and (**c, c'**) mean that the fish can turn right or left equally, the direction being determined by the other function in the product, $\hat{h}(\phi_{ij})$ and $g(\psi_{ij})$, respectively. Formulas indicate the sign and the parity of the interaction function. (**a, a'**) Odd contribution of the function of the viewing angle $g(\psi_{ij})$ to the attraction: same intensity but opposite sign depending on which left/right side the neighboring fish is located. (**b, b'**) Positive even contribution of $\hat{g}(\psi_{ij})$ to the alignment: same intensity and same sign independently of the relative position of the neighbor. (**c, c'**) Positive even contribution of $h(\phi_{ij})$ to the attraction: same value independently of the relative orientation of both fish. (**d, d'**) Odd contribution of $\hat{h}(\phi_{ij})$ to the alignment: same intensity but opposite sign depending on the sign of the relative orientation of both fish

This corresponds to specific situations, an example of which is given later in the case of *H. rhodostomus*.

Limiting the number of types of interactions to two simplifies both the characterization of the interaction functions and the description of the resulting behavior. This way, repulsion can be described as a *negative attraction*, as it happens when $f(d_{ij}) < 0$ at short distances, or as an alignment, from which it is mathematically indistinguishable.

Figure 6.6 shows three situations in which the change of direction of a fish can be interpreted as a repulsion. At time t^n, the fish i turns away from fish j by an angle $\delta\phi_i^n$, whose sign is the opposite to the sign of ψ_{ij}, and the same as the sign of ϕ_{ij}. This can be explained by (1) an attraction function $f(d)$ that is negative at

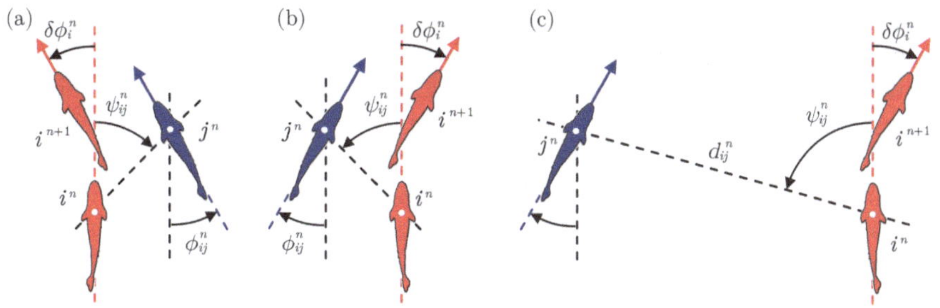

Fig. 6.6 Repulsion interpreted as alignment. Focal fish i position (in red) before and after reacting to a neighbor j (in blue). Index n indicates time iteration. (**a, b**) Typical scenarios where fish i turns away from j by an angle $\delta\phi_i^n$ such that $\text{sign}(\delta\phi_i^n) = -\text{sign}(\psi_{ij})$, suggesting a repulsion interaction. In (**a**), $\psi_{ij} < 0$ and $\phi_{ij} > 0$, so $g(\psi_{ij}) < 0$ and $h(\phi_{ij}) > 0$. As $\delta\phi_i^n > 0$, $f(d) < 0$ so it is indeed a *negative attraction*, the opposite as in Fig. 6.5a. Scenario (**b**) is a specular image of (**a**): $\psi_{ij} > 0$ and $\phi_{ij} < 0$, so $g(\psi_{ij}) > 0$ and $h(\phi_{ij}) > 0$. As $\delta\phi_i^n < 0$, we have $f(d) < 0$ as before. The angle turned is also such that $\text{sign}(\delta\phi_i^n) = \text{sign}(\phi_{ij})$, so the interaction in (**a, b**) can be viewed as an alignment as in Fig. 6.5d'. In (**a**), $\hat{g}(\psi_{ij}) > 0$ and $\hat{h}(\phi_{ij}) > 0$, so, assuming that $\hat{f}(d) > 0$, we have $\delta\phi_{\text{Ali}} > 0$. In (**b**), $\hat{g}(\psi_{ij}) > 0$ and $\hat{h}(\phi_{ij}) < 0$, so $\delta\phi_{\text{Ali}} < 0$, provided $\hat{f}(d) > 0$. The interaction in scenario (**c**) is more difficult to interpret as a repulsion because, although it is identical to (**b**), the distance between fish is much greater. It is therefore better described as alignment, and we extend this explanatory choice to scenarios (**a, b**)

short distances or (2) an alignment interaction, as in Fig. 6.5d', that dominates the attraction.

According to these definitions, the so-called short-range repulsion found in [25] for $d_{ij} < 2.5$ cm (less than one body length) can in fact be viewed as alignment.

We also emphasize that the interaction itself should not be confused with its effect: attraction increases cohesion (reduces distance between individuals), while alignment increases polarization (how much fish are aligned in the same direction). This distinction prevents the misconception that "alignment can be achieved through attraction and repulsion," which should be correctly stated as "polarization can be achieved independently of alignment."

Reconstruction Procedure

The reconstruction procedure then goes as follows.

Using the notation $f(d_i) = f_i$, $g(\psi_j) = g_j, \ldots, \hat{l}(\text{v}_m) = l_m$, (6.6) becomes a system of $I \times J \times K \times M$ equations with $D = 2(I + J + K + M)$ unknowns,

$$f_i \, g_j \, h_k \, l_m + \hat{f}_i \, \hat{g}_j \, \hat{h}_k \, \hat{l}_m = \delta\phi_{ijkm}. \tag{6.11}$$

Systems with more equations than unknowns are called *overdetermined systems* and rarely have a solution. One approach to assess this issue consists in minimizing the error with which the overdetermined system is satisfied, by a candidate solution that is updated iteratively. It is expected that natural phenomena give rise to iterative processes that provide information at each iteration, extracting it from the entangled average values of $\delta\phi$ and distributing it in a structured form among the pairwise interaction functions. Once the error function is smaller than a given tolerance, the iterative process is stopped and the solution is considered found. In the experimental data analyzed here, the error with which Eq. (6.11) is satisfied is very small. Thus, the essential information contained in $\delta\phi$ has been extracted, and its structure disentangled in two pairwise interaction functions, although some information remains hidden because the error is not zero; this corresponds, for example, to mixed or nonlinear components such as $\psi\phi$, $\sin(\psi)\sin(\phi)$, ψ^2, etc.

A typical error function is the sum of the weighted squared residuals,

$$\Delta(\mathbf{X}) = \sum_{i=1}^{I}\sum_{j=1}^{J}\sum_{k=1}^{K}\sum_{m=1}^{M} \epsilon_{ijkm}(f_i\, g_j\, h_k\, l_m + \hat{f}_i\, \hat{g}_j\, \hat{h}_k\, \hat{l}_m - \delta\phi_{ijkm})^2, \tag{6.12}$$

where $\mathbf{X}$ is the following vector of dimension D,

$$\begin{aligned}\mathbf{X} = (&f_1, \ldots, f_I, g_1, \ldots, g_J, h_1, \ldots, h_K, l_1, \ldots, l_M,\\ &\hat{f}_1, \ldots, \hat{f}_I, \hat{g}_1, \ldots, \hat{g}_J, \hat{h}_1, \ldots, \hat{h}_K, \hat{l}_1, \ldots, \hat{l}_M),\end{aligned}$$

and ϵ_{ijkm} is the number of data in the $ijkm$-box.

Weighting the regions with more data with ϵ_{ijkm} allows to preserve the structure of the dataset and the correlations between variables, resulting from the dynamics.

The minima of $\Delta(\mathbf{X})$ are obtained by finding the zeros of the gradient of Δ. Finding the zeros of a function in several dimensions is done using a descent method or other iterative methods. Here we use an iterative method that, thanks to the hypothesis of variable separability, takes advantage of the linearity of the equation in each component, allowing each one to be explicitly expressed in terms of the others. This relation is found as follows.

The gradient of Δ is zero if $\partial\Delta/\partial f_1 = 0$, $\partial\Delta/\partial f_2 = 0, \ldots, \partial\Delta/\partial l_M = 0$. For the component f_i, the partial derivative of Δ is

$$\frac{\partial\Delta}{\partial f_i} = 2\sum_{j=1}^{J}\sum_{k=1}^{K}\sum_{m=1}^{M} \epsilon_{ijkm}\, g_j\, h_k\, l_m(f_i\, g_j\, h_k\, l_m + \hat{f}_i\, \hat{g}_j\, \hat{h}_k\, \hat{l}_m - \delta\phi_{ijkm}),$$

so the condition $\partial\Delta/\partial f_i = 0$ is equivalent to

$$f_i = -\frac{\displaystyle\sum_{j=1}^{J}\sum_{k=1}^{K}\sum_{m=1}^{M} \epsilon_{ijkm}\, g_j h_k l_m(\hat{f}_i\hat{g}_j\hat{h}_k\hat{l}_m - \delta\phi_{ijkm})}{\displaystyle\sum_{j=1}^{J}\sum_{k=1}^{K}\sum_{m=1}^{M} \epsilon_{ijkm}(g_j h_k l_m)^2}. \tag{6.13}$$

The corresponding equations for the other components are derived in the same way and all have the same structure where each component is given by an explicit combination of the other components and the known values $\delta\phi_{ijkm}$ and ϵ_{ijkm}.

The resulting set of explicit equations for each component can be written in the following form:

$$\begin{aligned}
f_1 &= R_1(g_1, \ldots, g_J, h_1, \ldots, \hat{h}_K, \hat{l}_1, \ldots, \hat{l}_M), \\
f_2 &= R_2(g_1, \ldots, g_J, h_1, \ldots, \hat{h}_K, \hat{l}_1, \ldots, \hat{l}_M), \\
\vdots &= \vdots \\
f_I &= R_I(g_1, \ldots, g_J, h_1, \ldots, \hat{h}_K, \hat{l}_1, \ldots, \hat{l}_M), \\
g_1 &= R_{I+1}(f_1, \ldots, f_I, h_1, \ldots, \hat{h}_K, \hat{l}_1, \ldots, \hat{l}_M), \\
\vdots &= \vdots \\
\hat{g}_J &= R_{I+J}(f_1, \ldots, f_I, h_1, \ldots, \hat{h}_K, \hat{l}_1, \ldots, \hat{l}_M), \\
\vdots &= \vdots \\
\hat{l}_M &= R_D(f_1, \ldots, f_I, g_1, \ldots, \hat{g}_J, \hat{h}_1, \ldots, \hat{h}_K)).
\end{aligned}$$

This system can be written in a compact form as $\mathbf{X} = \mathbf{R}(\mathbf{X})$, where $\mathbf{R}$ is a function from $\mathbb{R}^D$ to $\mathbb{R}^D$, so that finding the minima of Δ reduces to find a fixed point $\mathbf{X_0}$ of the function $\mathbf{R}$, such that $\mathbf{R(X_0)} = \mathbf{X_0}$.

Under certain conditions,[1] fixed points can be found with a simple iterative method of the form $\mathbf{X_{n+1}} = \mathbf{R(X_n)}$, where n is the iteration number. As the dimension of the system D is very large, and in order to improve the stability of the recursion dynamics, it is convenient to use a relaxation method,

$$\mathbf{X_{n+1}} = \lambda\mathbf{X_n} + (1 - \lambda)\mathbf{R}(\mathbf{X_n}), \tag{6.14}$$

where $\lambda \in (0, 1)$ is the weight of the previous iteration in the value of the new iteration, averaged with what would have been the new iteration.

At the end, the procedure provides the values of each interaction function at the center of each box: $\{f_i\}_{i=1}^I$, $\{g_j\}_{j=1}^J$, ..., $\{\hat{l}_m\}_{m=1}^M$, and analytical forms of each one-variable function can be obtained, based on physical principles and specific observations. For example, the intensity of social interactions is expected to decay with the distance between fish d. Physical considerations show that this decay is like $1/d$ for large d [25], something that needs to be integrated in the analytical expressions of $f(d)$ and $\hat{f}(d)$. Angular functions are especially suitable to be described by Fourier series of the appropriate parity. The next section shows a

[1] The function $\mathbf{R}$ must be contractive and the initial value of the iterations must be sufficiently close to a fixed point.

specific case where these analytical expressions are provided for each interaction function.

Extension of the Model to Larger Groups
The analytical expressions of the pairwise interaction functions provided by the procedure allow us to evaluate at each instant the *influence* that each neighboring fish exerts on a focal fish. In groups of more than two fish, it is possible to rank the neighbors according to the magnitude of this influence, defined as the absolute contribution of the neighbor to the change of direction of the fish, $|\delta\phi_{ij}^{\mathrm{S}}|$, and determine the number and identity of the neighbors necessary to collect all the relevant social information to make a decision at the next kicking time. In *H. rhodostomus*, we found that paying attention to simply the two *most influential neighbors* is sufficient to reproduce the observed behavioral patterns in groups of five fish [31].

6.3 Reconstruction of Social Interactions in the Species *Hemigrammus rhodostomus*

The model proposed in Sect. 6.2 is based on the assumption that social interactions are combined in an additive form, Eq. (6.4), on the observation of a physical property of symmetry, Eq. (6.7), and on the hypothesis that the two functions $\delta\phi_{\mathrm{Att}}$ and $\delta\phi_{\mathrm{Ali}}$ accounting for the attraction and alignment interactions depend on the state variables d_{ij}, ψ_{ij}, and ϕ_{ij} in a decoupled way; Eqs. (6.9)–(6.10). Step 4 in Fig. 6.1 provides the shape and the intensity of the interaction functions, but do not determine their analytical expressions. Here, we describe this last step for the specific case of pairs of *H. rhodostomus* swimming in a circular tank.

Making use of the properties of parity described in the previous section, the interaction functions can be rewritten from Eqs. (6.9)–(6.10) as follows:

$$\delta\phi_{\mathrm{Att}}(d_{ij},\psi_{ij},\phi_{ij}) = F_{\mathrm{Att}}(d_{ij})\, O_{\mathrm{Att}}(\psi_{ij})\, E_{\mathrm{Att}}(\phi_{ij}), \tag{6.15}$$

$$\delta\phi_{\mathrm{Ali}}(d_{ij},\psi_{ij},\phi_{ij}) = F_{\mathrm{Ali}}(d_{ij})\, E_{\mathrm{Ali}}(\psi_{ij})\, O_{\mathrm{Ali}}(\phi_{ij}), \tag{6.16}$$

where function names "O" and "E" stand for "odd" and "even," respectively.

The six unknown interaction functions are then tabulated on a grid with typically 30 boxes per function, leading to effectively 180 fitting parameters, a number much smaller than the number of kicks experimentally recorded (2×10^5 in our experiments with *H. rhodostomus*). Then these parameters are determined by minimizing the corresponding error function (see Sect. 6.2).

Figure 6.7a, c, e shows the six sets of scattered dots resulting from the reconstruction procedure and for which analytical approximations have to be determined. These approximations are not a mere interpolation of nodes; instead, they are again

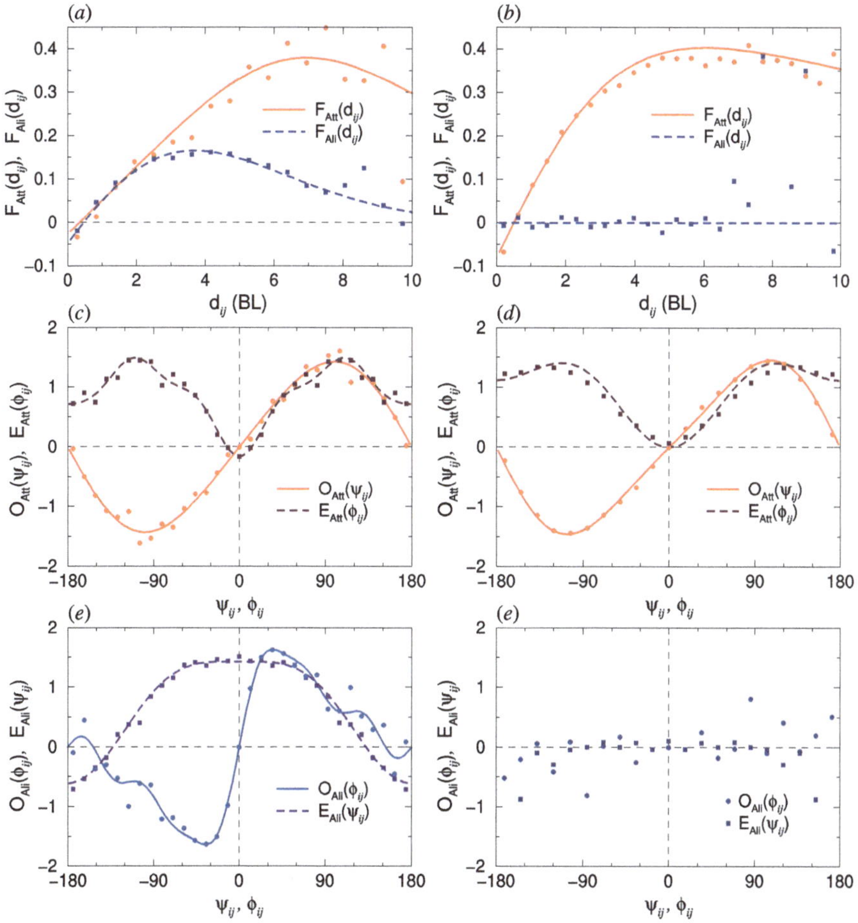

Fig. 6.7 (**a**, **c**, **e**) Discrete values (dots) extracted from the experimental data with the procedure described in Sect. 6.3 and analytical approximations (solid lines) used to interpolate these values and describe the social interactions of (**a**, **c**) attraction and (**a**, **e**) alignment in pairs of *H. rhodostomus* swimming in a circular tank of radius 25 cm. (**b**, **d**, **f**) Discrete values (dots) corresponding to a dataset produced by the numerical simulations of the model using the interaction functions defined by the analytical functions depicted with solid lines, without alignment interaction, $F_{Ali} \equiv 0$

based on the observation of the physical phenomenon and borrowed from very specific expressions that convey a precise physical meaning.

Let us observe the shape of the scattered dots in Fig. 6.7a. One can notice that both the intensity of attraction and alignment exhibit similar shapes: they have low (even negative) values when the distance between individuals is short, increase to a maximum, and finally decay to zero. The maximum value for attraction is higher than that for alignment, it is reached at larger distances between individuals, and it decays slower compared to the alignment.

Table 6.1 Parameter values for the attraction and alignment strength functions in *H. rhodostomus*

Species	BL (m)	γ_{Att}	d_{Att} (m)	l_{Att} (m)	γ_{Ali}	d_{Ali} (m)	l_{Ali} (m)	α
H. rhodostomus	0.03	0.026	0.01	0.27	0.047	0.014	0.14	1.7

These features can be converted into mathematical expressions as follows:

$$F_{\mathrm{Att}}(d) = \gamma_{\mathrm{Att}} \left(\frac{d}{d_{\mathrm{Att}}} - 1\right) \left[1 + \left(\frac{d}{l_{\mathrm{Att}}}\right)^4\right]^{-1}, \tag{6.17}$$

$$F_{\mathrm{Ali}}(d) = \gamma_{\mathrm{Ali}} \left(\frac{d}{d_{\mathrm{Ali}}} - 1\right) \exp\left[-\left(\frac{d}{l_{\mathrm{Ali}}}\right)^{\alpha}\right], \tag{6.18}$$

where γ_{Att} and γ_{Ali} are the (dimensionless) intensities of the attraction and alignment interactions, d_{Att} and d_{Ali} are the distance below which attraction and alignment change sign and become repulsion and anti-alignment respectively, and l_{Att} and l_{Ali} are the ranges of each interaction (the higher the value, the longer the range) .

Having dimensionless factors γ_{Att} and γ_{Ali} facilitates direct comparison between intensities of social interactions and with the intensity of the interaction with obstacles $\delta\phi_{\mathrm{w}}$ and the spontaneous decision term $\delta\phi_{\mathrm{R}}$ in Eq. (6.4).

Table 6.1 shows the parameter values corresponding to specific experiments carried out with *H. rhodostomus*. Note that the values have been adapted from those in [25], according to the slightly more accurate expressions used here.

The four angular functions are more easy to reproduce, by using simple Fourier expansions of the appropriate parity. Keeping a reduced number of modes, we found the following ones sufficiently accurate to fit the scattered dots of Fig. 6.7c, e:

$$\begin{aligned}
O_{\mathrm{Att}}(\psi) &= 1.41 \sin(\psi)[1 - 0.173 \cos(\psi)],\\
E_{\mathrm{Att}}(\phi) &= 0.895[1 - 0.38 \cos(\phi) - 0.576 \cos(2\phi) - 0.1 \cos(4\phi)\\
&\qquad - 0.1 \cos(5\phi)],\\
E_{\mathrm{Ali}}(\psi) &= 0.67[1 + 1.52 \cos(\psi) - 0.39 \cos(2\psi)],\\
O_{\mathrm{Ali}}(\phi) &= 1.41 \sin(\phi)[1 + 1.45 \cos(\phi) + 0.37 \cos(2\phi) + 0.48 \cos(3\phi)\\
&\qquad - 0.04 \cos(4\phi) + 0.2 \cos(5\phi)].
\end{aligned}$$

Following step 5 of our procedure (Fig. 6.1), these analytical expressions are implemented in the model derived in Sect. 6.2. Then, numerical simulations are performed and compared with the known experiments, and predictions are made that must be verified a posteriori. The agreement of our numerical simulations with the experimental observations is shown in the PDF of the main observables characterizing the collective behavior of pairs of individuals, in Fig. 6.3c, d, e.

Attraction and alignment are of the same order of magnitude at the distance of separation fish more frequently are, $d_{ij} \approx 2\,\text{BL}$ (Fig. 6.3). Beyond this distance, attraction is stronger than alignment and reaches a maximum at ~0.38, approximately twice higher than that of alignment, and over a larger range of ~7 BL, again approximately twice the one of alignment. Due to the high attraction, situations where fish are far from each other are rare, resulting in a limited data availability for long distances and a higher uncertainty in the scattered dots. However, the figure suggests that alignment decreases to zero at approximately 10 BL, whereas attraction remains quite high.

In *H. rhodostomus*, a fish i is subject to a stronger attraction when the other fish j is at its right or left side ($|O_{\text{Att}}|$ reaches its highest values when $\psi_{ij} \approx \pm 90°$; see Fig. 6.7c) and moves more or less perpendicular to it (E_{Att} is higher when $\phi_{ij} \approx 90$–$100°$, Fig. 6.7c). Alignment is stronger when the other fish is in front (E_{Ali} is higher when $|\psi_{ij}| < 80°$) and vanishes when fish are already almost aligned ($O_{\text{Ali}} \approx 0$ when $|\phi_{ij}| < 30°$; see Fig. 6.7e) or when they are practically anti-aligned ($|\phi_{ij}| > 120°$).

Notably, the two even angular functions are negative in some interval. For the attraction function, this interval is small ($E_{\text{Att}} < 0$ if $|\phi_{ij}| < 15°$) and corresponds to a weak repulsion when fish are practically aligned. In turn, the alignment function has a larger negative interval ($E_{\text{Ali}} < 0$ if $|\psi_{ij}| > 130°$), suggesting a specific behavior, such as the initiation of a U-turn maneuver when the other fish is behind.

Another benefit of having an analytical model is that one can explore what happens in different theoretical scenarios and analyze particular cases. High polarization at the group level can be achieved by combining attraction and repulsion at the individual level, without the need of alignment. Our method allows us not only to disentangle the combined effects of attraction and alignment interactions when they both exist, but also to detect the absence of alignment from the analysis of the trajectories. To demonstrate this capability of the method, we removed the alignment interaction from the *H. rhodostomus* model and produced simulated trajectories. We then reconstructed the interaction functions using this dataset, finding that the procedure is able to reveal the absence of alignment interaction.

Figure 6.7b, d, f shows the four analytical functions used in the model $F_{\text{Att}}(d)$, $O_{\text{Att}}(\psi)$, $E_{\text{Att}}(\phi)$, and $F_{\text{Ali}}(d) \equiv 0$, with solid lines, and the six sets of scattered dots found by the reconstruction procedure, $\{f^i_{\text{Att}}\}_{i=1}^I$, $\{O^j_{\text{Att}}\}_{j=1}^J$, $\{E^k_{\text{Att}}\}_{k=1}^K$, $\{f^i_{\text{Ali}}\}_{i=1}^I$, $\{O^j_{\text{Ali}}\}_{j=1}^J$, $\{E^k_{\text{Ali}}\}_{k=1}^K$. Not only the three functions of the attraction interaction are well reproduced, but the function for the intensity of alignment is also accurately captured, showing a null value except at long distances where the precision is lower due to the smaller number of data, as the fish are close together because of the attraction interaction.

Minimizing the Error Between Data and a Product Form

Let us consider a $N \times M$ two-dimensional array a_{ij}, for instance, built after averaging some quantity obtained from experiments on a $N \times M$ grid. We look for the best approximation of a_{ij} under a product of separable form, $a_{ij} \approx x_i \times y_j$,

$i = 1, \ldots, N$, $j = 1, \ldots, M$ (see the system (6.11), in the more general case when more than two dimensions are involved). Note that under the product form, the number of unknown variables is only $N + M$, compared to the much larger size $N \times M$ of the original array of data a_{ij}.

In order to quantify how accurately the product describes the data, we define the quadratic error function $\Delta : \mathbb{R}^{N+M} \to \mathbb{R}$ defined by

$$\Delta(x_1, \ldots, x_N, y_1, \ldots, y_M) = \sum_{i=1}^{N} \sum_{j=1}^{M} \epsilon_{ij} (x_i y_j - a_{ij})^2, \tag{6.19}$$

where ϵ_{ij} is the number of data that falls in the box ij. For systems that derive from physical and biological phenomena, it is fundamental to modulate the contribution of each cell to give a higher weight to the more frequent values and to preserve actual correlations arising between the variables.

The goal is thus to find a set of values $x_1, \ldots, x_N, y_1, \ldots, y_M$ that minimizes the error Δ, a positive quantity. If $\Delta = 0$, then $x_i y_j = a_{ij}$ for all i and j, and the original data have exactly a product form.

Finding the minima of a function requires finding the zeros of its derivative, which, in high dimensions, is the *gradient vector* $\nabla \Delta$, given by the partial derivatives of Δ with respect to its components.

Differentiating (6.19) with respect to, e.g., x_k, gives

$$\frac{\partial \Delta}{\partial x_k} = \sum_{i=1}^{N} \sum_{j=1}^{M} \frac{\partial}{\partial x_k} \Big[\epsilon_{ij} (x_i y_j - a_{ij})^2 \Big] = 2 \sum_{j=1}^{M} \epsilon_{ij} y_j (x_k y_j - a_{kj}), \tag{6.20}$$

where we have used the fact that $x_i y_j$ does not depend on x_k if $i \neq k$. Then,

$$\frac{\partial \Delta}{\partial x_k} = 2 x_k \sum_{j=1}^{M} \epsilon_{ij} y_j^2 - 2 \sum_{j=1}^{M} \epsilon_{ij} y_j a_{kj}, \tag{6.21}$$

and the conditions $\partial \Delta / \partial x_i = 0$ and $\partial \Delta / \partial y_j = 0$ for all i and j can be rewritten as

$$x_i = \frac{\sum_{j=1}^{M} \epsilon_{ij} y_j a_{ij}}{\sum_{j=1}^{M} \epsilon_{ij} y_j^2}, \quad i = 1, \ldots, N, \qquad y_i = \frac{\sum_{i=1}^{M} \epsilon_{ij} x_i a_{ij}}{\sum_{i=1}^{M} \epsilon_{ij} x_i^2}, \quad j = 1, \ldots, M. \tag{6.22}$$

The result is a system of $N + M$ equations and unknowns that can be solved with different methods. In the case of systems of high dimensions arising in social interaction analysis, iterative methods are often used.

6.4 Conclusions

Behavioral biology has recently evolved into a 'big data science,' driven by advances in imaging and tracking technologies that produce massive datasets on animal trajectories [32–34]. These innovations have transformed the way we observe and describe both individual and collective animal behaviors, strongly enhancing the diversity and precision of available data. In mathematical terms, these data, once processed, are trajectories that represent the movement patterns of individuals. These data are ideally suited for data-based modeling, which can uncover how local interaction rules generate collective motion and other group-level properties across species (swarms of insects, schools of fish, flocks of birds, herds of ungulates, or human crowds) [1, 3, 35, 36].

We have presented a general procedure that, in the framework of a mathematical model, allows to reconstruct these interactions and unveil the mechanisms at play. This procedure includes several steps. Once the experimental trajectories are obtained, individual and collective state variables are defined and a mathematical model is designed. The structure of the model depends on the specific kind of movement. Intermittent "burst-and-coast" swimming mode is adequately modeled by means of a discrete-time model, which returns the heading angle changes after each burst and calculate the effects of the environment for the next step. The model includes the equation of motion of each individual interacting with its environment (obstacles and other individuals) and considers relevant variables such as the instantaneous distance and orientation with respect to the wall and to other individuals. Assuming a spontaneous motion with no directional preference, each interaction component contributes additively, assuming that the effects of the interactions with wall and other individuals were independent.

We focused on the reconstruction of the social interactions between individuals, considering them as pairwise interactions. A crucial observation is that these interactions are symmetric: the intensity of the interaction cannot depend on whether a neighbor is on one side or the other (given by the viewing angle ψ_{ij}) or misaligned in one direction or the other (given by the relative heading angle ϕ_{ij}). This symmetry imposes a constraint on the angular functions and reduces the types of interactions to two: attraction and alignment. By making the strong hypothesis that the interaction functions are products of single-variable functions, the aforementioned observation allows us to determine the parity of each function. In the attraction interaction, the function of the viewing angle is odd, whereas the function of the relative heading angle is even. In turn, it is the opposite in alignment interaction: the function is even on ψ_{ij} and odd on ϕ_{ij}. All other types of interactions can be reduced to these two basic forms. For instance, the angular variation observed when a fish is repelled by another individual (e.g., to avoid collision) corresponds to a short-range alignment interaction. Except for their parity, the unknown interaction functions are not predetermined, and there are no assumptions about their form or intensity.

Then each function is tabulated on a grid, with values at each point serving as fitting parameters. These parameters are determined by minimizing the mean squared error between the model's predictions and the experimental data. This minimization is performed using iterative methods or gradient descent. Once determined, the interaction functions are fitted and represented by simple analytical forms with the extracted shape and intensity, yielding an explicit model that can be implemented numerically and studied mathematically [37].

Indeed, the model reproduces the experimental results and can be used to predict behaviors in new situations like changes of illumination in the environment or swimming in narrow corridors. Moreover, the procedure is robust against noise in the data and can measure spontaneous fluctuations in speed and heading angles. It requires minimal computational power and provides general insights into the interactions of the studied species. Additionally, it avoids issues associated with multidimensional force maps, which can be distorted by correlations among variables [24].

The main limitations of this procedure are the assumptions of additive contributions and the product of single-variable functions. However, this approach allows for a reduced number of fitting parameters while still capturing the relevant features of complex interactions and yielding an accurate and predictive model.

Ultimately, this methodology results in a concise and explicit model that helps understand and explain various experimental features and collective behaviors, with a predictive power that multidimensional force maps cannot offer.

Acknowledgments G.T., R.E. and C.S. were supported by Agence Nationale de la Recherche (ANR-20-CE45- 0006-1). G.T. also gratefully acknowledges the Indian Institute of Science for the support via the Infosys Visiting Chair Professor at the Centre for Ecological Sciences, IISc, Bengaluru. The funders had no role in the study design, data collection and analysis, decision to publish, or manuscript preparation.

References

1. Camazine, S., Deneubourg, J.-L., Franks, N., Sneyd, J., Theraulaz, G., Bonabeau, E.: Self-Organization in Biological Systems. Princeton Studies in Complexity. Princeton University Press, Princeton (2001)
2. Sumpter, D.J.T.: Collective Animal Behavior. Princeton University Press, Princeton (2010)
3. Lopez, U., Gautrais, J., Couzin, I.D., Theraulaz, G.: From behavioural analyses to models of collective motion in fish schools. Interface Focus **2**(6), 693–707 (2012)
4. Vicsek, T., Zafeiris, A.: Collective motion. Phys. Rep. **517**, 71–14 (2012)
5. Deutsch, A., Theraulaz, G., Vicsek, T.: Collective motion in biological systems. Interface Focus **2**, 689–692 (2012)
6. Strandburg-Peshkin, A., Farine, D.R., Crofoot, M.C., Couzin, I.D.: Habitat and social factors shape individual decisions and emergent group structure during baboon collective movement. Elife **6**, e19505 (2017)

7. Branson, K., Robie, A.A., Bender, J., Perona, J., Dickinson, M.H.: High-throughput ethomics in large groups of Drosophila. Nat. Methods **6**, 451 (2009)
8. Straw, A.D., Branson, K., Neumann, T.R., Dickinson, M.H.: Multi-camera real-time three-dimensional tracking of multiple flying animals. J. R. Soc. Interface **8**, 395–409 (2011)
9. Dell, A.I., Bender, J.A., Branson, K., Couzin, I.D., de Polavieja, G.G., Noldus, L.P.J.J, et al.: Automated image-based tracking and its application in ecology. Trends Ecol. Evol. **29**(7), 417–428 (2014)
10. Pérez-Escudero, A., Vicente-Page, J., Hinz, R.C., Arganda, S., de Polavieja, G.G.: idTracker: tracking individuals in a group by automatic identification of unmarked animals. Nat. Methods **11**, 743–748 (2014)
11. Walter, T., Couzin, I.D.: TRex, a fast multi-animal tracking system with markerless identification, and 2D estimation of posture and visual fields. Elife **10**, e64000 (2021)
12. Ballerini, M., Cabibbo, N., Candelier, R., Cavagna, A., Cisbani, E., Giardina, I., et al.: Empirical investigation of starling flocks: a benchmark study in collective animal behaviour. Anim. Behav. **76**, 201–215 (2008)
13. Francisco, F.A., Nührenberg, P., Jordan, A.: High-resolution, non-invasive animal tracking and reconstruction of local environment in aquatic ecosystems. Mov. Ecol. **8**, 27 (2020)
14. Haalck, L., Mangan, M., Wystrach, A., Clement, L., Webb, B., Risse, B.: CATER: combined animal tracking & environment reconstruction. Sci. Adv. **9**, eadg2094 (2023)
15. Lukeman, R., Li, Y.-X., Edelstein-Keshet, L.: Inferring individual rules from collective behavior. Proc. Natl. Acad. Sci. USA **107**(28), 12576 (2010)
16. Gautrais, J., et al.: Deciphering interactions in moving animal groups. PLoS Comput. Biol. **8**(9), e1002678 (2012)
17. Heras, F.J.H., Romero-Ferrero, F., Hinz, R.C., de Polavieja, G.G.: Deep attention networks reveal the rules of collective motion in zebrafish. PLoS Comput. Biol. **15**, e1007354 (2019)
18. Zienkiewicz, A.K., Ladu, F., Barton, D.A.W., Porfiri, M., Di Bernardo, M.: Data-driven modelling of social forces and collective behaviour in zebrafish. J. Theor. Biol. **443**, 39–51 (2018)
19. Katz, Y., Tunstrøm, K., Ioannou, C.C., Huepe, C., Couzin, I.D.: Inferring the structure and dynamics of interactions in schooling fish. Proc. Natl. Acad. Sci. USA **108**(46), 18720–18725 (2011)
20. Herbert-Read, J.E., Perna, A., Mann, R.P., Schaerf, T.M., Sumpter, D.J.T., Ward, A.J.W.: Inferring the rules of interaction of shoaling fish. Proc. Natl. Acad. Sci. USA **108**(46), 18726–18731 (2011)
21. Pettit, B., Perna, A., Biro, D., Sumpter, D.J.T.: Interaction rules underlying group decisions in homing pigeons. J. Roy. Soc. Interface 10(89), 20130529 (2013)
22. Mudaliar, R.K., Schaerf, T.M.: An examination of force maps targeted at orientation interactions in moving groups. PLoS One **18**, e0286810 (2023)
23. Puy, A., Gimeno, E., Torrents, J., Bartashevich, P., Miguel, M.C., Pastor-Satorras, R., Romanczuk, P.: Selective social interactions and speed-induced leadership in schooling fish. Proc. Natl. Acad. Sci. USA **121**(18), e2309733121 (2024)
24. Escobedo, R., Lecheval, V., Papaspyros, V., Bonnet, F., Mondada, F., Sire, C., Theraulaz, G.: A data-driven method for reconstructing and modelling social interactions in moving animal groups. Philos. Trans. R. Soc. B **375**(1807), 20190380 (2020)
25. Calovi, D.S., Litchinko, A., Lecheval, V., Lopez, U., Pérez-Escudero, A., Chaté, H., Sire, C., Theraulaz, G.: Disentangling and modeling interactions in fish with burst-and-coast swimming reveal distinct alignment and attraction behaviors. PLoS. Comput. Biol. **14**(1), e1005933 (2018)
26. Xue, T., Li, X., Lin, G., Escobedo, R., Han, Z., Chen, X., et al.: Tuning social interactions' strength drives collective response to light intensity in schooling fish. PLoS Comput. Biol. **19**(11), e1011636 (2023)
27. Partridge, B.L.: Internal dynamics and the interrelations of fish in schools. J. Comput. Physiol. **144**, 313–325 (2009)

28. McKee, A., Soto, A.P., Chen, P., McHenry, M.J.: The sensory basis of schooling by intermittent swimming in the rummy-nose tetra (*Hemigrammus rhodostomus*). Proc. R. Soc. B **287**, 20200568 (2020)
29. Ashraf, I., Van Wassenbergh, S., Verma, S.: Burst-and-coast swimming is not always energetically beneficial in fish (*Hemigrammus bleheri*). Bioinspir Biomim **16**, 016002 (2021)
30. Li, G., Ashraf, I., François, B., Kolomenskiy, D., Lechenault, F., Godoy-Diana, R., Thiria, B.: Burst-and-coast swimmers optimize gait by adapting unique intrinsic cycle. Commun. Biol. **4**, 40 (2021)
31. Lei, L., Escobedo, R., Sire, C., Theraulaz, G.: Computational and robotic modeling reveal parsimonious combinations of interactions between individuals in schooling fish. PLoS Comput. Biol. **16**(3), e1007194 (2020)
32. Reiser, M.: The ethomics era? Nat. Methods **6**, 413–414 (2009)
33. Bentley, R.A., O'Brien, M.J., Brock, W.A.: Mapping collective behavior in the big-data era. Behav. Brain Sci. **37**(1), 63 (2014)
34. Geissmann, Q., García Rodríguez, L., Beckwith, E.J., Gilestro, G.F.: Rethomics: An R framework to analyse high-throughput behavioural data. PLoS ONE **14**(1), e0209331 (2019)
35. Sumpter, D.J.T.: The principles of collective animal behaviour. Philos. Trans. R. Soc. B **361**, 5–22 (2006)
36. Moussaïd, M., Garnier, S., Theraulaz, G., Helbing, D.: Collective information processing in swarms, flocks and crowds. Topics Cogn. Sci. **1**, 469–497 (2009)
37. Wang, W., Escobedo, R., Sánchez, S., Sire, C., Han, Z., Theraulaz, G.: The impact of individual perceptual and cognitive factors on collective states in a data-driven fish school model. PLoS Comput. Biol. **18**, e1009437 (2022)

Chapter 7
Spatial Distributions for Animal Movement Processes

Christen H. Fleming, William F. Fagan, Jesse M. Alston, Gayatri Anand, Nozomu Hirama, Erika Y. Lin, Ricardo Martinez-Garcia, Michael J. Noonan, and Justin M. Calabrese

7.1 Introduction

Spatial distributions, $p(\mathbf{r})$, represent a relative frequency, p, at a specified location, $\mathbf{r} = (x, y)$ in two dimensions. Here, we cover a number of biologically relevant spatial distributions, and related concepts, that derive from movement processes—including home ranges, territories, and corridors (Table 7.1). We pay special attention to the frequent conflation between prediction and estimation and the commonly misapplied biological concepts of "space use" and "utilization"; and we

---

C. H. Fleming (✉) · N. Hirama · E. Y. Lin
University of Central Florida, Orlando, FL, USA
e-mail: christen.fleming@ucf.edu; nozomu.hirama@ucf.edu; erika.lin@ucf.edu

W. F. Fagan · G. Anand
University of Maryland, College Park, MD, USA
e-mail: bfagan@umd.edu; gayatri9@umd.edu

J. M. Alston
University of Arizona, Tuscon, AZ, USA
e-mail: jmalston@arizona.edu

R. Martinez-Garcia
Center for Advanced Systems Understanding (CASUS), Helmholtz-Zentrum Dresden-Rossendorf (HZDR), Görlitz, Germany

ICTP South American Institute for Fundamental Research & Instituto de Física Teórica, Universidade Estadual Paulista - UNESP, São Paulo, Brazil
e-mail: r.martinez-garcia@hzdr.de

M. J. Noonan
University of British Columbia, Kelowna, BC, Canada
e-mail: michael.noonan@ubc.ca

L. Giuggioli, P. K. Maini (eds.), *The Mathematics of Movement*,
https://doi.org/10.1007/978-3-032-00873-2_7

Table 7.1 Spatial distributions covered in this chapter.

Section	Distribution	Key References
7.2	Occurrence	[1, 2]
7.3	Range	[3–5]
7.3.2	Population Range	[6]
7.3.3	Corridor	[7]
7.3.4	Territory	[8]
7.4.1	Encounter	[9]
7.4.2	Revisitation	This work

delineate stronger linkages between important biological concepts and well-defined mathematical objects.

We start our coverage in Sect. 7.2 with distributions that interpolate unknown locations within a dataset and proceed to distributions that extrapolate future (or past) locations in Sect. 7.3. This order is anti-chronological, yet pedagogical, and we hope it will make the distinction between these two classes of distributions obvious in retrospect. Finally, in Sect. 7.4, we consider other biologically interesting spatial distributions that are derived from movement processes, but do not directly relate to the expected locations of the individual movers.

7.2 Occurrence Distributions: Where Has the Animal Gone?

In this section, we consider the task of the interpolative prediction of unknown locations. In the language of probability theory, random variables, such as the location process, $\mathbf{r}(t)$, are *predicted* at times, t, where $\mathbf{r}(t)$ is unknown. Predictions can be of the future, the past, or the present—before, after, or within the sampling period of the data. When the data contain substantial measurement error, predictions can meaningfully coincide with the sampled times.

By far, the most ubiquitous class of model employed for the interpolative prediction of animal movement data consists of Gaussian stochastic processes (GSPs) [1, 10, 11], where the location process of an individual mover is entirely characterized by its mean function

$$\mathrm{E}[\mathbf{r}(t)] = \boldsymbol{\mu}(t)\,, \tag{7.1}$$

J. M. Calabrese
Center for Advanced Systems Understanding (CASUS), Helmholtz-Zentrum Dresden-Rossendorf (HZDR), Görlitz, Germany

Department of Ecological Modelling, Helmholtz Centre for Environmental Research-UFZ, Leipzig, Germany

Department of Biology, University of Maryland, College Park, MD, USA
e-mail: j.calabrese@hzdr.de

and autocovariance function

$$\mathrm{COV}[\mathbf{r}(t), \mathbf{r}(t')] = \sigma(t, t') . \tag{7.2}$$

We will refer to the true distributions as p and their Gaussian approximations as $p_{\mathcal{N}}$. For a mathematical overview of the most commonly used GSPs in movement ecology, see [12, 13]. These include the "Brownian motion" (BM) process[15], the spatially constrained "Ornstein–Uhlenbeck" (OU) process [16], the continuous-velocity "integrated Ornstein–Uhlenbeck" (IOU) process [17], and the spatially constrained and continuous-velocity "Ornstein–Uhlenbeck Foraging" (OUF) processes [18].

Obviously, a realistic movement process will not be Gaussian, but for the purpose of interpolation, our model only needs to capture an appropriate number of continuous derivatives to obtain asymptotic efficiency [19]. So, for instance, IOU motion can be asymptotically efficient on finely sampled movement data that show obvious signs of finite speeds and directed motion, while Brownian motion can be asymptotically efficient on coarser data without these features.

For a GSP model, the prediction of the unknown location $\mathbf{r}(t)$ at some time t is then given by the conditional density

$$p_{\mathcal{N}}(\mathbf{r}(t)|\mathrm{DATA}) = \frac{p_{\mathcal{N}}(\mathbf{r}(t), \mathrm{DATA})}{p_{\mathcal{N}}(\mathrm{DATA})} , \tag{7.3}$$

$$= \mathcal{N}(\mathbf{r}(t)|\mathbf{M}(t), \boldsymbol{\Sigma}(t)) , \tag{7.4}$$

with a conditional mean vector, $\mathbf{M}(t) = \mathrm{E}[\mathbf{r}(t)|\mathrm{DATA}]$, and conditional covariance matrix, $\boldsymbol{\Sigma}(t) = \mathrm{COV}[\mathbf{r}(t)|\mathrm{DATA}]$ given by the Kriging relations [1, 20], or, equivalently, by the Kalman filter [13, 21], for a dataset consisting of a set of locations, $\mathbf{r}(t_i)$, tracked at times, t_i, indexed by $i \in \{1, 2, 3, \cdots, n\}$. Figure 7.1 provides a one-dimensional example. $\mathbf{M}(t)$ and $\boldsymbol{\Sigma}(t)$ can be formally calculated by applying Bayes' theorem to the joint density, $p_{\mathcal{N}}(\mathbf{r}(t), \mathrm{DATA})$, and marginal density, $p_{\mathcal{N}}(\mathrm{DATA})$, which are both multivariate normal distributions straightforwardly constructed from the mean (7.1) and autocovariance (7.2) functions evaluated at the sampled times and some time t [1]. Both cumulants of the conditional density, $\mathbf{M}(t)$ and $\boldsymbol{\Sigma}(t)$, are, therefore, functions of the data and of the model's mean and autocovariance functions. R packages that can efficiently calculate these probabilistic interpolations include `crawl` [11, 22], with an IOU model, `aniMotum` [23], with an integrated BM model, and `ctmm` with a selected continuous-time movement model [24, 25].

7.2.1 Derivation

The *occurrence distribution* is focused on representing the individual's movement path during the sampling period, when we can interpolate the movement path. Most

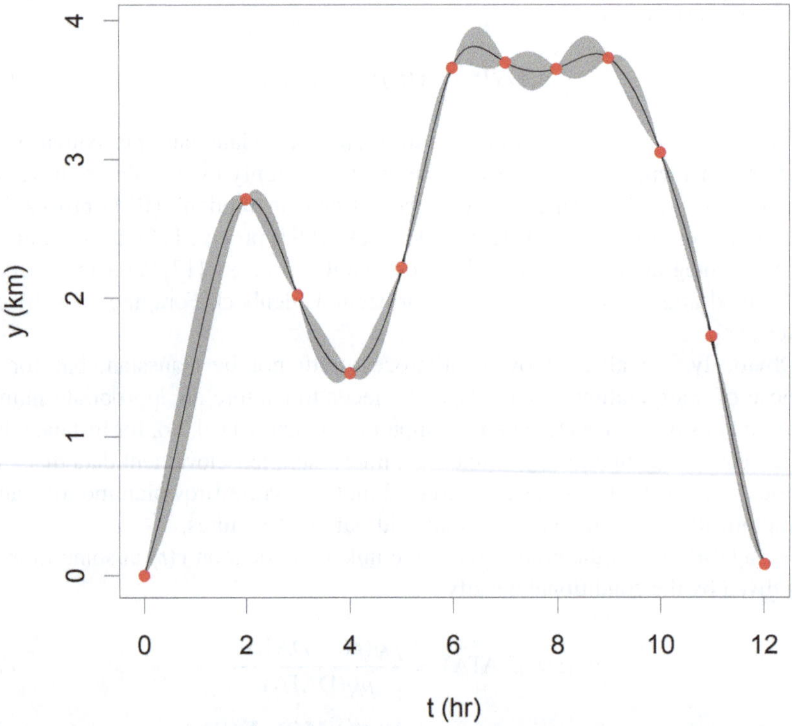

Fig. 7.1 Latitudinal predictions of 12 hours of African buffalo (*Syncerus caffer*) [14], with location data in red, best predicted locations in black, and prediction intervals in gray

simply, the occurrence distribution is calculated from the posterior density of a random time within the sampling period:

$$p_{\mathrm{OD}}(\mathbf{r}|t_1, t_n) = \frac{1}{t_n - t_1} \int_{t_1}^{t_n} p_{\mathcal{N}}(\mathbf{r}(t)|\mathrm{DATA})\, dt \,, \tag{7.5}$$

where t_1 and t_n denote the first and last sampled times. The result is a smooth distribution centered on our best predictions of the movement path, $\mathbf{M}(t)$, with dispersions given by our uncertainties, $\boldsymbol{\Sigma}(t)$. Figure 7.2 provides an example. R packages that can calculate occurrence distributions include `move` [26], with a BM model, and `ctmm` with a selected continuous-time movement model [24, 25].

7.2.2 *Space and Resource Utilization*

In the limit of finely (and accurately) sampled data, the occurrence distribution becomes singular on the true movement path, and in this limit, the occurrence

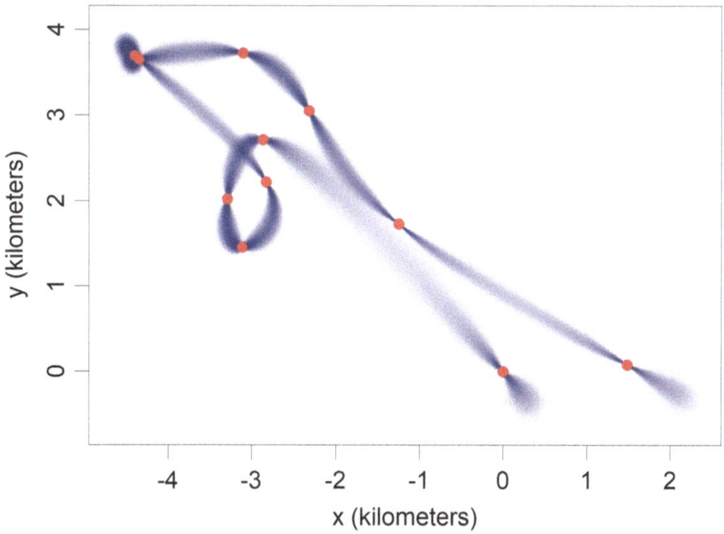

Fig. 7.2 An occurrence distribution, in blue, conditioned on 12 hours of African buffalo tracking data in red [14]

distribution—as an estimator of the unknown movement path—is asymptotically consistent. More generally, the coverage area of the occurrence distribution is entirely composed of uncertainties and is a reflection of our ignorance of where the mover was located during the sampling period. $p_{\mathrm{OD}}(\mathbf{r})$ does not represent a relative frequency of space use—only one trajectory was realized, and the true space use is measure zero without assigning some intrinsic positive area to the mover. In fact, if a small area, dA, is assigned to the mover, then the total area used by the mover between the times t_1 and t_n is simply given by the distance that mover traveled

$$A(t_1, t_n) = dA \int_{t_1}^{t_n} |\dot{\mathbf{r}}(t)|\, dt + O\Big(dA^2\Big)\,, \tag{7.6}$$

to first order in dA. It is, therefore, inaccurate to refer to the occurrence distribution as a space-use estimator.

On the other hand, the occurrence distribution is closely related to resource utilization, as for any static spatial field, $R(x, y)$, which could represent a potential resource, the average resource value sampled by the tracked animal during the observation period can be estimated by

$$\bar{R}(t_1, t_n) = \iint R(x, y)\, p_{\mathrm{OD}}(x, y|t_1, t_n)\, dx\, dy\,, \tag{7.7}$$

rather than the simple average of the sampled times, $R(x(t_i), y(t_i))$. The advantage of using the occurrence distribution to estimate resource utilization is that it accounts for location error, weights all times evenly in the case of irregular sampling, and habitats with poor satellite reception will be better sampled by the interpolation. However, the occurrence-distribution-based estimator is adding some statistical error with its Gaussian interpolation, and so for higher quality data, the sample mean is likely to be more efficient.

7.2.3 Asymptotic Behavior

Without loss of generality, let us consider a stationary movement process for $\mathbf{r}(t)$, with stationary distribution $p(\mathbf{r}(t)) = p(\mathbf{r})$, that is evenly sampled, with time steps $\Delta t = t_{i+1} - t_i$, for some period of time, $T = t_n - t_1$. It is important to note that $p_{\text{OD}}(\mathbf{r}|t_1, t_n)$ does *not* generally converge to the true stationary distribution, $p(\mathbf{r})$, in most asymptotic limits. If $\Delta t > 0$, then $p_{\mathcal{N}}(\mathbf{r}(t)|\text{DATA})$ is not asymptotically consistent for all times of one realization of a non-Gaussian processes, though it is asymptotically efficient when considering many realizations of a time. On the other hand, $\lim_{\Delta t \to 0} p_{\text{OD}}(\mathbf{r}|t_1, t_n)$ is a singular distribution with zero coverage area. Therefore, it is only the case that the occurrence distribution becomes the stationary distribution in the particular limit

$$\lim_{\Delta t \to 0} \lim_{T \to \infty} p_{\text{OD}}(\mathbf{r}|t_1, t_n) = p(\mathbf{r}) \,, \tag{7.8}$$

where the limits do *not* commute. Moreover, if the time step, Δt, is small, then the rate of convergence is slow and a larger sampling period, T, is required to approach convergence. On the other hand, if Δt is large, then $\lim_{T \to \infty} p_{\text{OD}}(\mathbf{r})$ will be appreciably different from the true stationary distribution, $p(\mathbf{r})$, and more similar to the Gaussian distribution used for interpolation, $p_{\mathcal{N}}(\mathbf{r})$. Therefore, in practice, the occurrence distribution only serves as a reasonable approximation of the stationary distribution, $p(\mathbf{r})$, when Δt takes an intermediate "Goldilocks" value that will not be known *a priori* [2]. This is in contrast to kernel density estimation (KDE), which is asymptotically consistent in the more general limit of $\lim_{T \to \infty}$ with $\Delta t > 0$, and autocorrelated kernel density estimation (AKDE), which is asymptotically consistent with an unimpeded rate of convergence even if Δt is arbitrarily small [4].

7.2.4 Information Gaps

When relying on a Gaussian stochastic process model, large gaps in the data will contain times far from those of the data, wherein correlations with $\mathbf{r}(t)$ will be weak, and the resulting conditional predictions resolve to the unconditioned Gaussian model, $p_{\mathcal{N}}(\mathbf{r}(t)) = \mathcal{N}(\mathbf{r}|\boldsymbol{\mu}(t), \boldsymbol{\sigma}(t, t))$. Such predictions contain no information beyond the Gaussian model and only serve to oversmooth the occurrence distribution. Moreover, if using the uniformly weighted occurrence disribution (7.5), then arbitrarily large gaps in the data will cause arbitrary oversmoothing until the occurrence distribution itself is nothing more than the Gaussian model.

For regularly sampled data, the conventional solution to this problem is to skip gaps in the data that are larger than some multiple, m, of the sampling interval, Δt. The obvious issues with this criterion are that m is subjectively chosen and a regular sampling interval, Δt, may not exist. [24] considered gap skipping rules based on the autocorrelation timescales, which resolves the latter issue, but still remains unsatisfactory.

More generally, we may consider both of these stop-gap solutions as weighted occurrence distributions of the form

$$p_{\mathrm{OD}}(\mathbf{r}) = \frac{\int w(t)\, p_{\mathcal{N}}(\mathbf{r}(t)|\mathrm{DATA})\, dt}{\int w(t')\, dt'}, \tag{7.9}$$

where, more ideally, the weighting function, $w(t)$, should be a positive function with the limits

$$w(t) = 0 \qquad \text{iff} \qquad p_{\mathcal{N}}(\mathbf{r}(t)|\mathrm{DATA}) = p_{\mathcal{N}}(\mathbf{r}(t)), \tag{7.10}$$

$$w(t) = 1 \qquad \text{iff} \qquad p_{\mathcal{N}}(\mathbf{r}(t)|\mathrm{DATA}) = \delta(\mathbf{r}(t) - \mathbf{M}(t)), \tag{7.11}$$

so that only the informative predictions are included. Therefore, the weighting function should be a dissimilarity measure of the form

$$w(t) = D[\, p_{\mathcal{N}}(\mathbf{r}(t)|\mathrm{DATA}) \mid p_{\mathcal{N}}(\mathbf{r}(t))\,]. \tag{7.12}$$

We close this section by noting that the Hellinger distance

$$H[p|q]^2 = \frac{1}{2} \iint \left(\sqrt{p(x,y)} - \sqrt{q(x,y)}\right)^2 dx\, dy \tag{7.13}$$

satisfies these limits, though so does any positive power of the Hellinger distance. So, while many such weighting functions do exist, their functional dependence should be optimized for an estimation task.

7.3 Range Distributions: Where Will The Animal Go?

> [...] that area traversed by the individual in its normal activities of food gathering, mating, and caring for young. Occasional sallies outside the area, perhaps exploratory in nature, should not be considered as in part of the home range. W. H. Burt (1943) [27]

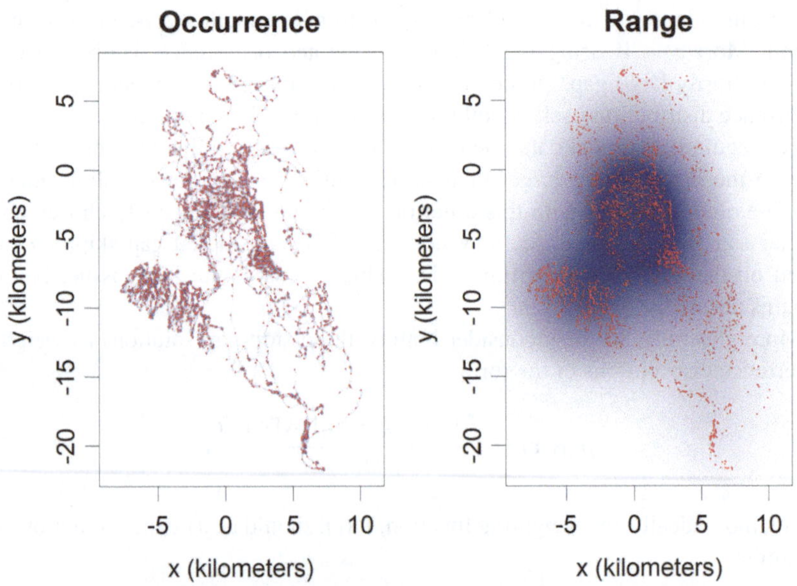

Fig. 7.3 Left panel: The same occurrence distribution as in Fig. 7.2 in blue, but calculated over all 5 months of buffalo tracking data in red [14]. Right panel: The stationary distribution of the same buffalo in blue

For any stochastic process, $\mathbf{r}(t)$, we can consider the marginal probability density $p(\mathbf{r}(t))$ as designating the potential range of movement at time t. If $\mathbf{r}(t)$ is also a stationary process, then $p(\mathbf{r}(t)) = p(\mathbf{r})$ is the same stationary distribution for all times t. If $\mathbf{r}(t)$ describes a range-resident mover, then the stationary distribution is the most obvious mathematical concept that directly relates to Burt's notion of an animal's "home range." The "areas traversed" can be identified with the coverage areas of the stationary distribution, and "occasional sallies" are rare events in the tails of the stationary distribution. An example is given in Fig. 7.3.

If we can decompose a non-stationary movement process into a set of behaviors, B, with each behavior exhibiting stationary dynamics that the mover can switch between [28]—such as a resting behavior and an active behavior—then we can consider the steady-state distribution

$$p_{\mathrm{SS}}(\mathbf{r}) = \sum_{\mathrm{B}} P(\mathrm{B})\, p_{\mathrm{B}}(\mathbf{r}) \,, \tag{7.14}$$

where $P(\mathrm{B})$ is the steady-state probability of being in state B, and $p_{\mathrm{B}}(\mathrm{r})$ is the stationary distribution of state B. The home range would then correspond to a suite of different range-resident behavioral distributions, while non-resident behaviors may also have associated stationary distributions and ranges. In general, we consider weighted marginal distributions, where the weights correspond to behaviors or periods of time, and we refer to these distributions as *range distributions*. For instance, a migratory individual might have a well-defined summer range, winter range, and migratory corridor—all of which could be combined into an annual range. The range distribution is conceptually meaningful, because it is the distribution used to calculate expectation values that correspond to its associated behaviors and periods of time.

Range distributions are most commonly estimated non-parametrically, though parametric estimates are not unheard of [8, 29, 30]. The most statistically efficient known method for non-parametric distribution estimation is kernel density estimation [31, 32], whereby the true distribution is approximated by a weighted average of normalized kernels, $\kappa(\mathbf{r}|\mathbf{H})$, placed upon the data:

$$\hat{p}(\mathbf{r}) = \sum_i w(t_i)\,\kappa(\mathbf{r}-\mathbf{r}(t_i)|\mathbf{H}(t_i))\,, \tag{7.15}$$

where $w(t_i)$ denotes the weights, which can prevent bias due to oversampled times, and $\mathbf{H}(t_i)$ denotes the bandwidth matrices, which control the spread of the kernels. In general, the bandwidth and weights are optimized to minimize an asymptotic approximation to the "mean integrated square error" (MISE) between the density estimate (7.15) and the true density, $p(\mathbf{r})$:

$$\mathrm{MISE}(\mathbf{w}, \mathbf{H}) = \mathrm{E}\left[\iint \left(\hat{p}(\mathbf{r}|\mathbf{w}, \mathbf{H}) - p(\mathbf{r})\right)^2 d^2\mathbf{r}\right]. \tag{7.16}$$

Different asymptotic approximations to the MISE then correspond to different bandwidth optimizers, but all good bandwidth optimizers achieve an asymptotically optimal order of convergence [31, 32]. For autocorrelated processes, there is, so far, only one known asymptotically optimal bandwidth optimizer, which is the autocorrelated Gaussian reference function bandwidth optimizer of AKDE [4, 33, 34]. The vast majority of modern animal tracking datasets are substantially autocorrelated, and non-parametric distribution estimates that neglect this property are known to underestimate the spread of the range distribution [35]

7.3.1 "Space Use" and Non-stationary Predictions

Movement behaviors may be subject to transient dynamics, especially if driven by time-dependent covariates [36]. For instance, the marginal distribution $p(\mathbf{r}(t))$ may contract in the dry season and expand in the wet season, and yet the dry-season, wet-season, and annual range distributions all make sense, conceptually. In contrast to occurrence distributions, range distributions do *predict* a frequency of space use, but not the space *used* during the sampling period. In the wet-season/dry-season example, the wet-season range distribution will be larger than the annual range, because the annual range is the weighted average of the larger wet-season range and the smaller dry-season range, and yet the individual animal will have *used* more space over the year than over the wet season. A range distribution only predicts space use over the long term—in the ergodic regime—and only if the corresponding behaviors and conditions assumed by that distribution persist. The sizes of each distribution are to be compared as one would compare variances and not in proportion to the amount of data collected.

7.3.2 Population Ranges

Estimating space use at the population level is a fundamental problem that is at the intersection of spatial ecology and wildlife management, with many real-world applications relating to the designation of protected areas for species conservation [37]. Individual-level range distributions can be generalized to population distributions, by considering the posterior density of a random individual in the population

$$p(\mathbf{r}) = \frac{1}{P} \sum_{k=1}^{P} p_k(\mathbf{r}) \,, \tag{7.17}$$

where P is the abundance of the population and $p_k(\mathbf{r})$ is the range distribution of individual k. In empirical applications, only a small subset of the population, $S \ll P$, will be tracked, and, therefore, one of the challenges in estimating distribution (7.17) is to make inferences about the population, P, from the sample, S.

Traditional population range estimation techniques have relied the dual assumptions of independently sampled movement tracks, ignoring temporal autocorrelation, and a large number of sampled individuals, S, ignoring individual variation. These methods have been found to be critically sensitive to the number of individuals sampled from the population [38–40]. Here we describe the mathematical underpinnings of newly developed methods in [6] that can account for both temporal autocorrelation and individual variation.

In most situations, the abundance will be a large number, and only a sample of individuals will be tracked, and so it becomes useful to model the individual distributions

$$p_k(\mathbf{r}) = p(\mathbf{r}|\boldsymbol{\theta}_k) \,, \tag{7.18}$$

as having random parameters $\boldsymbol{\theta}_k$ that are distributed according to a population-level parameter distribution

$$\boldsymbol{\theta} \sim p(\boldsymbol{\theta}|\boldsymbol{\Theta}) \,, \tag{7.19}$$

with population-level parameters, $\boldsymbol{\Theta}$, so that inference can be made with the hierarchical distribution

$$p(\mathbf{r}|\boldsymbol{\Theta}) = \int p(\mathbf{r}|\boldsymbol{\theta})\, p(\boldsymbol{\theta}|\boldsymbol{\Theta})\, d\boldsymbol{\theta}, \tag{7.20}$$

which accounts for variation in individual-level space use when scaling up to the population level. The population range distribution can then be estimated non-parametrically using a population-weighted kernel density estimate of the form:

$$\hat{p}(\mathbf{r}) = \sum_{k=1}^{P} W_k \sum_{i=1}^{n_k} w_k(t_i)\, \kappa(\mathbf{r}-\mathbf{r}_k(t_i)|\mathbf{H}_k) \;, \tag{7.21}$$

where W_k represents the individual weights to prevent bias due to differential sampling intensity between individuals. The bandwidths and weights can be optimized by minimizing the MISE between the density estimate (7.21) and the true density (7.20). An example of population range estimated using this method is given in Fig. 7.4.

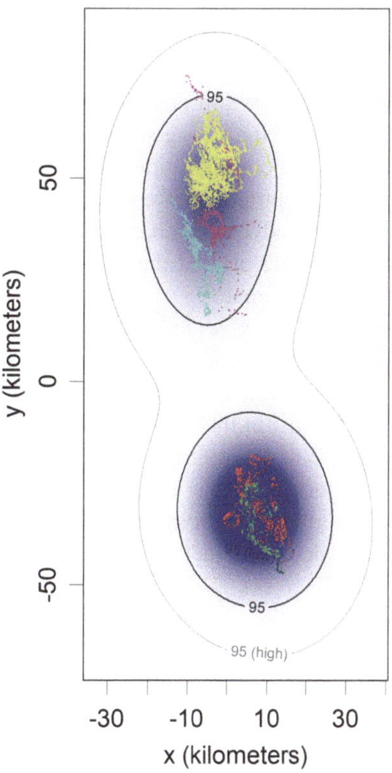

Fig. 7.4 The population range distribution (blue) estimated using tracking data from six African buffalos [14], with different colors per individual. The point estimate of the 95% coverage area is denoted by the solid black line, while the gray lines correspond to the 95% confidence intervals on the magnitude of that area

7.3.3 Corridors

Ecological corridors maintain connectivity within fragmented natural habitats by enabling animals to move between patches [41] and at larger scales (i.e., migratory corridors) [42]. Mathematically and functionally, the corridor itself can be defined as the marginal distribution of locations for trajectories that connect a set of patches or distributions—including population distributions in the meta-population context. Therefore, the corridor satisfies the definition of a range distribution that is not well modeled as a stationary distribution, but has a very specific conditioning on its end points. This definition is also consistent with resistance-circuit models of connectivity that liken the landscape to a surface of variable resistivity, in which populations move as the electric current flows between terminal points from a voltage difference [43]. However, the resistance-circuit model does not provide a likelihood for an individual animal's stochastic trajectory, and so there is no direct way to fit this model to the movement data that would best inform its parameters.

The most commonly applied method of estimating corridors to date has been to calculate occurrence distributions conditioned on Brownian motion, which produces an average of Brownian bridges that probabilistically interpolate the data. However, as described in Sect. 7.2, the width of the occurrence distribution corresponds to our uncertainty in where the animal traveled through the corridor and not the physical width of the corridor itself. All else being equal, the width of an occurrence distribution will vanish in the limit of finely and accurately sampled data, whereas the corridor width is a bio-physical parameter that exists independent of the data. Because of the strong sampling dependence inherent to occurrence distributions, it has been a common practice to subjectively tune the underlying model parameters until the output occurrence distribution widths satisfy expert visual inspection, and so [7] has advocated abandoning occurrence distributions in favor of an expert-tuned linear buffer around the data, which produces a more honest and less convoluted corridor estimate.

Now, we give some early results from [44] on identifying the proper target parameter for corridor estimation. If we consider any cross section of a migratory corridor, then the 95% corridor width should capture, on average, 95% of all migrations, in that 95% of the migrants will pass through the 95% corridor width, and 5% of the migrants will pass around the 95% corridor width. Non-parametric corridor estimation can, therefore, be reduced to running along the corridor and calculating the cross-sectional variation or distribution at each point. A simple example of this method is given in Fig. 7.5.

7.3.4 Territories

Most simply, a territory can be defined as "any defended area" [46]. However, there is great variation in the underlying mechanisms of territoriality, both within and

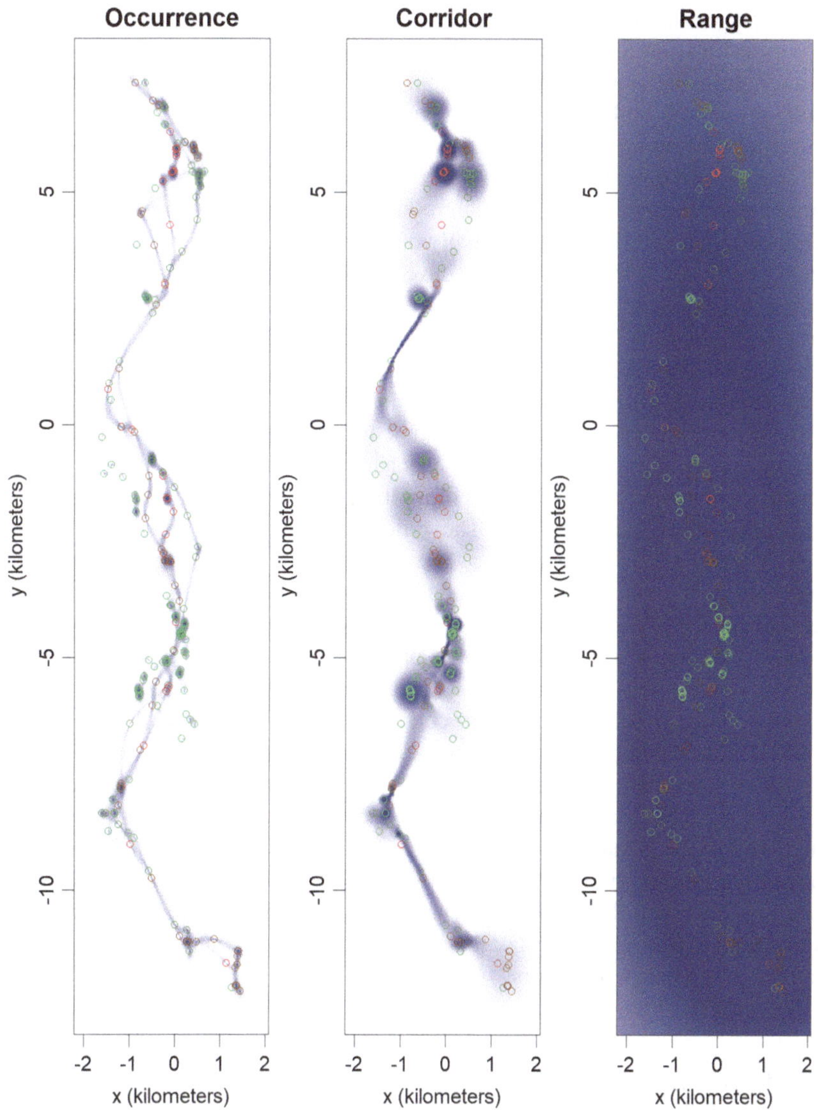

Fig. 7.5 Tracking data on multiple jaguar (*Panthera onca*) dispersals through a single corridor [45], where each color represents a different trek through the corridor. Left panel: The occurrence distribution, which is an interpolation of where the jaguar is thought to have traveled. The variation in dispersals is clearly not captured, because of the high sampling rate. Middle panel: A cross-sectional KDE method that targets the appropriate corridor width. Right panel: An AKDE range estimate that improperly models the movement process as being stationary and results in an excessively wide distribution

across species, as well as the dependencies on defended resources, which makes territory estimation a complex problem [47]. In particular, even distinguishing the territory from the stationary distribution, or home range, is a largely unresolved challenge, in both biological and mathematical contexts.

Territory mapping is a common method used to estimate territories in some bird species. An observable behavior such as the direct exclusion of intruders from hummingbird feeding territories or aggressive calls in many songbird species around their breeding territories is first spatially mapped based on field recordings [48, 49]. A range estimate, often a minimum convex polygon, can then be drawn to indicate an area that contains the recorded occurrences of said territorial behavior. This is thus referred to as the potential territory of an individual. While this incorporates the direct exhibited evidences of territoriality, the heavy dependence of these methods on observable field data limits the applicability of this approach to only a small subset of territorial species that must also be in a realistically observable range.

Using only animal tracking data, mechanistic home-range analysis (MHRA) estimates the stationary distributions of multiple individuals exhibiting intraspecific competition and exclusion [50]. The rationale behind this approach is that an animal's territory can be modeled as a more restricted version of its natural home range due to the avoidance of surrounding tracked conspecific territories. Thus, mechanistic home-range analysis serves as a flexible method to estimate territory size as a function of external territorial pressures. The time-dependent probability densities of the tracked individuals, $p_k(\mathbf{r}, t)$, are described by a system of diffusion equations with advection away from competing conspecific densities or pressures [50, 51]. This territorial pressure can be modeled to suit the territorial behavior of a species of interest. Repulsive scent marks, for example, can be modeled as accumulating from probability density and decaying exponentially in time. The key feature of MHRA is that it distinguishes territorial behavior from an animal's natural home-range fidelity via the consideration of external pressures. With case-by-case modeling of territorial pressures, mechanistic home ranges can produce reasonable estimates of territorial space use in various species without direct observational mapping of territorial behaviors.

7.4 Other Spatial Distributions

7.4.1 Encounter Distributions

Given the individual stationary distributions, $p_k(\mathbf{r})$, of two individuals indexed by k, where in space are these individuals likely to encounter each other? Individuals will be more likely to encounter one another at locations where both of their individual probabilities of space use are high, and vice versa. An historically common approach to identifying the area over which encounters between individuals k and l might occur has been to delineate the area of intersection between the respective coverage areas of $p_k(\mathbf{r})$ and $p_l(\mathbf{r})$ at a coverage level of interest (e.g., 50% or

95%). Though widespread in practice, this approach is only valid if $p_k(\mathbf{r})$ and $p_k(\mathbf{r})$ are uniform distributions. Realistically speaking, free-ranging animals exhibit nonuniform patterns of space use. A more generally valid approach to identifying the spatial distribution of encounter probabilities between individuals k and l works from the joint probability distribution for individuals that move independently (up to their encounter):

$$p_{kl}(\mathbf{r}_k, \mathbf{r}_l) = p_k(\mathbf{r}_k)\, p_l(\mathbf{r}_l)\,. \tag{7.22}$$

Next, we change coordinates to the mean, $\bar{\mathbf{r}} = {}^1\!/_2\,(\mathbf{r}_k + \mathbf{r}_l)$, and difference, $\boldsymbol{\Delta} = \mathbf{r}_k - \mathbf{r}_l$, location vectors:

$$p_{kl}(\bar{\mathbf{r}}, \boldsymbol{\Delta}) = p_k(\bar{\mathbf{r}} + {}^1\!/_2\,\boldsymbol{\Delta})\, p_l(\bar{\mathbf{r}} - {}^1\!/_2\,\boldsymbol{\Delta})\,. \tag{7.23}$$

The probability of these two individuals being within some distance $|\boldsymbol{\Delta}| < d$ of each other is then given by the marginal distribution:

$$P_{kl}(d) = \iint_{\mathbb{R}^2} d^2\bar{\mathbf{r}} \int_{-\pi}^{+\pi} d\theta \int_0^{d^2} d\Delta^2\, p_{kl}(\bar{\mathbf{r}}, \boldsymbol{\Delta})\,, \tag{7.24}$$

$$= \iint_{\mathbb{R}^2} d^2\bar{\mathbf{r}} \int_{-\pi}^{+\pi} d\theta \int_0^{d^2} d\Delta^2\, p_k(\bar{\mathbf{r}} + {}^1\!/_2\,\boldsymbol{\Delta})\, p_l(\bar{\mathbf{r}} - {}^1\!/_2\,\boldsymbol{\Delta})\,, \tag{7.25}$$

where the difference vector, $\boldsymbol{\Delta}$, was decomposed into polar coordinates, Δ and θ. We are generally interested in encounters that occur at small distances, d, relative to the broad scales of the distributions, and to second order in d, we have the encounter probability:

$$P_{kl}(d) = \left(2\pi \iint p_k(\mathbf{r})\, p_l(\mathbf{r})\, d^2\mathbf{r}\right) d^2 + O(d^4)\,. \tag{7.26}$$

This relation for encounter probabilities now informs us as to the origin of each spatial contribution, and so the spatial distribution of where these encounters take place must be given by the normalized density [9]:

$$\mathrm{CDE}_{kl}(\mathbf{r}) = \frac{p_k(\mathbf{r})\, p_l(\mathbf{r})}{\iint p_k(\mathbf{r}')\, p_l(\mathbf{r}')\, d^2\mathbf{r}'}\,, \tag{7.27}$$

for small encounter distances, d.

For some biological systems, it may be more informative to consider encounters between groups of interest (e.g., males and females, predators and prey, and inter-group encounters). The CDE can be expanded beyond individual-level encounters k and l to encounters between groups K and L. Making the same assumptions as before, the group-level CDE between individuals in group K and those in group L is given by the normalized density:

$$\text{CDE}_{KL}(\mathbf{r}) = \frac{\sum_{k\in K}\sum_{l\in L} p_k(\mathbf{r})\, p_l(\mathbf{r})}{\sum_{k'\in K}\sum_{l'\in L} \iint p_{k'}(\mathbf{r}')\, p_{l'}(\mathbf{r}')\, d^2\mathbf{r}'} \,. \tag{7.28}$$

In summary, (7.28) relies on three key assumptions to maintain analytical tractability: (i) stationarity in the movement processes; (ii) that encounters are local events; and (ii) that movement is uncorrelated across individuals (outside of said encounters). The simplicity of (7.28) allows for the immediate application of a plug-in estimator, whereby the CDE is estimated from estimates of the individual range distributions, $\hat{p}_k(\mathbf{r})$. Because many animals will exhibit non-stationarity, have perceptual ranges that allow for non-local encounters, and exhibit cross-correlated movement, future work on relaxing these assumptions is a promising avenue for future research.

7.4.2 Revisitation Distributions

While range distribution represents proportion of time spent in an area, biologists are also interested in the number of times that an area is revisited, or recursions to such areas, which distinguishes between a small number of long visits and a large number of short visits [52]. Like encounters, revisits are ordinarily based on a threshold radius (R) that determines what constitutes a unique visit. That is, time spent within the radius counts as one visit, while leaving and re-entering the radius counts as two revisits. Revisits are difficult to estimate, because of sensitivity to the threshold radius and because naive algorithms have an $O(n^2)$ computational cost, where n is the number of sampled locations. Moreover, there are further complications that are not normally considered, such as the fact that revisits cannot be well estimated with coarse data, because an uncorrelated or Brownian motion process supported by such data can realize an infinite number of revisits between any two sampled locations.

7.4.2.1 Derivation

Consider an individual moving at a constant velocity $\mathbf{v}$, from location $\mathbf{r}$. The time, dt, that the individual spends within a small radius R for that visit is

$$dt(\text{visit}) = \frac{R}{v} \,, \tag{7.29}$$

and the expected proportion of time, T, spent within a radius R of location $\mathbf{r}$ is

$$\frac{dt(\text{total})}{T} = \pi\, R^2\, p(\mathbf{r}) \,, \tag{7.30}$$

where $p(\mathbf{r})$ denotes the stationary distribution of the process. Therefore, the expected number of revisits, N_R, is given by

$$N_\mathrm{R} = \frac{dt(\text{total})}{dt(\text{visit})} = \pi\, R\, v\, p(\mathbf{r})\, T\,, \tag{7.31}$$

and the frequency of revisitation, F_R, is given by

$$F_\mathrm{R} = \frac{dN_\mathrm{R}}{dT} = \pi\, R\, v\, p(\mathbf{r})\,. \tag{7.32}$$

Finally, for small radii, R, it suffices to consider

$$\frac{dF_\mathrm{R}}{dR} = \pi\, v\, p(\mathbf{r})\,, \tag{7.33}$$

which is proportional to the revisitation rate.

Given the joint distribution of position and velocity, $p(\mathbf{r},\mathbf{v})$, the expected revisitation rate can then be calculated by the marginal expectation

$$\frac{dF_\mathrm{R}}{dR} = \pi \iint v\, p(\mathbf{r},\mathbf{v})\, d^2\mathbf{v}\,, \tag{7.34}$$

which is straightforwardly estimated via non-parametric regression methods. In particular, this formula has the same target parameter as the numerator in the standard Nadaraya–Watson kernel regression for calculating $\mathrm{E}[v|\mathbf{r}]$. That is, given a weighted kernel density estimator of the form

$$\hat{p}(\mathbf{r}) = \sum_{i=1}^{n} w(t_i)\, \kappa(\mathbf{r}-\mathbf{r}(t_i))\,, \tag{7.35}$$

then the same weights and kernels provide a revisitation-rate estimator of

$$\frac{d\hat{F}_\mathrm{R}}{dR}(\mathbf{r}) = \pi \sum_{i=1}^{n} w(t_i)\, v(t_i)\, \kappa(\mathbf{r}-\mathbf{r}(t_i))\,, \tag{7.36}$$

whereas the expected speed at a given location would be conventionally estimated via

$$\hat{\mathrm{E}}[v|\mathbf{r}] = \frac{\sum_{i=1}^{n} w(t_i)\, v(t_i)\, \kappa(\mathbf{r}-\mathbf{r}(t_i))}{\sum_{j=1}^{n} w(t_j)\, \kappa\big(\mathbf{r}-\mathbf{r}(t_j)\big)}\,, \tag{7.37}$$

for comparison.

Given that different numbers of revisitations occur at different locations, we can consider the spatial distribution of revisitations, which is simply given by normalizing the revisitation-rate prediction surface

$$p_{\mathrm{R}}(\mathbf{r}) = \frac{\iint d^2\mathbf{v}\, v\, p(\mathbf{r},\mathbf{v})}{\iint d^2\mathbf{r}' \iint d^2\mathbf{v}'\, v'\, p(\mathbf{r}',\mathbf{v}')}, \tag{7.38}$$

and can be estimated via

$$\hat{p}_{\mathrm{R}}(\mathbf{r}) = \frac{\sum_{i=1}^{n} w(t_i)\, v(t_i)\, \kappa(\mathbf{r}-\mathbf{r}(t_i))}{\sum_{j=1}^{n} w(t_j)\, v(t_j)}. \tag{7.39}$$

From this perspective, it is easier to see that the revisitation distribution is simply the speed-weighted location distribution.

7.5 Summary

Spatial distributions are a natural product of animal locations that are tracked through time, and knowledge of that time dependence allows for a multitude of meaningful spatial distributions to be formulated. The natural generalization of the stationary distribution for non-stationary movement processes is the *range distribution*, which, in other contexts, might be referred to as the equilibrium, steady-state, or regulated distribution. The range distribution is predictive of out-of-sample locations during a particular set of behavioral states or in a particular temporal or seasonal context. For a range-resident individual, the range distribution is the distribution that most closely corresponds to the biological notion of an animal's "home range." This range-distribution concept can be naturally extended to corridors, territories, and populations.

In contrast to the range distribution, the *occurrence distribution* is the posterior prediction of where the individual was at a random time within the sampling period. Although these two distributions—range and occurrence—have been frequently conflated in the literature [2], their derivations here should provide clear mathematical insight into how distinct these two distributions are, via their respective extrapolative and interpolative targets.

Another example of conflation between range and occurrence exists in corridor estimation, where occurrence distributions have also been historically applied toward depicting the *corridor distribution*. Again, the interpolative target of the occurrence distribution makes it poorly suited for predicting out-of-sample migration paths. Though, in fairness and in contrast to conflation with the home range, an appropriate target distribution for corridors has only been recently identified [44].

A less mathematically resolvable debate exists in the concept of the *territory*, in how the territory can be estimated from movement data, and in how the territory differs from the home range. Following the definition of a range distribution, the territory could be defined as the territorial range distribution—the range distribution associated with territorial behaviors. Importantly, the range distribution of a territorial individual can be properly modeled and estimated without modeling the territory as a distinct subset of the home range that is defended, but rather by modeling territorial pressure and its response [8].

The spatial distributions of location-dependent events can also be considered, and one of the more interesting examples is the conditional *encounter distribution*, which predicts where encounters are likely to take place between movers, given their respective distributions. While first derived in [9], our derivation here avoids any notion of a phenomenological encounter-rate parameter.

Finally, we have introduced the spatial distribution of revisits, which we have termed the *revisitation distribution*, and we have shown it to be equivalent to the speed-weighted location distribution. This distribution cannot be calculated from stochastic-process models that do not exhibit finite speeds, such as Brownian motion, because fractal motion can produce an infinite number of revisits within a finite duration of time. We believe that this distribution may find use in corridor estimation, where the amount of time spent in an area may not be as important as how often an animal needs to move through an area.

Funding Acknowledgement This work was partially funded by the Center of Advanced Systems Understanding (CASUS), which is financed by Germany's Federal Ministry of Education and Research (BMBF) and by SMWK with tax funds on the basis of the budget approved by the Saxon State Parliament.

RMG also received partial support from FAPESP through ICTP-SAIFR 2021/14335-0.

WFF acknowledges funding support from NSF-NNA 212727.

References

1. Fleming, C.H., Fagan, W.F., Mueller, T., Olson, K.A., Leimgruber, P., Calabrese, J.M.: Estimating where and how animals travel: an optimal framework for path reconstruction from autocorrelated tracking data. Ecology **97**, 3 (2016). https://doi.org/10.1890/15-1607.1
2. Alston, J.M., Fleming, C.H., Noonan, M.J., Tucker, M.A., Silva, I., Folta, C., Akre, T.S.B., Ali, A.H., Belant, J.L., Beyer, D., Blaum, N., Böhning-Gaese, K., de Paula, R.C., Dekker, J., Drescher-Lehman, J., Farwig, N., Fichtel, C., Fischer, C., Ford, A.T., Janssen, R., Jeltsch, F., Kappeler, P.M., LaPoint, S.D., Markham, A.C., Medici, E.P., Morato, R.G., Nathan, R., Olson, K.A., Patterson, B.D., Petroelje, T.R., Ramalho, E.E., Rösner, S., Santos, L.G.O., Schabo, D.G., Selva, N., Sergiel, A., Spiegel, O., Ullmann, W., Zieba, F., Zwijacz-Kozica, T., Wittemyer, G., Fagan, W.F., Müller, T., Calabrese, J.M.: Clarifying space use concepts in ecology: range vs. occurrence distributions (2022). bioRxiv:2022.09.29.509951
3. Worton, B.J.: Kernel methods for estimating the utilization distribution in home-range studies. Ecology **70**(1), 164–168 (1989). https://doi.org/10.2307/1938423
4. Fleming, C.H., Fagan, W.F., Mueller, T., Olson, K.A., Leimgruber, P., Calabrese, J.M.: Rigorous home-range estimation with movement data: a new autocorrelated kernel-density estimator. Ecology **96**, 1182–1188 (2015)
5. Silva, I., Fleming, C.H., Noonan, M.J., Alston, J., Folta, C., Fagan, W.F., Calabrese, J.M.: Autocorrelation-informed home range estimation: a review and practical guide. Methods Ecol. Evol. **13**(3), 534–544 (2022)
6. Gayatri, A., Fleming, C.H., Krishnan, A.G., Lamb, C.T., Medici, E.P., Prugh, L.R., Calabrese, J.M., Fagan, W.F.: Estimating population range distributions from animal tracking data (in preparation)

7. Merkle, J.A., Lowrey, B., Wallace, C.F., Hall, L.E., Wilde, L., Kauffman, M.J., Sawyer, H.: Conserving habitat for migratory ungulates: how wide is a migration corridor? J. Appl. Ecol. **60**(9), 1763–1770 (2023)
8. Moorcroft, R.P., Lewis, A.M.: Mechanistic Home Range Analysis. Princeton University Press, New Jersey (2006)
9. Noonan, M.J., Martinez-Garcia, R., Davis, G., Crofoot, M.C., Kays, R., Hirsch, B.T., Caillaud, D., Payne, E., Sih, A., Sinn, D.L., Spiegel, O., Fleming, C.H., Calabrese, J.M.: Estimating encounter location distributions from animal tracking data. Methods Ecol. Evol. **12**(7), 1158–1173 (2021)
10. Horne, J.S., Garton, E.O., Krone, S.M., Lewis, J.S.: Analyzing animal movements using Brownian bridges. Ecology **88**(9), 2354–63 (2007)
11. Johnson, D.S., London, J.M., Lea, M.-A., Durban, J.W.: Continuous-time correlated random walk model for animal telemetry data. Ecology **89**(5), 1208–1215 (2008)
12. Fleming, C.H., Subaşı, Y., Calabrese, J.M.: A maximum-entropy description of animal movement. Phys. Rev. E **91**, 032107 (2015)
13. Fleming, C.H., Sheldon, D., Gurarie, E., Fagan, W.F., LaPoint, S., Calabrese, J.M.: Kálmán filters for continuous-time movement models. Ecol. Inform. **40**, 8–21 (2017)
14. Cross, P.C., Bowers, J.A., Hay, C.T., Wolhuter, J., Buss, P., Hofmeyr, M., du Toit, J.T., Getz, W.M.: Data from: nonparameteric kernel methods for constructing home ranges and utilization distributions. Movebank Data Repository (2016). https://doi.org/10.5441/001/1.j900f88t
15. Einstein, A.: Über die von der molekularkinetischen Theorie der Wärme geforderte Bewegung von in ruhenden Flüssigkeiten suspendierten Teilchen. Annalen der Physik **322**(8), 1521–3889 (1905)
16. Uhlenbeck, G.E., Ornstein, L.S.: On the theory of the Brownian motion. Phys. Rev. **36**(5), 823–841 (1930)
17. Hesse, C.H.: The one-sided barrier problem for an integrated Ornstein-Uhlenbeck process. Commun. Stat. Stoch. Models **7**(3), 447–480 (1991)
18. Fleming, C.H., Calabrese, J.M., Mueller, T., Olson, K.A., Leimgruber, P., Fagan, W.F.: From fine-scale foraging to home ranges: a semi-variance approach to identifying movement modes across spatiotemporal scales. Am. Nat. **183**(5), E154–E167 (2014)
19. Stein, M.L.: Asymptotically efficient prediction of a random field with a misspecified covariance function. Ann. Stat. **16**(1), 55–63 (1988)
20. Krige, D.G.: A statistical approach to some basis mine valuation problems on the Witwatersrand. J. Chem. Metal Mining Soc. S. Africa **52**, 119–139 (1951)
21. Kalman, R.E.: A new approach to linear filtering and prediction problems. J. Basic Eng. **82**(1), 35–45 (1960)
22. Johnson, D.S.: `crawl`: fit continuous-time correlated random walk models to animal movement data (2008). https://CRAN.R-project.org/package=crawl
23. Jonsen, I.D., Grecian, W.J., Phillips, L., Carroll, G., McMahon, C., Harcourt, R.G., Hindell, M.A., Patterson, T.A.: `aniMotum`, an `R` package for animal movement data: Rapid quality control, behavioural estimation and simulation. Methods Ecol. Evol. **14**(3), 806–816 (2023)
24. Fleming, C.H., Calabrese, J.M.: `ctmm`: continuous-time movement modeling (2015). https://CRAN.R-project.org/package=ctmm
25. Calabrese, J.M., Fleming, C.H., Gurarie, E.: `ctmm`: an `R` package for analyzing animal relocation data as a continuous-time stochastic process. Methods Ecol. Evol. **7**(9), 1124–1132 (2016)
26. Kranstauber, B., Smolla, M., Scharf, A.K.: `move`: visualizing and analyzing animal track data (2012). https://CRAN.R-project.org/package=move
27. Burt, W.H.: Territoriality and home range concepts as applied to mammals. J. Mammal. **24**(3), 436–352 (1943)
28. Michelot, T., Blackwell, P.G.: State-switching continuous-time correlated random walks. Methods Ecol. Evol. **10**, 637–649 (2019)
29. Horne, J.S., Garton, E.O., Rachlow, J.L.: A synoptic model of animal space use: simultaneous estimation of home range, habitat selection, and inter/intra-specific relationships, Ecol. Model. **214**(2–4), 338–348 (2008)

30. Alston, J.M., Fleming, C.H., Kays, R., Streicher, J.P., Downs, C.T., Ramesh, T., Reineking, B., Calabrese, J.M.: Mitigating pseudoreplication and bias in resource selection functions with autocorrelation-informed weighting. Methods Ecol. Evol. **14**(2), 643–654 (2022)
31. Silverman, B.W.: Density Estimation for Statistics and Data Analysis. Chapman & Hall, Boca Raton (1986)
32. Turlach, B.A.: Bandwidth Selection in Kernel Density Estimation: A Review. CORE and Institut de Statistique (1993)
33. Fleming, C.H., Calabrese, M.J.: A new kernel-density estimator for accurate home-range and species-range area estimation. Methods Ecol. Evol. **8**(5), 571–579 (2017)
34. Fleming, C.H., Sheldon, D., Fagan, W.F., Leimgruber, P., Mueller, T., Nandintsetseg, D., Noonan, M.J., Olson, K.A., Setyawan, E., Sianipar, A., Calabrese, J.M.: Correcting for missing and irregular data in home-range estimation. Ecol. Appl. **28**(4), 1003–1010 (2018)
35. Noonan, M.J., Tucker, M.A., Fleming, C.H., et al.: A comprehensive analysis of autocorrelation and bias in home range estimation. Ecol. Monographs **89**(2), e01344 (2019). https://doi.org/10.1002/ecm.1344
36. Michelot, T., Blackwell, P.G., Matthiopoulos, J.: Linking resource selection and step selection models for habitat preferences in animals. Ecology **100**(1), e02452 (2019)
37. Pe'er, G., Tsianou, M.A., Franz, K.W., Matsinos, Y.G., Mazaris, A.D., Storch, D., Kopsova, L., Verboom, J., Baguette, M., Stevens, V.M., et al.: Toward better application of minimum area requirements in conservation planning. Biol. Conserv. **170**, 92–102 (2014)
38. Gutowsky, S., Leonard, M., Conners, M., Shaffer, S., Jonsen, I.: Individual-level variation and higher-level interpretations of space use in wide-ranging species: an albatross case study of sampling effects. Front. Marine Sci. **2**, 93 (2015)
39. Sequeira, A., Heupel, M., Lea, M., Eguíluz, V., Duarte, C., Meekan, M., Thums, M., Calich, H., Carmichael, R., Costa, D., et al.: The importance of sample size in marine megafauna tagging studies. Ecol. Appl. **29**, e01947 (2019)
40. Soanes, L., Arnould, J., Dodd, S., Sumner, M., Green, J.: How many seabirds do we need to track to define home-range area? J. Appl. Ecol. **50**, 671–679 (2013)
41. Scharf, A., Belant, J., Beyer, D., Wikelski, M., Safi, K.: Habitat suitability does not capture the essence of animal-defined corridors. Movement Ecol. **6**, 18 (2012)
42. Merkle, J., Gage, J., Sawyer, H., Lowrey, B., Kauffman, M.: Migration Mapper: identifying movement corridors and seasonal ranges for large mammal conservation. Methods Ecol. Evol. **13**(11), 2397–2403 (2022)
43. McRae, B., Dickson, B., Keitt, T., Shah, V.: Using circuit theory to model connectivity in ecology, evolution, and conservation. Ecology **89**(10), 2712–2724 (2008)
44. Lin, E., et al.: Estimating corridors: what is the target distribution? (in preparation)
45. Morato, R.G., et al.: Jaguar movement database: a GPS-based movement dataset of an apex predator in the Neotropic. Ecology, **99**(7), 1691–1691 (2018)
46. Noble, G.K.: The role of dominance in the social life of birds. Auk **56**(3), 263–273 (1939)
47. Powell, R.A.: Animal home ranges and territories and home range estimators. In: Boitani, L., Fuller, T. (eds.) Research Techniques in Animal Ecology: Controversies and Consequences, 2nd edn., pp. 65–110. Columbia University Press, New York (2000)
48. Hixon, M.A., Food production and competitor density as the determinants of feeding territory size. Am. Naturalist **115**(4), 510–530 (1980)
49. Gibbons, D.W., Gregory, R.D.: Birds. In: Sutherland, J.W. (ed.) Ecological census techniques: A handbook, 2nd edn., pp. 308–350. Cambridge University Press, Cambridge (2006)
50. Moorcroft, P.R., Lewis, M.A.: Mathematical analysis of the conspecific avoidance model. In: Mechanistic Home Range Analysis, pp. 67–78. Princeton University Press, New Jersey (2006)
51. Bateman, A.W., Lewis, M.A., Gall, G., Manser, M.B., Clutton, Brock, H.-T.: Territoriality and home-range dynamics in meerkats, Suricata suricatta: a mechanistic modelling approach. J. Animal Ecol. **84**(1), 260–271 (2014)
52. Bracis, C., Bildstein, K.L., Mueller, T.: Revisitation analysis uncovers spatio-temporal patterns in animal movement data. Ecography **41**(11), 1801–1811 (2018)

Chapter 8
An Information Theory Treatment of Animal Movement Tracks

Wayne M. Getz

8.1 Introduction

At its most fundamental level, the movement track $\mathcal{T}$ of an animal over a landscape is a time series of position relocation points. Movement track relocation data conceal a movement behavior narrative that, once revealed, provides insights into and a deeper knowledge of the ecological aspects of an individual's life history [1]. Extracting this knowledge requires, first, a rigorous decoding to reveal the movement elements underpinning the relocation data. Second, it requires an understanding of how these movement elements are influenced by both the internal state (age, sex, health, physiology) of an individual and its external surroundings (landscape, climatic, and population ecological factors).

These movement elements themselves cannot be identified by their mechanical signature (i.e., fundamental movement elements in the sense described in [2]) because such information is not coded in relocation data alone, but requires video, accelerometer, or other body-parts movement data for identification [3]. Rather, they can be replaced by statistically derived movement elements computed from a fixed number of relocation data steps, using a recent approach described in [4]. This approach permits a coding theory of the relocation data to be formulated, as developed here, though the approach requires relatively high-frequency relocation data that have an inter-point interval of, at most, seconds rather than minutes (i.e., frequencies in excess of 0.1 Hz).

W. M. Getz (✉)
Department Environmental Science, Policy and Management, University of California, Berkeley, CA, USA

School of Mathematics, Statistics & Computer Science, University of KwaZulu-Natal, Durban, South Africa
e-mail: wgetz@berkeley.edu

L. Giuggioli, P. K. Maini (eds.), *The Mathematics of Movement*,
https://doi.org/10.1007/978-3-032-00873-2_8

Decoding a movement track to read the story it encodes involves a process called path segmentation [5]. This process relies on time series segmentation methods such as behavioral change point analysis (BCPA) [6–11] and hidden Markov methods (HMM) [12–16] that produce variable length segments. Typically, such segments are much longer than the statistical elements discussed in [4]; and we note that if a code is not based on fixed-length segments, then ambiguity arises regarding the number of elements used to code any given segment of track. Additionally, BCPA and HMM methods, as mentioned by [10], may not be the best way to identify rare events, such as "parturition, migration initiation, and juvenile dispersal."

The decoding process is generally not applied directly to the relocation data, but first these data are transformed into a time series of step lengths (SL) and turning angles (TA) [11, 17, 18]. From these, other quantities can be extracted, such as the individual's persistent and angular velocities [6]. These quantities can then be used to identify intervals of time over which they vary in characteristic ways. Through clustering or other category generating procedures [19, 20], segments with particular characteristic variations can be interpreted as expressing some identifiable behavioral movement mode, such as walking, resting, or feeding.

BCPA has proven to be a potent segmentation technique on sub-minute data. For example, as an illustration of its accuracy at this level of resolution, [9] used BCPA to segment relocation data collected every 10 seconds from four sheep grazing at an experimental site in Norway. To reduce global navigation satellite systems errors, though, they smoothed their data to obtain a relocation point every 1 min, then applied a 30 point (i.e., half hour) moving window to generate both persistent and angular velocities at each point. This approach allowed them to identify four types of segments using Ward's agglomerative hierarchical clustering method [21]: "foraging," "resting," "walking," and "other." The segments they identified were all of variable length to a resolution of 1 minute. After ground-truthing a number of their segments, [9] found that their method scored an average classification accuracy of around 80% across their movement modes.

For very large, high-resolution relocation datasets, BCPA may be relatively computationally expensive and also have problems converging on a solution, particularly if one seeks to identify four or more movement behavioral modes. An alternative approach based on Shannon coding theory can be taken. This approach is computationally much more manageable than BCPA and also provides a rigorous theory for measuring and comparing the information content (entropy) of the movement track segmentation process.

In 1948, Claude Shannon presented his mathematical theory of communication in a two-part article published in the Bell System Technical Journal [22, 23]. This theory can be applied to any sequential strings of symbols used to code information, whether they be strings of electronic bits, nucleic acids, numbers and letters, or animal movement track segments. In the latter case, however, the segments underlying the analysis (i.e., the lowest set of coding symbols) need to be standardized to all be of the same size to avoid confusion between single and compound segments of relocation data.

Table 8.1 Hierarchical segmentation of an animal track relocation time series $\mathcal{T}^{\text{loc}}$ collected at a frequency f HZ; names and resolutions provided here are a more precise rendering of previous presentations [18, 25] (see footnotes)

Acronym (symbol)	Number of types	Resolution	Comment
StaME* (σ_i)	$i = 1, \cdots, n$	μf secs	n clusters of segments and μ steps
			per segment are method parameters
CAM† (κ_c)	$c = 1, \cdots, k$	$m\mu f$ secs	k clusters of words and m StaMEs
			per word are method parameters
BAM‡	TBD**	sub-diel	homogeneous: string of same CAMs
		but variable	heterogeneous: characteristic CAM mix
DAR§	TBD††	24-hrs	bottom up: strings of BAMs
		fixed	top down: e.g., geometry of DARs
LiMP¶	depends on	multi-day to	LiMP specific DARs condition on
	age & env.‡‡	seasonal	internal and external factors
LiT ‖	syndromic	individual's	LiMP specific DARs condition on
	move. types§§	life span	internal and external factors

*Statistical movement elements, formerly called metaFuMEs
†Canonical activity mode segments, formerly called "short duration" or homogeneous CAMs
‡Behavioral activity mode segments, formerly called "long duration" or heterogeneous CAMs
§Diel activity routine segments
¶Life-time movement phase segments, including dispersal, migration, and nurturing young
‖Life-time track, including dispersal, migration, and nurturing young
**To be determined using biological change point analysis [6, 7] and hidden Markov methods [12, 14, 15]
††May be extracted using appropriate clustering methods [24]
‡‡To be determined using appropriate statistical methods [26, 27]
§§For discussions on movement syndrome types, see [28, 29]

In the material that follows, we review an approach to segmenting movement tracks using elements that are μ steps long and, after clustering and identifying cluster centroids, provides a set of n statistical movement elements (StaME set $\mathcal{S}$, elements σ_i, $i = 1, \cdots, n$). The approach also segments movement tracks using elements referred to as "words" that are $m\mu$ steps long. After clustering these words, there centroids provide a set of k canonical activity modes (CAM set $\mathcal{K}$, elements κ_c, $c = 1, \cdots, k$). The smaller StaMEs may then serve as set of m symbols for analyzing the information content of the larger CAMs that, in turn, can be used to probe the information content of subdiel, diel, and supra-diel segments of animal movement tracks (Table 8.1).

Our formulation also allows us to compare coding efficiencies across different methods for clustering and generating StaMEs and CAMs. This includes the effects of the number of steps μ selected for the StaME building blocks of movement tracks, the number n of symbol types identified, the number m of smallest segments used to build words, and the number k of CAMs used to generate larger interpretable movement behavior segments, such as behavioral activity modes and diel activity routines [18, 24].

8.2 Segmentation Hierarchy

A hierarchical scheme for the segmentation of tracks has been discussed in considerable depth elsewhere [4, 18, 25]. An updated summary is provided in Table 8.1. Originally, [2] proposed that, at its most basic level, a movement track $\mathcal{T}$ can be conceptualized as a sequence of fundamental movement elements (FuMEs; referred to as FMEs in [2]). These elements when strung together either homogeneously (or one type), or in characteristic mixes of several types, produce canonical activity mode (CAMs) segments. [2] pointed out that "with most current data, it is not possible to construct distributions of building block movement elements in terms of FuMEs" (they used the acronym FME in [2]): essentially, as previously mentioned, FuMEs are unobservable in relocation data alone and companion video or accelerometer data are likely needed to identify the start of each new FuME. This subsequently led to the concept of a set $\mathcal{S}$ of statistical movement elements (StaMEs), formerly referred to as a metaFuMEs [18] to be used in place of FuMEs as a set of lowest level movement building block elements or coding symbols. In the formulation here, the set $\mathcal{S}$ is regarded as an underlying set of symbols that can be used to encode the movement track relocation data time series $\mathcal{T}^{\text{loc}}$ as a StaMEs time series $\mathcal{T}^{\sigma}$ (Table 8.2). This symbolic representation then allows an information theory toolbox to be used to analyze the information content and, ultimately, meaning of animal movement tracks [30].

The elements of $\mathcal{S}$, as discussed in some detail in [4], are generated from a segmentation of the $T+1$ point (0 included) relocation time series $\mathcal{T}^{\text{loc}}$ into a track segment time series $\mathcal{T}^{\text{seg}}$ consisting of $\lfloor T/\mu \rfloor$ ($\lfloor \cdot \rfloor$ means round down the integer below) segments each containing μ steps (Task 1, Fig. 8.1). A cluster analysis of these segments is then used to place them into one of n similarly-shaped categories of segment types [4]. The centroids of these n segment-type clusters can then be extracted and treated as archetypal elements in a set $\mathcal{S}$ of symbolic segments σ_i

Table 8.2 Representation of movement tracks at different stages and resolutions of encoding the relocation data $\mathcal{T}^{\text{loc}}$ into StaMEs/symbols and CAMs (Fig 8.1)

Track (description)	Elements	Relevant Measures	Comment
$\mathcal{T}^{\text{loc}}$ (point location track)	$\big(t; (x_t, y_t)\big)$	$T+1$ points	relocation data
$\mathcal{T}^{\text{seg}}$ (segment track)	$(z; \text{seg}_z)$	μ-steps; $\lfloor \frac{T}{\mu} \rfloor$ segments	1st segmentation
$\mathcal{T}^{\sigma}$ (StaME track)	$(i; \sigma_i)$	$\lfloor \frac{T}{\mu} \rfloor H(\mathbf{p}^{\sigma})$ bits*	1st level centroids
$\mathcal{T}^{\text{wd}}$ (word track)	$(j; \text{wd}_j)$	mμ-steps; $\lfloor \frac{T}{m\mu} \rfloor$ words	2nd segmentation
$\mathcal{T}^{\kappa}$ (raw CAM track)	$(j; \kappa_c)$	$D_{\text{JS}}^{\text{ens}}\left(\mathbf{p}^{\mathcal{W}_1}, \cdots, \mathbf{p}^{\mathcal{W}_k}\right)^{\dagger}$	centroids of $\text{wd}_j \in \mathcal{W}_c$
$\mathcal{T}^{\text{CAM}}$ (rectified CAM track)	$(j; \kappa_c^{\star})^{+}$	$\lfloor \frac{T}{m\mu} \rfloor H(\mathbf{p}^{\text{CAM}})$ bits§	argmax c reassignment

$^{+}$The difference between raw (κ_c) and rectified ($\kappa_c^{\star}$) CAMs based on Eq. 8.14 (see Fig. 8.1)
*Information encoded into $\mathcal{T}^{\text{loc}}$ if symbol sequences are uncorrelated
†Divergence across the ensemble distributions $\mathbf{p}^{\mathcal{W}_c}$, $c = 1, \cdots, k$ (Eq. 8.26)
‡Avg. divergence of ω_ℓ distributions in $\mathcal{W}_c$ from their distribution in $\mathcal{W}$ itself
§Information that can be coded into $\mathcal{T}^{\text{loc}}$ if CAM sequences are uncorrelated

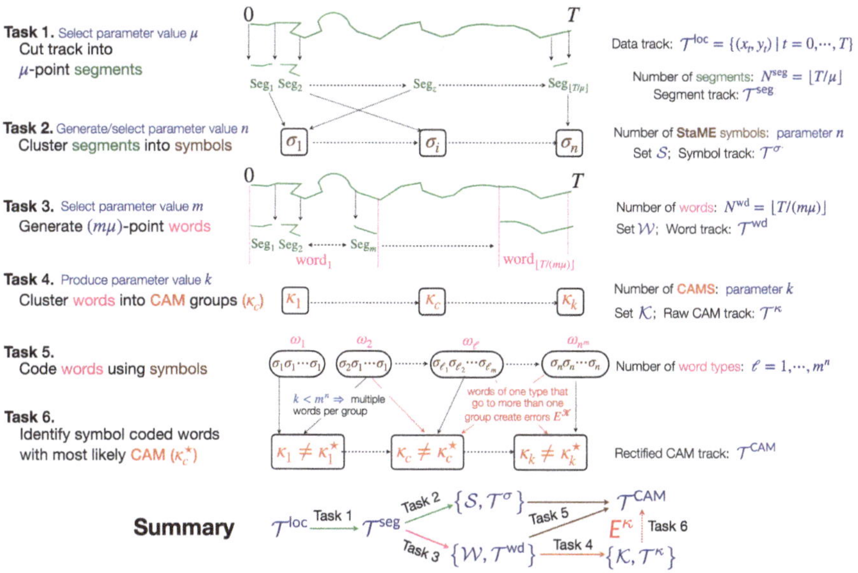

Fig. 8.1 A graphic depiction of tasks 1–6 required to code an animal movement relocation data series ($\mathcal{T}^{\mathrm{loc}}$) into a statistical movement elements (StaMEs) data series $\mathcal{T}^{\sigma}$ and a canonical activity modes (CAMs) data series $\mathcal{T}^{\mathrm{CAM}}$ and compute the coding accuracy E^{κ} of the process: i.e., implement the functor $\mathcal{M} : \mathcal{T}^{\mathrm{loc}} \mapsto \{\mathcal{T}^{\sigma}, \mathcal{T}^{\mathrm{CAM}}, E^{\kappa}\}$. Parameter values $\mu, n, m,$ and k need to be either a priori selected or determined during implementation of the clustering methods, which are highlighted in blue. Extracted objects are: segments in green (Tasks 1 and 3), symbols in brown (Task 2), words in fuchsia (Tasks 3–5), CAMs in orange (Tasks 4 and 6), and a coding accuracy measure in red (Task 6). With regard to the latter, the red dotted arrows between Tasks 5 and 6 indicate misassignment of a proportion of the N_ℓ words of type ω_ℓ ($\ell = 1, \cdots, n^m$; $N^{\mathrm{wd}} = \sum_{\ell=1}^{n^m} N_\ell$) as instances of particular CAMs assigned to $\kappa^{\star}_{c_1}$ when initially a member of $\mathcal{W}_{c_2}$ for $c_1 \neq c_2, c_1, c_2 \in \{1, \cdots, k\}$). A **summary** of the tasks and objects produced at each step of the process is provided at the bottom of this graphic. (Note: calligraphy letters in the graph and caption are the same symbol, but generated by different font sets)

that can be used to code a time series $\mathcal{T}^{\sigma}$ (Task 2, Fig. 8.1). The values μ and n themselves become tune-able parameters of the StaME/symbol identification and creation process, while the method used to cluster the segments into a set of n StaMEs becomes one of the arguments of the segmentation process when viewed as a functor $\mathcal{M}$ acting on the relocation data time series $\mathcal{T}^{\mathrm{loc}}$ (Table 8.2; also see Eq. 8.22 below).

The current concept of a CAM also needs to be treated more rigorously than before if it is to be integrated into an information theoretic framework. Specifically, in the formulation below, m μ-length segments are strung together into $m\mu$-step words (Task 3; Fig. 8.1, Tables 8.1 and 8.2) that then constitute a word set $\mathcal{W}$. The words in $\mathcal{W}$ can then be used to generate a word time series $\mathcal{T}^{\mathrm{wd}}$ of $\lfloor T/(m\mu)\rfloor$ elements (Table 8.2). A cluster analysis of the words in $\mathcal{W}$ is then used to identify k word clusters denoted by $\mathcal{W}_c$, $c = 1, \ldots, k$. The centroids of these k clusters can be used to generate the k-element raw CAM set $\mathcal{K}$ (Task 4, Fig. 8.1), with elements κ_c used to code track $\mathcal{T}^{\mathrm{wd}}$ into a raw CAM track $\mathcal{T}^{\kappa}$ (Table 8.2).

Additionally, one may translate each word into its representative m-symbol string and then identify which cluster most words that have the same m-symbol code belong to. If this is the word set $\mathcal{W}_c$, then all words with this particular m-symbol coding sequence will be identified with a rectified CAM type c, as represented by its archetype/ideal $\kappa_c^{\star}$ (Task 5, Fig. 8.1). The corresponding rectified CAM coding track is denoted by $\mathcal{T}^{\mathrm{CAM}}$. The proportion E^{κ} of these particular m-symbol words that had to be reassigned, when added for all $c = 1 \cdots, k$, provides a measure of the reliability of the CAM coding process. In a nutshell, the difference between the $\mathcal{T}^{\kappa}$ and $\mathcal{T}^{\mathrm{CAM}}$ time series is that the former arises from a cluster analysis of segments containing mμ steps and the latter assigns a CAM type $\kappa_c^{\star}$ to all words that are coded by the same m-symbol sequence of StaMEs $\sigma_i \in \mathcal{S}$. This distinction will be made more rigorous in the next section where we formalize the set of Tasks 1–6 (Fig. 8.1) that defines the functor $\mathcal{M} : \mathcal{T}^{\mathrm{loc}} \mapsto \left\{\mathcal{T}^{\sigma}, \mathcal{T}^{\mathrm{CAM}}, E^{\kappa}\right\}$ used to derive the StaME and rectified CAM sets and track codings, and the assignment error associated with the process.

The frequencies of StaMEs and CAMs respectively coded into the movement track representations $\mathcal{T}^{\sigma}$ and $\mathcal{T}^{\mathrm{CAM}}$ can be used to derive a measure of the entropy of these coding schemes if coding elements are not autocorrelated. If they are, then an autocorrelation analysis can be used to derive more accurate measures of the entropy of these coded tracks. Further, if CAMs are regarded as each consisting of a particular sequence of m StaMEs, then it becomes relatively simple to apply Shannon's Information Theory [22, 23] concepts to compare the information content per unit time of different tracks.

Although each of the CAMs will come to be identified in terms of some emblematic activity such as resting, random stepping, directed stepping, jogging/cantering, sprinting/galloping, individual CAMs will not constitute a complete behavioral repertoire of a particular kind. Rather, it will take either a homogeneous or a characteristic-heterogeneous mix of CAM strings of, in practice, variable length to then interpret the resulting multi-CAM segment as a particular kind of behavioral mode [18, 25]. For example, in the context of a homogeneous CAM string, a sequence of "high velocity, low turning angle" CAMs may be interpreted as a "beelining" behavioral activity mode (BAM). In the context of a characteristic mix of CAMs, a mix of stepping and resting CAMs may come to be ground-truthed as a foraging BAM. Since BAMs are generally of variable length, they currently constitute the types of movement mode structures identified using BCPA methods [6–11].

From a top-down viewpoint, BAMs are variable-length behavioral segments that switch intermittently during the course of animal's diurnal activity routine (DAR). The lengths and frequency of switching among BAMs depend on many factors including, as pointed out in our introduction, the internal state of an individual (e.g., its age, sex, health, physiology) and its external environment (e.g., landscape and vegetation structure of surroundings, weather variables, and presence of particular conspecifics, heterospecific competitors, and predators). The sequencing and switching probabilities of BAMs can be modeled by fitting an autoregressive time series (AR) CAM occurrence model to empirical movement path data $\mathcal{T}^{\mathrm{loc}}$ that has been segmented into a rectified CAM times series $\mathcal{T}^{\mathrm{CAM}}$. Such models can also be elaborated by including the effects of internal and external auxiliary variables on sequencing and switching rates.

Methods for fitting categorical variable AR(p) models to empirical data (here p denotes the time span of the correlated effect—but to begin it may be sufficient to fit an AR(1) model). Dependence of model parameters on auxiliary factors is likely to be important and models that deal with such situations can be used (e.g., [31]). Model parameters influencing CAM switch probabilities in generalized AR(p) models are likely to be time-dependent, particularly with regard to BAMs that occur at characteristic times of the day or night within the diel cycle (e.g., resting, heading to water, heading home, and feeding). Seasonal effects are also likely to be a factor and can be treated under the rubric of segmenting the lifetime track (LiT) of individuals into lifetime movement phases (LiMPs), as they may pertain, for example, to breeding, hibernation, and migration cycles.

8.3 Two-Level Track Segmentation

An animal movement track $\mathcal{T}$ can viewed from a number different perspectives. In a two-level hierarchical segmentation of the time series $\mathcal{T}^{\mathrm{loc}}$ of relocation points $(t; x_t, y_t)$, $t = 0, \cdots, T$, derived time series, as previously mentioned, can be viewed as (Fig. 8.1, Table 8.2): (i) a time series $\mathcal{T}^{\mathrm{seg}}$ of building-block segments, where each segment consists of μ consecutive steps; (ii) a time series $\mathcal{T}^{\sigma}$ of segments labeled by their StaME (symbol) type, as it arises from a clustering of segments in $\mathcal{T}^{\mathrm{seg}}$ into n distinct categories; (iii) a time series $\mathcal{T}^{\mathrm{wd}}$ of word segments, where each segment consists of mμ consecutive steps; (iv) a time series $\mathcal{T}^{\kappa}$ of raw CAMs that arise from clustering (mμ)-step length words into k CAM categories; and v) a time series $\mathcal{T}^{\mathrm{CAM}}$ of rectified CAMs that arises from reassigning words, based on their m-length string of StaME coding symbols σ_i to the CAM where such words are most likely to occur in the first place. Once this segmentation has been accomplished, then different methods can be used to parse the rectified CAM time series $\mathcal{T}^{\mathrm{CAM}}$ into variable length BAMs, fixed time periods DARS, and multi-day lifetime movement phases (LiMPs) using autoregressive modeling, BCPA, and HMM methods. These various coding schemes for $\mathcal{T}$ then allow one to use Shannon information theory to estimate coding entropy and error rates and compare the coding accuracy of different coding schemes.

In summary, the six tasks needed to implement the functor $\mathcal{M} : \mathcal{T}^{\mathrm{loc}} \mapsto \{\mathcal{T}^{\sigma}, \mathcal{T}^{\mathrm{CAM}}, E^{\kappa}\}$ are (Fig. 8.1):

Task 1. *Movement Path Segmentation:* This task requires that we segment the movement track relocation dataset

$$\mathcal{T}^{\text{loc}} = \{(t; x_t, y_t)|t = 0, \cdots, T\} \tag{8.1}$$

into segments of size μ (i.e., each consists of μ steps) to obtain a movement track representation $\mathcal{T}^{\text{seg}}$ of $N^{\text{seg}} = \lfloor T/\mu \rfloor$ segments ($\lfloor \bullet \rfloor$ means round down to the integer value): i.e.,

$$\mathcal{T}^{\text{seg}} = \big\{(z; \text{seg}_z)|\ \text{seg}_z = \big((x_{\mu(z-1)}, y_{\mu(z-1)}), \cdots, (x_{\mu z}, y_{\mu z})\big),\ z = 1, \cdots, N^{\text{seg}}\big\}. \tag{8.2}$$

Task 2. *Segment Clustering:*
This task produces a set StaMEs

$$\mathcal{S} = \{\sigma_i | i = 1, \cdots, n\} \tag{8.3}$$

that can be used as a set of symbols to recode the set of $T + 1$ relocation points $\mathcal{T}^{\text{loc}}$ as a times series $\mathcal{T}^{\sigma}$ of $\lfloor T/\mu \rfloor$ symbols (Table 8.2). The process of creating the set $\mathcal{S}$ requires that we cluster the segments making up the track $\mathcal{T}^{\text{seg}}$ using either shape (typically, unsupervised machine learning [32]) methods or vector methods that use summary statistics as variables defining each segment [19, 33]. Shape methods my be set up to ignore rotations and reflections when movement on a landscape is unlikely to be directionally and rotationally biased. Biases enter when individuals have a proclivity to move latitudionally or longitudinally at a local scale (e.g., in neotropical birds [34]), or clockwise or anticlockwise [35, 36]. Vector methods in the past, using summary statistics, have included the statistical means of persistence and angular velocities [6, 9], the means and standard deviations of the speed (i.e., step length) and absolute turning angle of each segment, and sometimes even the net displacement (distance between the two end points) of each segment [4, 33].

Once the set of symbols S has been constructed—where, for convenience, symbols are numbered by size (i.e., by step-length/velocity averaged across their μ consecutive steps) so the size(σ_i) $\geq$ size(σ_{i+1}), $i = 1, \cdots, n-1$—then the segmented data series $\mathcal{T}^{\text{seg}}$ can be coded using these symbols to obtain the symbol time series

$$\mathcal{T}^{\sigma} = \{(z; \sigma_{i_z})|z = 1, \cdots, N^{\text{seg}}\}. \tag{8.4}$$

We can now count the number of times that each symbol σ_i appears in $\mathcal{T}^{\sigma}$ to obtain the vector

$$\mathbf{p}^{\sigma} = \left(p_1^{\sigma} \cdots, p_n^{\omega}\right)', \quad \text{where } p_i^{\sigma} \text{ is the proportion of } \sigma_i \in \mathcal{T}^{\sigma}, i = 1, \cdots, n \tag{8.5}$$

of the distribution of symbols in $\mathcal{T}^{\sigma}$.

Task 3. *Word Generation:*
In the same way that $\mathcal{T}^{\text{loc}}$ can be parsed into N^{seg} segments each consisting of μ steps, we can parse $\mathcal{T}^{\text{loc}}$ into N^{wd} words each consisting of m consecutive

segments providing us with a next level of segmentation in which segments contain mμ consecutive steps. In this case, we obtain a time series of $N^{\text{wd}} = \lfloor T/(m\mu) \rfloor$ words

$$\mathcal{T}^{\text{wd}} = \big\{(j;\, \text{wd}_j | \text{wd}_j = \big((x_{m\mu(j-1)}, y_{m\mu(j-1)}),\cdots, (x_{\text{m}\mu j}, y_{m\mu j})\big),\ j=1,\cdots, N^{\text{wd}}\big\}. \tag{8.6}$$

After this segmentation is complete, we can gather these segments into a set of words:

$$\mathcal{W} = \big\{\text{wd}_j \in \mathcal{T}^{\text{wd}} | j = 1, \cdots, N^{\text{wd}}\big\}. \tag{8.7}$$

Task 4. *Direct Word Clustering and CAM Generation:*
The set $\mathcal{W}$ of the words in the time series $\mathcal{T}^{\text{wd}}$ can also be clustered to obtain a set of objects that we call raw canonical activity modes (CAMs). Thus, after clustering the N^{wd} words in $\mathcal{T}^{\text{wd}}$ into k different categories, using some appropriate vector or shape method, we obtain k subsets $\mathcal{W}_c \subset \mathcal{W}$, $c = 1, \cdots, k$, each with N_c^{wd} words, where $\sum_{c=1}^{k} N_c^{\text{wd}} = N^{\text{wd}}$. We can either regard each word segment ω_ℓ ($\ell = 1, \cdots, m^n$) in $\mathcal{W}_c$ as an instance of the CAM κ_c in the set

$$\mathcal{K} = \{\kappa_c \,|\, c = 1, \cdots, k\}, \text{ and } \mathbf{p}^\kappa = \left(p_1^\kappa, \cdots, p_k^\kappa\right)', \quad \text{where } \ p_c^\kappa = \frac{N_c^{\text{wd}}}{N^{\text{wd}}}, \tag{8.8}$$

and κ_c is the centroid of all $\omega_\ell \in \mathcal{W}_c$.

Once all the words ω_ℓ, $\ell = 1, \cdots, m^n$, have been assigned to some particular raw CAM group represented by the centroid κ_c of the word cluster $\mathcal{W}_c$, then our movement track $\mathcal{T}^{\text{wd}}$ can be rewritten as the raw CAM time series

$$\mathcal{T}^\kappa = \big\{(j; \kappa_c) | \text{ whenever wd}_j \in \mathcal{W}_c\big\}. \tag{8.9}$$

Task 5. *Symbolic Word Coding:*
By comparing the way segments in the two time series $\mathcal{T}^\sigma$ and $\mathcal{T}^{\text{wd}}$ have been labeled, we can time match each of the words $\text{wd}_j \in \mathcal{T}^{\text{wd}}$ with its corresponding coding sequence to obtain its representation

$$\text{wd}_j \equiv \sigma_{j_1}\sigma_{j_2}\cdots\sigma_{j_m},\ j = 1, \cdots, N^{\text{wd}}. \tag{8.10}$$

The number of different types of words that exist when m symbols are strung together to code a word is m^n. Thus, we can generate the set Ω of all possible words that can be constructed from the set of m symbols $\sigma_r \in \mathcal{S}$ (Eq. 8.3). The number of words in Ω is thus n^m and

$$\Omega = \{\omega_\ell | \ell = 1, \cdots, n^m\}, \text{ where } \omega_\ell \equiv \sigma_{\ell_1}\sigma_{\ell_2}\cdots\sigma_{\ell_m} \text{ and } \sigma_{\ell_r} \in \mathcal{S} \text{ for all } r = 1, \cdots, m. \tag{8.11}$$

For convenience, the words in Ω are partially ordered so that the largest word is $\omega_1 \equiv \sigma_1 \cdots \sigma_1$, the smallest $\omega_{m^n} \equiv \sigma_n \cdots \sigma_n$ with the rest of the words going approximately from largest to smallest (in terms of the average step size $\equiv$ velocity taken across their consecutive steps) using an ordered numbering scheme, as discussed in [33]. Once all these words have been numbered, we can generate the proportions

$$\mathbf{p}^\omega = \left(p_1^\omega \cdots, p_{m^n}^\omega\right)' \text{ where } p_\ell^\omega \text{ is the proportion of } \omega_\ell \in \mathcal{W}, \ell = 1, \cdots, m^n. \tag{8.12}$$

Task 6. *Assign Coded Words to Rectified CAM:*
Beyond generating $\mathcal{T}^\kappa$ directly from $\mathcal{T}^{\text{loc}}$, we can identify different word types ω_ℓ with specific CAMs κ_c using some appropriate rule. For example, if we define $\mathbf{p}^{\mathcal{W}_c}$ as the vector of the proportions of ω_ℓ words in the word clusters $\mathcal{W}_c$: i.e.,

$$\mathbf{p}^{\mathcal{W}_c} = \left(p_1^{\mathcal{W}_c} \cdots, p_{m^n}^{\mathcal{W}_c}\right)', \quad c = 1, \cdots, k, \tag{8.13}$$

then we can apply a rule, such as the following, to assign the various word types ω_ℓ to particular rectified CAM types $\kappa_c^\star$

$$\omega_\ell \in \kappa_c^\star \text{ whenever } c = \underset{c=1,\cdots,k}{\text{argmax}} \{p_\ell^{\mathcal{W}_c}\}, \quad \ell = 1, \cdots, m^n. \tag{8.14}$$

The star is used to denote the fact that the words ω_ℓ that are identified with $\kappa_c^\star$ are different from the words $\omega_\ell \in \mathcal{W}_c$, whose centroid is our archetype/ideal $\kappa_c \in \mathcal{K}$ (Eq. 8.8). In short, the archetypes κ_c arise from a cluster analysis of a segmentation of $\mathcal{T}^{\text{loc}}$ using segments of size mμ. On the other hand, the archetypes $\kappa_c^\star$ arise from a reassortment of the clusters $\mathcal{W}_c$ using rule Eq. 8.14. If there is more than one maximum value when applying rule Eq. 8.14, then we can select among them at random or using some additional cluster-related criterion.

Once a rule such as Eq. 8.14 has been applied to associate the ℓ-th word position in $\mathcal{T}^\kappa$ with a particular $\kappa_c^\star$, then our time series and proportional occurrence can be recast as

$$\mathcal{T}^{\text{CAM}} = \left\{(j; \kappa_c^\star) | \text{ whenever wd}_j \text{ is assigned to } \kappa_c \text{ using Eq. 8.14}\right\} \tag{8.15}$$
$$\text{with } \mathcal{K}^\star = \{\kappa_c^\star \,|\, c = 1, \cdots, k\} \text{ and } \mathbf{p}^{\kappa^\star} = \left(p_1^{\kappa^\star} \cdots, p_k^{\kappa^\star}\right).$$

In general, we cannot expect $\mathcal{T}^\kappa$ and $\mathcal{T}^{\text{CAM}}$ to represent the same sequence of CAMs. Errors will occur, and the proportion of errors that occur can be determined by computing the proportion of locations j where the CAM designations in $\mathcal{T}^\kappa$ and $\mathcal{T}^{\text{CAM}}$ do not match up. This error computation can be either be computed by directly comparing the CAM sequences in $\mathcal{T}^{\kappa}$ and $\mathcal{T}^{\text{CAM}}$ or using the proportion of misclassifications that arise in computing the CAM assignment using rule Eq. 8.14. In the latter case, it is clear that the proportion of times that word$_j$, when coded as word type ω_ℓ, is correctly assigned to CAM κ_c is $\left(\max_c\{p_j^{\mathcal{W}_c}\}/\sum_{c=1}^{k} p_j^{\mathcal{W}_c}\right)$: proportional misassigned to another CAM is 1 minus this quantity. Thus, summing over all misassignments of ω_ℓ, taking into account the frequency $N^{\omega_\ell}/N^{\text{wd}}$ of these words in $\mathcal{W}$ we obtain the percentage error rate

$$E^{\kappa} = 100 \sum_{\ell=1}^{n^m} \frac{N^{\omega_\ell}}{N^{\text{wd}}} \left(1 - \frac{\max_c\{p_\ell^{\mathcal{W}_c}\}}{\sum_{c=1}^{k} p_\ell^{\mathcal{W}_c}}\right) \%. \tag{8.16}$$

8.4 Information Entropy of Movement Track Segments

Information theory, as developed by Claude Shannon [22, 23] in the late 1940s, posits a set of n symbols $\mathcal{S}$ (Eq. 8.3) that when strung together sequentially encodes information at a basic rate determined by the probability p_i^{σ} of encountering symbol σ_i at randomly selected locations along an encoded string. Under these circumstances, the information coding rate of the symbols in $\mathcal{S}$ is determined by its $\log_2$ entropy $H(\mathbf{p}^{\sigma})$ [37]

$$H^{\sigma} \equiv H\left(\mathbf{p}^{\sigma}\right) = -\sum_{i=1}^{n} p_i^{\sigma} \log_2 p_i^{\sigma} \quad \text{(bits)}. \tag{8.17}$$

The units of entropy, when cast in the information theoretic $\log_2$ context, are "bits" of information with a single bit the rate at which information can be coded just using uncorrelated sequences of 0's and 1's with equal frequency (i.e., $-2 \times \frac{1}{2} \log_2 \left(\frac{1}{2}\right) =$ 1 bit).

Similarly, for the distribution $\mathbf{p}^{\omega}$ defined in Eq. 8.12, we can compute the entropy of the word coding set Ω in the context of the complete word string $\mathcal{T}^{\text{wd}}$ as

$$H_0^{\omega} \equiv H\left(\mathbf{p}^{\omega}\right) = -\sum_{\ell=1}^{n^m} p_\ell^{\omega} \log_2 p_\ell^{\omega}, \tag{8.18}$$

where the subscript 0 is used to denote the fact that this compution assumes no autocorrelation among consecutive words. If consecutive words are autocorrelated, however, then the coding rate must account for the conditional probability $p_{\ell'|\ell}^{\omega}$ of the next word being $\omega_{\ell'}$ when the previous word is ω_ℓ. In this case, if $p_{\ell\ell'}^{\omega}$ is the joint probability for the co-occurrence of the word sequence ω_ℓ and $\omega_{\ell'}$ independent of the order in which they occur, then the information content is modified to

$$H_1^{\omega} = -\sum_{\ell=1}^{m^n}\sum_{\ell'=1}^{m^n} p_{\ell\ell'}^{\omega} \log_2 \frac{p_{\ell\ell'}^{\omega}}{p_{\ell}^{\omega}}. \tag{8.19}$$

This formulation accounts for only the first level of autocorrelation when higher orders of correlation may occur, as can be assessed by fitting an autoregressive p-term model AR(p) [31] to a word-encoded empirical movement track $\mathcal{T}^{\text{wd}}$.

Finally, the set of CAMs has its own entropy, which can be computed in terms of the frequency of CAM types $\kappa_c^\star$ in the track $\mathcal{T}^{\kappa^\star}$ when CAM occurrence is not autocorrelated (which, in animal movement ecology, it generally will be). In this case, it follows from Eq. 8.15 that

$$H_0^{\kappa^\star} = -\sum_{c=1}^{k} p_c^{\kappa^\star} \log_2 p_c^{\kappa^\star} \quad \text{and} \quad H_1^{\kappa^\star} = -\sum_{c=1}^{k} p_{cc'}^{\kappa^\star} \log_2 \frac{p_{cc'}^{\kappa^\star}}{p_c^{\kappa^\star}}. \tag{8.20}$$

8.5 Information Content and Method Comparisons

The word assignment error E^{κ} (Eq. 8.16) and entropy measure $H^{\kappa^\star}$ of a set of rectified CAMs extracted from a movement track $\mathcal{T}^{\text{loc}}$ using the approach illustrated in Fig. 8.1 provide coding scheme measures whereby the performance of two different CAM extraction functors that use different clustering methods and parameter values $\{\mu, n, m, k\}$ (i.e., segment size, StaME number, word size, and CAM number) can be compared. The coding potential of a set of CAMs, however, also needs to account for the size of the CAMs themselves and not just the entropy H_0^{κ} of the CAM set $\mathcal{K}$.

First, we expect that the number of StaMEs groups that can be clearly separated increases with the segment size parameter μ. If we ignore orientation, then for $\mu = 2$ only the size (i.e., step length) of StaMEs can vary. If $\mu = 3$, then both size and turning angle can vary: apart from size, some StaMEs may have highly acute (almost reversed) turning angles, others closer to right angles, and still others with more obtuse or even rather small turning angles. The latter is a characteristic of what is called persistent movement. As μ increases beyond 3, StaME shapes can become more varied with saw-tooth-like elements, circular elements or combinations thereof, and absolute size (i.e., step-size/velocity averaged across all relocation points in the segment) coming into the mix. The more CAMs that are identified, the greater the potential coding rate since the maximum coding rate of a set of k rectified CAMs occurs when $p_c^{\kappa^\star} = 1/k$, $c = 1, \cdots, k$ (Eq. 8.8), and is given by

$$H^{\kappa^\star}_{\max} = -k(1/k)\log_2(1/k) = \log_2 k. \tag{8.21}$$

For a track containing $T + 1$ relocation points (i.e., including the point 0), however, the number of CAMs of size $m\mu$ that can be used to code this track is $N^{\text{wd}} = \lfloor T/(\text{m}\mu)\rfloor$ so that the the maximum amount of information that can be coded into track $\mathcal{T}^{\text{CAM}}$ is $N^{\text{wd}} H_0^{\kappa^\star}$ bits (Table 8.2). This, of course, does not account for the fact that CAM sequences may be autocorrelated, in which case $N^{\text{wd}} H_1^{\kappa}$ bits (Eq. 8.20) provides a better estimate of the amount of information that can be coded into a track of length T when segmented into CAMs that each include mμ steps.

Essentially, the task of producing the best (among a comparative group of parameter and clustering method approaches) rectified CAMs from relocation data can be expressed in terms of a family of functors $\mathcal{M}_a$, $a = 0, 1, 2, \cdots,$. The arguments of these functors are the parameters $\{\mu, n, m, k\}$ and the methods selected to cluster the segments into StaMEs (denoted CM(n,approach$^\sigma$)) and the words into CAMs (denoted CM(k,approach$^\kappa$)). Following our earlier notation, our family of functors are mappings of a track $\mathcal{T}^{\text{loc}}$ onto a set of StaMEs $\mathcal{S}$, rectified CAMs $\mathcal{K}^\star$, and error measure E^κ all indexed by an approach designator a:

$$\mathcal{M}_a : \mathcal{T}_a^{\text{loc}} \mapsto \{\mathcal{T}_a^{\sigma}, \mathcal{T}_a^{\text{CAM}}, E^{\kappa_a}\}, \quad a = 1, 2, \cdots, \tag{8.22}$$
$$\text{where } \mathcal{M}_a(\mathcal{T}_a) \equiv \mathcal{M}\big(\mathcal{T}_a; \mu_a, \text{CM}(n_a,\text{approach}_a^{\sigma}), m_a, \text{CM}(k_a, \text{approach}_a^{\kappa})\big).$$

Beyond using the entropy measures $H^{\kappa_a^\star}$ (Eq. 8.20) and misassignment error rates E^{κ_a} (Eq. 8.16) of different functors a to compare various approaches and identify those with high coding and low error rates, we can compare the CAM distributions $\mathbf{p}^{\kappa^\star}$ in sets $\mathcal{K}_a^\star$ that have the same number k of CAMs using the information theoretic Kullback–Leibler and Jensen–Shannon divergence measures [37]. The Kullback–Leibler divergence measure allows us to compute the "distances" (divergences) between two distributions. Specifically, if $\mathbf{p}^1 = \big(p_1^1, \cdots, p_k^1\big)$ and $\mathbf{p}^2 = \big(p_1^2, \cdots, p_k^2\big)$ are any two discrete distributions with k bins, then the Kullback–Leibler measure of the divergence of distribution $\mathbf{p}^2$ from distribution $\mathbf{p}^1$ is

$$D_{\text{KL}}(\mathbf{p}^1 : \mathbf{p}^2) = \sum_{c=1}^{k} p_c^1 \log_2\left(\frac{p_c^1}{p_c^2}\right). \tag{8.23}$$

This measure is not symmetric with respect to the distributions $\mathbf{p}^1 = \big(p_1^1, \cdots, p_k^1\big)$ and $\mathbf{p}^2 = \big(p_1^2, \cdots, p_k^2\big)$ and is not defined if any of the elements of $\mathbf{p}^2 = \big(p_1^2, \cdots, p_k^2\big)$ are zero. The Jensen–Shannon measure solves these issues by first defining the mixture distribution $\mathbf{p}^{\text{mix}}$ of $\mathbf{p}^1$ and $\mathbf{p}^2$ as follows: if N_1 and N_2 are the sample sizes of distributions $\mathbf{p}^1$ and $\mathbf{p}^2$, then

$$\mathbf{p}^{\mathrm{mix}} = \left(p_1^{\mathrm{mix}}, \cdots, p_k^{\mathrm{mix}}\right), \text{ where } p_c^{\mathrm{mix}} = \frac{N_1 p_c^1}{N_1 + N_2} + \frac{N_2 p_c^2}{N_1 + N_2}, \quad c = 1, \cdots, k. \tag{8.24}$$

In terms of this mixed distribution, the Jensen–Shannon divergence is then defined by

$$D_{\mathrm{JS}}(\mathbf{p}^1, \mathbf{p}^2) = \frac{1}{2}\Big(D_{\mathrm{KL}}(\mathbf{p}^1 : \mathbf{p}^{\mathrm{mix}}) + D_{\mathrm{KL}}(\mathbf{p}^2 : \mathbf{p}^{\mathrm{mix}})\Big). \tag{8.25}$$

Clearly, this measure is now symmetric with respect to distribution $\mathbf{p}^1$ and $\mathbf{p}^2$. Also, because proportions are nonnegative, $p_c^{\mathrm{mix}} = 0$ if and only if $p_c^1 = 0$ and $p_c^2 = 0$, the problem of dividing by 0 can be avoided: just remove bin c itself from the comparison.

In the context of the raw CAM set $\mathcal{K}$, we may want to evaluate the difference between any one of the distributions $\mathbf{p}^{\mathcal{W}_c}$ $c = 1, \cdots, k$, and the before-cluster distribution $\mathbf{p}^{\omega}$ of words in Ω by computing $D_{\mathrm{KL}}(\mathbf{p}^{\mathcal{W}_c} : \mathbf{p}^{\omega})$. More informatively, however, the Jensen–Shannon divergence measure can be extended to estimate that average divergence across the ensemble of discrete distributions $\mathbf{p}^{\mathcal{W}_c}$ by generating a mixed distribution $\mathbf{p}^{\mathrm{mix}}$ using a version of Eq. 8.24 extended from 2 to k distributions. Accounting for the relative size p_c^{κ} of each of the subsets $\mathcal{W}_c$ (Eq. 8.8), the resulting Jensen–Shannon ensemble divergence is then given by

$$\begin{aligned} D_{\mathrm{JS}}^{\mathrm{ens}}(\mathbf{p}^{\mathcal{W}_1}, \cdots, \mathbf{p}^{\mathcal{W}_k}) &= \sum_{c=1}^{k} p_c^{\kappa} D_{\mathrm{KL}}(\mathbf{p}^{\mathcal{W}_c}, \mathbf{p}^{\mathrm{mix}}) \\ &= \sum_{c=1}^{k} p_c^{\kappa} \sum_{\ell=1}^{n^m} p_\ell^{\mathcal{W}_c} \log_2 p_\ell^{\mathcal{W}_c} - \sum_{\ell=1}^{n^m} p_\ell^{\mathrm{mix}} \log_2 p_\ell^{\mathrm{mix}} \\ &= H^{\omega}(\mathbf{p}^{\mathrm{mix}}) - \sum_{c=1}^{k} p_c^{\kappa} H^{\omega}(\mathbf{p}^{\mathcal{W}_c}). \end{aligned} \tag{8.26}$$

Thus, the Jensen–Shannon divergence, used to measure the coding efficacy of track segmentation into raw CAMs, comes down to the difference between the entropy of i) the word distribution before CAM clustering (i.e., $H(\mathbf{p}^{\mathrm{mix}})$) and ii) a weighted average of the entropies of the k clusters after clustering. The error rate E^{κ} (Eq. 8.16) can then used to assess the error of assigning raw to rectified CAMs.

8.6 Discussion

The formulation provided here only considers information that can be extracted from the relocation time series $\mathcal{T}^{\mathrm{loc}}$ itself, whereas these data can be highly context-dependent. The movement of individuals can be both individual-specific and noisy

[38]. It is likely to be affected by the internal state of an individual including hunger [39], thirst [40], age and sex [24], nutritional state [41], health [42], not to mention an individual's "personality" or syndromic movement type [29, 43, 44]. It is also affected by the external state and structure of an individual's immediate [45] and remembered [46–48] environment, landscape structure, and resource distribution [49]. Immediate external factors include current weather, the presence of allies, competitors, and predators, as well as a perception of danger from enemies result in a so-called "landscape of fear" [50]).

When auxiliary data such as accelerometer, heart rate, skin temperature, and barometric readings are collected from an individual along with GPS location data [3, 17, 51], then such data can be incorporated into the formulation to fine-tune the informational aspects of movement behavior. The most obvious way to do this is to condition the occurrence and autocorrelations associated with probabilities and sequencing of StaMEs within words and CAMs within BAMs and to incorporate these changes into the various informational and divergence measures presented here.

At the BAM level, the motivational state of animal, though influenced by the internal and external variables mentioned above, will also be conditioned by diel-cycle time parameters (e.g., when to sleep, rest, search for food, and feed). In addition, when an animal commutes on the landscape, its direction and speed of heading are likely to be influenced by navigational beacons that relate to an individual's knowledge of its landscape [46–48]. Such complexities can only be considered once suitable data exist that link an individual's behavior to such beacons, but the next step remains to see how useful the concepts presented here are when applied to real data. These data, of course, must be of sufficient resolution to render the identification of StaMEs a useful enterprise [25, 52], rather than the enterprise of directly identifying BAMs using BCPA methods on data that has a resolution of minutes [9] rather than the needed range of seconds or tens of seconds.

The informational approach presented here to segmenting movement tracks of animals, as a way of identifying various factors that influence the structure of animals movement path, provides a rigorous basis for comparing movement tracks of individuals as a function of internal, external, and idiosyncratic factors that ultimately determine the movement behavior and trajectory of individuals. The theory and methods articulated provide for the first time a way to rigorously address such questions as "Does the information content of an individual increase as it learns to navigate its landscape?" or "Is the information content of an individual reduced when it is infected with parasites?"

At the cusp of the "big data" information age in movement ecology [1], it is only fitting that we have a way of coding movement tracks and assessing and comparing the efficacy of these codes across different parameter sets and clustering algorithms. Further, as other kinds of data derived from various types of inertial sensors (e.g., gyroscopes, magnetometers, accelerometers) are collected, such data can be used to supplement location data in characterizing segments of fixed size (either time-wise or step-wise).

Auxiliary data may also become an increasingly important augmentation of relocation time series collected at relatively high frequencies in the context of compensating for observation errors associated with GPS location data. These location data currently have a CEP (circular error probable) of a few meters, depending on the particular hardware used to collect the data [9]. Averaging across μ steps, however, will considerably reduce this error [53]. The addition of other data types, such as accelerometer data, is likely to increase the statistical robustness of identifying the StaMEs that underpin different types of movement, particular relatively low accelerative (e.g. gliding or striding) versus relatively high accelerative (e.g., jumping or swooping) modes of movement.

The method presented in this chapter holds no matter what type of movement data are used to define a movement track, provided suitable AI clustering methods can be applied to the identification of a robust set of StaMEs. Thus, performance-measure-supported method provides a way forward to implementing a fine-scale structural analyses of animal movement tracks in the decades to come.

Acknowledgments I would like to thank Luca Giuggioli and Ran Nathan for inviting me to participate in the program "Mathematics of Movement: an interdisciplinary approach to mutual challenges in animal ecology and cell biology" held at the Isaac Newton Institute for Mathematical Sciences in 2023 and where part of the work on this chapter was undertaken and supported by the EPSRC Grant Number EP/R014604/1. WG acknowledges funding from A Starker Leopold Endowed Chair in Wildlife Biology at the University of California, Berkeley.

I would like to thank Varun Sethi for constructive discussions regarding the development of the approach taken here and for taking the time to provide comments on this chapter, as well as Richard Salter and Orr Spiegel for discussions over the past year regarding animal track segmentation. In addition, and in approximate chronological order, I have benefited over the past 20 years from the many discussions I have had with the following students and colleagues who have influenced my thinking within the field of spatial and movement ecology: Norman Owen-Smith, Jessica Redfern, Peter Baxter, Chris Wilmers, Paul Cross, Jamie Lloyd-Smith, George Wittemyer, Sadie Ryan, Shirli Bar-David, David Saltz, Ran Nathan, Craig Tambling, Leo Polansky, Andy Lyons, Wendy Turner, Steve Bellan, Alex Boettiger, Richard Salter, Werner Kiliam, Royi Zidon, Scott Fortmann-Roe, Miri Tsalyuk, Briana Abrams, Dana Seidel, Eric Dougherty, Colin Carlson, Krti Tallam, Jason Blackburn, Wayne Linklater, Yen-Hua Huang, and Ludovica Luisa Vassat.

References

1. Nathan, R., Monk, C.T., Arlinghaus, R., Adam, T., Alós, J., Assaf, M., Baktoft, H., Beardsworth, C.E., Bertram, M.G., Bijleveld, A.I., et al.: Big-data approaches lead to an increased understanding of the ecology of animal movement. Science **375**(6582), eabg1780 (2022)
2. Getz, W.M., Saltz, D.: A framework for generating and analyzing movement paths on ecological landscapes. Proc. Natl. Acad. Sci. **105**(49), 19066–19071 (2008)
3. Nathan, R., Spiegel, O., Fortmann-Roe, S., Harel, R., Wikelski, M., Getz, W.M.: Using tri-axial acceleration data to identify behavioral modes of free-ranging animals: general concepts and tools illustrated for griffon vultures. *Journal of Experimental Biology*, 215(6):986–996, 2012.
4. Getz, W.M., Salter, R., Sethi, V., Cain, S., Spiegel, O., Toledo, S.: The statistical building blocks of animal movement simulations. Movement Ecol. **12**, 67 (2024)

5. Edelhoff, H., Signer, J., Balkenhol, N.: Path segmentation for beginners: an overview of current methods for detecting changes in animal movement patterns. Movement Ecol. **4**(1), 21 (2016)
6. Gurarie, E., Andrews, R.D., Laidre, K.L.: A novel method for identifying behavioural changes in animal movement data. Ecol. Lett. **12**(5), 395–408 (2009)
7. Chen, J., Gupta, A.K.: Parametric Statistical Change Point Analysis: With Applications to Genetics, Medicine, and Finance. Springer, Berlin (2011)
8. Gurarie, E., Bracis, C., Delgado, M., Meckley, T.D., Kojola, I., Wagner, C.M.: What is the animal doing? Tools for exploring behavioural structure in animal movements. J. Anim. Ecol. **85**(1), 69–84 (2016)
9. Teimouri, M., Indahl, U.G., Sickel, H., Tveite, H.: Deriving animal movement behaviors using movement parameters extracted from location data. ISPRS Int. J. Geo-Inform. **7**, 78 (2018)
10. Gundermann, K.P., Diefenbach, D.R., Walter, W.D., Corondi, A.M., Banfield, J.E., Wallingford, B.D., Stainbrook, D.P., Rosenberry, C.S., Buderman, F.E.: Change-point models for identifying behavioral transitions in wild animals. Movement Ecol. **11**(1), 65 (2023)
11. Thompson, P.R., Harrington, P.D., Mallory, C.D., Lele, S.R., Bayne, E.M., Derocher, A.E., Edwards, M.A., Campbell, M., Lewis, M.A.: Simultaneous estimation of the temporal and spatial extent of animal migration using step lengths and turning angles. Movement Ecol. **12**(1), 1 (2024)
12. Franke, A., Caelli, T., Hudson, R.J.: Analysis of movements and behavior of caribou (Rangifer tarandus) using hidden Markov models. Ecol. Model. **173**(2–3), 259–270 (2004)
13. Langrock, R., King, R., Matthiopoulos, J., Thomas, L., Fortin, D., Morales, J.M.: Flexible and practical modeling of animal telemetry data: hidden Markov models and extensions. Ecology **93**(11), 2336–2342 (2012)
14. Michelot, T., Langrock, R. and Patterson, T.A.: movehmm: an r package for the statistical modelling of animal movement data using hidden Markov models. Methods Ecol. Evol. **7**(11), 1308–1315 (2016)
15. Zucchini, W., MacDonald, I.L., Langrock, R.: Hidden Markov Models for Time Series: An Introduction Using R. Chapman and Hall/CRC (2016)
16. Pohle, J., Langrock, R., Van Beest, F.M., Schmidt, N.M.: Selecting the number of states in hidden Markov models: pragmatic solutions illustrated using animal movement. J. Agric. Biol. Environ. Stat. **22**(3), 270–293 (2017)
17. Seidel, D.P., Dougherty, E., Carlson, C., Getz, W.M.: Ecological metrics and methods for GPS movement data. Int. J. Geograph. Inform. Sci. **32**(11), 2272–2293 (2018)
18. Getz, W.M.: A hierarchical path-segmentation movement ecology framework. Ecol. Process. **11**(1), 1–15 (2022)
19. Kassambara, A.: Practical Guide to Cluster Analysis in R: Unsupervised Machine Learning, vol. 1. Sthda (2017)
20. Jaeger, A., Banks, D.: Cluster analysis: a modern statistical review. Wiley Interdiscip. Rev. Comput. Stat. **15**(3), e1597 (2023)
21. Murtagh, F. and Legendre, P.: Ward's hierarchical agglomerative clustering method: which algorithms implement ward's criterion? J. Classif. **31**(3), 274–295 (2014)
22. Shannon, C.E.: A mathematical theory of communication. Bell Syst. Tech. J. **27**, 379–423 (1948)
23. Shannon, C.E.: A mathematical theory of communication. Bell Syst. Tech. J. **27**, 623–656 (1948)
24. Luisa Vissat, L., Cain, S., Toledo, S., Spiegel, O. and Getz, W.M.: Categorizing the geometry of animal diel movement patterns with examples from high-resolution barn owl tracking. Movement Ecol. **11**(1), 1–20 (2023)
25. Getz, W.M.: An animal movement track segmentation framework for forecasting range adaptation under global change. Front. Ecol. Evol. **11**, 1171169 (2023)
26. Owen-Smith, N., Goodall, V.: Coping with savanna seasonality: comparative daily activity patterns of African ungulates as revealed by GPS telemetry. J. Zool. **293**(3), 181–191 (2014)
27. Klarevas-Irby, J.A., Farine, D.R.: Diel patterns of movement reveal temporal strategies during dispersal. Anim. Behav. **207**, 119–129 (2024)

28. Abrahms, B., Seidel, D.P., Dougherty, E., Hazen, E.L., Bograd, S.J., Wilson, A.M., Weldon McNutt, J., Costa, D.P., Blake, S., Brashares, J.S., et al.: Suite of simple metrics reveals common movement syndromes across vertebrate taxa. Movement Ecol. **5**(1), 1–11 (2017)
29. Spiegel, O., Leu, S.T., Bull, C.M. and Sih, A.: What's your move? Movement as a link between personality and spatial dynamics in animal populations. Ecol. Lett. **20**(1), 3–18 (2017)
30. Mayner, W.G., Marshall, W., Albantakis, L., Findlay, G., Marchman, R., Tononi, G.: Pyphi: a toolbox for integrated information theory. PLoS Comput. Biol. **14**(7), e1006343 (2018)
31. Biswas, A., Guha, A.: Time series analysis of categorical data using auto-mutual information. J. Stat. Plan. Inference **139**(9), 3076–3087 (2009)
32. Alloghani, M., Al-Jumeily, D., Mustafina, J., Hussain, A., Aljaaf, A.J.: A systematic review on supervised and unsupervised machine learning algorithms for data science. In: Supervised and Unsupervised Learning for Data Science, pp. 3–21 (2020)
33. Varun Sethi, Richard Spiegel, Orrand Salter, Shlomo Cain, Sivan Toledo, and Wayne M Getz. An information theory assessment of animal track hierarchical segmentation methods (2024). bioRxiv
34. Jahn, A.E., Cueto, V.R., Fontana, C.S., Guaraldo, A.C., Levey, D.J., Marra, P.P., Ryder, T.B.: Bird migration within the neotropics. Auk **137**(4), ukaa033 (2020)
35. Jacobs, J., Kahana, M.J., Ekstrom, A.D., Mollison, M.V., Fried, I.: A sense of direction in human entorhinal cortex. Proc. Natl. Acad. Sci. **107**(14), 6487–6492 (2010)
36. Narazaki, T., Nakamura, I., Aoki, K., Iwata, T., Shiomi, K., Luschi, P., Suganuma, H., Meyer, C.G., Matsumoto, R., Bost, C.A., et al.: Similar circling movements observed across marine megafauna taxa. Iscience **24**(4), 102221 (2021)
37. Manning, C., Schutze, H.: Foundations of Statistical Natural Language Processing. MIT Press, Cambridge (1999)
38. Shaw, A.K.: Causes and consequences of individual variation in animal movement. Movement Ecol. **8**(1), 12 (2020)
39. Hansen, M.J., Schaerf, T.M., Ward, A.J.: The effect of hunger on the exploratory behaviour of shoals of mosquitofish Gambusia holbrooki. Behaviour **152**(12–13), 1659–1677 (2015)
40. Owen-Smith, N., Hopcraft, G., Morrison, T., Chamaillé-Jammes, S., Hetem, R., Bennitt, E., Van Langevelde, F.: Movement ecology of large herbivores in African savannas: current knowledge and gaps. Mammal Rev. **50**(3), 252–266 (2020)
41. Bazazi, S., Romanczuk, P., Thomas, S., Schimansky-Geier, L., Hale, J.J., Miller, G.A., Sword, G.A., Simpson, S.J., Couzin, I.D.: Nutritional state and collective motion: from individuals to mass migration. Proc. Roy. Soc. B: Biol. Sci. **278**(1704), 356–363 (2011)
42. Binning, S.A., Shaw, A.K., Roche, D.G.: Parasites and host performance: incorporating infection into our understanding of animal movement. Integrative Comparat. Biol. **57**(2), 267–280 (2017)
43. Spiegel, O., Leu, S.T., Sih, A., Godfrey, S.S., Bull, C.M.: When the going gets tough: behavioural type-dependent space use in the sleepy lizard changes as the season dries. Proc. Roy. Soc. B: Biol. Sci. **282**(1819), 20151768 (2015)
44. Hertel, A.G., Niemelä, P.T., Dingemanse, N.J., Mueller, T.: A guide for studying among-individual behavioral variation from movement data in the wild. Movement Ecol. **8**, 1–18 (2020)
45. Wittemyer, G., Northrup, J.M., Bastille-Rousseau, G.: Behavioural valuation of landscapes using movement data. Philos. Trans. Roy. Soc. B **374**(1781), 20180046 (2019)
46. Fagan, W.F., Lewis, M.A., Auger-Méthé, M., Avgar, T., Benhamou, S., Breed, G., LaDage, L., Schlägel, U.E., Tang, W.-W., Papastamatiou, Y.P., et al.: Spatial memory and animal movement. Ecol. Lett. **16**(10), 1316–1329 (2013)
47. Polansky, L., Kilian, W., Wittemyer, G.: Elucidating the significance of spatial memory on movement decisions by African savannah elephants using state–space models. Proc. Roy. Soc. B: Biol. Sci. **282**(1805), 20143042 (2015)
48. Abrahms, B., Hazen, E.L., Aikens, E.O., Savoca, M.S., Goldbogen, J.A., Bograd, S.J., Jacox, M.G., Irvine, L.M., Palacios, D.M., Mate, B.R.: Memory and resource tracking drive blue whale migrations. Proc. Natl. Acad. Sci. **116**(12), 5582–5587 (2019)

49. Thurfjell, H., Ciuti, S., Boyce, M.S.: Applications of step-selection functions in ecology and conservation. Movement Ecol. **2**(1), 4 (2014)
50. Bleicher, S.S.: The landscape of fear conceptual framework: definition and review of current applications and misuses. PeerJ **5**, e3772 (2017)
51. de Müllenheim, P.-Y., Chaudru, S., Emily, M., Gernigon, M., Mahé, G., Bickert, S., Prioux, J., Noury-Desvaux, B., Le Faucheur, A.: Using GPS, accelerometry and heart rate to predict outdoor graded walking energy expenditure. J. Sci. Med. Sport **21**(2), 166–172 (2018)
52. Getz, W.M., Salter, R., Vissat, L.L.: Simulation applications to support teaching and research in epidemiological dynamics. BMC Med. Educ. **22**(1), 632 (2022)
53. Signer, J., Fieberg, J., Avgar, T.: Animal movement tools (amt): R package for managing tracking data and conducting habitat selection analyses. Ecol. Evol. **9**(2), 880–890 (2019)

Chapter 9
Some Statistical Mechanics Techniques for Ecology: Random Walks for Epidemics Spread and Kinetic Equations for Patchiness

V. M. Kenkre

9.1 Introduction

Statistical Mechanics provides tools for analysis and description, and indeed an investigative approach, to phenomena in a variety of disciplines. This chapter is an attempt to show examples in the study of ecology. One of the phenomena is the spread of epidemics, and the principal technique employed is specialized random walks. Let me begin with a few preliminary words about the role Statistical Mechanics plays in theoretical science. Mechanics is that part of physics that describes motion. When the moving bodies under investigation are enormous in number, not merely thousands or trillions but *zillions* if we may use such a curious word, as in the case of atoms in a solid or molecules in a liquid or gas, the discipline is called Statistical Mechanics because it obviously requires statistical methods and procedures. Although these are the origins of Statistical Mechanics, the introduction of powerful methods and techniques developed under its wings by brilliant expositors such as Boltzmann[1] have made it a rich and mighty subject

[1] Boltzmann is one of the patriarchs of Statistical Mechanics and often considered by many the father of the subject. In a blend of humility and pride I recall what my thesis advisor, Max Dresden, said to me during my PhD thesis defense in 1971: "Boltzmann's star student was Ehrenfest, his was Uhlenbeck who supervised my thesis. Don't ever forget your illustrious lineage." Needless to say, I never have. I feel honored to carry this information in my mind with delight although I can take absolutely no credit for it. I must confess I have occasionally used the connection to lure brilliant students into my research lair.

V. M. Kenkre (✉)
Department of Physics and Astronomy, University of New Mexico, Albuquerque, NM, USA

L. Giuggioli, P. K. Maini (eds.), *The Mathematics of Movement*,
https://doi.org/10.1007/978-3-032-00873-2_9

on its own account. It has taken the form of a language or a mode of thinking, and its special characteristics are used in the description of a multitude of physical phenomena ranging from gas dynamics and the science of materials to virology and the control of traffic.

This chapter is based on, and summarizes in brief, two invited lectures presented in the summer of 2023 at the Conference/Workshop in Cambridge whose proceedings constitute the present volume. The title of the first was *Infection Transmission among Rodents during a Random Walk: Exact Model Calculations* while that of the second was *Masked Bacteria in a Petri dish: How to Wield the Fisher Equation without Sweat*. The next six sections of this chapter refer to the first talk. The following five refer to the second. A concluding section gives a summary and also alludes passingly to a third lecture that I was scheduled to present but unable to give for extraneous reasons.

9.2 A Mechanistic Aproach for Infection Transmission in the Hantavirus Epidemic

Various details, calculational and pictorial (graphical), may be found by the reader in the original publications cited in the body of this chapter. What served its purpose well in the lecture that was presented at the Cambridge conference was a brief story of the logic of how the theory was constructed. Because I think that it will also be useful here, I begin with it below.

The story took place more than a couple of decades ago during a 1-day workshop I arranged at the University of New Mexico in Albuquerque. Two leading observational biologists, Terry Yates and Bob Parmenter, who had been part of the group that discovered the Hantavirus epidemic in the Four Corners area of the USA decades ago, were specifically asked to describe to us the *smoking guns* of the epidemic. By that curious phrase we mean, of course, salient features that might guide a serious theorist in the construction of an appropriate description of the epidemic. The two biologists instructed us about various characteristic of the Hantavirus scenario including refugia of infection (places where the Hantavirus infection persists in focused spatial regions in off-periods of the epidemic), the influence of precipitation as caused by storms and rainfall on the timing and intensity of the epidemic, about the link between the virus causing the epidemic and the carriers of the disease that happen to be deer mice (Peromyscus Maniculatus), and a number of other relevant characteristics. Two eager theorists, Guillermo Abramson from Argentina and myself, pounced upon the information and, guided closely by the experimental observations, constructed a simple theory of the epidemics spread whose essence is described in the next section.

9.3 The Kinetic Equation Approach: Fisher Equation with Internal States

The basic idea emerging from what we learnt from the biologists is that the Hantavirus epidemic is spread by the motion of the mice on the terrain, the encounters between infected mice and uninfected mice, the so-called *susceptible* mice, being the cause of the spread. The natural easy choice for our theory that kept the observations carefully in mind was to take as our point of departure two coupled equations for the densities of the two kinds of mice. The motion of the mice was considered to be simple random walks and was incorporated in the equations as simple diffusion of the density of mice. Terms were added that described the birth and the death of the mice and their interaction with the environment in the simplest possible manner. The equations took the form

$$\begin{aligned}\frac{\partial M_I}{\partial t} &= -c\, M_I - \frac{M_I M}{K} + a\, M_S M_I + D\,\nabla^2 M_I,\\ \frac{\partial M_S}{\partial t} &= b\, M - c\, M_S - \frac{M_S M}{K} - a\, M_I M_S + D\,\nabla^2 M_S. \end{aligned} \tag{9.1}$$

Here M is the density of the mice (or rodents that are responsible for carrying the Hantavirus), the suffix I or S denoting that they are infected or uninfected (susceptible). Without suffix the symbol represents the sum $M_I + M_S$. The last term on the right-hand side of the two equations describes their motion on the terrain, visualized as being simple random walks with diffusion constant D. The other terms are characterized by parameters c, b, $1/K$, and a. They represent, respectively, death, birth, competitive interaction with the environment, and aggressive encounters of infected animals with uninfected ones that increase the number of the former and reduce in consequence the latter. [2] A number of observable features that characterize the Hantavirus are built into these equations such as that no mice are ever born with infection, that infection does not impair motion or health of the mice (unlike in a case such as plague), and that there is no category of recovered rodents.

The approach [1] was clearly of coupled kinetic equations of the Boltzmann type formulated in what Ehrenfest termed (in the context of gas molecules) the μ space that describes the interacting mice populations. A mathematical way of regarding the starting point is thus in terms of a Fisher-KPP equation [2, 3] with internal states. Relative to original simplistic descriptions of epidemic dynamics, these equations can be said to treat physical space (in which the virus-carrying animals move) as a feature of the system from the very beginning: the approach follows lines similar to Boltzmann's description of gas dynamics. Various nuances that go into

[2] Like the other terms in the equation, the competitive interaction term also follows standard usage in the literature. It is represented as a decrease in the population proportional to the product of the density of the mice under consideration (susceptible or infected) with the density of all the mice (total), the proportionality constant being the environment parameter $1/K$.

the construction of the theory will be appreciated by the reader if reference is made to the original papers such as [1] or the expository book [4].

The theory met with reasonable success, at least semiquantitatively, in that it addressed a number of observations adequately. Chief among them were three: temporal correlations between precipitation such as from El Niño and the timing of the spread of the Hanta epidemic; spatial phenomenon of the formation of refugia; and the onset of traveling waves of infection that emanate from infection centers. The interested reader will find these described in [1, 4–7].

An important undertaking was the quantitative determination of the parameters of the theory, the principal among them being D, the diffusion constant of the mice. A careful study was undertaken of copious data available for the motion of mice collected from mark-recapture observations in two locations, in New Mexico and in Panama [8, 9], for which my theoretical collaborators were Luca Giuggioli and Guillermo Abramson. The basic reasoning behind the analysis was Einstein's relationship between the mean square displacement and the time, which are related in the well-known linear proportionality controlled by D in ordinary random walks. Satisfactory results were obtained including the fact that the deduced magnitudes for D were different in the two locations appropriate to the fact that New Mexico is a desert and Panama has much more vegetation.

9.4 Observational Input Leading to a Modification and Extension

Although the extraction of the diffusion constant of mice in the two sites studied thus worked well, a couple of surprises arose. In both cases, a plot of the time dependence of the observed mean square displacement showed its initial linear rise with time as expected, but then a puzzling saturation to a constant value. The values of D deduced from the slope of the initial rise were satisfactory, but why did the saturation occur? The saturation implies the existence of a characteristic distance in the system. Two different possibilities suggest themselves. One stems from the character of the observational technique used (finite extension of the region in which the traps to capture mice are placed in the terrain) and the other from the biological propensity of each animal to roam preferentially in one spatial location and return repeatedly to that location, its home.

An ingenuous method of analysis was used to disentangle the interfering effects of the two sources and thereby to eliminate the first of these lengths from consideration (see [4, 10]), but the biological length had to be dealt with. We have to contend here not with simple random walks but *tethered random walks*, each mouse attracted to a different location while it moves otherwise randomly during its interactions with the other.

The fascinating problem of how to extend the AK theory of [1] to incorporate the tethering required by home range considerations was approached in three different ways (see [4] for a detailed explanation), but the most reasonable solution was to study a simple model of two mice, one infected and one susceptible, tethered to *two separate* homes. That is the solution that will be explained in the following. The approach is simple in concept. Despite its simplicity, it has the advantage that, once the system is set up, exact *analytical* calculations can be performed. They lead directly and explicitly to the time development of infection. The difficult but highly interesting problem the analysis entails is described in the following two sections along with a somewhat unexpected result that emerges from the calculations.

9.5 Model and Exact Calculation

The previous sections have brought us to the point where one of the two main contents of this chapter begins. The approach we will take to understand the transmission of Hantavirus infection from an infected mouse (that we will label by the suffix 1) to an uninfected mouse (that we will label by the suffix 2), both of them executing random walks while being tethered (attracted) to *separate locations*, their homes, is to abandon our earlier μ-space description of the AK equations for a simple but exact two-mice analysis [11].

We denote by $\boldsymbol{R}_1$ and $\boldsymbol{R}_2$ the locations of the respective mice and assume as is appropriate from Hantavirus data that the diffusion constant for both is the same D. The focus of our study is the joint probability density $P(\boldsymbol{r}_1, \boldsymbol{r}_2, t)$ that the infected mouse is at $\boldsymbol{r}_1$ and the uninfected one at $\boldsymbol{r}_2$ at time t. A moment's reflection should make clear that it vanishes when the uninfected mouse acquires infection. We take this process to occur at a rate controlled by the parameter C. While one can easily specify a non-zero range of interaction for the passage of infection (and we have done so explicitly in the published articles [4, 12], we initially formulate the possibility of infection passing to the susceptible animal when the two animals, regarded as point particles, occupy the *same* location. Our point of departure is the following evolution equation for $P(\boldsymbol{r}_1, \boldsymbol{r}_2, t)$:

$$\begin{aligned} \frac{\partial P}{\partial t} &= \nabla_1 \cdot [\gamma\,(\boldsymbol{r}_1 - \boldsymbol{R}_1)\,P] + \nabla_2 \cdot [\gamma\,(\boldsymbol{r}_2 - \boldsymbol{R}_2)\,P] \\ &\quad + D\left(\nabla_1^2 + \nabla_2^2\right) P - \delta(\boldsymbol{r}_1 - \boldsymbol{r}_2) C P. \end{aligned} \tag{9.2}$$

The vector notation makes clear that we have set up the equation as applicable for arbitrary number of dimensions (which we will denote by s). The crucial new element forced upon us by the field observations explained in the previous section is the saturation of the mean square displacement at large times. It stems from the attraction of the two mice to their homes. The representative parameter that controls this attraction is γ. Like the diffusion constant D, it is taken to be the same for the two mice, infected and susceptible, although the locations of the two homes are explicitly different from each other: $\boldsymbol{R}_1$ is different from $\boldsymbol{R}_2$.

Before describing the analytical procedure to solve this equation exactly for $P(\boldsymbol{r}_1, \boldsymbol{r}_2, t)$, let me clarify how we can use that solution to get the physical information we desire. Once we have solved for $P(\boldsymbol{r}_1, \boldsymbol{r}_2, t)$ explicitly, we can get, by integrating it over all space, the infection probability $\mathcal{I}(t)$. That quantity is our central focus:

$$\mathcal{I}(t) = 1 - \int_{-\infty}^{\infty} \int_{-\infty}^{\infty} d^s r_1 d^s r_2 \; P(\boldsymbol{r}_1, \boldsymbol{r}_2, t). \tag{9.3}$$

The infection curve ($\mathcal{I}$ plotted against the time t) contains all the information we seek. To facilitate an understanding of the behavior of the entire curve, let us extract from it a single number that describes succintly the rate at which infection occurs throughout the terrain. As will be seen below, the infection curve has an aspect similar to that of an exponential rise. Inspired by this observation, we can define a function $\tilde{\alpha}(\epsilon)$ in the Laplace domain, where ϵ is the Laplace variable, and tildes denote Laplace transforms through

$$\tilde{\mathcal{I}}(\epsilon) = \frac{\tilde{\alpha}(\epsilon)}{\epsilon\,(\epsilon + \tilde{\alpha}(\epsilon))}. \tag{9.4}$$

In order to appreciate why this α is introduced, notice that $\mathcal{I}(t)$ would indeed be an exponential rise if $\tilde{\alpha}(\epsilon)$ were independent of the Laplace variable ϵ in Eq. (9.4).

As will be seen in the ensuing explicit expressions, the ϵ-dependence of $\tilde{\alpha}(\epsilon)$ arises primarily (and fundamentally) from the motion of the mice. If, however, there would be no dependence of $\tilde{\alpha}(\epsilon)$ on the Laplace variable, the infection curve would be precisely an exponential rise $1 - \exp(-\alpha t)$. It is straightforward then to seek a single parameter α to denote the infection rate by taking the Markoffian limit of $\tilde{\alpha}(\epsilon)$:

$$\alpha \equiv \lim_{\epsilon \to 0} \tilde{\alpha}(\epsilon).$$

The single quantity α and the full time dependence $\mathcal{I}(t)$ are the products of our exact analysis. Expressions relating these two to the underlying parameters of our system such as D and C, as well as the relevant graphs, will be shown in the next section. We now complete the elucidation of how the solution to $P(\boldsymbol{r}_1, \boldsymbol{r}_2, t)$ is found, given our starting equation (9.2) for its evolution.

Mathematically regarded, the starting equation is of a type that should be amenable to standard Green function methods if the propagator solutions of the equation with $C = 0$ are known. The seeming many-particle nature of the homogeneous problem (in the system considered, the mice are two in number) can be addressed via a procedure I used in the context of excitons that annihilate mutually when close to one another [13]. What that method entails is to consider a space of twice the number of dimensions as the one in which the two mice move (for instance, 6 in a realistic case and 2 in an illustrative 1-d case) and to perform the calculations for capture *along the diagonal* in this space, that is where $\boldsymbol{r}_1 = \boldsymbol{r}_2$.

All that then remains is the study of the propagators of the Smoluchowski equation that describes the homogeneous problem.

If we denote the propagator (Green function) of the homogeneous multi-dimensional Smoluchowki equation by the symbol $\Pi(\boldsymbol{r}_1, \boldsymbol{r}_1^0, \boldsymbol{r}_2, \boldsymbol{r}_2^0, t)$, where the superscript 0 represents initial locations, we can write down at once the solution of the homogeneous problem ($C = 0$) for any initial locations of the two mice as

$$\eta(\boldsymbol{r}_1, \boldsymbol{r}_2, t) = \int_{-\infty}^{\infty} \int_{-\infty}^{\infty} d^s r_1^0 d^s r_2^0 \; \Pi(\boldsymbol{r}_1, \boldsymbol{r}_1^0, \boldsymbol{r}_2, \boldsymbol{r}_2^0, t) P(\boldsymbol{r}_1^0, \boldsymbol{r}_2^0, 0). \tag{9.5}$$

In the presence of infection, as a consequence of the linearity of the equation of evolution, we can further write for the full inhomogeneous situation,

$$\begin{aligned} P(\boldsymbol{r}_1, \boldsymbol{r}_2, t) =& \eta(\boldsymbol{r}_1, \boldsymbol{r}_2, t) \\ & - C \int_0^t dt' \int_{-\infty}^{\infty} d^s r'1 \; \Pi(\boldsymbol{r}_1, \boldsymbol{r}'1, \boldsymbol{r}_2, \boldsymbol{r}'1, t - t') P(\boldsymbol{r}'1, \boldsymbol{r}'1, t'). \end{aligned} \tag{9.6}$$

As it stands, this is by no means a solution for $P(\boldsymbol{r}_1, \boldsymbol{r}_2, t)$ given that the unknown $P(\boldsymbol{r}'1, \boldsymbol{r}'1, t')$ also lies on the right-hand side. However, if we Laplace transform this equation, set $\boldsymbol{r}_1 = \boldsymbol{r}_2$, and integrate over all space of the appropriate number s of dimensions, we obtain

$$\begin{aligned} \int_{-\infty}^{\infty} d^s r_1 \; \tilde{P}(\boldsymbol{r}_1, \boldsymbol{r}_1, \epsilon) =& \int_{-\infty}^{\infty} d^s r_1 \; \tilde{\eta}(\boldsymbol{r}_1, \boldsymbol{r}_1, \epsilon) \\ & - C \int_{-\infty}^{\infty} d^s r'1 \int_{-\infty}^{\infty} d^s r_1 \; \tilde{\Pi}(\boldsymbol{r}_1, \boldsymbol{r}'1, \boldsymbol{r}_1, \boldsymbol{r}'1, \epsilon) \\ & \times \tilde{P}(\boldsymbol{r}'1, \boldsymbol{r}'1, \epsilon). \end{aligned} \tag{9.7}$$

Motivated by the averaging procedure of the so-called ν-function analysis introduced in capture problems [14, 15] (for a relatively recent review and application, see [16]), and assisted by the observation that the integral of $\tilde{\Pi}(\boldsymbol{r}_1, \boldsymbol{r}'1, \boldsymbol{r}_1, \boldsymbol{r}'1, \epsilon)$ over the entire domain of $\boldsymbol{r}_1$ (i.e., all space) appearing in Eq. (9.7) is independent of $\boldsymbol{r}'1$, we introduce the symbol $\tilde{\nu}(\epsilon)$ to denote that integral,

$$\tilde{\nu}(\epsilon) = \int_{-\infty}^{\infty} d^s r_1 \; \tilde{\Pi}(\boldsymbol{r}_1, \boldsymbol{r}'1, \boldsymbol{r}_1, \boldsymbol{r}'1, \epsilon), \tag{9.8}$$

and succeed in obtaining, in the Laplace domain, an *explicit* solution for the joint probability (density) that the two animals occupy the same position,

$$\int_{-\infty}^{\infty} d^s r'1 \; \tilde{P}(\boldsymbol{r}'1, \boldsymbol{r}'1, \epsilon) = \frac{\tilde{\mu}(\epsilon)}{1 + C\tilde{\nu}(\epsilon)}. \tag{9.9}$$

In light of Eq. (9.3), which yields the infection curve $\mathcal{I}$ simply from the left side of Eq. (9.9), we can be said to have completely solved our problem in a formal sense: $\mathcal{I}$ is determined in terms of the motion quantities ν and μ.

The two motion quantities mentioned are key to the analysis and have a simple physical meaning. The first of these, $\nu(t)$ in the time domain, is the probability (density) that the locations of the two animals coincide (whatever that location) if at a time t earlier their locations also coincided. The second key quantity, $\mu(t)$, whose Laplace transform is

$$\tilde{\mu}(\epsilon) = \int_{-\infty}^{\infty} d^s r'1\ \tilde{\eta}(\boldsymbol{r}'1, \boldsymbol{r}'1, \epsilon)$$
$$= \int_{-\infty}^{\infty} d^s r'1 \int_{-\infty}^{\infty}\int_{-\infty}^{\infty} d^s r_1^0 d^s r_2^0\ \tilde{\Pi}(\boldsymbol{r}'1, \boldsymbol{r}_1^0, \boldsymbol{r}'1, \boldsymbol{r}_2^0, \epsilon) P(\boldsymbol{r}_1^0, \boldsymbol{r}_2^0, 0), \tag{9.10}$$

is the probability (density) that the two animals occupy the same location at the present time (whatever that location) if at a time t earlier they occupied locations as per the *given initial condition* of the problem. Both refer to the homogeneous situation, i.e., the problem without infection ($C = 0$). They are integrals (over the s-dimensional space) of the two-particle joint probability density and have the dimensions of reciprocal length raised to s. The rest of the calculation is straightforward. Knowledge of the propagators of the system (generally in the presence of constraining potentials to represent the attraction to the homes) gives ν and, in combination with the given initial conditions, yields μ. The two, together with Eq. (9.9), provide all that is necessary to obtain the infection probability and the nuances of its behavior. The final explicit expression that gives the infection curve $\mathcal{I}(t)$ quite generally in terms of C and the motion quantities ν and μ is explicit in the Laplace domain:

$$\tilde{\mathcal{I}}(\epsilon) = \frac{1}{\epsilon}\left[\frac{\tilde{\mu}(\epsilon)}{(1/C) + \tilde{\nu}(\epsilon)}\right]. \tag{9.11}$$

Correspondingly, the infection *memory* introduced in Eq. (9.4) is given in the Laplace domain as

$$\tilde{\alpha}(\epsilon) = \frac{\epsilon\tilde{\mu}(\epsilon)}{(1/C) + \tilde{\nu}(\epsilon) - \tilde{\mu}(\epsilon)}, \tag{9.12}$$

with the result that the effective overall rate of transmission of infection has a suggestive form

$$\alpha = \frac{\mu(\infty)}{(1/C) + (1/\mathcal{M})}. \tag{9.13}$$

Here $\mu(\infty)$ is the infinite-time limit of $\mu(t)$ and the motion parameter $\mathcal{M}$ is the reciprocal of the difference in the 0 to ∞ time integrals of $\nu(t)$ and $\mu(t)$:

$$1/\mathcal{M} = \int_0^\infty dt[\nu(t) - \mu(t)]. \tag{9.14}$$

The effective rate α emerges as a product of the probability in the steady state that the two mice occupy the same position, inclusive of whatever the given initial condition, and a combined rate involving the contact (encounter) parameter and the motion parameter. In the contact limit $C << \mathcal{M}$, α equals $C\mu(\infty)$ and in the motion limit $\mathcal{M} << C$, it equals $\mathcal{M}\mu(\infty)$.

9.6 Primary Results and Discussion

It is time now to extract physics (as physicists are wont to say) from our mathematical solutions obtained both for α, the compact effective rate at which infection occurs over the terrain, and for the full infection curve $\mathcal{I}(t)$. Restricting the analysis to mice moving in 1-dimensional space ($s = 1$) is amply adequate to understand most of the primary features. The explicit expressions for $\nu(t)$ and $\mu(t)$ are [11]

$$\nu(t) = \frac{e^{-\frac{H^2}{8D\mathcal{T}(t)}(1-e^{-\gamma t})^2}}{\sqrt{8\pi D\mathcal{T}(t)}}; \quad \mu(t) = \frac{e^{-\frac{H^2}{8D\mathcal{T}(t)}}}{\sqrt{8\pi D\mathcal{T}(t)}}. \tag{9.15}$$

Here h_1 and h_2 are the locations of the homes of the two mice and $H = h_1 - h_2$ is the distance between them. The well-known Smoluchowski time

$$\mathcal{T}(t) = \frac{1 - e^{-2\gamma t}}{2\gamma} \tag{9.16}$$

is the same at short times as ordinary time t but saturates at large times to $1/2\gamma$ to represent the equilibration of the random walking mouse to its being near its home. The Smoluchowski time is controlled solely by the attraction strength γ.

In order to appreciate what our analytical solution adds to our understanding of the phenomena we have been investigating, let us begin with the simpler quantity α. Figure 9.1 describes its behavior. How it depends on the elemental capture rate C_1 (the suffix 1 reminds that this is a 1-dimensional result) is shown on the left side of the graph while how it depends on the strength of attraction of the mice to their homes is shown on the right side.

The left panel is expected. The variation of α is linear in C_1 for small values of the latter but becomes independent of it for large values. The latter situation is one that the motion, controlled by D, determines completely the effective rate of infection. This comes from the faster of the two rates controlling the combined process dropping out of the picture and is well-known from electrical network

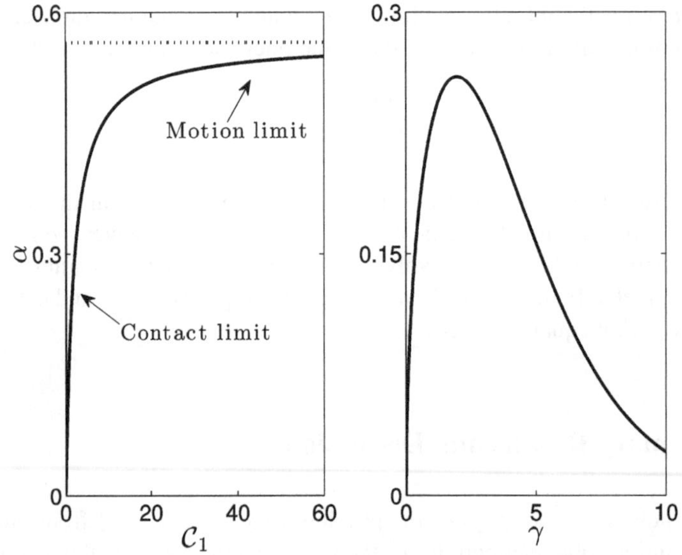

Fig. 9.1 Dependence of α that describes the overall rate at which infection is transmitted, plotted as a function of the contact rate C_1 (scaled to $2D/H$) in the left panel and as a function of γ, the strength of attraction of the mice to their homes, in the right panel. Both α and γ are scaled to $1/\tau_H$. See the text for the meaning of τ_H. The left panel shows an expected result that α is linear in the contact rate for small values of the latter but saturates to the motion-limited value (0.56 in this example) for large values. The right panel contains a surprising result. It shows the non-monotonicity effect on infection: as confinement steepness γ increases, the infection rate α rises to a peak and decreases for larger γ. For the right panel, C_1 in units of $2D/H$ is 15. Reprinted from Fig. 2 of ref. [11]; copyright (2014) with permission from Springer Nature

theory. The saturation of α at high values of the capture rate (for instance infinite) means that the phenomenon is now motion-limited. In mathematical parlance, this is the domain of first passage time analyses [17, 18].

The right part of the graph shows an unexpected and noteworthy result. Having introduced, as a result of field observations, the attraction of the individual mice to their homes of strength controlled by γ, we naturally ask is the transmission of infection *helped* or *hindered* by increasing γ with all other parameters held constant. The answer is a humorous "yes": it can do either. We have here a non-monotonic dependence of the process of infection on γ. Inspect the right panel of the plot. It starts with an increase in α with increasing γ. However, a peak is attained and then α *decreases* with further increase in γ.

Figure 9.2 depicts the same unexpected and meaningful finding in the context of the entire infection curve. The caption should be self-explanatory. Note that the quantity τ_H, used to scale the time t, equals $H^2/2D$, the time required for either animal to traverse diffusively the distance H between the two homes.

Elementary considerations establish that the width of the steady-state distribution of the Smoluchowski random walker arising from the opposing tendencies of

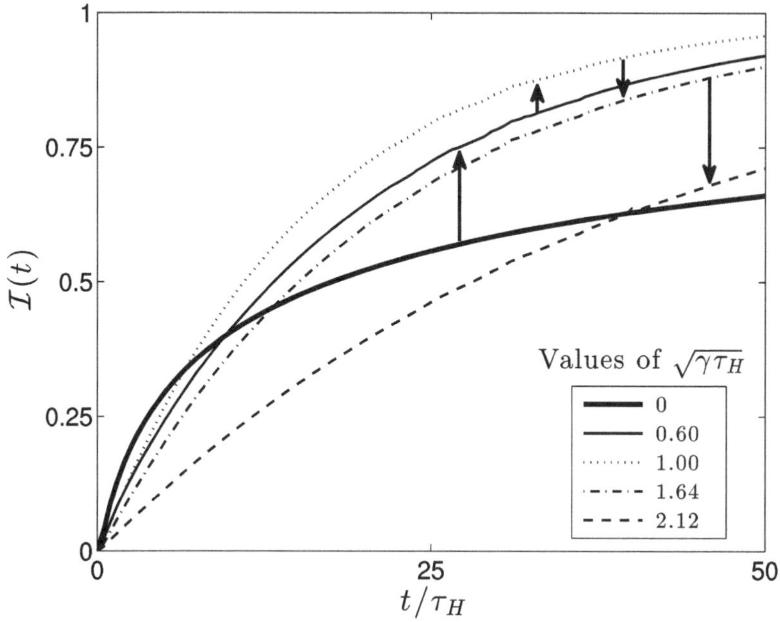

Fig. 9.2 Variation of the infection curve $\mathcal{I}(t)$ with change in γ, the steepness of the potential confining the animals to their home ranges. The value shown for each line in the legend is of $\sqrt{\gamma\tau_H}$. This quantity equals H/σ, the ratio of the inter-home distance to the steady-state Smoluchowski width. Time is scaled to τ_H; C_1 scaled to $2D/H$ is taken to equal 0.3. Starting with the unconfined case (γ=0), that corresponds to the analytic solution Eq. (9.17) below, an increase in γ makes infection more effective for small values of γ *but less effective* for larger values. Reprinted from Fig. 1 of ref. [11]; copyright (2014) with permission from Springer Nature

being attracted to the home (controlled by γ) and of wandering away from home (controlled by D) happens to be $\sqrt{2D/\gamma}$. We will call this distance as σ. Let us keep D and the distance H between the two homes constant. Begin by inspecting the case of no attraction ($\gamma = 0$) depicted by the thickest solid curve in the graph. Increment the strength of attraction γ. The upward arrow shows that infection spreads faster. A further increase shows the same tendency. However, if one increases γ more, the downward arrow shows that the tendency reverses itself and infection now spreads *slower*. This is shown in one more case in the figure.

This is the same phenomenon as represented by the initial rise and subsequent fall of α in the right panel of Fig. 9.1. This behavior is obviously a new effect predicted by our analysis. It is also observed if the diffusion constant of the animals is varied keeping the potential steepness constant. It arises from the interplay of three quantities, the diffusion constant D, the steepness γ and the inter-center distance H, which here is also the distance between the initial locations of the animals. For a given value of H, changes in D or γ exhibit the phenomenon. Varying H does not: maximum transmission occurs when $H = 0$, i.e., when the animals do not have to move to find each other for the infection to be propagated. The key parameter is $\gamma \tau_H = H^2\gamma/2D$ that is nothing other than $(H/\sigma)^2$: for a given H, optimum transmission of infection occurs when the parameter equals 1, particularly in the contact-limited case. More generally, the critical value is different from 1. By the term contact-limited, we mean the case that C_1 is much smaller than the corresponding motion parameter. In such a case, the contact process of infection when the animals meet, rather than the motion, determines the overall infection event.[3]

A remarkable side result that I found during my 1-dimensional analysis of the problem described was an analytic expression for the infection curve in the time domain in the absence of attraction ($\gamma = 0$). The result is explicit in terms of the complementary error function

$$I(t) = \text{erfc}\left(\sqrt{\frac{\tau_H}{4t}}\right) - e^{\left(\sqrt{\frac{\tau_H}{4t}}+\frac{t}{\theta}\right)}\text{erfc}\left(\sqrt{\frac{\tau_H}{4t}} + \sqrt{\frac{t}{\theta}}\right), \tag{9.17}$$

where the parameter θ is given by $8D/(\pi C_1^2)$. This infection curve is the thickest solid line in Fig. 9.2. To obtain Eq. (9.17), one first shows that ν and μ for this case of no attraction are given by

$$\nu(t) = \frac{1}{\sqrt{8\pi Dt}}; \quad \mu(t) = \frac{1}{\sqrt{8\pi Dt}}e^{-\frac{H^2}{8Dt}}. \tag{9.18}$$

One then uses their Laplace transforms in the expression for the infection curve.

Needless to say, this result is only marginally relevant to our problem of transmission of infection (in the presence of home ranges of mice), but it is analytically satisfying to recover it. It also makes contact with totally unrelated reaction diffusion contexts. See [16, 17, 19, 20].

9.7 Further Studies of Interacting Random Walkers

The original 1-dimensional analysis of our problem may be found in [11]. There is a lot of further work that has been done in related directions, one notable example being the calculations that Sugaya performed in collaboration with me [4, 12] on extensions for higher dimensions. They are not trivial generalizations of a straight-

[3] At the Cambridge workshop, in talks following the presentation summarized here, this non-monotonic effect with respect to confinement strength was spoken about by several speakers and arrived at in different ways. We had discovered it not only in the context of interacting walkers as presented in [11] but even earlier [21] as an inherent feature of the Smoluchowski equation for a *single* random walker. The interested reader will find helpful discussions of its physical origin in that earlier publication including the link to the relevant variation of the equilibrium probablility distribution of the Smoluchowski equation with changes in γ.

forward nature. The reciprocal of the motion parameter $\mathcal{M}$ in Eq. (9.14) involves an improper integral calculable in 1-dimension but not in higher dimensions if mouse encounters are taken to occur at *a point*, i.e., a region of zero dimensions. Subtle analysis is necessary about the relative dimensionality of the space in which mouse motion occurs and of that in which the transmission (encounter) phenomenon occurs. The interested reader will find all that has been shown here explicitly for 1-dimension generalized to higher dimensions with careful examination of all these details. Treated there is also the effect of a finite radius of interaction. When such a finite radius is considered, the possibility of a new phenomenon arises whereby the peak value of the effective rate of transmission, α, can exhibit a sudden resonance increase when the radius of interaction matches the distance between the homes of the two mice. We have used the terms "resonance" and "home sharing" to label this effect. I have not included those useful calculations in this chapter for want of space.

9.8 An Ecological Problem on a Different Spatial Scale: Bacteria

The following sections have a different theme as their focus: not the transmission of infection in an epidemic but patchiness of resources and concomitant abrupt transitions. If we return to Eq. (9.1), add the two evolution equations for the susceptible and infected mice and regard the resulting single equation as for the evolution of the total mice density M, we have the Fisher-KPP equation, which here we will write with different symbols:

$$\frac{\partial u(X,t)}{\partial t} = Au(X,t) - Bu^2(X,t) + D\frac{\partial^2 u(X,t)}{\partial x^2}. \tag{9.19}$$

Corresponding to Eq. (9.1), we have taken A to represent $(b-c)$ and B to represent $1/K$ and considered a single dimension where the position is denoted by X. The density is now $u(X,t)$ rather than M because we are interested not in mice but initially entities on a much different spatial scale, viz., bacteria in a Petri dish.

This so-called Fisher-KPP equation [2, 3] has had a venerable history and describes the time evolution of entities that perform random walks as they move in space while simultaneously increasing their population through a Malthusian growth limited by the nonlinear term quadratic in their density. In the absence of motion, i.e., if $D = 0$, Eq. (9.19) is the well-known logistic equation, whereas if $A = 0 = B$, it is the linear diffusion equation. The reader will find intriguing descriptions of that equation and its applications to ecology in various places in the literature [4, 22, 23].

Two questions had been interesting investigators particularly occupied with bacterial dynamics in Petri dishes. One was whether the Fisher-KPP equation, often used to describe such dynamics, was valid at all, i.e., in accord with experiment, or simply represented the desire of eager theorists to impose their favorite model on the interpretation of observations. The other, independently, centered on understanding what effects patchiness of resources had on the survival and motion of bacteria. The latter investigation followed a curious experimental situation that was in fashion at that time. Ultraviolet light was made to fall on bacteria in a Petri dish leading to their death except under a mask that protected a limited region. The mask was then moved, and the time evolution of the population studied.

It is obviously more simple to ask questions concerning bacterial extinction if the mask is static. With this thought, in collaboration with Marcelo Kuperman, I asked such a question as well as one dealing with the more general issue of whether the use of the Fisher-KPP equation is justified in bacterial contexts. The resulting publication is [24]. Useful considerations emerged in our study of the second issue for which I refer the interested reader to the published article.

To investigate the effects of patchiness of resources, we considered Eq. (9.19) and asked what would be happen if A in it were to have a constant positive value only within a distance of length L (the extent of the mask in the Petri dish) and that the ultraviolet light shining on the Petri dish were lethal enough to make the bacterial population zero outside it. At long times, when steady state has been attained, the time derivative on the left side of the equation vanishes and leaves us with the task of solving a second-order spatial differential equation with the quadratic term in u. The differential equation is to be solved with the boundary condition that u vanishes at the edges of the mask and beyond.

9.9 Description of the Theory and Onset of Transition

Long struggles that the reader can witness in the paper cited [24] led Kuperman and me to show that the steady state $u(x, \infty)$ is proportional to the square of the *cd* elliptic function. We could also obtain

$$L = L_{cr} = \pi\sqrt{D/A} \tag{9.20}$$

as the relation between the parameters of the equation and the critical value L_{cr} of the mask length. If the mask length is less than this critical value, the bacterial population vanishes *abruptly* everywhere under the mask. On the basis of values for D and A known from experiments, we made a prediction that L_{cr} would be of the order of 0.5 cm.

A year or so after our prediction, an experimental verification of our prediction was published by Perry [25], the critical length having been found to be 0.8 cm for *E. Coli*. Our analysis appeared thus to be vindicated.

While we were pleased with this reported verification of what we had thought were our original calculations, we also learnt around this time, quite to our embarrassment,[4] that both the elliptic function solution and the entire effect of the abrupt transition had been discovered decades earlier than our calculation by Skellam [26].

This Skellam result can be understood quite simply by recalling the Einstein result for the mean square displacement that equates the square of the distance covered on the average by a random walker with the product of $2D$ and t, the time traversed. If π in the relation is replaced by a slightly different coefficient, viz., 2, we understand Skellam's finding to mean that if the bacteria can diffuse within the growth time $1/A$ from within to outside the mask where they are destroyed by the ultraviolet light, extinction occurs. It is interesting that, although it is non-zero B that causes the abrupt transition, B drops completely out of the expression for the critical length L_{cr}.

Perusal of the literature also shows additional insights into the Skellam problem offered by Ludwig and collaborators [27] during their study of the spatial patterning of spruce budworm. If one does not impose drastic boundary conditions of vanishing u at the borders of the mask but merely takes the growth rate to have a negative value $-A_1$ outside the mask (but a positive value A inside the mask), one obtains a generalization of Skellam's critical condition given above:

$$L_{cr} = 2\sqrt{D/A}\,\arctan(\sqrt{A_1/A}). \tag{9.21}$$

If A_1 is taken to be infinite, ensuring that u is zero everywhere outside the mask, this Ludwig result reduces precisely to the Skellam result, given that the tangent of $\pi/2$ is infinity.

The simplicity of the Ludwig analysis is also worth appreciating. It is not based on hefty calculations with elliptic functions arising from nonlinear equations. It uses, instead, the clever idea that, if extinction is the phenomenon under consideration, the term proportional to u^2 can be safely neglected. This is so because it is an order of magnitude smaller than terms linear in u. One then has a linear equation to solve, no elliptic function struggles to contend with, and matters are utterly simple. Armed with this cunning approach given to us by Ludwig et al., let us analyze two new scenarios in the following two sections. The first is the gradual spatial variation of resources at the border of the mask. The second has to do with multiple masks.

9.10 Effect of Gradual Spatial Variation of Resources

Let us assume that the growth rate is the constant A within a length L and the constant $-A_1$ outside the mask but that the transition in its value at the mask border is gradual rather than precipitous. This is obviously a better representation of

[4] We could not even offer an excuse to ourselves for not having known of the earlier publication because physicists have little contact with the journal Biometrika: we found that the Skellam result appears even in a book [28] that had escaped our notice.

experiment. Using the suffixes 1, 2, and 3 for the respective regions I, II, and III in Fig. 9.3, we have

$$\frac{\partial u_1(X,t)}{\partial t} = D\frac{\partial^2 u_1(X,t)}{\partial X^2} + Au_1(X,t) - Bu_1^2(X,t),$$
$$\frac{\partial u_2(X,t)}{\partial t} = D\frac{\partial^2 u_2(X,t)}{\partial X^2} + A\left(1 - \frac{2X-L}{2R}\right)u_2(X,t)$$
$$-Bu_2^2(X,t),$$
$$\frac{\partial u_3(X,t)}{\partial t} = D\frac{\partial^2 u_3(X,t)}{\partial X^2} - A_1 u_3(X,t) - Bu_3^2(X,t). \quad (9.22)$$

To investigate into the extinction condition, it is not necessary to solve either this nonlinear partial differential equation or the corresponding nonlinear ordinary differential equation for the steady state $u(X)$ obtained by setting the time derivative equal to zero. It suffices to argue that, because our interest lies only in the extinction phenomenon, i.e., the vanishing of $u(X)$, we can eliminate the bilinear terms of the density, and match solutions of the ensuing *linear* equation across the interfaces of the three regions I, II, and III (see Fig. 9.3).

The critical value of L signifying extinction, that results for *any positive* B, can therefore be calculated precisely by considering the linear equation obtained by putting B equal to zero. Defining the dimensionless position x and the quantities l and r, all obtained by dividing X, $L/2$, and R respectively by the diffusion length $\sqrt{D/A}$, we focus on the linear steady-state counterpart of Eq. (9.22), with the notation $\xi = \sqrt{A_1/A}$:

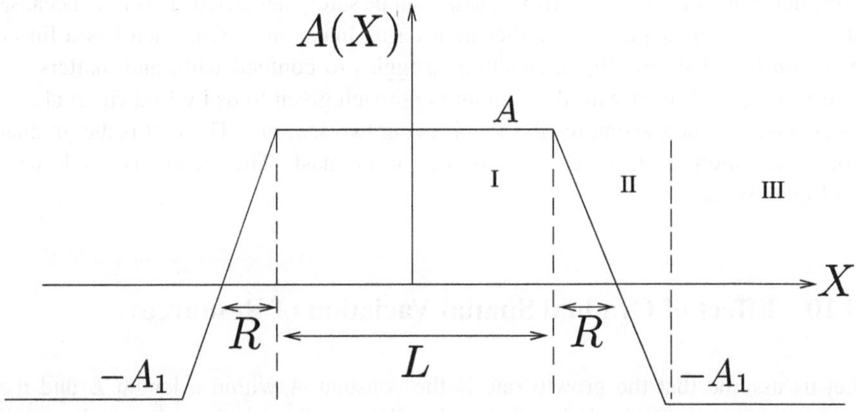

Fig. 9.3 The spatial dependence of the growth rate $A(X)$ around the mask in the Petri dish. It is constant with value A in a segment of length L that represents the mask. It undergoes a linear decrease to the value $-A_1$ on both sides. Adapted from Fig. 1 of ref [29]; copyright (2011) with permission from Elsevier

$$\frac{d^2u_1(x)}{dx^2} + u_1(x) = 0,$$
$$\frac{d^2u_2(x)}{dx^2} + \left(1 - \frac{x-l}{r}\right) u_2(x) = 0,$$
$$\frac{d^2u_3(x)}{dx^2} - \xi^2 u_3(x) = 0. \tag{9.23}$$

The first of these equations is solved trivially in terms of trigonometric functions, the third in terms of exponentials, and the second in terms of Airy functions.

Specifically, using the fact that the system has symmetry about $x = 0$, the solution in the three regions is given by

$$u_1(x) = E_1 \cos(x),$$
$$u_2(x) = E_2\, Ai\left[-r^{2/3}\left(1 - \frac{x-l}{r}\right)\right] + B_2\, Bi\left[-r^{2/3}\left(1 - \frac{x-l}{r}\right)\right],$$
$$u_3(x) = E_3\, e^{-\xi x}, \tag{9.24}$$

where Ai and Bi are Airy functions and E's and B_2 are constants to be determined from the boundary conditions.

Let us introduce the symbol $\eta = r^{1/3}$ for notational convenience, recall that we have defined $\xi = \sqrt{A_1/A}$ and denote by p their product, i.e. $p = \xi\eta$. We now match the logarithmic derivative of the solutions across the interfaces and arrive at the expression for the critical length L_{cr} of the mask for extinction:

$$\tan l = \left(\frac{1}{\eta}\right) \times \left(\frac{Ai'(p^2)Bi'(-\eta^2) - Ai'(-\eta^2)Bi'(p^2) + p\left[Ai(p^2)Bi'(-\eta^2) - Bi(p^2)Ai'(-\eta^2)\right]}{Ai(-\eta^2)Bi'(p^2) - Ai'(p^2)Bi(-\eta^2) + p\left[Ai(-\eta^2)Bi(p^2) - Ai(p^2)Bi(-\eta^2)\right]}\right). \tag{9.25}$$

Primes denote derivatives. The normalized extinction length is $l = L_{cr}/\sqrt{D/A}$ and is seen to equal the arctangent of the right-hand side if the latter is positive, and zero otherwise.

Equation (9.25) is a result Kumar and I obtained [29] as a generalization of the result of Ludwig et al. (9.21) for the case of a gradual variation of A around the mask edge. The crucial parameters in our expression are η, ξ, and their product p. The first of these measures the transition region relative to the diffusion length. The second measures the death rate outside the segment, relative to the growth rate inside it.

If the transition region vanishes ($R = 0$), we take the limit $\eta \to 0$ in Eq. (9.25) and recover the Ludwig formula [27] given in Eq. (9.21). Simulations reported in

[29] show excellent agreement between the numerical results from the nonlinear equation (whatever the value of B) and the Kumar–Kenkre analytic result from the linear extinction analysis (that has no B). The agreement vindicates the theoretical procedure we have explained.

9.11 Two-Mask Idea, Analytical Approach

The original experiment in the bacterial context [30, 31] involved measuring critical velocities of *moving* masks. The single stationary mask idea based on the Skellam or Ludwig concept was much simpler but uses for its interpretation [24, 25], a number of inputs that suffer from a certain degree of uncertainty. For instance, the values of D and A are not known with precision [32]; furthermore, the extent to which phenomena such as signaling [25] that occur in bacterial movement might affect the outcome of the experiment is also quite unclear.

Absolute values of critical lengths are consequently difficult to obtain with an acceptable degree of accuracy. If it were possible to find a variety of systems with differing values of D and A, one might be able to be more sure of the agreement of experiment and theory. Such variation in system parameters D and A is not possible in practice.

What is possible, however, is to introduce a variation in experimentally controllable parameters by placing *two masks* on the Petri dish each of length L as before but separated by a distance $2d$ as shown in Fig. 9.4. This significantly improved version of the stationary mask experiment was proposed by Kenkre and Kumar [33]. It focuses on the *interdependence of two critical lengths*. The essential feature of that proposal is to have two controllable lengths, a favorable length associated with a livable region for the bacteria, and a hostile length associated with an unlivable region. The double control possible in the proposed experiment opens the prediction space from a single value to an infinity of values. The variation of the critical value of one of the lengths when the other is changed presents an entire *relationship* that is directly observable rather than a single value and therefore could lead to a much cleaner and trustworthy interpretation of the observations. The theoretical analysis

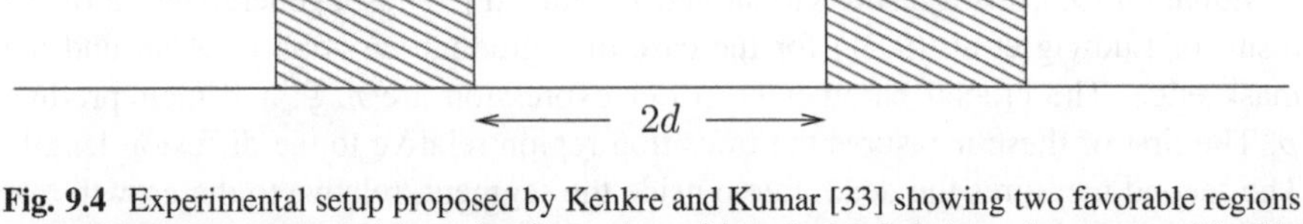

Fig. 9.4 Experimental setup proposed by Kenkre and Kumar [33] showing two favorable regions each of length L separated by a hostile region of length $2d$. It brings about a considerable improvement by focusing on the interdependence of the critical values of L and d at which extinction occurs. Modified with permission from Fig. 1 of ref [33]; Copyright (2008) National Academy of Sciences

is based on the neglect of the terms quadratic in u as earlier, a procedure justified by the fact that we focus only on extinction.

What we seek is the solutions of

$$\frac{d^2u(x)}{dx^2} + \alpha^2 u(x) = 0, \tag{9.26}$$

under the two masks, and of

$$\frac{d^2u(x)}{dx^2} - \alpha_1^2 u(x) = 0, \tag{9.27}$$

outside the masks, with $\alpha^2 = A/D$ and $\alpha_1^2 = A_1/D$. The masks happen to occupy the space from $x = \pm d$ to $x = \pm(d + L)$. We match the logarithmic derivative of the solution at the outer and the inner boundaries. The consequence is [33]

$$\tan[\alpha(d + L) - \phi] = \alpha_1/\alpha,$$
$$\tan(\alpha d - \phi) = -(\alpha_1/\alpha)\tanh \alpha_1 d.$$

All we now need to do is to eliminate ϕ from these two equations. The result is an *entire relationship* that the experimentally controllable parameters L and d must satisfy for extinction to occur:

$$L = \sqrt{\frac{D}{A}}\left\{\arctan\sqrt{\frac{A_1}{A}} + \arctan\left[\sqrt{\frac{A_1}{A}}\tanh\left(d\sqrt{\frac{A_1}{D}}\right)\right]\right\}. \tag{9.28}$$

Equation (9.28) is at the center of the Kenkre–Kumar proposal to verify the theory in light of experiments on bacteria in Petri dishes. Notice how, in light of the behavior of the hyperbolic tangent, L in Eq. (9.28) reduces to two expressions known earlier in two extreme limits: Ludwig et al.'s single mask value [27] of L when d attains infinite values, and *half* that value when d vanishes. In the former case, the two masks are so far (infinitely far) apart from each other that only one mask needs be considered. In the latter case, i.e., if $d = 0$, the two masks are next to each other, i.e., there is only a single mask of twice the width.

It is also useful to rewrite the transition relation in the compact form

$$\mathcal{L} = \arctan \xi + \arctan(\xi \tanh \delta\xi). \tag{9.29}$$

The favorable distance L and hostile distance d have been written normalized to the growth diffusion length $g = \sqrt{D/A}$ as $\mathcal{L} = L/g$ and $\delta = d/g$, respectively. The depletion parameter $\xi = \sqrt{A_1/A}$, the square root of the ratio of the destruction rate A_1 and the growth rate A, has already been defined in the previous section.

Careful numerical analysis based on the full nonlinear equations of motion underlying the bacterial dynamics in the presence of two masks was performed in

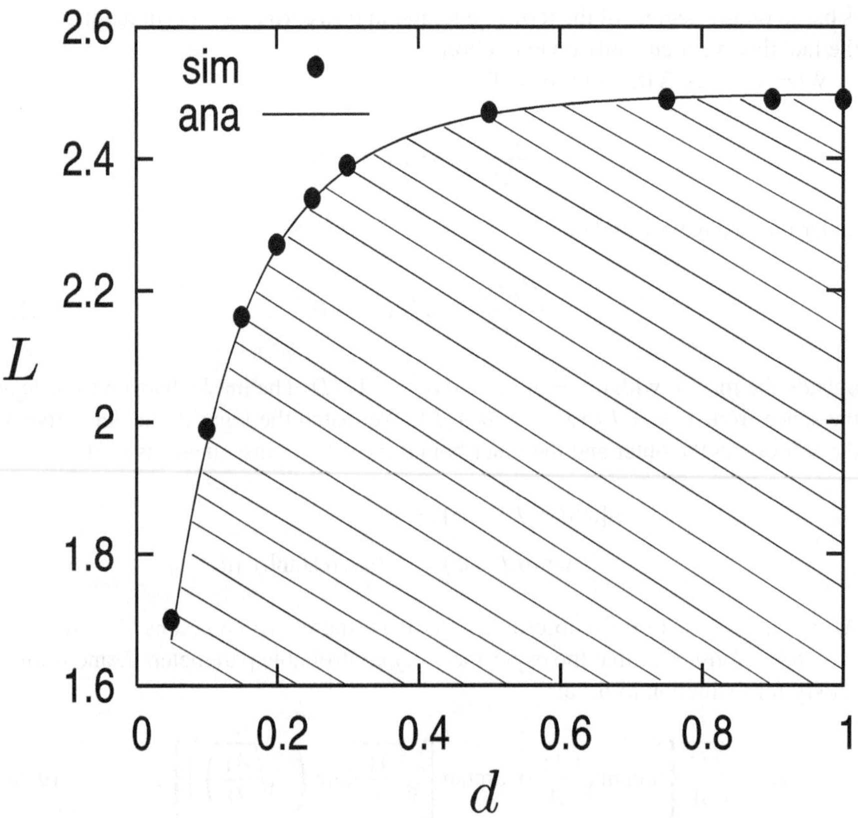

Fig. 9.5 Two-mask proposal: our analytic predictions for the dependence of critical values of the favorable length L and the hostile length d, respectively, from Eq. (9.28) along with numerical simulations. The lengths are expressed as dimensionless ratios to the diffusion length $\sqrt{D/A}$. For the purposes of this plot, $A_1/A = 9$ (see text). Numerical solution of the starting equations for values of B shows results to be independent of the (positive) value of B used and produces the filled circles for the onset of extinction. Note the excellent agreement with the analytic curve. The shaded region represents corresponding pairs of values of L and d that lead to extinction. Modified with permission from Fig. 2 of ref [33]; Copyright (2008) National Academy of Sciences

[33] and resulted in the solid circles shown in Fig. 9.5 below. They indicate the onset of the transition. The shaded portion of the L-d parameter space corresponds to extinction. It is remarkable that the line predicted analytically by our theoretical calculations, Eq. (9.28), agrees so well with the numerical results.

It is easy to understand, on the basis of a qualitative argument, the shape and tendency of the results of the numerical solutions of the nonlinear equation. Bacteria diffuse from within the mask to the harsh region and die if they reach that region. Small values of the mask width or large values of the intermask distance result in extinction. The extinction effect is worsened by an increase in the intermask distance for small values of that distance, but the effect saturates for larger values. Hence, the saturation in the curve.

All these features should be easy to test experimentally in the proposed setup. Quantitative comparison with the predictions of the Fisher equation is possible in terms of an exact analytic theory of the interplay of the favorable and hostile distances as reflected in the transition. The Fisher-KPP equation, whose *numerical* solution leads one to set out the separation curve between the extinction region and the rest in Fig. 9.4, cannot be solved exactly by analytic means. What makes the proposed comparison with experiment significant is that, nevertheless, *the separation curve itself can be obtained analytically.*

A number of useful analytical results emerge from extensions of the analysis to multiple masks placed on the Petri dish as also to *circular Petri dishes* and to the inclusion of chemotaxis effects. The interested reader is referred to [4, 33].

9.12 Approximate Approach Leading To Refugia Formation

These ideas can be used also to understand the very *emergence* of refugia in the Hantavirus context. A brief description is given here.

We return to the first of the equations set forth in this book for the dynamics of infected and susceptible mice, Eq. (9.1) but rewrite it differently in terms of equations for the *total* mice density M and the infected mice density M_I. In the steady state, when the time derivatives have vanished, we have

$$D\frac{d^2M(x)}{dx^2} + (b-c)M(x) = \frac{M^2(x)}{K(x)},$$

$$D\frac{d^2M_I(x)}{dx^2} + \mathcal{A}(x)M_I(x) = aM_I^2(x). \tag{9.30}$$

We have reemployed the original symbols: b and c are birth and death rates taken constant as earlier, but the environment parameter is taken to be explicitly space-dependent and denoted by $K(x)$. We have also introduced the expression

$$\mathcal{A}(x) = \left[-c + \left(a - \frac{1}{K(x)}\right)M(x)\right] \tag{9.31}$$

in the equation for the infected mice density.

Our goal is to understand whether patchiness in the environment parameter $K(x)$, representing the presence and absence of vegetation and water within the landscape, can give rise to abrupt extinction in M_I. If it can, it would be natural to identify regions where infection persists with observed refugia. We are investigating here abrupt transitions in $M_I(x)$, not in $M(x)$. The role of the equation for the total mice density $M(x)$ is simply to provide $\mathcal{A}(x)$ that goes into the equation for $M_I(x)$.

Let us first observe that in that equation for $M(x)$, if the diffusion coefficient D were to vanish, the steady-state solution for the total density would be $M(x) = K(x)(b-c)$, the other possible solution, $M(x) = 0$, being an unstable zero. If we take the environment parameter $K(x)$ to be a step function with constant values K_p and K_b respectively inside and outside the region of large resources that we suspect might become a refugium, $M(x)$ would also appear as a step function with respective values $K_p(b-c)$ and $K_b(b-c)$. Kumar et al. [7] showed that in the presence of D, we can employ a linearization approximation and obtain

$$M(x) = K_b(b-c) + \Delta \sinh\left(\frac{\alpha L}{2}\right) e^{-\alpha|x|}, \quad \text{for } |x| > L/2,$$

$$M(x) = K_p(b-c) - \Delta e^{-\alpha L/2} \cosh(\alpha x), \quad \text{for } |x| < L/2. \tag{9.32}$$

The first solution holds outside the extent L of the refugium while the second applies inside the refugium, the difference $(K_p - K_b)(b-c)$ in the carrying capacities inside and outside is denoted by Δ, and $\alpha = \sqrt{(b-c)/D}$ is the reciprocal of the diffusion length of the mice, i.e., is the distance traversed by the random walking mice in the effective growth time $1/(b-c)$. It has been clearly demonstrated in [7] that this is an excellent approximation for a small diffusion coefficient but produces departures for large D. This is so because it is diffusion that makes $M(x)$ depart from the step functionshape of $K(x)$.

The strategy followed in [7] was to plot the $\mathcal{A}(x)$ obtained from the conjunction of Eqs. (9.31) and (9.32), verify from the plot that it can be represented by a step function and apply the Ludwig analysis that we have been using to the $M_I(x)$ equation to extract a prescription for extinction.

This program turns out successful and yields explicit expressions relating the extent of the favorable region and other parameters of the problem. We refer the interested reader to [4] as well as to the original publication [7] for all details including the connection to the so-called "island theory" in the ecological literature.

Why is the theory of this infection problem much more difficult than that of the bacterial population transition problem analyzed in the preceding sections in this chapter? The answer is as follows. Ludwig arguments allow us to neglect the bilinear term at the end of the second (but not of the first) of the above two equations. Our interest lies in the transition in the infection here so, at the transition, we may certainly neglect the nonlinearity in $M_I(x)$ since the infected density tends to zero. However, at that transition, the total density does *not* vanish; the term proportional to its square in the first of the equations cannot be neglected.

9.13 Concluding Remarks

Two ecological areas have been examined in this chapter in the context of the application of the methods of Statistical Mechanics to observed phenomena: transmission of infection and the spread of epidemics on the one hand and patchiness and the onset of refugia on the other. A third area for which I had been scheduled to make a presentation (that could not take place) was the application of the effective medium theory. Before presenting a summary of the two topics discussed in this chapter, I will mention without detailed elucidation what the effective medium theory is and point out a few references.

In the analysis presented here, the key motion quantity, the diffusion constant D of the rodents or bacteria performing random walks, is a constant quantity: it does not vary over the terrain. What if there were reason to include a D that varies in space and does so in a random fashion? One of the theoretical approaches employed in the literature for such a scenario consists of the effective medium theory. The procedure appears to have originated in the engineering literature and has had a venerable history of applications. The basic idea is that the given spatially disordered system is replaced (for the purposes of analysis) by a spatially ordered counterpart that can be tackled by standard techniques for solution such as Fourier transforms. The price one pays for the replacement is the introduction of memory functions or time non-locality into the equations of motion. How the memory functions can be obtained through a practical prescription of translating given information about the probability distribution ρ into time non-local but spatially invariant motion parameters has been explained in full detail in [34] and in the expository Chapters 13 and 14 of a recent book [35]. Here I will only state the results for a simplified system where the original motion is as described by a Master equation for the probabilities P on a 1-dimensional lattice of sites m, $m+1$, etc.:

$$\frac{dP_m(t)}{dt} = F_{m+1}[P_{m+1}(t) - P_{m-1}(t)] + F_m[P_{m-1}(t) - P_{m-1}(t)]. \tag{9.33}$$

Notice that the probabilites are time-dependent, but the rates F are independent of time yet vary from site to site. The variation is probabilistic and described via a distribution $\rho(f)$. What this means is that a transition rate F_m can have any positive value f with probability density $\rho(f)$ normalized such that $\int_0^\infty \rho(f)df = 1$.

It should be evident that, as it stands, it is impossible to solve this problem analytically. As the sources I have cited show, however, it is possible to replace this equation by one for the equivalent ordered counterpart

$$\frac{dP_m(t)}{dt} = \int_0^t ds\mathcal{F}(t-s)[P_{m+1}(s) + P_{m-1}(s) - 2P_m(s)], \tag{9.34}$$

whose solution does a fine job of describing the motion of the disordered system. Notice that in Eq. (9.34) there is no spatial variation of the effective nearest-neighbor rate (actually memory function) of transfer $\mathcal{F}$. Its time dependence is obtained by solving in the Laplace domain

$$\int_0^\infty df \frac{\rho(f)}{f + \tilde{\xi}(\epsilon, \tilde{\mathcal{F}})} = \frac{1}{\tilde{\xi}(\epsilon, \tilde{\mathcal{F}}) + \tilde{\mathcal{F}}(\epsilon)}. \tag{9.35}$$

Here $\tilde{\xi}$ is a function of the variables ϵ and $\tilde{\mathcal{F}}$, its form being

$$\tilde{\xi}(\epsilon, \tilde{\mathcal{F}}) = \frac{\epsilon}{4}\left(1 + \sqrt{1 + \frac{\tilde{\mathcal{F}}(\epsilon)}{\epsilon}}\right). \tag{9.36}$$

Despite the formidable appearance of the task, it is possible to carry out the solution for simple representative forms of the probability distribution $\rho(f)$ and to verify the results via numerical solutions. Such a program has been carried out in [34] and explained also in a book I published recently [35]. It is hoped that the strategy of the effective medium approach will be used by future investigators into ecological problems that are characterized by random parameters such as the diffusion constant.

Let us now summarize what we have done in the rest of this chapter.

Investigations of the transmission of Hantavirus infection occupied us in the first part and of patchiness in resources leading to abrupt transitions in the second part. These were selected for their ecological relevance as well as their inherent mathematical significance. The first part was based on the solution of an exactly soluble model of tethered random walks. The primary entity used was the Smoluchowski equation for a representative entity describing an infected and a susceptible mouse that interacts and transmits infection. A number of ideas are worth noting: the representation of the interaction through a double-space description as a defect problem and the Green function approach, the ν-function averaging procedure to solve the problem exactly, and the explicit use of Smoluchowski equation characteristics. A new and unexpected result that emerged was the non-monotonicity of the rate of transmission of infection on the underlying parameters.

The second part of the chapter was based on the Fisher-KPP equation (diffusion with logistic nonlinearity) and its form with internal states employed in [1]. See Eq. (9.1) at the beginning of this chapter. We invoked the Skellam formula for bacteria in Petri dishes, augmented it via the Ludwig analysis, and applied the approach to a variety of ecological contexts on different spatial scales. In the case of bacteria in a Petri dish, we analyzed experiments in such a dish with a single mask, showed that gradual variation of resources at the mask edges can be treated analytically via Airy functions if the variation is linear in space, carried out a significantly new proposal with two masks designed to remove uncertainties in the correspondence between theory and experiment, and mentioned multiple-mask analysis and a number of other related calculations. They included, especially, an understanding of how Hantavirus infection refugia can occur from environmental resources that vary on the terrain. Although the last mentioned study necessitated the use of an approximation procedure, the procedure was strategically constructed and led even to the so-called island theory [4, 7] known in ecological literature.[5]

I hope that readers will find this description useful for their own studies of complex phenomena with the help of the tools of Statistical Mechanics.

Acknowledgments Many friends and collaborators have aided my efforts to develop the approach and apply the techniques described in this article. I acknowledge the late Max Dresden and the late Elliott Montroll for introducing me to the essentials of the techniques a half century ago. I recall the excitement and fun I had learning and working together with Guillermo Abramson, Marcelo Kuperman, Luca Giuggioli, Kathrin Spendier, and Satomi Sugaya as also many others including, especially, Niraj Kumar, Bob Parmenter, Ziya Kalay, and Paul Parris. I also thank Bob Knox for his unerring editorial eye and comments that inspired me to improve this manuscript.

My debt to Luca Giuggioli, who started out as a student I advised, is central in the context of this chapter. It makes me happy to express it. Not only did he collaborate on some aspects of the science explained here, but he has the lion's share of making possible presenting the lectures at the Cambridge conference—first by insisting that I accept the invitation despite health issues and then facilitating my movements while in Cambridge. His kind assistance has included even helping me reacquire my lost skills at latex without which it would have been impossible for me to publish this chapter. I acknowledge the economic support of EPSRC grant EP/R014604/1 for my participation at the program on the Mathematics of movement: An interdisciplinary approach to mutual challenges in animal ecology and cell biology.

References

1. Abramson, G., Kenkre, V.M.: Spatio-temporal patterns in the Hantavirus infection. Phys. Rev. E **66**(5), 011912 (2002)
2. Fisher, R.A.: The wave of advance of advantageous genes. Ann. Eugenics **7**(4), 355–369 (1937)
3. Kolmogorov, A.N., Petrovsky, N.G., Piskunov, N.S.: Study of the diffusion equation with growth of the quantity of matter and its application to a biology problem. Bull. Moscow State Univ. Ser. A. Math. Mech. **1**(6), 1–26 (1937); translated and reprinted in P. Pelce, Dynamics of Curved Fronts (Academic Press, San Diego, 1988)
4. (Nitant) Kenkre, V.M., Giuggioli, L.: Theory of the Spread of Epidemics and Movement Ecology of Animals: An Interdisciplinary Approach Using Methodologies of Physics and Mathematics. Cambridge University Press, Cambridge (2021)
5. Aguirre, M.A., Abramson, G., Bishop, A.R., Kenkre, V.M.: Simulations in the mathematical modeling of the spread of the Hantavirus. Phys. Rev. E **66**, 041908 (2002)
6. Abramson, G., Kenkre, V.M., Yates, T.L., Parmenter, R.R.: Traveling waves of infection in the Hantavirus epidemics. Ecol. Complexity **65**, 519–534 (2003)
7. Kumar, N., Parmenter, R.R., Kenkre, V.M.: Extinction of refugia of hantavirus infection in a spatially heterogeneous environment. Phys. Rev. E **82**(1), 011920 (2010)

[5] Hoping that it would not be presumptuous or redundant to offer advice to beginning investigators who might chance to read this chapter, I might make two suggestions. One should not merely present one's solutions in their bare form but express them in terms of dimensionless ratios of physical quantities: typically, this greatly enhances our understanding of what the results obtained. And one should not be daunted by the toughness of a mathematical problem during these applications: there are always useful approximation procedures that it is possible to invent in order to forge ahead. Several examples of both these strategies are available in this chapter.

8. Giuggioli, L., Abramson, G., Kenkre, V.M., Suzán, G., Marcé, E., Yates, T.L.: Diffusion and home range parameters from rodent population measurements in Panama. Bull. Math. Biol. **67**(5), 1135–1149 (2005)
9. Abramson, G., Giuggioli, L., Kenkre, V.M., Dragoo, J.W., Parmenter, R.R., Parmenter, C.A., Yates, T.L.: Diffusion and home range parameters for rodents: *Peromyscus maniculatus* in New Mexico. Ecol. Complexity **3**, 64–70 (2006)
10. Giuggioli, L., Abramson, G., Kenkre, V.M., Parmenter, R.R., Yates, T.L.: Theory of home range estimation from displacement measurements of animal populations. J. Theor. Biol. **240**, 126–135 (2006)
11. Kenkre, V.M., Sugaya, S.: Theory of the transmission of infection in the spread of epidemics: interacting random walkers with and without confinement. Bull. Math. Biol. **76**(12), 3016–3027 (2014)
12. Sugaya, S., Kenkre, V.M.: Analysis of transmission of infection in epidemics: confined random walkers in dimensions higher than one. Bull. Math. Biol. **80**(12), 3106–3126 (2018)
13. Kenkre, V.M.: Theory of exciton annihilation in molecular crystals. Phys. Rev. B **22**, 2089–2098 (1980)
14. Kenkre, V.M.: A theoretical approach to exciton trapping in systems with arbitrary trap concentration. Chem. Phys. Lett. **93**(3), 260–263 (1982)
15. Kenkre, V.M., Parris, P.E.: Exciton trapping and sensitized luminescence: a generalized theory for all trap concentrations. Phys. Rev. B **27**(6), 3221 (1983)
16. Spendier, K., Kenkre, V.M.: Analytic solutions for some reaction-diffusion scenarios. J. Phys. Chem. B **117**(49), 15639–15650 (2013)
17. Redner, S.: A Guide to First-Passage Processes. Cambridge University Press, Cambridge (2001)
18. Bray, A.J., Majumdar, S.N., Schehr, G.: Persistence and first-passage properties in nonequilibrium systems. *Advances in Physics*, 62(3):225–361, 2013.
19. Abramson, G., Wio, H.S.: Time behaviour for diffusion in the presence of static imperfect traps. Chaos, Solitons Fractals **6**, 1–5 (1995)
20. Carslaw, H.S., Jaeger, J.C.: Conduction of Heat in Solids. Clarendon Press, Oxford (1959)
21. Spendier, K., Sugaya, S., Kenkre, V.M.: Reaction-diffusion theory in the presence of an attractive harmonic potential. Phys. Rev. E **88**, 062142 (2013)
22. Murray, J.D.: Mathematical Biology: I. An Introduction, 3rd edn. Springer, Berlin (2003)
23. Murray, J.D. Mathematical Biology: II. Spatial Models and Biomedical Applications, 3rd edn. Springer, Berlin (2003)
24. Kenkre, V.M., Kuperman, M.N.: Applicability of the fisher equation to bacterial population dynamics. Phys. Rev. E **67**(5), 051921 (2003)
25. Perry, N.: Experimental validation of a critical domain size in reaction–diffusion systems with escherichia coli populations. J. Roy. Soc. Interface **2**(4), 379–387 (2005)
26. Skellam, J.G.: Random dispersal in theoretical populations. Biometrika **38**, 196–218 (1951)
27. Ludwig, D., Aronson, D.G., Weinberger, H.F.: Spatial patterning of the spruce budworm. J. Math. Biol. **8**(3), 217–258 (1979)
28. Kot, M.: Elements of Mathematical Ecology. Cambridge University Press, Cambridge (2001)
29. Kumar, N., Kenkre, V.M.: Effects of gradual spatial variation in resources on population extinction: analytic calculations for abrupt transitions. Phys. A: Stat. Mech. Appl. **390**(2), 257–262 (2011)
30. Lin, A.L.: Resonant chemical oscillations: pattern formation in reaction-diffusion systems. In: Kenkre, V.M., Lindenberg, K. (eds) AIP Conference Proceedings on Modern Challenges in Statistical Mechanics: Patterns, Noise, and the Interplay of Nonlinearity and Complexity, vol. 658, pp. 315–328. American Institute of Physics, Melville (2003)
31. Lin, A.L., Mann, B.A., Torres-Oviedo, G., Lincoln, B., Käs, J., Swinney, H.L.: Localization and extinction of bacterial populations under inhomogeneous growth conditions. Biophys. J. **87**(1), 75–80 (2004)
32. Mann, B.: Spatial Phase-Transitions in Bacterial Growth. Ph.D. Thesis, University of Texas, Austin, TX, USA, 2001

33. Kenkre, V.M., Kumar, N.: Nonlinearity in bacterial population dynamics: proposal for experiments for the observation of abrupt transitions in patches. Proc. Natl. Acad. Sci. USA **105**(48), 18752–18757 (2008)
34. Kenkre, V.M., Kalay, Z., Parris, P.E.: Extensions of effective-medium theory of transport in disordered systems. Phys. Rev. E **79**, 011114 (2009)
35. (Nitant) Kenkre, V.M.: Memory Functions, Projection Operators, and the Defect Technique. Springer, Berlin (2021)

Chapter 10
Modelling the Movements of Organisms: From Cell Migration to Bumblebee Flights to Foraging Sea Turtles

Rainer Klages, Norberto Lucero-Azuara, Vijay Kumar, Alessia Gentili, Perla A. R. Torres, Christophe Eizaguirre, Giorgio Volpe, and Vladimir V. Palyulin

10.1 Introduction

Movements of organisms living at very different spatio-temporal scales, from migrating in the microworld to foraging at the surface of the Earth, exemplify the intrinsic complexity of nature. Figure 10.1 depicts three examples: White blood cells *(Neutrophil granulocytes)* crawling along a chemical gradient [1]; a bumblebee *(Bombus terrestris)* foraging in an artificial carpet of flowers in a laboratory experiment [2]; and a loggerhead sea turtle *(Caretta caretta)* swimming off the coast of West Africa [3]. All these different paths display what one may call random-looking movement patterns. Especially within the past decade, such patterns have increasingly been recorded experimentally with big data techniques for a huge variety of organisms, from bacteria [4] to insects [5], fishes, birds, mammals [6, 7] and even humans [8], to name only a few.

R. Klages (✉)
Centre for Complex Systems, School of Mathematical Sciences, Queen Mary University of London, London, UK

London Mathematical Laboratory, London, UK
e-mail: r.klages@qmul.ac.uk

N. Lucero-Azuara
Centre for Complex Systems, School of Mathematical Sciences, Queen Mary University of London, London, UK

V. Kumar · V. Palyulin
CAIT, Skolkovo Institute of Science and Technology, Bolshoy Bulwar, Moscow, Russia

A. Gentili · G. Volpe
Department of Chemistry, University College London, London, UK

P. A. R. Torres · C. Eizaguirre
School of Biological and Behavioural Sciences, Queen Mary University of London, London, UK

L. Giuggioli, P. K. Maini (eds.), *The Mathematics of Movement*,
https://doi.org/10.1007/978-3-032-00873-2_10

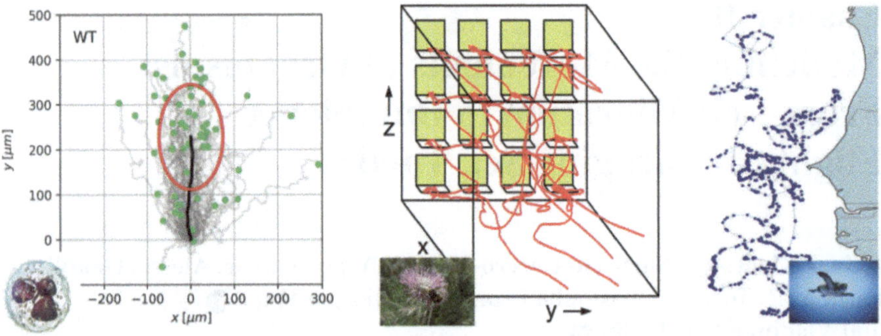

Fig. 10.1 Three examples of movement patterns across very different scales. Left: 60 neutrophil cells moving along a chemical gradient applied parallel to the y axis. The bold (black) vertical line depicts the mean position of the whole cell ensemble up to 30 min, the bold (red) ellipse the mean squared distance travelled after this time (from [1]). Centre: Part of the flight trajectory of a bumblebee foraging in an arena of artificial flowers (from [2]. Right: Path of a loggerhead sea turtle foraging in front of the coast of West Africa [9]. The insets show schematics or photos of these three organisms

Understanding these movement patterns by constructing mathematical models from data provides a fundamental challenge. The first approach for modelling the random-looking migration of organisms employed simple random walks [10, 11]. Depending on whether one was looking at the migration of organisms in the microworld or the movements of animals on macroscopic scales, this basic idea of a stochastic description of movement branched out in different directions. For example, for modelling cell migration, Langevin equations (LEs) have been used, inspired by the similarly random-looking Brownian motion of a tracer particle in a fluid [12]. Only more recently, it was found that these equations have to be generalised by including long-term memory to reproduce the persistence of cellular paths [13]. Constructing mathematical models for microswimmers like bacteria or sperm cells led yet to a different type of generalised Langevin dynamics and associated stochastic models, driven by the need to especially capture the self-propelled motion of these biological agents. The field of active particles emerged [4, 14, 15], related to the more general framework of active matter [16–18]. On the macroscale, the analysis of albatross flights led to introducing Lévy dynamics into the arsenal of stochastic methods for understanding the movements of organisms [19–21]. Lévy walks (LWs) and flights belong to an advanced class of stochastic processes exhibiting diffusive properties beyond Brownian motion, called anomalous diffusion [22–24]. In parallel, the new framework of movement ecology was formulated, which endeavours to understand organismic movement in view of important real-life biological properties like biomechanical generation of movement, perception and interaction with the environment [25–27].

The key problem which we will address in this book chapter is that, as far as biological dynamics is concerned, the theory of active particles has especially been applied to understand organismic movement on microscales. Macroscopic motion of animals, on the other hand, has rather been modelled inspired by the stochastic theory of passively diffusing particles, either Brownian ones, as largely applied in movement ecology, or anomalously diffusing ones, as in the approach formulated by anomalous stochastic theory. In our view, the three disciplines of movement ecology, active matter and anomalous diffusion need to be suitably combined to arrive at a coherent description of movements of organisms at all scales. The framework that we develop here to achieve such a theoretical description is based on Langevin dynamics, however, generalised in a particular way which blends fundamental elements of these different three fields. We will introduce to this approach as follows: The next Sect. 10.2 briefly outlines the fields of movement ecology, active matter (viz. active particles) and anomalous diffusion by focussing on key stochastic models and their connections among each other. In Sect. 10.3, we demonstrate how to construct generalised LEs from statistical experimental data analysis of various types of biological movements. Working in a fixed Cartesian frame, we reveal features of active motion and anomalous dynamics in generalised LEs constructed for migrating cells, bumblebee flights and foraging sea turtles. Inspired by movement ecology, we then switch the perspective to a stochastic modelling in a frame comoving with a biological entity. Applied to data of bumblebee flights, this approach yields a generalised LE combining basic features of three generic active (Brownian) particle models. As we argue in Sect. 10.4, these findings suggest to use generalised Langevin dynamics in a comoving frame for modelling the movements of organisms. In the same section, we outline experimental applications of this framework and elaborate on the theoretical foundation of our generalised LE approach to formulate it on a rigorous mathematical basis. In the final Sect. 10.5, we briefly discuss the promise, main open questions and biological implications related to this approach in view of providing a general framework for modelling organismic movement at all scales.

10.2 Movement Ecology Meets Active Matter and Anomalous Dynamics

The approach that in the following we put forward for modelling the movements of organisms is based on cross-linking three fundamental fields of research. Briefly outlining them by focussing on important movement models used in these different research areas will set the scene for the precise formulation of our theory and its meaning in terms of the associated physical, mathematical and biological properties. We start with the most biological scientific field, moving down to more theoretical physical and mathematical formulations.

Movement ecology is a rather new discipline that endeavours to understand the movements of individual organisms, especially on larger scales, in view of interactions of the biological dynamics with the environment [25, 27]. While movement ecology is, to quite some extent, driven by experimental biologists [6, 7], it has developed a mathematical framework in both statistical data analysis and probabilistic modelling by applying state space [28, 29], hidden Markov [28, 30] or random walk [26, 31] models to experimental data. For our approach, the concept of correlated random walks (CRWs) [32] developed within movement ecology is especially important. In two dimensions, this stochastic model describes the movement of an organism in terms of its step length, i.e. distance travelled per time step, and turning angle, both defined in a comoving frame [15, 31, 33]. Both variables are sampled independently and identically distributed at any time step from two associated probability distributions. In the simplest case, the step length may be chosen as constant while the turning angle is drawn from a non-uniform probability distribution, e.g. a Gaussian wrapped on the circle. Notably, the non-uniformity of the angular distribution models a persistence of the motion, i.e. a preference to move in the same direction, as is often observed in organismic movement, hence the notion of CRWs. Nevertheless, the whole stochastic process as defined above in the comoving frame is still Markovian, i.e. memoryless. CRWs have been widely applied to describe animal movement [31].

In contrast to movement ecology, the field of *active matter*, which developed in parallel over the past two decades, focusses on understanding the collective dynamics and the resulting material properties of a large number of interacting living agents at all scales, from swarms of bacteria to cytoskeleton gels in living cells, to schools of fish and flocking birds [16–18, 34]. Single entities of active matter can in turn be modelled as *active particles* [4, 14, 15, 35]. In detail, in contrast to the passive Brownian motion of a tracer particle in a fluid, which emerges from collisions with the surrounding fluid molecules, active biological motion is generated by an agent itself, due to taking up energy from the environment and converting it into self-propulsion [14, 15, 35]. This phenomenon is formulated mathematically most conveniently in terms of Langevin dynamics. For a passive Brownian particle in a fluid, the famous LE reads [36, 37]

$$m\dot{\mathbf{v}} = -\eta\mathbf{v} + \boldsymbol{\xi}(t), \tag{10.1}$$

with velocity $\mathbf{v} = \dot{\mathbf{x}}$ for a particle of mass m at position $\mathbf{x}$ at time t. The force on the right-hand side is decomposed into a viscous damping term with (Stokes) friction coefficient η while the random kicks of the microscopic surrounding particles are modelled by Gaussian white noise $\boldsymbol{\xi}$. Physically, this equation generalises Newton's second law to describing random-looking Brownian motion by decomposing the force into a deterministic and a stochastic component. Accordingly, the LE is sometimes called 'Newton's Law of stochastic physics'. Mathematically, it defines a stochastic differential equation generating an Ornstein–Uhlenbeck (OU) process for the velocity of the particle. The resulting Brownian motion is a Gaussian stochastic process characterised by a mean squared displacement that grows linearly in the long time limit, $\langle x^2 \rangle \sim t$ $(t \to \infty)$, where the angular brackets denote an ensemble average, which yields what is called normal diffusion.

For modelling a self-propelled active particle, this LE has been modified in different ways. First, [14, 15, 35, 38], the constant friction η has been replaced by a speed-dependent friction coefficient,

$$m\dot{\mathbf{v}} = -\eta(s)\mathbf{v} + \boldsymbol{\xi}(t). \tag{10.2}$$

The simplest example is the linear Schienbein–Gruler friction defined by $\eta(s) = \eta_0(1 - s_0/s)$, where $s = |\mathbf{v}|$ denotes the speed of the particle, yielding $-\eta(s)\mathbf{v} = -\eta_0(s-s_0)\mathbf{e_v}$ with $\mathbf{e_v} = \mathbf{v}/s$ as the unit vector of the velocity. A non-linear version is the Rayleigh–Helmholtz friction, $\eta(s) = \eta_0(s^2/s_0^2-1)$, corresponding to $-\eta(s)\mathbf{v} = -\eta_0(s^2 - s_0^2)s\mathbf{e_v}/s_0^2$ [14, 15]. The idea underlying this modification is that an active particle can take up energy from the environment by converting it into kinetic energy while it can also lose energy to the environment, depending on the sign of the friction term. Such an active particle aims at stabilising itself to move with a self-generated speed s_0 for which the active friction term is zero. This model has been called an *active Brownian particle* (ABP) in the original literature starting from the 1990s [15, 38].

A second type of active particle [4, 39] has been introduced by using an overdamped LE for which there is no particle acceleration anymore. Here the self-propulsion is achieved by a velocity term that is driven by angular noise,

$$\dot{x} = s_0 \cos\phi + \xi_x(t)\ , \ \ \dot{y} = s_0 \sin\phi + \xi_y(t), \tag{10.3}$$

where ϕ is the polar angle of the (here and in the following two-dimensional) velocity vector $\boldsymbol{v} = (v_x, v_y)$ fixed at constant speed, s_0, and governed by rotational diffusion,

$$\dot{\phi} = \xi_\phi(t), \tag{10.4}$$

where ξ_ϕ is wrapped Gaussian white noise. That is, the orientation ϕ is generated by a Wiener process on the circle. This model has also been called an ABP in the literature, with some authors reserving this term for that particular model only [4].

A third generic type of active particle was also defined in terms of an overdamped LE [40, 41],

$$\dot{\mathbf{x}} = \mu_{\mathrm{mob}}\boldsymbol{F}(\mathbf{x}) + \boldsymbol{\xi}_{\mathrm{c}}(t)\ , \tag{10.5}$$

where $\boldsymbol{F}(\mathbf{x})$ holds for a force and μ_{mob} for the mobility of the particle. Here the Gaussian noise $\boldsymbol{\xi}_{\mathrm{c}}$ is assumed to be coloured, typically in terms of exponential correlations [40],

$$\langle\boldsymbol{\xi}_{\mathrm{c}}(0)\boldsymbol{\xi}_{\mathrm{c}}(t)\rangle \sim \exp(-t/\tau), \tag{10.6}$$

which due to the non-Markovian memory yields active, persistent motion. In the case of a number of interacting active particles, the force is generated from a putative interaction potential [42]. If the exponentially correlated Gaussian noise is obtained from solving the OU equation (10.1) [41], this third model is called an active OU particle [42].

These three different models have in common that, inspired by the theory of passive Brownian motion, they are all formulated in terms of (underdamped or overdamped) generalised Langevin dynamics. However, in contrast to the LE (10.1) for a passive particle, they all model self-propulsion by including persistence that is very different from simple constant friction. In each of the three cases, the precise type of active persistence is entirely different. Furthermore, in all three cases, the fluctuation–dissipation relation is broken, as is typically found in active matter systems, since active particles operate far from equilibrium [16, 17]. On the other hand, in analogy to passive Brownian motion, all three types of active particle dynamics again exhibit normal diffusion in the long time limit, with the particles' persistence leading to an enhanced diffusion coefficient. On this basis, we classify all these three generic models of active particles that have been widely studied in the literature [4, 14, 15, 34, 35, 38–40] as ABP models, more specifically as ABPs type I, II and III, in (chronological) order of appearance.

Apart from this broad class of ABPs, there is another basic type of active particles describing run-and-tumble dynamics [43, 44]. For *Escherichia coli* bacteria under chemotaxis, it was observed [43] that these microswimmers perform sequences of long 'runs' in one direction, interrupted by periods of time where the bacteria 'tumble' by choosing a new direction, seemingly at random. This class of active motion can be modelled by using the overdamped LE (10.3) amended by the different equation for the dynamics of the polar angle ϕ [45]

$$\dot{\phi} = \Sigma_i \Delta\phi_i \delta(t - T_i) \,. \tag{10.7}$$

Here $\Delta\phi_i$ denotes the turning angle at times T_i when the particle tumbles, with both quantities drawn independently and identically distributed from a uniform distribution on the circle, respectively from some other (e.g. Poisson) distribution. In this particular sense, we have an ABP type IV with biological activity in terms of a specific tumble time distribution. However, there are other modelling approaches for run-and-tumble dynamics deviating from LEs [34, 40, 44]. We will not focus on run-and-tumble motion but will nevertheless come back to it in Sect. 10.4. This attempt of categorising active particle dynamics is of course not complete. There exist, for example, heterogeneous ABPs [46], anisotropic ABPs in three dimensions [47] as well as purely deterministic, nonlinear active particle models [48, 49] going beyond these four basic types.

At this point, there is already an important, although somewhat trivial, observation to be made. Namely, ABP II is intimately related to the CRW model of movement ecology introduced above in the sense that the self-propulsion term of ABP II, discretised in time, corresponds to a CRW model with a non-uniform wrapped Gaussian turning angle distribution. This is evident by noticing that the first terms in ABP II Eqs. (10.3), (10.4) represent the CRW model, defined within the comoving frame, transformed back to Cartesian coordinates by using time-continuous stochastic differential equations. Interestingly, ABP II is non-Markovian in the Cartesian frame, as the velocity autocorrelation function decays exponentially [50]. Hence, we have a Markovian representation of ABP II in the comoving frame while there is a non-Markovian one in the Cartesian frame. We will come back to this interesting interplay between different frames of reference for defining stochastic processes in Sect. 10.4.

Mathematically, both movement ecology and active matter, the latter especially in view of ABPs, are rooted to quite some extent in stochastic theory, combined with statistical data analysis. In 1904, Sir Ronald Ross inspired [10] Karl Pearson to model the movements of organisms by a random walk in the plane, which led to a famous series of papers published by Pearson in 1905 and 1906 [11, 51]. Therein he defined a two-dimensional random walk in the comoving frame in complete analogy to setting up CRWs, with the only simplification that in addition to a constant step length, the turning angle is drawn from a uniform distribution on the circle. The paradigm of modelling biological movements as simple types of Brownian motion prevailed for almost a century. In parallel to the development of active matter theory, in the 1990s, this idea was questioned by experimental findings [19, 52] suggesting that albatrosses perform LWs characterised by *anomalous diffusion*, where in contrast to Brownian motion, the mean squared displacement increases nonlinearly with time, $\langle \boldsymbol{x}^2 \rangle \sim t^\alpha$ $(t \to \infty)$ with $\alpha \neq 1$. For the albatross, an exponent $\alpha > 1$ was found corresponding to superdiffusion; motion with $\alpha < 1$ is called subdiffusion [22–24, 37]. A two-dimensional LW can be easily formulated within the CRW framework by sampling the step lengths from an α-stable Lévy distribution [24] and assuming a uniform distribution of turning angles. As an additional constraint, the speed is held constant, thus inducing the very same power law distribution of corresponding flight times [53].

We remark that the hypothesis that organisms perform LWs is discussed controversially in the literature [19, 21, 54–57]. This relates to criticism of the underlying theory [21, 58] as well as to problems with the experimental data analysis, as it is difficult to argue for power laws when they stretch over a decade only and the methods of data analysis are not robust [55, 56]. On the other hand, this line of work indicates that biological dynamics may need to be described by stochastic processes that are more non-trivial than normal diffusive (active or passive) Brownian motion by exhibiting anomalous stochastic dynamics, thus going beyond the framework of both conventional movement ecology and active particles. Anomalous diffusion defines a very active field of research, mainly driven by mathematicians and theoretical physicists, who explore the diffusive properties of non-Markovian, non-Gaussian stochastic processes beyond Brownian motion [22–24, 37]. Such dynamics have been experimentally observed in a wide variety of real-world systems [20, 24]. Apart from Brownian motion and LWs, there are many other advanced stochastic models available for understanding biological movement, from intermittent processes where organisms switch between different modes of motion [59, 60] to other generic anomalous stochastic models like continuous-time random walks [61] or fractional Brownian motion [62]. More recently, anomalous diffusion has also been considered for modelling ABPs [63].

10.3 Data-Driven Construction of Generalised Langevin Equations for Modelling the Movements of Organisms

In the first part of this section, we will analyse movement data of three different types of organisms with the goal to construct generalised LEs in the Cartesian frame for them. We will first illustrate this approach for bumblebee flights by then applying it to cell migration and sea turtle foraging. Employing the very same theoretical framework, we will reveal non-trivial active and anomalous movement properties for these organisms. In the second part, bumblebee flights will be reconsidered by constructing from data an underdamped generalised LE in the comoving frame for them. We will show that this equation combines the characteristic features of the first three generic types of ABPs dynamics discussed in the previous section.

10.3.1 The Cartesian Frame

The very simple model that here we put forward for analysing movement data is the overdamped LE, obtained from the underdamped LE (10.1) in the limit of large friction where there is no particle acceleration anymore,

$$\dot{\mathbf{x}} = \boldsymbol{\xi}_o(t) . \tag{10.8}$$

Here $\boldsymbol{\xi}_o(t)$ denotes the noise of the overdamped equation. This model sits between time-discrete random walks and time-continuous underdamped LEs. A simple mathematical example is the Wiener process, for which the velocities are drawn from Gaussian white noise. As $\mathbf{v} = \dot{\mathbf{x}}$, in this equation, the distribution function of the noise $\rho(\boldsymbol{\xi}_o)$ matches to the velocity distribution function $\rho(\mathbf{v})$ of the moving organism, and the autocorrelation function of the noise $\langle \boldsymbol{\xi}_o(0)\boldsymbol{\xi}_o(t)\rangle$ is identical to the velocity autocorrelation function $\langle \mathbf{v}(0)\mathbf{v}(t)\rangle$ of the dynamics (here by assuming stationarity). Knowing both quantities fully defines this model, hence it can be constructed by extracting them from the data. This very elementary stochastic model, defined in the Cartesian frame, provides only a first approach and will be substantially refined in the next Sect. 10.3.2. In the following, we will nevertheless show that this simplistic framework can already yield a lot of insight into the origins of organismic movement. Most prominently featured by this model is the velocity autocorrelation function, which represents memory in the dynamics. An example is fractional Brownian motion, which can be generated by using for $\rho(\boldsymbol{\xi}_o)$ fractional (power law correlated) Gaussian noise [23, 62]. However, allowing arbitrary correlations obtained from data goes beyond other standard models such as Brownian motion, CRWs or LWs.

Starting from Eq. (10.8), we will first construct a stochastic model for bumblebee flights tracked in a laboratory experiment [64], as illustrated in Fig. 10.1(middle). A single bumblebee (*Bombus terrestris*) was flying in a cube of 75 cm side length

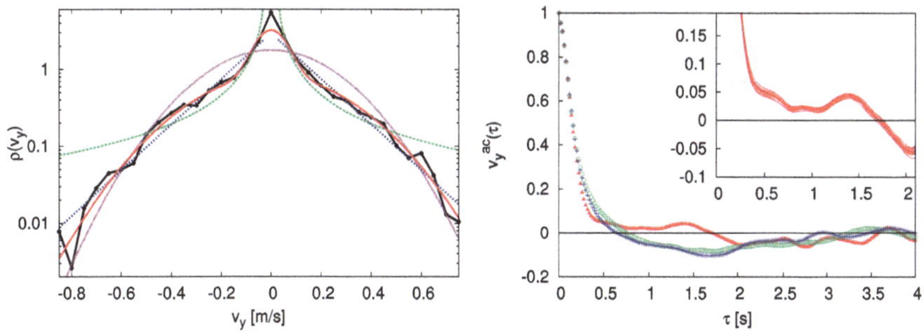

Fig. 10.2 Left: Semi-logarithmic plot of the velocity distribution $\rho(v_y)$ parallel to the y axis in the experimental set-up illustrated in Fig. 10.1(middle) (black crosses) for a single bumblebee in the spider-free phase (1). Included are lines for maximum likelihood fits with (from bottom to top in left corner) a single Gaussian (violet dotted), a Gaussian mixture (red line), an exponential (blue dotted) and a power law (green dashed). Right: Velocity autocorrelation function Eq. (10.9) for bumblebee velocities v_y in the three different phases of the experiment as described in the text: (1) red triangles, no spiders; (2) green circles, spiders; (3) blue crosses, spiders, memory test. The effect of the presence of spiders on the bumblebee flights is clearly seen, even in the memory test. The inset displays the resampled autocorrelation for stage (1) confirming that the positive autocorrelations are not a numerical artifact. Figures are from Ref. [2]

with a vertical grid of artificial yellow flowers on the wall in the back. Each flower was equipped with a landing platform and a source of nectar, replenished after consumption. The goal of the experiment was to analyse how bumblebee flights change under predation risk. For this purpose, four of the 16 flowers were randomly equipped with spider images, and 30 bumblebees were trained by a specific mechanism to avoid any flower with a spider image. The flight of a single bumblebee in this arena was recorded by two high frame rate (0.02 s) cameras yielding on average 6000 data points per bumblebee trajectory. In the following, we will analyse three phases of the experiment, which are (1) spider-free foraging; (2) foraging under predation risk; (3) a memory test with the same setting as (2) 1 day later; see Ref [64] for full details of the experiment.

First, we discuss the velocity distributions in phase (1) by focussing on $\rho(v_y)$ horizontally parallel to the wall, see the coordinate system in Fig. 10.1(middle), which turned out to be symmetric. The other ones along x and z were found to be asymmetric, due to the setup of the experiment or gravity, respectively. Applying maximum likelihood fits together with information criteria $\rho(v_y)$ shown in Fig. 10.2(left) was found to be best fitted by a mixture of two Gaussians representing two different flight modes, one with larger variance far away from the flowers and another one with smaller variance close to them, as confirmed by an additional spatial data analysis [2]. Hence, a bumblebee flies with on average smaller velocities when approaching a flower and with larger ones farther away from it. This corresponds to spatio-temporal intermittent motion, suitably adapted to this environment by exhibiting different flight modes when searching for food sources, respectively accessing them [59, 60].

Very surprisingly, the velocity distribution of phase (1) did not change in phases (2) and (3) under predation risk [2]. One might have expected that the bumblebees are getting more nervous in the presence of enemies, leading to flights with larger velocities, but this was not the case. Instead, the whole adaptation in the dynamics to environmental predation risk is contained in the velocity autocorrelation functions. Shown in Fig. 10.2(right) are the normalised velocity correlations horizontally parallel to the wall,

$$v_y^{\mathrm{ac}}(\tau) = \sigma^{-2} \left\langle (v_y(t) - \mu)(v_y(t+\tau) - \mu) \right\rangle_{\mathrm{e},t} , \tag{10.9}$$

where μ and σ^2 denote the mean and the variance of the corresponding velocity distribution of v_y, respectively, and the subscripts of the angular brackets denote a combined time-ensemble average over all bumblebees and over time. The time dependence of this quantity yields important information about persistence, memory and overall the topology of a bumblebee path. One can see that in the spider-free phase (1), the correlation decay remains positive for about 1.5 s representing persistence, that is, the bumblebee keeps flying in the same direction it came from. In contrast, for phases (2) and (3) under predation risk, the decay immediately becomes negative, which implies anti-persistence: The bumblebees quickly reverse the velocities by making more turns. This reflects the predation threat, as the flights are much less directed than in phase (1), i.e. more curvy, so the bumblebees are more careful. However, this change is only represented in the topology of the path, not in the velocity distributions, perhaps because the bumblebees find it more difficult to change their speeds.

In summary, after statistical analysis of experimental bumblebee data guided by Eq. (10.8), for an overdamped generalised LE modelling bumblebee flights under predation risk, the velocities are sampled from the mixture of two Gaussians identified by fitting the data in Fig. 10.2(left), correlated in time by using (suitably simplified versions of) the velocity autocorrelation functions shown in Fig. 10.2(right), depending on the phase one wishes to model. This stochastic model corresponds to a generalised version of ABP III Eq. (10.6), because the noise is non-trivially correlated, hence non-Markov and active. In addition the noise distribution is a bit more complicated than a simple Gaussian reflecting spatio-temporal intermittency. The coupling to the environment, in the spirit of movement ecology, can qualitatively be reproduced by another, related LE model, where changing the correlation function under predation risk is modelled by an interactive force term, as shown in Ref. [2]. A more refined model for bumblebee flights will be built in Sect. 10.3.2, therein defined fully analytically.

Along the same lines and again driven by experimental data analysis, an overdamped LE (10.8) can be constructed for cell chemotaxis [1]. Shown in Fig. 10.1(left) are the paths of 60 wild-type murine neutrophil cells recorded over $T = 30$ min with a sampling interval of 5 s yielding 361 data points per track. For these cells, the velocity distributions were found to be simple Gaussians [1]. Figure 10.3(top) displays the associated velocity autocorrelation functions, here

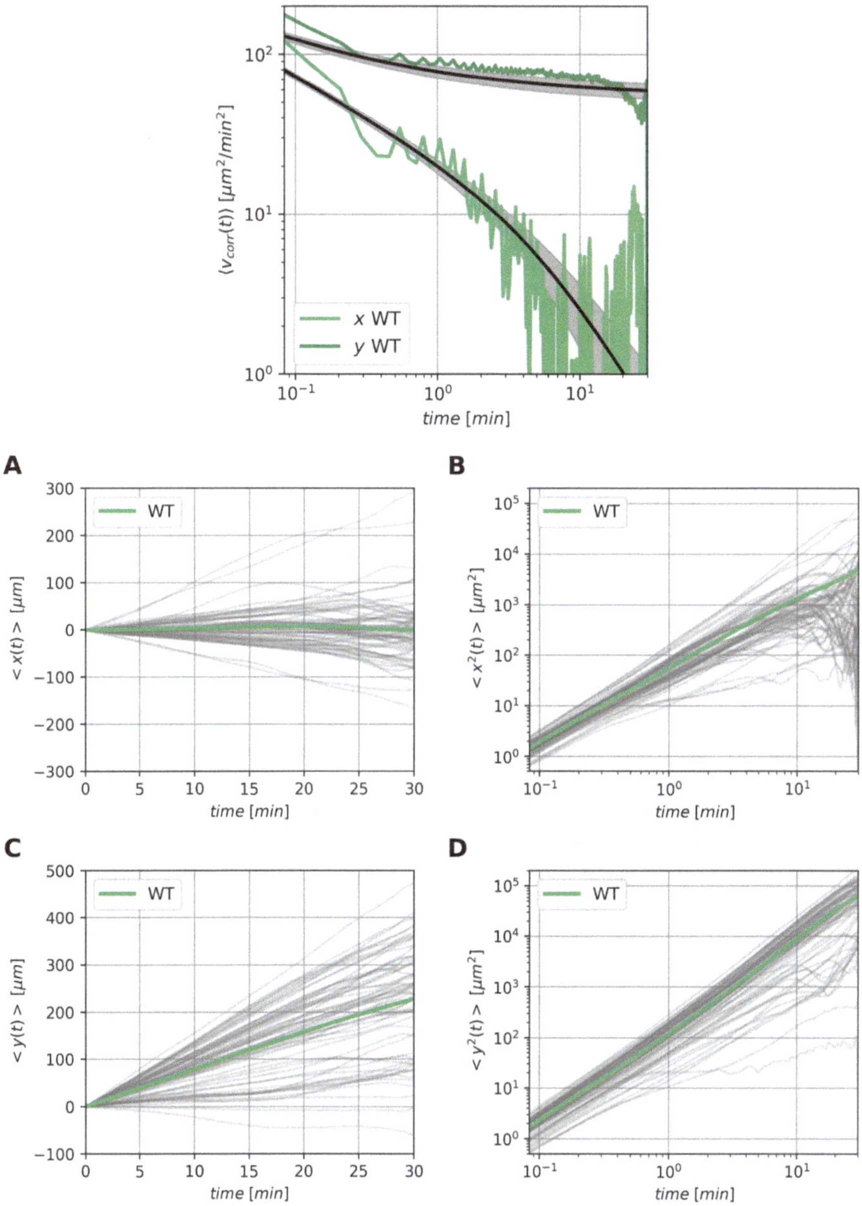

Fig. 10.3 Top: Velocity autocorrelation functions along x and y for wild-type neutrophils performing chemotaxis parallel to the y-axis. Thick green lines show the combined time-ensemble averaged correlation function obtained from data for the x and y direction in bright and dark colors, respectively. The fits discussed in the text are included as bold black lines together with a grey area of uncertainty ($\pm$ one standard deviation). Bottom: Corresponding first and second moments of positions along x and y. Thin grey lines show time-averaged moments of single cells. Thick green lines display the combined time-ensemble average of the corresponding moment. Figures are from Ref. [1]

denoted by $\langle v_{\mathrm{corr}}(t)\rangle$, defined again by a combined time and ensemble average but not normalised and without subtracting the drift. The fits represent, along y, pure power law correlation decay, $\langle \xi_y(t)\xi_y(t')\rangle \sim \tau^{2H-2}, \tau = |t-t'|, 0 < H < 1$, and along x the same with power law tempering, $\langle \xi_x(t)\xi_x(t')\rangle \sim \tau^{2H-2}(1 + \tau/\tau^*)^{-\mu}, \mu, \tau^* \geq 0$. However, as Fig. 10.1(left) indicates, here we have motion with an average non-zero drift along the chemical gradient in the y direction, which deserves further analysis. Fig. 10.3(bottom), showing first and second moments of the position distributions along x and y as functions of time, demonstrates that indeed there is a drift linear in time along y with $v_{\mathrm{d}} \simeq 7.4\,\mu\mathrm{m/min}$ while there is no drift along the x direction perpendicular to the chemical gradient. These results suggest a modelling of this cell chemotaxis in terms of the generalised overdamped LE

$$\dot{x} = \xi_x(t),\ \dot{y} = v_{\mathrm{d}} + \xi_y(t). \tag{10.10}$$

The dynamics along y represents (biased) fractional Brownian motion [23, 62]. Along the x-axis, there is no drift, but the correlations are power law tempered. These results reflect the symmetry breaking by the chemical gradient and long-term persistence induced by it. Accordingly, this dynamics yields anomalous (super) diffusion both along x and y as shown in Fig. 10.3(bottom) by the second moment obtained from the data, with exponents α changing between 1.2 and 1.5 along x, due to the tempering, and $\alpha \sim 1.55$ along y. Note that the associated (anomalous) diffusion coefficients, i.e. the prefactors in the mean squared displacement, are more than an order of magnitude larger along y than along x yielding a much faster diffusion parallel to the gradient. The same analysis has been performed for three other cell types that have been chemically modified, revealing subtly different chemotactic dynamics in terms of the resulting LEs, which helps to understand the impact of specific chemical modifications on chemotaxis, as well as identifying cell deficiencies [1].

In summary, for neutrophil cells we have constructed a generalised overdamped LE from data that again may be associated with ABP III, as the noise is non-trivially correlated; hence, the dynamics is again non-Markovian and active, though Gaussian. In contrast to the previous bumblebee model, however, these cells display anomalous (super) diffusion, which demonstrates the need to go beyond Markovian normal diffusive Brownian motion models to understand organismic movement [13, 65, 66].

Finally, as a third example of data-driven application of the overdamped LE framework based on Eq. (10.8), we present results for the off-shore foraging of loggerhead sea turtles *(Caretta caretta)*. A sample path is depicted in Fig. 10.1(right). These turtles are listed as endangered on the IUCN red list; hence, learning about their foraging activities is important for improving their protection. The length of a typical loggerhead sea turtle in this population of Cabo Verde is around 75–85 cm of carapace. The largest individuals can weight up to 135 kg and some may live up tp 60 years in the wild. If threatened, in short burst, they can swim up to 35 km/h. The group of turtles studied in this experiment [67] nested on the Cabo Verde islands from which they swim in directed paths to the coastline of West Africa, where they forage. For this experiment, 23 turtles have been tagged since 2011 using satellite-relayed ARGOS systems so that their positions could be recorded at high accuracy [67]. The following results [9] are for a female turtle named Mokamba, where around 1350 data points have been obtained over 400 days from August 2012 to September 2013.

A crucial difference to the previous two organisms, whose positions have been determined with high constant frame rates in a lab, is that these turtles have been tracked via satellites in the wild. As signals could only be emitted with different quality when a turtle was close to the surface, and sometimes the satellite was not available to receive the signal, all turtle data are recorded with varying precision and at irregular time intervals. This poses a severe problem to any reliable statistical data analysis. First, the turtle data had to be cleaned by a specific procedure for removing, e.g. unrealistic or duplicate data points. Then the data had to be interpolated, as statistical averages can only be constructed based on data sets sampled at constant time intervals. Here we chose the most straightforward way of linear interpolation by resampling the interpolated data with a constant time interval, given by the average of the underlying irregular time intervals; details of the full data analysis are given in Ref. [9].

First, we extracted the velocity distributions from the data in a Cartesian coordinate system aligned to longitude and latitude, where according to Fig. 10.1(right) the y-axis is on average almost parallel to the coastline. It was found that both velocity distributions are best fitted by symmetric Gaussians for small velocities while they showed exponential tails for large velocities, in combination matching well to a Gamma distribution, related to what was reported for albatross flights [55]. No significant deviations between velocities along x and y were detected in the data. This corresponds to spatio-temporal intermittency in the turtle movements as an alternation between two different movement modes, similar to our previous results for bumblebee flights. However, here the intermittency occurs as a random-looking switching between foraging (diving, searching, small velocities) and travelling to the next foraging spot (larger velocities). Most interesting were, again, the velocity autocorrelation functions, computed by time averaging along the trajectory. The one for v_x is shown in Fig. 10.4(left); the one for v_y looks similar, but the oscillations are a bit more irregular. Both correlation functions are best fitted by the functional form $\langle v_x(0)v_x(t)\rangle_t = a\exp(-bt) + c\cos(dt)$, where a, b, c, d are fit parameters. This corresponds to exponential decay for very short times while the long-time dynamics exhibits persistent oscillations with a certain period. Figure 10.4(right) displays the corresponding mean squared displacement $\langle x^2(t)\rangle_t$ computed from the data, again via time averaging. One can see that on time scales up to 100–200 h, the turtle moves superdiffusively with an exponent of approximately $\alpha \simeq 1.7$ while for longer times the graph crosses over to subdiffusion, eventually leading to localisation.

Based on these experimental findings, we simulated an overdamped LE where the noise was sampled from a Gaussian (ignoring the exponential tails) with the

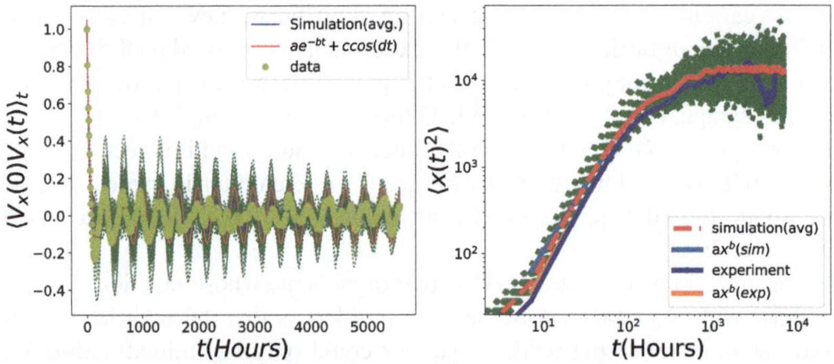

Fig. 10.4 Left: Velocity autocorrelation function along x for a loggerhead sea turtle foraging off the coast of West Africa. The filled olive circles show experimental data, the thin red line with constant oscillation amplitude our best fit to this data, the green dots results from individual simulations of our stochastic model and the blue line with varying amplitude the ensemble averaged simulation result. Right: Corresponding mean squared displacement along x. The wiggled blue line shows experimental data, the overlapping straight orange line is a power law fit to the data for short times, the green dots result from individual simulations of our stochastic model, the (mostly hidden) dashed red line depicts our ensemble averaged simulation result and the turquois line matching the red line at short times yields a power law fit to the simulation results. Figures are from Ref. [9]

correlation decay as identified above, amended by boundary conditions. One can observe in Fig. 10.1(right) that the turtle swimming is bounded from the right by the continental shelve, and on the left, there is also some less rigorous boundedness, likely associated with deeper and less productive (food-rich) waters. We modelled this qualitatively by adding to the LE a force generated by a harmonic potential with slightly asymmetric prefactors to the left and to the right. The numerical results are compared to the experimental data in Fig. 10.4 showing excellent agreement. What one can learn from this stochastic model is that, first, again there is biological activity along the lines of ABP III, as the noise is non-Markovian correlated. Second, the turtles move superdiffusively for shorter times. This is directly related to the strong, very slow periodic oscillations seen in the correlation function: the average half period of one oscillation is about 200 h and thus matches approximately to the cross-over time to subdiffusion seen in the mean squared displacement. Hence, the turtle swims in a rather directed way superdiffusively before it turns around due to hitting the boundaries of the chosen foraging region, which induces subdiffusion (eventually crossing over to localisation on longer time scales; not seen here but for other turtles [9]). The average distance swam during 200 h calculated from the mean squared displacement is approximately 83 km. This matches in order of magnitude to the longitudinal width of the turtles' foraging region, which is hard to estimate but lies around 100–200 km. All these details explain the large scale 'looping' seen in the trajectory displayed in Fig. 10.1(right). That the oscillations in the velocity autocorrelations do not decay may be considered as an unusually strong sign of biological activity. One might speculate that the turtles have developed this looping as an optimal foraging strategy to efficiently search this region. It was checked that the looping is not generated by ocean currents, nor do the turtles exhibit chirality, as often observed for microswimmers [4]. However, there is some coupling to the chlorophyll concentration that is a proxy for food sources, which is ignored in this modelling.

To summarise this section, we have applied the rather basic approach of a generalised overdamped LE to explain the experimentally recorded movements of three very different types of organisms. In all three cases, the stochastic model constructed from data showed that the most significant information is not contained in the associated probability distributions, which is in sharp contrast to the optimal foraging hypothesis based on LWs [19]. Instead, important details characterising the biological movements are captured by the velocity autocorrelation function. That these quantities deserve more attention when analysing movement data has already been emphasised in Ref. [2]. In all three cases, the observed dynamics turned out to be non-Markovian correlated, which may be understood as biological activity along the lines of ABP III. In two cases, the associated diffusion was furthermore anomalous. The constructed models were also to some extent capable to reproduce couplings with the environment in the spirit of movement ecology.

While this straightforward approach of stochastic model building seems to be very promising for further experimental data analysis of organismic movement by already providing significant insight into the emergence of the observed biological dynamics, it is theoretically somewhat a bit naive. In the next part, we will show that one can do much better by drawing inspiration from movement ecology, which suggests to analyse data in the comoving frame.

10.3.2 The Comoving Frame

The CRW model of movement ecology introduced in Sect. 10.2 entails to analyse movement data in a frame attached to the centre of mass of an organism cotranslating and corotating with its movements, which we call the *comoving frame*. This frame is biologically very well motivated, as it seems natural to think of higher-dimensional movements of organisms in terms of step length and turning angle. On the other hand, while a CRW already exhibits persistence, as discussed in Sect. 10.2, it still belongs to the simple class of Markovian overdamped random walks and thus looks somewhat artificial in order to capture the full spectrum of organismic movement. For that reason, in Ref. [33], the conventional CRW model was generalised by formulating underdamped non-Markovian dynamics in a comoving frame. Inspiration was taken from underdamped Langevin dynamics, see Eq. (10.1), which suggested to write down ad hoc the following two Langevin-type equations of motion in the comoving frame:

$$\dot{\beta} = h(\beta, s) + \xi_{\beta;s}(t), \tag{10.11}$$

$$\dot{s} = g(\beta, s) + \xi_{s;\beta}(t). \tag{10.12}$$

In the spirit of a CRW, here β denotes the *turning angle* defined by integration over the angular velocity of the orientation ϕ, $\beta(t) = \int_{t-\Delta t}^{t} \dot{\phi} dt$ [33]. Note that ϕ refers to the polar angle of the velocity vector $\mathbf{v}$ of the moving organism defined in the Cartesian frame, as introduced in Eq. (10.3). Likewise, s holds for the speed of the moving organism, i.e. the radial component of the velocity in the Cartesian frame, cf. Eq. (10.2). In summary, Eqs. (10.11), (10.12) are based on an interplay between three different coordinate frames: (1) The usual fixed Cartesian frame in which overall the three ABP models of Sect. 10.2 were defined; (2) the polar frame in which the speed of ABP I [15] and the velocity of ABP II were defined; and (3) the comoving frame, which refers to a *self-consistent* definition of the dynamics of a moving agent in terms of the turning angle β and its speed s only, i.e. without any explicit reference to the Cartesian or polar frames (1) or (2). Examples of such dynamics are the CRW of movement ecology and the Langevin-type equations introduced above. The tricky part, crucially distinguishing the comoving from the polar frame, is to obtain a self-consistent equation for the turning angle variable, as for the example of the OU process we will show in Sect. 10.4.

To conclude the discussion of the new model Eqs. (10.11), (10.12), both equations are underdamped featuring acceleration of β and s on the left-hand side, which is generated by splitting the dynamics into deterministic drift and stochastic diffusion terms on the right-hand side. While this very basic structure corresponds to the one of the ordinary LE (10.1), here the functional forms of these four terms, including the autocorrelation decay of the two noises, are per se completely general and need to be determined, e.g. by experimental data analysis. For this reason, we use the more general notation $h(\beta, s)$ and $g(\beta, s)$ for the drift terms, even though at least for the speed, we would expect the respective term to boil down to some type of friction, cf. $\eta(s)$ in ABP I Eq. (10.2), as indeed will be shown in the following.

We now demonstrate that Eqs. (10.11), (10.12) can be used to analyse movement data by standard methods of drift-diffusion time series analysis. As an example, we construct a stochastic model for 'free' bumblebee flights by applying this framework to the experimental data discussed in Sect. 10.3.1, where here we only consider flights far away from the flower carpet in the spider-free phase (1). The analysis was based on 51 trajectories with approximately 49,000 data points [33]. Plotting the normalised drift vector field of (s, β), respectively the deterministic terms of Eqs. (10.11), (10.12) extracted from data, shows that the cross-dependencies of $h(\beta, s)$ on s and of $g(\beta, s)$ on β are weak [33]. This suggests to split the vector field into

$$\dot{\beta} = h(\beta) + \xi_{\beta;s}(t), \tag{10.13}$$

$$\dot{s} = g(s) + \xi_{s;\beta}(t). \tag{10.14}$$

Figure 10.5 depicts results for both simplified drift terms, extracted from data and with fits to the data. As can be seen in Fig. 10.5(top), $h(\beta) \simeq -k\beta$ with $k \approx 1/\Delta t$.

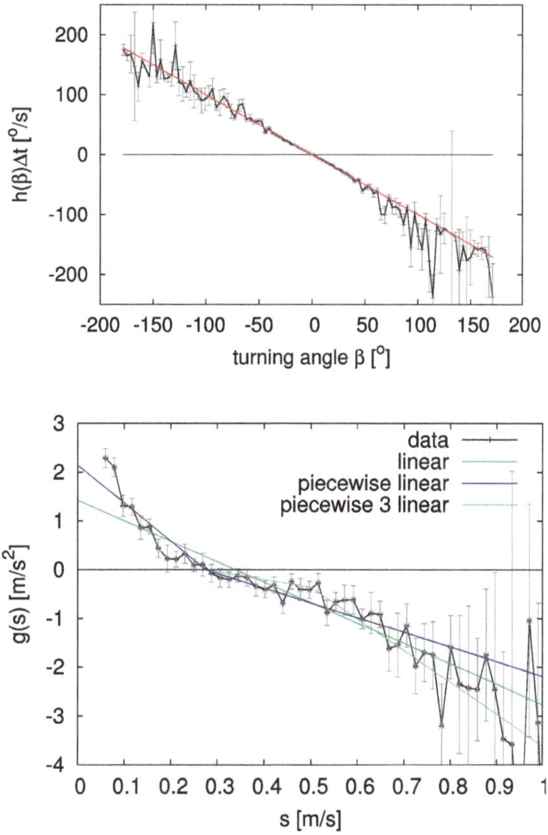

Fig. 10.5 Top: Drift coefficient for the turning angle β. The drift $h(\beta)$ obtained from bumblebee data (wiggled black line) is linear in β, as demonstrated by the fitted straight red line. Bottom: Drift coefficient for the speed s. Here the drift $g(s)$ obtained from bumblebee data (black wiggled line) has been approximated by piecewise linear functions from one to three pieces (blue, green, cyan, right corner from top to bottom). 95% confidence intervals are shown in grey. Figures are from Ref. [33]

Integrating Eq. (10.13) with respect to Δt yields $\beta(t) = \tilde{\xi}_{\beta;s}(t)$ [33]. Hence, for the angular dynamics, we are almost back to a CRW where the turning angles are sampled from some probability distribution, except that here we allow them to be correlated both with respect to s and in time. Determining analogously the drift term $g(s)$ in the speed equation (10.14) from data suggests simple (piecewise) linear approximations as suitable fit functions, cf. Fig. 10.5(bottom). For an analytical model, we choose the one consisting of two pieces,

$$g(s) \approx (s - s_0) \cdot \begin{cases} -\eta_1, s < s_0, \\ -\eta_2, s \geq s_0, \end{cases} \tag{10.15}$$

with $\eta_1 > \eta_2 > 0$ and a preferred speed s_0. We see that this deterministic drift term constructed from data yields an expression that, interestingly, can be understood as an active, speed-dependent friction coefficient, which is somewhat in-between the Schienbein–Gruler and Rayleigh–Helmholtz forms discussed for $\eta(s)$ in ABP I, Eq. (10.2).

We now take on the two noise terms in Eqs. (10.11), (10.12). First, it was found that the turning angle distribution $\rho_s(\beta)$ at each speed is best approximated by a wrapped Gaussian [33]. However, the standard deviation σ_β turned out to be strongly speed-dependent, with exponential decay in s fitting best [33]. This matches to naive biological reasoning: At larger speeds, it is more difficult for a bumblebee to make frequent turns. It remains to determine the correlation decay of the turning angle dynamics, where the autocorrelation function is defined in analogy to Eq. (10.9). Figure 10.6(top) shows that it follows a steep power law. This analysis yields a stochastic process for the turning angle dynamics defined by

$$\beta = \tilde{\xi}_{\beta;s}(t),\ \tilde{\xi}_{\beta;s} \overset{d}{\sim} \mathcal{N}(0, \sigma_\beta(s)),\ \sigma_\beta(s) = c_1 e^{-c_2 s} + c_3, \tag{10.16}$$

with c_1, c_2, c_3 as fit parameters, where the turning angle noise is power law correlated.

Second, the noise for the speed changes can be computed from Eq. (10.12), rewritten as $\xi_{s;\beta}(t) = \dot{s} - g(s)$. This noise obtained from data showed no dependence on β and was best fitted with a Gaussian [33]. The correlations, defined in the same way as in Eq. (10.9), turned out to be anti-persistent, as depicted in Fig. 10.6(bottom), which was best approximated by a difference between two exponential terms, $\mathrm{acf}_{\xi_s}(\tau) \approx ae^{-\lambda_1\tau} + (1-a)e^{-\lambda_2\tau}$, $\lambda_1, \lambda_2 > 0$. This completes Eq. (10.14) for the speed dynamics.

In summary, we obtain as a full stochastic bumblebee flight model

$$\beta = \tilde{\xi}_{\beta;s}(t), \tag{10.17}$$

$$\dot{s} = -\eta_i(s - s_0) + \xi_s(t),\ i = 1, 2, \tag{10.18}$$

where $\tilde{\xi}_{\beta;s}$ is speed-dependent, power law correlated wrapped Gaussian noise defining the turning angle β at each time t. The speed dynamics is governed by an underdamped LE with a speed-dependent piecewise linear friction coefficient and exponentially anti-correlated Gaussian noise ξ_s. As shown in detail in Ref. [33], this fully analytically defined model, which is based on quite some approximations, works surprisingly well to reproduce the experimental bumblebee flight data it was constructed from.

There are some important conclusions to be drawn from this part, which we summarise as follows. We have shown that starting from the (underdamped) generalised LE (10.11), (10.12) formulated in the comoving frame for both the turning angle and the speed dynamics, one can construct a corresponding stochastic model from data for 'free' (modulo the experimental setup of the cage) bumblebee flights. Interestingly, the speed equation contains a speed-dependent friction term right in the spirit of ABP I. The angular dynamics features a wrapped Gaussian turning angle distribution as for a standard CRW and the associated ABP II. Finally, both the angular and the speed dynamics exhibit correlated, non-Markovian noise

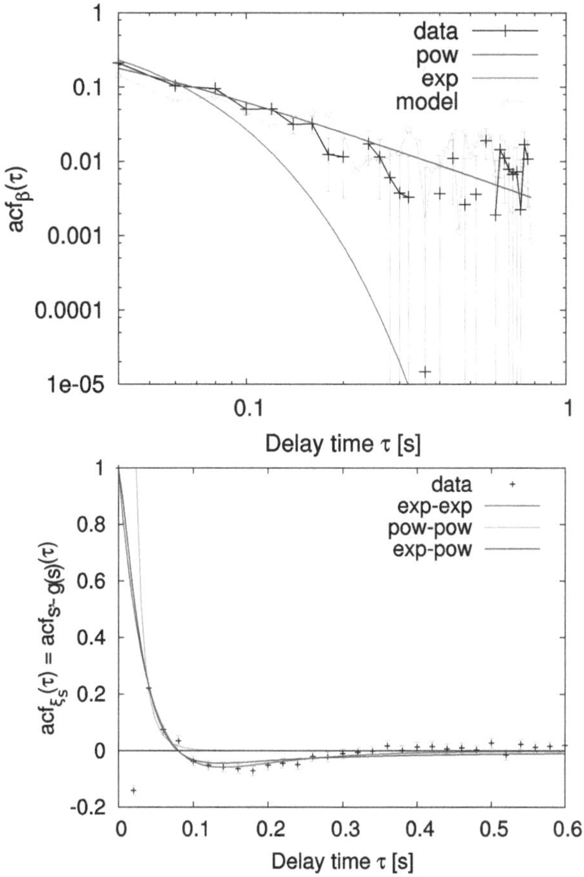

Fig. 10.6 Top: Autocorrelation function for turning angles β. Experimental data (black crosses) is shown together with an exponential (magenta curved line) and a power law (blue straight line) fit, supplemented by the large-lag standard error (grey). The green circles depict the autocorrelation extracted from the simulated data Eqs. (10.17), (10.18). Bottom: Autocorrelation of the non-deterministic speed changes ξ_s from experimental data (dots) with two times the large-lag standard error (grey) and three fits (top left corner left to right): Difference of two exponentials (red), exponential and power law (blue) and of two power laws (green). The outlier at $t \sim 0.02$ s is a discretisation artefact due to the finite resolution of the data. Figures are from Ref. [33]

terms, thus displaying the main property of ABP III. We have thus constructed from experimental data a movement model for bumblebee flights that contains the characteristic features of all three generic ABP models I–III discussed before. When Ref. [33] was published in 2013, only activity related to ABP I was pointed out but not the cross-links to ABP II (invented earlier) and III (invented later), with the field of active particles still being in rapid development during that time. Whether this model generates anomalous diffusion was not tested. There is no mechanism for such dynamics in the Gaussian noise distributions, but the turning angle correlations decay as a power law, similar to the fractional Brownian motion stochastic model for the angular dynamics studied in [63]. Therein it was shown, however, that such dynamics only displays transient, no long-term anomalous diffusion. More general conclusions based on this analysis are drawn in the following section.

10.4 A Newtonian Equation for Modelling the Movements of Organisms

How do you, reader, walk in a plane? You make a step per time interval by choosing a direction and a step length, and you repeat the process. Typically, you include some memory for steering where you want to go. But this dynamics is per se generated by yourself solely in the comoving frame. Of course, you may wish to navigate, given environmental conditions (moving in a room full of objects, going shopping, etc.). But we do not compute our intrinsic movements in a fixed Cartesian frame by then translating these results into the active movement outlined above, nor is there any reason why other living organisms should do so. Representing passive, environmentally generated movement mathematically in a fixed Cartesian frame makes perfect sense for modelling physical phenomena like Brownian motion of a tracer particle in a fluid, where the fluid is contained in some region and the tracer is driven passively by the surrounding fluid molecules. However, any living organism is self-propelled by 'biological noise' *internally* generated by the organism itself, *and thus in the comoving frame*, not originating *externally* in some fixed Cartesian frame. The latter should rather model external environmental noise on top of internal biological activity. For these biophysical reasons, we argue that the right frame to work in for understanding the movements of organisms is primarily the comoving one, not the fixed Cartesian one. This was already amply realised by Sir Ronald Ross in 1904 when he suggested to Karl Pearson to model mosquito flights by using step lengths and turning angles. Pearson rephrased this idea in his 1905 papers in terms of the 'drunken sailor problem', where he defined the random walk of a drunken sailor in the comoving frame, not in the Cartesian one [51]. In that sense, the CRW formulated within movement ecology is a straightforward generalisation of Ross' and Pearson's ideas to a sailor who is not fully drunk by exhibiting some persistence. In contrast, the physically and mathematically inspired approaches of active particles and anomalous diffusion naturally use the Cartesian frame, reflecting their origin of describing passive motion and that stochastic processes are defined and analysed way easier in these coordinates. Accordingly, higher-dimensional formulations typically consist of straightforward component-wise vectorial extensions of one-dimensional stochastic processes in a Cartesian frame. Note also that, as we remarked in Sect. 10.2, a CRW can be formulated as a Markov process in the comoving frame while it exhibits non-Markovian properties in the Cartesian frame. To what extent this applies to other stochastic processes is an interesting open question, possibly also of some biological relevance.

The problem of representing stochastic processes in different frames of reference is exemplified by a recent experiment, whose aim was to synthetically generate different types of anomalous diffusion. For this purpose, the dynamics of a super-paramagnetic colloidal particle was finely tuned by using magnetic fields. These fields were changed in time according to specific protocols defined in the comoving frame, which were designed to reproduce generic types of anomalous diffusion [68]. This experimental approach raised the question of how to define anomalous stochastic processes like LWs and fractional Brownian motion in the comoving frame. For LWs, the answer is to quite some extent known but, interestingly, ambiguous: In Ref. [53], it was shown that there exist three different LW models in two dimensions, two defined in the Cartesian frame and one in the comoving frame. What we introduced as LWs in Sect. 10.2 is actually only the version in the comoving frame, called uniform LW. It turns out that all three LW models exhibit different diffusive properties [53]; hence, they cannot be transformed into each other. This is due to the non-Markovianity of the LW dynamics. For other processes, such as two-dimensional Cartesian fractional Brownian motion, so far no corresponding formulation in the comoving frame is known; one may doubt whether there is any exact one, given the inherent non-Markovianity of this process.

The above discussion suggests that working in different reference frames requires very different definitions of stochastic processes. This problem has been solved for the Markovian OU process [69], defined in Eq. (10.1) for the velocity in the Cartesian frame (where in the following, we explicitly spell out the noise strength as $\sqrt{2D}$ with velocity diffusion coefficient D). Transforming the Cartesian OU process into the comoving frame requires to use polar coordinates for the velocities, as in ABP II Eq. (10.3), i.e. $v_x = s\cos\phi$, $v_y = s\sin\phi$. However, since both the associated speed s and the orientation ϕ are stochastic variables, we encounter an Itô-Stratonovich problem for defining the corresponding stochastic equations in the comoving frame. The Itô equations for (ϕ, s) were already derived in Ref. [70],

$$\dot{\phi} = \frac{1}{s}\sqrt{2D}\xi_\phi(t), \tag{10.19}$$

$$\dot{s} = \left(-\frac{\eta}{m}s + \frac{D}{s}\right) + \sqrt{2D}\xi_s(t), \tag{10.20}$$

with the two noises defined as

$$\xi_\phi(t) = -\xi_x(t)\sin\phi + \xi_y(t)\cos\phi,\ \xi_s(t) = \xi_x(t)\cos\phi + \xi_y(t)\sin\phi, \tag{10.21}$$

where ξ_x, ξ_y are the two corresponding Gaussian white noises in the Cartesian frame. One can show that the resulting two noises ξ_ϕ, ξ_s are also white Gaussian [70]. Equations (10.19), (10.20) can be solved numerically, confirming that they reproduce Cartesian OU dynamics [69]. However, in order to obtain a self-consistent description of OU in the comoving frame, we need an equation for the turning angle β. This can be accomplished starting from $\beta \simeq \dot{\phi}\delta t$ and by calculating the functional form of the turning angle distribution from the right-hand side of Eq. (10.19). One can show that in a steady state, the speed generated by Eq. (10.20) is distributed according to a Rayleigh distribution with exponentially decaying autocorrelations. Remarkably, the wrapping onto the circle entailed in the calculation yields turning angles that are uncorrelated. Hence, this wrapping, which is a crucial operation to generate self-consistent angular dynamics in the comoving frame, marks a highly non-trivial step of loss of memory. Along these lines one finds for the turning angle distribution

$$\rho(\beta) = \frac{\sqrt{2D\delta t}}{2\pi} \sum_{k=-\infty}^{\infty} \frac{\sigma_{\xi,\phi}\,|k|}{\sigma_{\rho,s}} K_1 \left(\frac{\sqrt{2D\delta t}\,\sigma_{\xi,\phi}\,|k|}{\sigma_{\rho,s}} \right) e^{ik\beta}, \tag{10.22}$$

with $\sigma_{\xi,\phi}$ the standard deviation of the Gaussian noise ξ_ϕ in Eq. (10.19) and $\sigma_{\rho,s}$ the one of the Rayleigh distribution for the speed obtained from solving Eq. (10.20), where $K_1(z)$ is the modified Bessel function of the second kind. $\rho(\beta)$ looks overall similar to a wrapped Gaussian, respectively a von Mises distribution, but shows deviations in detail [69]. One thus arrives at a self-consistent formulation of a Cartesian OU process transformed into the comoving frame in the form of [69]

$$\beta = \xi_\beta(t),\ s = \tilde{\xi}_s(t), \tag{10.23}$$

where ξ_β is white noise with the functional form of $\rho(\beta)$ while $\tilde{\xi}_s$ is exponentially correlated Rayleigh noise. Alternatively, the equation for the speed could be replaced by Eq. (10.20) with Gaussian white noise for ξ_s.

In summary, in this section, we first argued that for biophysical reasons, organismic movement models should ideally generically be defined in the comoving frame. We then demonstrated that this change of perspective has profound implications on the corresponding description of movement in terms of stochastic processes. While a Markovian OU process can still be transformed exactly from the Cartesian into the comoving frame, this is not possible anymore for non-Markovian LWs and appears unlikely for other non-Markovian processes like fractional Brownian motion. The resulting Eqs. (10.23) for OU in the comoving frame illustrate the framework provided by Eqs. (10.11), (10.12), which define generalised Langevin dynamics in the comoving frame. Equations (10.17), (10.18), on the other hand, demonstrated that a comoving generalised LE can also be constructed from experimental data. Similarly, this theoretical framework contains CRWs and LWs as special cases by furthermore capturing all characteristic features of ABP I–III. It also includes run-and-tumble dynamics if described by what we called ABP IV. These findings substantiate the purely heuristic formulation of Eqs. (10.11), (10.12) as a promising candidate for what one may call a Newtonian equation for modelling the movements of organisms, in the form of a generalised LE in the comoving frame.

This theoretical framework raises some important open questions. First, both the bumblebee model and the OU process in the comoving frame yield turning angle dynamics that does not involve a first-order time derivative as proposed in the original Eq. (10.11). Upon further scrutiny, one may argue that such a first-order differential equation for the turning angle models a kind of 'jerky' angular acceleration that goes beyond classical Newtonian dynamics. At present, it is not clear whether such a higher-order modelling is needed for describing biological movements, or not. On the other hand, there is no reason why organismic movement should obey simple Newtonian-type dynamics. Second, OU dynamics in the comoving frame is represented by a non-uniform turning angle distribution that reminds of ABP II as well as by exponential speed correlations similar to APB III. Hence, one may question whether characterising biological activity based on the existence of non-uniform turning angle distributions or correlations only is sufficient, as these properties also characterise passive OU dynamics transformed into the comoving frame. This asks for a more robust, general classification of active biological motion. Along these lines, third, OU for a passive particle exhibits fluctuation–dissipation relations while for an active particle, they are typically expected to be broken. It is not clear, however, how fluctuation–dissipation relations look like for dynamics in a comoving frame. These appear to be important theoretical questions that require further investigation.

10.5 Conclusions

The LE provides a powerful paradigm, which has been widely used across different fields of science. Originally formulated to model the Brownian motion of a tracer in a fluid, this concept has been successfully applied in terms of geometric Brownian motion to economics and, more recently, to climate science for modelling the temperature of the Earth [71]. Here we argue that it may be similarly promising to apply a suitably generalised LE for modelling and understanding the movements of organisms. In order to do so, we combined key concepts and ideas extracted from movement ecology, active matter and anomalous diffusion. We first demonstrated that a very simplistic application of Langevin dynamics in terms of a generalised overdamped LE formulated in the Cartesian frame already proves useful to describe experimental results for foraging bumblebees, migrating cells and swimming sea turtles. In all these cases, the resulting LEs constructed from data contained specific features of ABPs, partially together with properties of anomalous diffusion. We then showed that a more general, and biophysically more meaningful, formulation of generalised Langevin dynamics can be achieved by using the comoving frame. Working out this idea for bumblebee flights we constructed a generalised underdamped LE from data in the comoving frame, combining the characteristic features of all three ABP models discussed before. We also outlined an application of this framework to a recent experiment synthetically generating anomalous diffusion. Theoretically, we illustrated this formalism by transforming the Markovian OU process defined in the Cartesian frame into a new set of stochastic equations defined self-consistently in the comoving frame. Whether the transformation of known stochastic processes from the Cartesian into the comoving frame is generally the right way to study organismic movement may be questioned. Rather, it looks as if this theoretical framework asks for a formulation, and study, of novel stochastic processes specifically defined in the comoving frame. Our central result is the generalised LE in the comoving frame Eqs. (10.11), (10.12), already put forward in Ref. [33], which we advertise as a promising approach to model the movements of organisms. This equation asks to be tested experimentally for movements of other organisms and laboratory-based studies with synthetic active matter (e.g., active colloids [68]). Such experiments can help to identify and formulate generic, potentially new stochastic processes for explaining other observed dynamics. The proposed theoretical framework, in turn, poses interesting new mathematical problems in terms of formulating stochastic processes in the comoving frame by studying their statistical physical properties.

In view of a biological meaning, the theory that we put forward here may relate in some sense to the work by Uexküll [72]. Along similar lines, the Cartesian and the comoving frames that we identified as being intimately associated with different representations of stochastic dynamics for modelling organismic movement may have direct biological interpretations in terms of what is called allocentric and egocentric reference frames [73]. The interplay between these two frames is in turn very important for understanding organismic navigation strategies like path integration [73, 74]. Interestingly, recent research provides evidence that these two different, biologically highly relevant frames are linked on a neural level in the brain of organisms to specific neurons called place and grid cells [74].

Finally, apart from movement ecology, active matter and anomalous diffusion, strictly speaking, another fourth research area comes into play within this framework, which is *nonlinear dynamics*. This is most easily seen in the ABP model I, which employs potentially nonlinear friction coefficients. More recently, there have been speculations about the origin of specific organismic movement patterns by nonlinear oscillatory dynamics in the brain [75, 76]. Along all these lines, the door seems wide open for new research to model and understand the movements of organisms from first principles.

Acknowledgments R.K. wishes to thank Friedrich Lenz, Aleksei Chechkin, Thomas Ings and Lars Chittka for joint work on bumblebee flights as well as Peter Dieterich, Albrecht Schwab and Aleksei Chechkin for joint work on cell migration. He also thanks Nicholas Watkins for many inspiring discussions on Langevin dynamics, Sergei Fedotov for making him aware of Ross' letter to Person and Frederike Hanke for interesting hints on biological literature. Special thanks go to Luca Giuggioli for inviting him to the INI Programme, and INI for the hospitality, supported by EPSRC grant EP/R014604/1. Giving three talks during his visits to INI triggered the thoughts that led to this book chapter. A.G. and G.V. acknowledge sponsorship for this work by the US Office of Naval Research Global (award no. N62909-18-1-2170).

References

1. Dieterich, P., Lindemann, O., Moskopp, M.L., Tauzin, S., Huttenlocher, A., Klages, R., Chechkin, A., Schwab, A.: Anomalous diffusion and asymmetric tempering memory in neutrophil chemotaxis. PLOS Comput. Biol. **18**, 1–26 (2022)
2. Lenz, F., Ings, T.C., Chittka, L., Chechkin, A.V., Klages, R.: Spatiotemporal dynamics of bumblebees foraging under predation risk. Phys. Rev. Lett. **108**, 098103/1–5 (2012)
3. Parthasarathy, A.S.P.: Statistical analysis of sea turtle foraging. Master's Thesis, Queen Mary University of London, 2021
4. Bechinger, C., Di Leonardo, R., Löwen, H., Reichhardt, C., Volpe, G. and Volpe, G.: Active particles in complex and crowded environments. Rev. Mod. Phys. **88**, 045006 (2016)
5. Wehner, R.: Desert Navigator: The Journey of an Ant. Harvard University Press, Cambridge (2020)

6. Kays, R., Crofoot, M.C., Jetz, W., Wikelski, M.: Terrestrial animal tracking as an eye on life and planet. Science **348**, aaa2478 (2015)
7. Nathan, R., et al.: Big-data approaches lead to an increased understanding of the ecology of animal movement. Science **375**, eabg1780 (2022)
8. Brockmann, D., Hufnagel, L., Geisel, T.: The scaling laws of human travel. Nature **439**, 462–465 (2006)
9. Kumar, V., Torres, P.A.R., Eizaguirre, C., Palyulin, V.V., Klages, R.: in preparation (2024)
10. Cranna, V.: Karl Pearson and Sir Ronald Ross, 2015. Library, Archive and Open Research Services blog
11. Pearson, K.: Mathematical contributions to the theory of evolution—a mathematical theory of random migration. Biometric Ser. **3**, 54 (1906)
12. Brückner, D.B. and Broedersz, C.P.: Learning dynamical models of single and collective cell migration: a review. Rep. Progr. Phys. **87**, 056601 (2024)
13. Klages, R Cell migration: Beyond Brownian motion. Biophys. J. **123**, 1167–1169 (2024)
14. Schweitzer, F.: Brownian Agents and Active Particles. Springer, Berlin (2003)
15. Romanczuk, P., Bär, M., Ebeling, W., Lindner, B., Schimansky-Geier, L.: Active Brownian particles. Eur. Phys. J. Spec. Top. **202**, 1–162 (2012)
16. Ramaswamy, S.: The mechanics and statistics of active matter. Annu. Rev. Condes. Matter Phys. **1**, 323–345 (2010)
17. Marchetti, M.C., Joanny, J.F., Ramaswamy, S., Liverpool, T.B., Prost, J., Rao, M., Simha, R.A.: Hydrodynamics of soft active matter. Rev. Mod. Phys. **85**, 1143–1189 (2013)
18. Bernheim-Groswasser, A., Gov, N.S., Safran, S.A., Tzlil, S.: Living matter: mesoscopic active materials. Adv. Mater. **30**, 1707028 (2018)
19. Viswanathan, G.M., da Luz, M.G.E., Raposo, E.P., Stanley, H.E.: The Physics of Foraging. Cambridge University Press, Cambridge (2011)
20. Zaburdaev, V., Denisov, S., Klafter, J.: Lévy walks. Rev. Mod. Phys. **87**, 483–529 (2015)
21. Klages, R.: Search for food of birds, fish and insects. book chapter. In: Bunde, A., Caro, J., Kaerger, J., Vogl, G. (eds.) Diffusive Spreading in Nature, Technology and Society, 2nd edn., pp. 53–74. Springer, Berlin (2023)
22. Metzler, R., Klafter, J.: The random walk's guide to anomalous diffusion: a fractional dynamics approach. Phys. Rep. **339**, 1–77 (2000)
23. Metzler, R., Jeon, J.-H., Cherstvy, A.G. and Barkai, E.: Anomalous diffusion models and their properties: non-stationarity, non-ergodicity, and ageing at the centenary of single particle tracking. Phys. Chem. Chem. Phys. **16**, 24128–24164 (2014)
24. Klages, R., Radons, G., Sokolov, I.M. (eds.): Anomalous transport: Foundations and Applications. Wiley-VCH, Berlin (2008)
25. Nathan, R., Getz, W.M., Revilla, E., Holyoak, M., Kadmon, R., Saltz, D. and Smouse, P.E.: A movement ecology paradigm for unifying organismal movement research. Proc. Natl. Acad. Sci. **105**, 19052 (2008)
26. Méndez, V., Campos, D., Bartumeus, F.: Stochastic Foundations in Movement Ecology. Springer, Berlin (2014)
27. Joo, R., Picardi, S., Boone, M.E., Clay, T.A., Patrick, S.C., Romero-Romero, V.S., Basille, M.: Recent trends in movement ecology of animals and human mobility. Mov. Ecol. **10**, 26 (2022)
28. Patterson, T.A., Parton, A., Langrock, R., Blackwell, P.G., Thomas, L., King, R.: Statistical modelling of individual animal movement: an overview of key methods and a discussion of practical challenges. AStA Adv. Stat. Anal. **101**, 399–438 (2017)
29. Auger-Méthé, M., Newman, K., Cole, D., Empacher, F., Gryba, R., King, A.A., Leos-Barajas, V., Mills Flemming, J., Nielsen, A., Petris, G., Thomas, L.: A guide to state–space modeling of ecological time series. Ecol. Monogr. **91**, e01470 (2021)
30. Glennie, R., Adam, T., Leos-Barajas, V., Michelot, T., Photopoulou, T., McClintock, B.T.: Hidden Markov models: Pitfalls and opportunities in ecology. Methods Ecol. Evol. **14**, 43–56 (2023)
31. Codling, E.A., Plank, M.J., Benhamou, S.: Random walk models in biology. J. R. Soc. Interface **5**, 813–834 (2008)

32. Kareiva, P.M., Shigesada, N.: Analyzing insect movement as a correlated random walk. Oecologia **56**, 234–238 (1983)
33. Lenz, F., Chechkin, A.V., Klages, R.: Constructing a stochastic model of bumblebee flights from experimental data. PLoS ONE **8**, e59036/1–7 (2013)
34. te Vrugt, M, Wittkowski, R.: A review of active matter reviews (2024). preprint arXiv:2405.15751
35. Schweitzer, F.: An agent-based framework of active matter with applications in biological and social systems. Eur. J. Phys. **40**, 014003 (2019)
36. Langevin, P.: Sur la théorie du mouvement brownien. C.R. Acad. Sci. (Paris) **146**, 530–533 (1908). see also the English translation in Am. J. Phys. **65**, 1079 (1997)
37. Coffey, W., Kalmykov, Y.P., Waldron, J.T.: The Langevin Equation. World Scientific, Singapore (2004)
38. Ebeling, W., Schweitzer, F., Tilch, B.: Active Brownian particles with energy depots modeling animal mobility. BioSystems **49**, 17–29 (1999)
39. Howse, J.R., Jones, R.A., Ryan, A.J., Gough, T., Vafabakhsh, R., Golestanian, R.: Self-motile colloidal particles: From directed propulsion to random walk. Phys. Rev. Lett. **99**, 048102 (2007)
40. Koumakis, N., Maggi, C., Di Leonardo, R.: Directed transport of active particles over asymmetric energy barriers. Soft Matter **10**, 5695–5701 (2014)
41. Szamel, G.: Self-propelled particle in an external potential: existence of an effective temperature. Phys. Rev. E **90**, 012111 (2014)
42. Martin, D., O'Byrne, J., Cates, M.E., Fodor, É., Nardini, C., Tailleur, J., Van Wijland, F.: Statistical mechanics of active Ornstein-Uhlenbeck particles. Phys. Rev. E **103**, 032607 (2021)
43. Berg, H.C.: E. Coli in Motion. Springer, New York (2008)
44. Schnitzer, M.J.: Theory of continuum random walks and application to chemotaxis. Phys. Rev. E **48**, 2553–2568 (1993)
45. Khatami, M., Wolff, K., Pohl, O., Ejtehadi, M.R., Stark, H.: Active Brownian particles and run-and-tumble particles separate inside a maze. Sci. Rep. **6**, 1–10 (2016)
46. Lemaitre, E., Sokolov, I.M., Metzler, R., Chechkin, A.V.: Non-Gaussian displacement distributions in models of heterogeneous active particle dynamics. New J. Phys. **25**, 013010 (2023)
47. Kurzthaler, C., Leitmann, S., Franosch, T.: Intermediate scattering function of an anisotropic active Brownian particle. Sci. Rep. **6**, 36702 (2016)
48. Valani, R.N., Paganin, D.M.: Attractor-driven matter. Chaos **33**, 023125 (2023)
49. Pikovsky, A.: Deterministic active particles in the overactive limit. Chaos **33**, 113114 (2023)
50. Basu, U., Majumdar, S.N., Rosso, A., Schehr, G.: Active Brownian motion in two dimensions. Phys. Rev. E **98**, 062121 (2018)
51. Pearson, K.: The problem of the random walk. Nature **72**, 294, 342 (1905)
52. Viswanathan, G.M., Afanasyev, V., Buldyrev, S.V., Murphy, E.J., Prince, P.A., Stanley, H.E.: Lévy flight search patterns of wandering albatrosses. Nature **381**, 413–415 (1996)
53. Zaburdaev, V., Fouxon, I., Denisov, S., Barkai, E.: Superdiffusive dispersals impart the geometry of underlying random walks. Phys. Rev. Lett. **117**, 270601 (2016)
54. Viswanathan, G.M., Buldyrev, S.V., Havlin, S., da Luz, M.G., Raposo, E.P., Stanley, H.E.: Optimizing the success of random searches. Nature **401**, 911–914 (1999)
55. Edwards, A.M., Phillips, R.A., Watkins, N.W., Freeman, M.P., Murphy, E.J., Afanasyev, V., Buldyrev, S.V., da Luz, M.G., Raposo, E.P., Stanley, H.E., Viswanathan, G.M.: Revisiting Lévy flight search patterns of wandering albatrosses, bumblebees and deer. Nature **449**, 1044–1048 (2007)
56. Pyke, G.H.: Understanding movements of organisms: it's time to abandon the Lévy foraging hypothesis. Meth. Ecol. Evol. **6**, 1 (2015)
57. Reynolds, A.M.: Current status and future directions of Lévy walk research. Biol. Open **7**, 03016 (2018)
58. Levernier, N., Textor, J., Bénichou, O. and Voituriez, R.: Inverse square Lévy walks are not optimal search strategies for $d \geq 2$. Phys. Rev. Lett. **124**, 080601 (2020)
59. Benhamou, S.: How many animals really do the lévy walk? Ecology **88**, 1962–1969 (2007)

60. Bénichou, O., Loverdo, C., Moreau, M., Voituriez, R.: Intermittent search strategies. Rev. Mod. Phys. **83**, 81–129 (2011)
61. Vilk, O., Orchan, Y., Charter, M., Ganot, N., Toledo, S., Nathan, R., Assaf, M.: Ergodicity breaking in area-restricted search of avian predators. Phys. Rev. X **12**, 031005 (2022)
62. Khadem, S.M.J., Klapp, S.H., Klages, R.: Search efficiency of discrete fractional Brownian motion in a random distribution of targets. Phys. Rev. Res. **3**, 023169 (2021)
63. Gomez-Solano, J.R., Sevilla, F.J.: Active particles with fractional rotational brownian motion. J. Stat. Mech.: Theor. Exp. **2020**, 063213 (2020)
64. Ings, T.C., Chittka, L.: Speed-accuracy tradeoffs and false alarms in bee responses to cryptic predators. Curr. Biol. **18**, 1520–1524 (2008)
65. Dieterich, P., Klages, R., Preuss, R., Schwab, A.: Anomalous dynamics of cell migration. PNAS **105**, 459–463 (2008)
66. Klimek, A., Mondal, D., Block, S., Sharma, P., Netz, R.R.: Data-driven classification of individual cells by their non-Markovian motion. Biophys. J. **123**, 1173–1183 (2024)
67. Roman Torres, P.A.: Environmental variation shapes life history traits of sea turtles: an ecological modelling approach. Ph.D. Thesis, Queen Mary University of London, 2022
68. Gentili, A., Klages, R., Volpe, G.: Anomalous diffusion of superparamagnetic walkers with tailored statistics (2025). preprint arXiv:2412.13960
69. Lucero-Azuara, N., Klages, R.: in preparation (2024)
70. Gardiner, C.: Stochastic Methods: A Handbook for the Natural and Social Sciences. Springer Series in Synergetics. Springer, Berlin (2009)
71. Watkins, N.W., Calel, R., Chapman, S.C., Chechkin, A., Klages, R., Stainforth, D.A.: The challenge of non-Markovian energy balance models in climate. Chaos **34**, 072105 (2024)
72. von Uexküll, J.: Umwelt und Innenwelt der Tiere. Springer, Berlin (1909)
73. Klatzky, R.L.: Allocentric and egocentric spatial representations: definitions, distinctions, and interconnections. book chapter. In: Freksa, C., Habel, C., Wender, K.F. (eds.) Spatial Cognition: An Interdisciplinary Approach to Representing and Processing Spatial Knowledge, pp. 1–17. Springer, Berlin (1998)
74. Edvardsen, V., Bicanski, A., Burgess, N.: Navigating with grid and place cells in cluttered environments. Hippocampus **30**, 220–232 (2020)
75. Abe, M.S.: Functional advantages of lèvy walks emerging near a critical point. PNAS **117**(39), 24336–24344 (2020)
76. Cheng, K.: Oscillators and servomechanisms in orientation and navigation, and sometimes in cognition. Proc. R. Soc. Lond. Ser. B, Biol. Sci. **289**(1974), 20220237 (2022)

Chapter 11
Spatial Competition and Repulsion: Pattern Formation and the Role of Movement

Cristóbal López, Eduardo H. Colombo, Emilio Hernández-García, and Ricardo Martinez-Garcia

11.1 Introduction

Collections of interacting particles are widely used to study complex systems, as they provide insights into diverse collective behaviors observed in various fields, such as the flocking of animals [1], spatial patterns in ecosystems, embryos and colloids [2–5], cell migration [6], and phase transitions in active matter [7]. Interactions among particles in these models occur via forces and/or dynamical rules, which can be of physical, biological, or social origins, and can operate either locally, involving a few individuals, or over a finite distance, influencing many individuals and leading to long-range aggregates [8, 9].

The formation of these particle aggregates or clusters, which may appear forming global periodically ordered patterns, is particularly significant for biological populations like vegetation systems and cell populations [8, 9]. The spatial self-organization has important ecological implications, producing feedbacks with biological functionalities that are fundamental for the biodiversity maintenance, population survival, and the interactions with environmental factors [10]. A com-

C. López (✉) · E. Hernández-García
Instituto de Física Interdisciplinar y Sistemas Complejos (IFISC), CSIC-UIB, Palma de Mallorca, Spain
e-mail: clopez@ifisc.uib-csic.es

E. H. Colombo
Center for Advanced Systems Understanding (CASUS) – Helmholtz-Zentrum Dresden-Rossendorf (HZDR), Görlitz, Germany

R. Martinez-Garcia
Center for Advanced Systems Understanding (CASUS), Helmholtz-Zentrum Dresden-Rossendorf (HZDR), Görlitz, Germany

ICTP South American Institute for Fundamental Research & Instituto de Física Teórica, Universidade Estadual Paulista - UNESP, São Paulo, Brazil

L. Giuggioli, P. K. Maini (eds.), *The Mathematics of Movement*,
https://doi.org/10.1007/978-3-032-00873-2_11

plete knowledge of the conditions that lead to these spatial structures and their feedback on the system dynamics is thus essential for the understanding of complex living systems.

In recent years, the authors have explored different models of pattern formation driven only by anti-crowding interactions through repelling forces and intraspecific competition (throughout this chapter, we will refer to the both types of interactions as competing) and examined how these patterns are affected by particles' movement. Traditionally, patterns in natural systems have been interpreted in terms of variations of the Turing mechanism [11], involving a combination of short-range attraction (facilitation) to start aggregation, and long-range repulsion (competition) to limit it spatially [8]. However, the pioneer works by Likos and colleagues demonstrated that repulsive-only interactions in colloidal solutions and polymers could form cluster crystals, where clusters of particles organize into a crystalline structure [2, 12]. In biological contexts, it has been shown that competitive-only mechanisms (spatially non-local) can give rise to vegetation pattern formation [13], and species clustering in ecological niche space can occur despite competitive exclusion [14, 15].

This chapter presents results and discussions on two phenomena: pattern formation through anti-crowding interactions (which prevent individuals from getting too close to each other), and the role of movement on the spatial structure. Two different models of motile interacting individuals are discussed: one where particles repel each other via forces deriving from a soft potential, and another based on a birth-death model with competition rules. These models are studied using two complementary approaches: (i) describing the particle dynamics, mainly based on off-lattice numerical simulations, and (ii) a continuum mathematical description using partial differential equations for the number density of particles. Despite their differences, both models form the same type of patterns (cluster crystals) and, most importantly, under similar properties for the interaction. However, the underlying physical and/or biological mechanisms are completely different. The chapter also examines the influence of the type of movement of the organisms on the spatial structures. It has been noted that movement patterns of some living organisms are consistent with Lévy flight behavior [16–18], which has been argued to be advantageous over standard Brownian motion in certain foraging strategies [19]. This advantage primarily stems from the occurrence of occasional long jumps. These findings (and many others) highlight the importance of the type of movement when studying the spatial self-organization of biological populations. However, the impact of Lévy-type diffusion on the properties of organism aggregates has not been extensively studied. Moreover, the importance of self-propelled motion (by which particles move consuming internal energy) is key to understand living systems [20]. We will briefly present some results concerning cluster formation with active (repelling) and Lévy particles.

The structure of this chapter is as follows. In Sect. 11.2, we present a comparative study of the pattern formation dynamics in the two discrete particle models. In Sect. 11.3, we analyze the impact of the type of motion on the spatial structures. Finally, Sect. 11.4 provides a summary and discussion.

11.2 Mechanisms of Pattern Formation for Repelling/Competing Individuals

Here we present a comparative study of the pattern formation dynamics in two types of interacting particle systems. The first system consists of Brownian particles interacting via a repulsive soft potential [21]. The second one is a set of Brownianly moving particles, with no forces among them but with a birth–death dynamics [22, 23]. For both models, we provide their mean-field density equation description to highlight the similarities in the interaction and diffusion conditions required to obtain spatial periodic patterns and discuss the different physical mechanisms leading to these patterns.

11.2.1 A System of Moving Individuals with Repulsive Forces

Let us consider a system of N Brownian particles in contact with a thermal bath, in the overdamped limit, contained in a periodic two-dimensional domain. They interact through a soft-core (i.e., a potential not divergent at $\mathbf{x} = 0$) repulsive two-body potential. The motion of the particles is given by

$$\dot{\mathbf{x}}_i = -\sum_{j=1}^{N} \nabla V(\mathbf{x}_i - \mathbf{x}_j) + \sqrt{2D}\, \boldsymbol{\xi}_i(t), \tag{11.1}$$

where the diffusion coefficient is $D = k_B T/\gamma$, with γ the friction coefficient, T the temperature of the bath, and Gaussian noise vectors $\boldsymbol{\xi}_i$ satisfying

$$\langle \boldsymbol{\xi}_i \rangle = 0\,,\ \langle \boldsymbol{\xi}_i(t)\boldsymbol{\xi}_j(t') \rangle = \mathbb{I}\delta_{ij}\delta(t - t'). \tag{11.2}$$

For the interaction potential, we consider the generalized exponential model of exponent α (GEM-α), parameterized with exponent α, widely used in many applications [12, 15]:

$$V(\mathbf{x}) = \epsilon \exp\left(-\left|\frac{\mathbf{x}}{R}\right|^{\alpha}\right)\,. \tag{11.3}$$

ϵ is an energy scale and R a characteristic interaction range. Numerical simulations (see Fig. 11.1) confirm that hexagonal patterns spontaneously appear if $\alpha > 2$ and the diffusivity is small enough. Patterns do not occur if $\alpha \leq 2$, independently of the value of the diffusion coefficient. For any α, large values of the diffusivities give rise to a homogeneous spatial distribution of particles.

This can be analytically understood by considering the Dean–Kawasaki (DK) equation [24] describing the system dynamics for the microscopic density of particles $\rho(\mathbf{x}, t) = \sum_{i=1}^{N} \delta(\mathbf{x} - \mathbf{x}_i(t))$. In fact, to reproduce the phenomenology

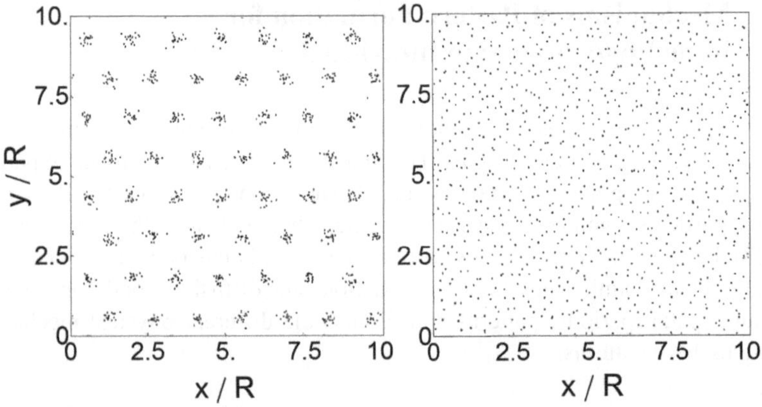

Fig. 11.1 Snapshot of the positions of $N = 1000$ particles at large times, moving according to Eq. (11.1), for $R = 0.1$, $\epsilon = 0.033$, and $L = 1$. Left plot is for the GEM-3 potential, i.e, $\alpha = 3$ and $D = 0.02$. The right plot is for GEM-1 ($\alpha = 1$) and $D = 0.005$. Modified with permission from [21]

presented here it is enough to consider its deterministic version, i.e., the one obtained by neglecting stochastic noise terms, which is appropriate for sufficiently large density of particles, and amounts to a kind of mean-field description:

$$\partial_t \rho(\mathbf{x}, t) = \nabla \cdot \left(\rho(\mathbf{x}, t) \int d\mathbf{x}' \nabla V(\mathbf{x} - \mathbf{x}') \rho(\mathbf{x}', t) \right) + D \nabla^2 \rho(\mathbf{x}, t). \quad (11.4)$$

The diffusion term results from the random motion of Brownian particles, while the potential term describes density advection by the local velocity due to repulsive forces. Note the conservative character of this equation (it is of continuity type) reflecting that the total number of particles in the system remains constant. Numerical simulations confirm that it properly describes the density of the particle system at large scales [21], in regimes where fluctuations can be neglected. Any constant density ρ_0 is a solution of Eq. (11.4). We perform a linear stability analysis around this homogeneous solution by considering a small perturbation $\rho(x, t) = \rho_0 + \exp(\lambda t + i\boldsymbol{k} \cdot \boldsymbol{r})$, which gives the following dispersion relation, or growth rate:

$$\lambda(k) = -k^2 \left[D + \rho_0 \hat{V}(k) \right], \quad (11.5)$$

with $\hat{V}(k)$ the Fourier transform of the interaction potential ($k = |\mathbf{k}|$). A positive value of $\lambda(k)$ for some k reveals a pattern-forming instability in which perturbations of periodicity $2\pi/k$ will grow on top of the homogeneous density ρ_0, leading to pattern formation with that periodicity. Note that a necessary condition for $\lambda(k) > 0$ is that the Fourier transform of the potential takes negative values for some values of k [23]. This explains why patterns can never be observed with, for example, GEM-

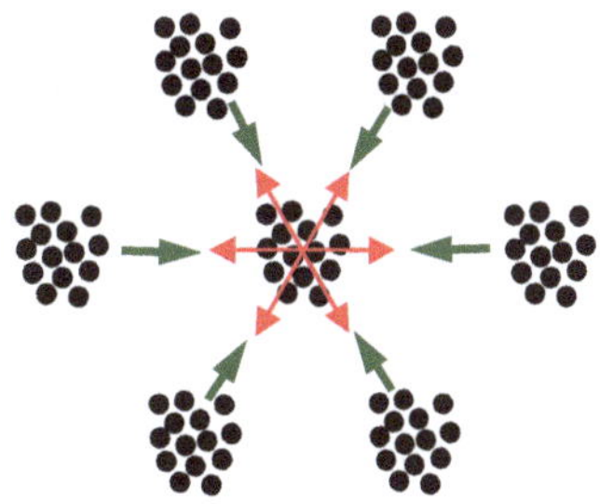

Fig. 11.2 Schematic representation of balance of forces between neighboring clusters (green) and neighboring particles inside a cluster (red)

1 or GEM-2 potentials, whose Fourier transforms are always positive. In general, the Fourier transform of the GEM-α potential is always positive if $\alpha \leq 2$. This is consistent with numerical simulations of the density equation, Eq. (11.4), which lead to a homogeneous state for $\alpha \leq 2$ [21], and analogously for the particle system. For $\alpha > 2$, however, the potential is more box-shaped, leading to Fourier transforms that may have negative values [15]. Note that for the derivation of Eq. (11.5) to hold, the soft character of the potential is crucial since otherwise, i.e., the hard-core case, its Fourier transform is not properly defined and another methodology is needed [25]. Also, in Eq. (11.5) we observe that the growth rate $\lambda(k)$ is always negative for large values of D, so that large diffusion coefficient inhibits pattern formation and makes the spatial distribution of particles homogeneous.

Equation (11.5) shows the mathematical conditions for the instability of the homogeneous solution in the continuum density description of the system of particles. A detailed study performed in [21] indicates that physically the formation of clusters comes from a balance between the internal repulsion of particles inside a cluster and the external repulsion from the particles in neighboring clusters (see Fig. 11.2) and mediated by the diffusion of the particles. Thus, when diffusion is small and repulsion from particles in neighboring clusters dominates the internal repulsion with other particles in the same cluster, aggregation tends to be enhanced.

11.2.2 A System of Birth–Death Particles

Let us consider a system of initially N_0 particles performing random Brownian motion with diffusivity D in a two-dimensional square with periodic boundary conditions. In addition, the particles undergo a birth–death dynamics, so that the total number of them is not constant, and there are $N(t)$ particles in the system at time t. The demographic events occur at stochastic Poisson times with the following rates per particle: (i) at rate β_0 one particle is chosen to die, i.e., to disappear from the system; (ii) particle i is chosen with rate λ_i to reproduce, with the offspring being placed at the same location as the parent. The important point is that the birth rate λ_i diminishes with the number of particles within a radius R from the focal particle i. This is a kind of competing dynamics since the birth rate of a given individual decreases with the number of others in its surroundings, which maybe due to competition for resources or mates. More explicitly, the rate λ_i at which a new particle is introduced in the system, exactly at the location of the parent particle i, is $\lambda_i = \lambda_0 - gN_R^i$, where λ_0 and g are constants, and N_R^i is the number of particles around i within a radius R.

We can consider more generally that this neighborhood dependence is weighted with the distance. The weight is expressed through an *influence* function that decays with the distance between the particles and the focal one $G(\mathbf{x}_i - \mathbf{x}_j)$, $j = 1, \ldots, N(t)$. In this way, particle i will give rise to the birth of a new particle at its position with a probability rate $\lambda_i = \lambda_0 - g \sum_{j \neq i} G(\mathbf{x}_i - \mathbf{x}_j)$. The previous case is recovered for a top-hat function $G(\mathbf{x})$, i.e., a constant if $|\mathbf{x}| < R$ and 0 otherwise.

Two transitions can be identified in the system:

1. An absorbing transition that differentiates between the absorbing phase, where all particles are dead (i.e. there are no particles in the system), and an active phase. In the active phase, at large times and high values of $\mu = \lambda_0 - \beta_0$ (the parameter driving the transition), the system contains many particles.
2. Depending on the shape of the influence function, G, and if the diffusion coefficient of the particles is sufficiently small, the particles may arrange into a hexagonal pattern, as illustrated in Fig. 11.3 for a top-hat influence function. The plot also shows the statistically homogeneous distribution that occurs when the diffusion coefficient is large.

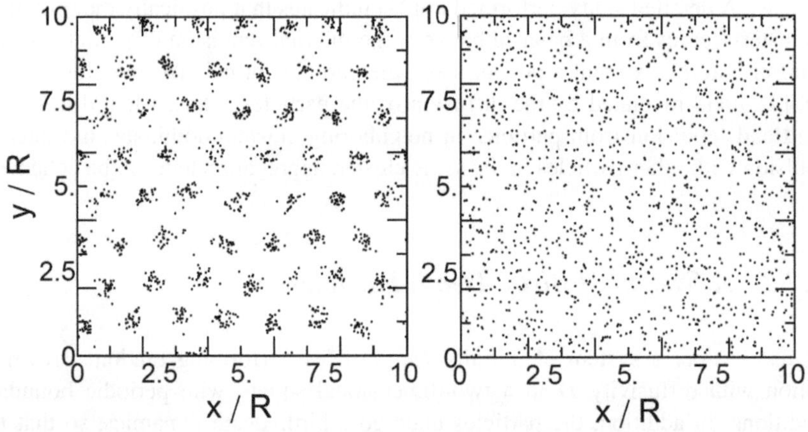

Fig. 11.3 Long-time spatial structures for the birth–death model with the top-hat influence function, i.e., birth rate $\lambda_i = \lambda_0 - gN_R^i$, where N_R^i is the number of neighbors within distance R from particle i. $R = 0.1$, system size $L = 1$, $g = 0.02$. Left: $D = 10^{-5}$, $\lambda_0 = 0.85$, and $\beta_0 = 0.15$, so that $\mu = \lambda_0 - \beta_0 = 0.7$. Right: $D = 10^{-4}$, $\lambda_0 = 0.95$, and $\beta_0 = 0.05$, so that $\mu = 0.9$. The largest diffusion coefficient destroys the spatial pattern. Modified with permission from [22]

Note that the pattern structure resembles that observed in the system of repulsive particles (Fig. 11.1). However, there are significant differences between the two models. The system of repelling particles is in equilibrium with a thermal bath and the total number of particles is conserved. In contrast, the birth–death system is a prototypical non-equilibrium system, characterized by the presence of an absorbing configuration, and the number of particles fluctuates and is not conserved (although its average reaches a constant value at long times).

We can further explore the analogies and differences by examining the deterministic macroscopic density equation for this system with an arbitrary influence function G. This equation was derived using field theoretical techniques in [22]. As before, we neglect here the stochastic terms so that a mean-field version of it is given by

$$\partial_t \rho(\mathbf{x}, t) = \rho(\mathbf{x}, t) \left(\mu - g \int d\mathbf{x}' G(\mathbf{x} - \mathbf{x}') \rho(\mathbf{x}', t) \right) + D\nabla^2 \rho(\mathbf{x}, t). \tag{11.6}$$

Here, the influence function is usually normalized so that $\int G(\mathbf{x}) d\mathbf{x} = 1$. Note that Eq. (11.6) is not a continuity equation as Eq. (11.4), indicating that the total number of particles is not conserved now. A linear stability analysis of Eq. (11.6) is also straightforward. Again considering a perturbation around the homogeneous solution, $\rho_0 = \mu/g$, we obtain the following growth rate

$$\lambda(k) = -(Dk^2 + \mu \hat{G}(k)), \tag{11.7}$$

where $\hat{G}(k)$ is the Fourier transform of the influence function. Note the similarity between this expression and Eq. (11.5). A necessary condition for pattern formation is that the Fourier transform of the influence function has negative values, which is the same condition required for the potential in the system of repelling particles (numerical simulations of the density equation, Eq. (11.6), confirming this scenario [22]). But the underlying mechanism is quite different. Patterns appear due to the presence of exclusion zones, which are spatial zones between the clusters (at distances larger than R and smaller than $2R$) that fall within the range of influence of two clusters. In these zones, particles face enhanced competition as they must compete with particles from both clusters. This phenomenon is similar to that observed in some nonlocal vegetation models, where exclusion areas were first identified [13]. See Fig. 11.4 for a visual explanation of this concept. Note that, when spacing is fixed by the interaction range, hexagonal patterns naturally result from self-organization due to competitive interactions. This configuration maximizes the exclusion area while maintaining spacing. However, more complex models can produce other structures like stripes [26].

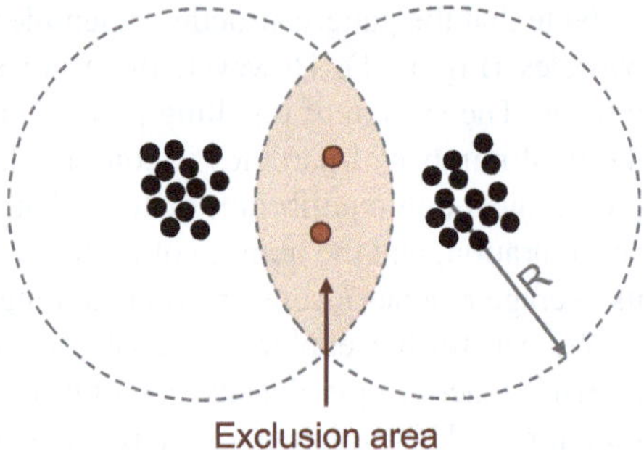

Fig. 11.4 Schematic representation of the formation of exclusion areas (orange shaded region), where individuals have to compete with two different clusters, whereas individuals in each patch compete only with individuals in its own patch

11.2.3 Discussion on the Influence Function

The models discussed in the previous section are phenomenological effective descriptions of populations of interacting individuals. Real interactions, specially in biological contexts, are often mediated by agents, visual or chemical signals, or substances (in general called mediators) whose temporal dynamics is usually faster than that of the population. A more fundamental description of the system should explicitly consider these signal dynamics, with the effective mathematical equations (e.g. Eq. (11.6)) derived through the adiabatical elimination of the signal dynamics. However, there is no systematic derivation of the population dynamics leading to arbitrary influence functions keeping the character of competitive-only interactions. In cases where such a derivation has been provided, the resulting equations did not have the characteristics necessary for pattern formation [26] (in vegetation systems), or, although leading to patterns, the influence functions lack of generality [27] (in the context of cell dynamics).

A new mechanism has been proposed that allows for the derivation of more general influence functions, including those with a shape similar to the GEM-α function [28, 29], in particular with cases leading to non-positive Fourier transforms. Briefly, this mechanism considers that mediators are released into the environment in the form of short pulses. The response to these pulse releases can significantly change the spatial stability of the population. In particular, in the limit of fast mediator dynamics, an effective description for the single population emerges, where the influence function is general enough and capable of leading to spatial instabilities.

11.3 Role of Motion on Spatial Structure

In the models of the previous section, despite their profound differences, we observed that the same ingredients are key for pattern formation: the spatial shape of the function (the potential or the influence) giving the particle interactions and a low enough diffusivity. Having focused on the role of interactions in the earlier sections, we will now discuss deeper the role of motility, considering two types: Lévy flights and active motion. First, we will analyze the impact of Lévy on the birth–death model, and then we will briefly examine the effects of active motion on the system of repulsive particles. Given the numerous analogies in the descriptions of both models, many of the results discussed for one system are applicable to the other.

As demonstrated both using the discrete particle dynamics and the instability analysis of the continuum description, clusters disappear and the population shows a uniform distribution for Brownian particles moving *very randomly*, i.e., when their diffusion coefficient is large. A natural question arises: what happens in the system if particles exhibit other types of motion? Of particular biological relevance is to consider particles performing Lévy flights or active motion.

Lévy flights result in movement patterns that are statistically different to Brownian motion. The displacement lengths have a power-law distribution resulting in the alternation of frequent, short displacements, with rare, long ones. Due to the weight of these long jumps, the variance of the displacement length is infinite, which translates into an infinite diffusion coefficient. Yet, one can still define a generalized diffusion coefficient, κ_γ, such that the typical displacement of an individual particle scales as $x \propto (\kappa_\gamma t)^{1/\gamma}$, where $0 < \gamma < 2$ is the Lévy index or anomalous exponent, which determines the type of motion. The smaller the value of γ, the more anomalous is the random walk (more frequent are long jumps). The question is whether the formally infinite diffusion coefficient destroys the spatial patterns. To analyze the impact of Lévy type of motion and spatial patterns we, as mentioned, will focus on the birth–death system (see [30]). We consider the same demographic processes as in Sec.11.2.2, but instead of Brownian motion, the particles perform Lévy flights.

As shown in [30], see also Fig. 11.5, clusters are not broken by the infinite diffusion coefficient of Lévy flights. However, large jumps do influence the characteristics of the patterns: clusters are less compact, and many single particles

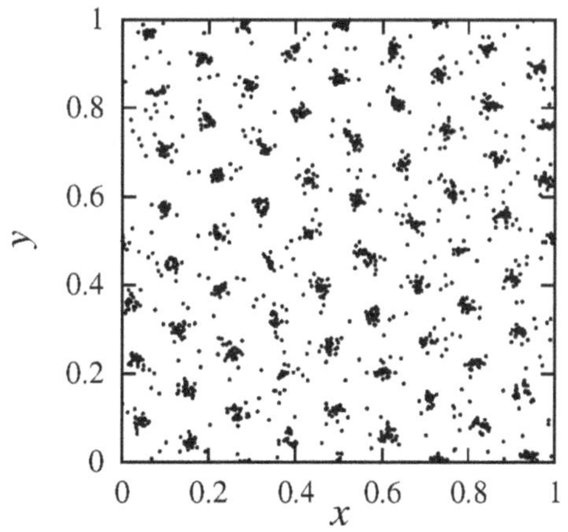

Fig. 11.5 Particle configuration at long times for particles interacting as in Fig. 11.3, but with Lévy motion with Lévy index $\gamma = 1$ and $\kappa_\gamma = 56 \times 10^{-5}$. Other parameters: $R = 0.1$, $L = 1$, $\lambda_0 = 1$, $\beta_0 = 0.1$, $g = 0.02$. Modified with permission from [30]

are found moving between them. The large-scale properties of the system can be derived from the mean-field density equation, which now, in terms of the fractional Laplacian ∇^γ, takes the form:

$$\partial_t \rho(\mathbf{x},t) = \rho(\mathbf{x},t)\left(\mu - g\int d\mathbf{x}' G(\mathbf{x}-\mathbf{x}')\rho(\mathbf{x}',t)\right) + \kappa_\gamma \nabla^\gamma \rho(\mathbf{x},t). \quad (11.8)$$

Doing a stability analysis similar to the previous cases, we obtain the following perturbation growth rate :

$$\lambda(k) = -(\kappa_\gamma k^\gamma + \mu \hat{G}(k)), \quad (11.9)$$

where $\hat{G}(k)$ is the Fourier transform of the influence function. There are no essential differences between Eqs. (11.9) and (11.7). As shown in [30] the Lévy index γ has only a small impact on the properties of the pattern.

One can conclude that the large-scale collective behavior of the system is more strongly influenced by the competitive interactions than by the type of motion performed by the particles. The long flights characteristics of Lévy movement do not accumulate their effect enough to break clusters. However, this is of course not always the case, as demonstrated by the consideration of active motion. [31, 32] explored a simple extension of the system of repulsive particles Eq. (11.1) by introducing an internal degree of freedom: the orientation of a self-propulsion speed. This results in a system of active Brownian particles interacting in pairs through a repulsive potential. Briefly, activity has two main effects on the distribution of particles (when the diffusivity of the particles is small, and thus, clusters form in the absence of activity): (a) For high values of the self-propulsion velocity, clusters are destroyed, and the steady-state becomes statistically homogeneous (the cluster crystal melts). This effect is similar to an enhanced effective diffusivity of the particles. (b) For intermediate values of the self-propulsion velocity, particle density within clusters is depleted, see Fig. 11.6, so that they reach a ring-like shape. This is a striking effect induced by the self-propelled motion.

11.4 Summary and Conclusions

We have presented a comparative study of two types of motile competing particle systems. Despite their significant differences, both of them exhibit similar spatial patterns that are induced by the same conditions for the interactions and the diffusivity of the individuals. However, the underlying physical or biological mechanisms differ substantially. We also analyzed the role of motion on patterns characteristics. In the case of Lévy flights, we concluded that their impact is less significant than that of the interactions. Competitive interactions are the main drivers of the global spatial structure of the system. However, striking impacts were

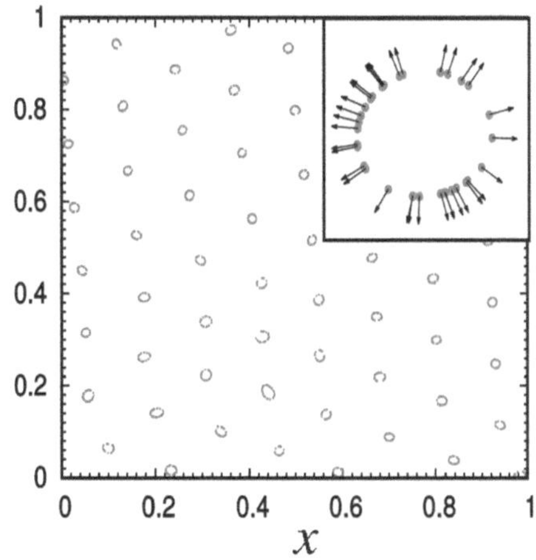

Fig. 11.6 Ring-shaped clusters appearing in a system of self-propelled repulsive particles. The inset shows the orientation of the self-propulsion velocity of the particles in a given cluster. Taken from [32]

observed when we considered active motion, for example, making single aggregates of particles become rings.

The examples we showed highlight an important observation in the literature on pattern formation studies: while analyzing the properties of structures can be challenging, it often does not reveal the underlying natural mechanisms or the functionalities of the individuals that are organized in such spatial distributions. However, numerous studies show that pattern formation is key to understand the factors leading to survival, the proper response to external factors, and ecological and evolutionary processes in biological dynamics. For instance, there is a known spatial pattern sequence to desertification in water-limited ecosystems, so that it may serve as early warning signal [33]. At much smaller scales, the spatial organization of cells is essential for accurately characterizing, forecasting, and controlling tumor evolution [34]. For non-neutral species, patterns may enable weaker competitors to survive in conditions that would otherwise lead to their extinction when spatial distribution is ignored [35]. In the context of the birth–death model we presented, it has been shown that when two otherwise identical types of particles compete, the type dispersing less and forming more compact clusters, due to its type of motion, is more likely to survive [36].

Regarding the interplay between spatial dispersion of organisms and their mobility, it is worth mentioning an emerging field known as proliferating active matter, which considers demographic dynamics in addition to the other factors. This field combines elements of the two models presented in this chapter, along with self-propulsion, and offers a natural framework for studying many biological systems [37].

In conclusion, we believe that gaining a deeper understanding of the feedbacks between spatial structure, interactions, and individual motility will provide valuable insights into the dynamics of living complex systems.

Acknowledgments C.L. acknowledges the warm hospitality at the Isaac Newton Institute for Mathematical Sciences (INI) during the program *Mathematics of movement: An interdisciplinary approach to mutual challenges in animal ecology and cell biology*, supported by EPSRC grant EP/R014604/1. He also acknowledges the Scultetus Visiting Scientist Program at the Center for Advanced Systems Understanding (CASUS) in Görlitz, Germany, where this work was completed. C.L. and E.H-G. acknowledge grant LAMARCA PID2021-123352OB-C32 funded by MCIN/AEI/10.13039/501100011033323 and FEDER "Una manera de hacer Europa"; and Grant TED2021-131836B-I00 funded by MICIU/AEI/10.13039/501100011033 and by the European Union "NextGenerationEU/PRTR." E.H-G. also acknowledges the Maria de Maeztu program for Units of Excellence, CEX2021-001164-M funded by MCIN/AEI/10.13039/501100011033. E.H.C. and R.M-G. are partially funded by the Center of Advanced Systems Understanding (CASUS), which is financed by Germany's Federal Ministry of Education and Research (BMBF) and by the Saxon Ministry for Science, Culture and Tourism (SMWK) with tax funds on the basis of the budget approved by the Saxon State Parliament.

References

1. Cavagna, A., Giardina, I.: Bird flocks as condensed matter. Annu. Rev. Conden. Matter Phys. **5**(1), 183–207 (2014)
2. Likos, C.N., Mladek, B.M., Gottwald, D., Kahl, G.: Why do ultrasoft repulsive particles cluster and crystallize? Analytical results from density-functional theory. J. Chem. Phys. **126**(22), 224502 (2007)
3. Rietkerk, M., van de Koppel, J.: Regular pattern formation in real ecosystems. Trends Ecol. Evol. **23**(3), 169–175 (2008)
4. Baker, R.E., Gaffney, E.A., Maini, P.K.: Partial differential equations for self-organization in cellular and developmental biology. Nonlinearity **21**(11), R251 (2008)
5. Borgogno, F., D'odorico, P., Laio, F., Ridolfi, L.: Mathematical models of vegetation pattern formation in ecohydrology. Rev. Geophys. **47**(1), RG1005 (2009)
6. Alert, R., Trepat, X.: Physical models of collective cell migration. Annu. Rev. Conden. Matter Phys. **11**(1), 77–101 (2020)
7. Cates, M.E., Tailleur, J.: Motility-induced phase separation. Annu. Rev. Conden. Matter Phys. **6**(1), 219–244 (2015)
8. Murray, J.D.: Mathematical Biology: II: Spatial Models and Biomedical Applications vol. 18. Springer, New York (2003)
9. Martínez-García, R., Tarnita, C., Bonachela, J.A.: Spatial patterns in ecological systems: from microbial colonies to landscapes. Emer. Topics Life Sci. **6**(3), 245–248 (2022)
10. Wiegand, T., Wang, X., Anderson-Teixeira, K.J., Bourg, N.A., Cao, M., Ci, X., Davies, S.J., Hao, Z., Howe, R.W., Kress, W.J., Lian, J., Li, J., Lin, L., Lin, Y., Ma, K., McShea, W., Mi, X., Su, S.-H., Sun, I-F., Wolf, A., Ye, W., Huth, A.: Consequences of spatial patterns for coexistence in species-rich plant communities. Nat. Ecol. Evol. **5**, 965–973 (2021)
11. Turing, A.: The chemical basis of morphogenesis. Philos. Trans. Roy. Soc. Lond. B **237**(641), 37–72 (1952)
12. Mladek, B.M., Gottwald, D., Kahl, G., Neumann, M., Likos, C.N.: Formation of polymorphic cluster phases for a class of models of purely repulsive soft spheres. Phys. Rev. Lett. **96**(4), 045701 (2006)
13. Martínez-García, R., Calabrese, J.M., Hernández-García, E., López, C.: Vegetation pattern formation in semiarid systems without facilitative mechanisms, Geophys. Res. Lett. **40**, 6143–6147 (2013)
14. Scheffer, M., Van Nes, E.H.: Self-organized similarity, the evolutionary emergence of groups of similar species. Proc. Natl. Acad. Sci. USA **103**(16), 6230–6235 (2006)

15. Pigolotti, S., López, C., Hernández-García, H.: Species clustering in competitive Lotka-Volterra models. Phys. Rev. Lett. **98**, 258101 (2007)
16. Metzler, R., Klafter, J.: The random walk's guide to anomalous diffusion: a fractional dynamics approach. Phys. Rep. **339**(1), 1–77 (2000)
17. Dieterich, P., Klages, R., Preuss, R., Schwab, A.: Anomalous dynamics of cell migration. Proc. Natl. Acad. Sci. USA **105**(2), 459–463 (2008)
18. iswanathan, G.M., Raposo, E.P., da Luz, M.G.E.: Lévy flights and superdiffusion in the context of biological encounters and random searches. Phys. Life Rev. **5**(3), 133–150 (2008)
19. Bartumeus, F., Peters, F., Pueyo, F.S., Marrasé, C., Catalán, J.: Helical Lévy walks: adjusting searching statistics to resource availability in microzooplankton. Proc. Natl. Acad. Sci. USA **100**(22), 12771–12775 (2003)
20. Marchetti, M.C., Joanny, J.F., Ramaswamy, S., Liverpool, T.B., Prost, J., Rao, M., Simha, R.A.: Hydrodynamics of soft active matter. Rev. Mod. Phys. **85**(3), 1143–1189 (2013)
21. Delfau, J.B., Ollivier, H., López, C., Blasius, B., Hernández-García, E.: Pattern formation with repulsive soft-core interactions: Discrete particle dynamics and Dean-Kawasaki equation. Phys. Rev. E **94**(4), 042120 (2016)
22. Hernández-García, E., López, C.: Clustering, advection, and patterns in a model of population dynamics with neighborhood-dependent rates. Phys. Rev. E **70**(1), 016216 (2004)
23. López, C., Hernández-García, E.: Fluctuations impact on a pattern-forming model of population dynamics with non-local interactions. Phys. D: Nonlinear Phenom. **199**, 223–234 (2004)
24. Dean, D.S.: Langevin equation for the density of a system of interacting Langevin processes. J. Phys. A: Math. Gener. **29**(24), L613 (1996)
25. Caprini, L., Hernández-García, E., López, C.: Cluster crystals with combined soft-and hard-core repulsive interactions. Phys. Rev. E **98**(5), 052607 (2018)
26. Martínez-García, R., Calabrese, J.M., Hernández-García, E., López, C.: Minimal mechanisms for vegetation patterns in semiarid regions. Philos. Trans. Roy. Soc. Lond. A: Math. Phys. Eng. Sci. **372**(2027), 20140068 (2014)
27. Mogilner, A.: Modelling spatio-angular patterns in cell biology. Ph.D. Thesis, University of British Columbia (1995)
28. Colombo, E.H., López, C., Hernández-García, E.: Pulsed interaction signals as a route to biological pattern formation. Phys. Rev. Lett. **130**, 058401 (2023)
29. Colombo, E.H., Martínez-García, R., Calabrese, J.M., López, C., Hernández-García, E.: Pulsed interactions unify reaction–diffusion and spatial nonlocal models for biological pattern formation. J. Stat. Mech. **2024**(3), 034001 (2024)
30. Heinsalu, E., Hernández-García, E., López, C.: Spatial clustering of interacting bugs: Lévy flights versus Gaussian jumps Europhys. Lett. **92**(4), 40011 (2010)
31. Delfau, J.B., López, C., Hernández-García, E.: Active cluster crystals. New J. Phys. **19**(9), 095001 (2017)
32. Caprini, L., Hernández-García, E., López, C., Marini Bettolo Marconi, U.: A comparative study between two models of active cluster crystals. Sci. Rep. **9**(1), 16687 (2019)
33. Kéfi, S., Rietkerk, M., Alados, C.L., Pueyo, Y., Papanastasis, V.P., ElAich, A., De Ruiter, P.C.: Spatial vegetation patterns and imminent desertification in Mediterranean arid ecosystems. Nature **449**(7159), 213–217 (2007)
34. Bellomo, N., Li, N.K., Maini, P.K.: On the foundations of cancer modelling: selected topics, speculations, and perspectives. Math. Models Methods Appl. Sci. **18**(4), 593–646 (2008)
35. Maciel, G.A., Martínez-García, R.: Enhanced species coexistence in Lotka-Volterra competition models due to nonlocal interactions. J. Theor. Biol. **530**, 110872 (2021)
36. Heinsalu, E., Hernández-García, E., López, C.: Clustering determines who survives for competing Brownian and Lévy walkers Phys. Rev. Lett. **110**(25), 258101 (2013)
37. Hallatschek, O., Datta, S.S., Drescher, K., Dunkel, J., Elgeti, J., Waclaw, B., Wingreen, N.S.: Proliferating active matter. Nat. Rev. Phys. **5**, 407 (2023)

Chapter 12
Number of Distinct and Common Sites Visited by N Independent Random Walkers

Satya N. Majumdar and Grégory Schehr

12.1 Introduction

In elementary set theory, two fundamental concepts are the *union* and the *intersection* of a number of N sets. While the union consists of all *distinct* elements of the collection of sets, the intersection consists of *common* elements of all the sets. These two notions appear naturally in everyday life: for example, the area of common knowledge or the whole range of different interests among the members of a society would define respectively its stability and activity. In a habitat of N animals, the union of the territories covered by different animals sets the geographical range of the habitat, while the intersection refers to the common area (e.g., a water body) frequented by all animals. In statistical physics, these two objects are modeled respectively by the number of distinct and common sites visited by N random walkers (RWs) on a d-dimensional hypercubic lattice. The knowledge about the number of distinct sites has applications ranging from the annealing of defects in crystals [1, 2] and relaxation processes [3–6] to the spread of populations in ecology [7, 8] or to the dynamics of web annotation systems [9].

Dvoretzky and Erdös [10] first studied the average number of distinct sites $\langle S_1(t)\rangle$ visited by a single t-step RW in d-dimensions, subsequently studied in [11–13]. Larralde et al. generalized this to N independent t-step walkers moving on a d-dimensional lattice [14]. They found three regimes of growth (early, intermediate,

S. N. Majumdar
Laboratoire de Physique Théorique et Modèles Statistiques (UMR 8626 du CNRS), Université Paris-Sud, Orsay Cedex, France
e-mail: satya.majumdar@universite-paris-saclay.fr

G. Schehr (✉)
Sorbonne Université, Laboratoire de Physique Théorique et Hautes Energies, CNRS UMR 7589, Paris, France
e-mail: schehr@lpthe.jussieu.fr

L. Giuggioli, P. K. Maini (eds.), *The Mathematics of Movement*,
https://doi.org/10.1007/978-3-032-00873-2_12

and late) for the average number of distinct sites $\langle S_N(t)\rangle$ as a function of time. These three regimes are separated by two N-dependent times scales [14]. In particular, they showed that in $d = 1$ and $t \gg \sqrt{\log N}$, $\langle S_N(t)\rangle \propto \sqrt{4D\,t\,\log N}$ where D is the diffusion constant of a single walker. Recently, Majumdar and Tamm [15] studied the complementary quantity, the number of common sites $C_N(t)$ visited by N walkers, each of t steps, and found analytically a rich asymptotic late time growth of the average $\langle C_N(t)\rangle$. Even though N and d are integers, and so is $C_N(t)$, one can analytically continue these results to real N and real d (for example, on fractal lattices with fractal dimensions typically non-integers). In the $(N-d)$ plane, the authors of Ref. [15] found three distinct phases separated by two critical lines $d = 2$ and $d_c(N) = 2N/(N-1)$, with $\langle C_N(t)\rangle \sim t^{\nu}$ at late times where $\nu = d/2$ (for $d < 2$), $\nu = N - d(N-1)/2$ [for $2 < d < d_c(N)$] and $\nu = 0$ [for $d > d_c(N)$] (see also [16]). In particular, in $d = 1$, $\langle W_N(t)\rangle \sim \sqrt{4Dt}$ with a N-dependent prefactor. While the mean number of distinct and common sites visited by N independent RW's is now well studied in all dimensions, computing their full distribution is highly nontrivial. Only recently, the full distributions of both $D_N(t)$ and $C_N(t)$ were computed exactly in $1d$, and an interesting link to extreme value statistics was established [17].

The rest of this book chapter is organized as follows. We will first provide a comprehensive and pedagogical introduction to computing the mean number of distinct and common sites $\langle D_N(t)\rangle$ and $\langle C_N(t)\rangle$ visited by N independent RWs in all dimensions. We will show that both $\langle D_N(t)\rangle$ and $\langle C_N(t)\rangle$ can be expressed in terms of a central quantity $p(\mathbf{x}, t)$ denoting the probability that a site $\mathbf{x}$ is visited by a single t-step walker, which starts at the origin. We will then analyze the scaling behavior of $p(\mathbf{x}, t)$ in all dimensions and use this to compute the asymptotic large t behavior of $\langle D_N(t)\rangle$ and $\langle C_N(t)\rangle$. Next we will focus on $d = 1$ and demonstrate how to compute the full distribution of $D_N(t)$ and $C_N(t)$ and also establish the interesting link to extreme value statistics. In particular, we will show that, for large N, the scaling form of the distributions of $\langle D_N(t)\rangle$ and $\langle C_N(t)\rangle$ can be expressed in terms of two well-known functions (namely Gumbel and Weibull) that appear in extreme value theory. Some more recent extensions of these techniques will also be discussed at the end. The results that we discuss in this book chapter have already been published elsewhere but, here, we gather all the results together with unifying notations and methods. We also take this opportunity to provide some more recent applications and some perspective for future works.

12.2 The Mean Number of Distinct and Common Sites in d Dimensions

We first consider a single random walker on a d-dimensional hyper cubic lattice. The walker starts at the origin. At each discrete time-step, the walker hops from a given site to anyone of the $2d$ neighbors with equal probability $1/(2d)$. Next we

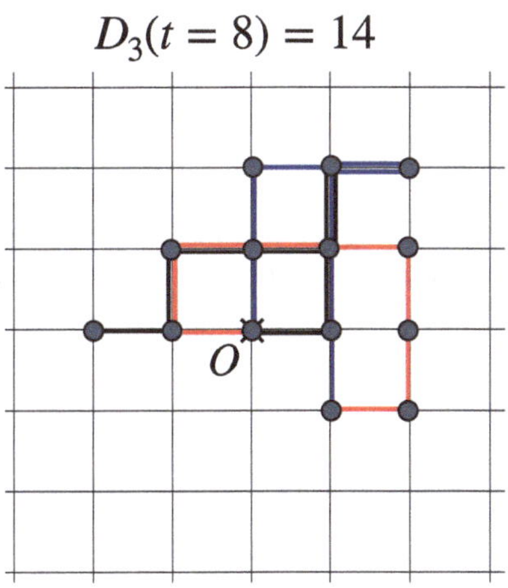

Fig. 12.1 Sketch of one realization of $N = 3$ lattice random walks starting from the origin, each being of duration $t = 8$ steps. On the panel we have marked with dots the distinct sites visited by the $N = 3$ random walkers, their number being $D_3(t = 8) = 14$. On the right panel we have marked instead the number of common sites visited by the $N = 3$ walkers, their number being $C_3(t = 8) = 3$

consider N such walkers, all starting at the origin and evolving independently up to t steps (see Fig. 12.1). We denote by $D_N(t)$ as the number of sites that are visited by at least one walker and by $C_N(t)$ the number of sites that have been visited by *all* walkers. Both $D_N(t)$ and $C_N(t)$ are random variables and fluctuate from sample to sample. Our main interest is to compute the statistics of $D_N(t)$ and $C_N(t)$. Note that for $N = 1$, we have $D_1(t) = C_1(t)$, but for $N > 1$, they are quite different. In this section, we will focus on the mean values $\langle D_N(t)\rangle$ and $\langle C_N(t)\rangle$.

We now show how both $\langle D_N(t)\rangle$ and $\langle C_N(t)\rangle$ can be expressed in terms of a central single walker quantity $p(\mathbf{x}, t)$ denoting the probability that the site $\mathbf{x}$ is visited by one walker up to time t. In order to establish this connection, we consider a single walker starting at the origin and, following Ref. [15], we introduce a binary variable

$$\sigma(\mathbf{x}, t) = \begin{cases} 1, & \text{if } \mathbf{x} \text{ has been visited by the walker up to time } t, \\ 0, & \text{otherwise.} \end{cases} \tag{12.1}$$

Then for a single walker, the random variable $D_1(t) = C_1(t)$ can be written as

$$D_1(t) = C_1(t) = \sum_{\mathbf{x}} \sigma(\mathbf{x}, t). \tag{12.2}$$

Taking the average on both sides, we get

$$\langle D_1(t)\rangle = \langle C_1(t)\rangle = \sum_{\mathbf{x}} p(\mathbf{x}, t), \tag{12.3}$$

where $p(\mathbf{x}, t) = \langle \sigma(\mathbf{x}, t) \rangle$ is the probability that the site $\mathbf{x}$ is visited by the walker within time t. Since the walker starts at the origin $\mathbf{x} = \mathbf{0}$

$$p(\mathbf{0}, t) = 1. \tag{12.4}$$

We now consider N such independent walkers. Using the fact that they all start at the origin, and the fact that they are independent, the probability that a site $\mathbf{x}$ is *not* visited by any one of the t-step walkers is simply $(1 - p(\mathbf{x}, t))^N$. Hence, the probability that the site $\mathbf{x}$ is visited by at least one of the walkers is $1-(1-p(\mathbf{x}, t))^N$. Summing over all the sites, we get

$$\langle D_N(t) \rangle = \sum_{\mathbf{x}} \left[1 - (1 - p(\mathbf{x}, t))^N \right]. \tag{12.5}$$

Similarly, the probability that a site $\mathbf{x}$ is visited by all the t-step walkers is $(p(\mathbf{x}, t))^N$. Consequently, summing over $\mathbf{x}$, we get

$$\langle C_N(t) \rangle = \sum_{\mathbf{x}} (p(\mathbf{x}, t))^N. \tag{12.6}$$

Thus, to compute $\langle D_N(t) \rangle$ and $\langle C_N(t) \rangle$, we need to compute one single central quantity, namely $p(\mathbf{x}, t)$. This can be computed using the standard generating function technique as follows. Let $F(\mathbf{x}, t)$ denote the first-passage probability to $\mathbf{x}$, i.e., $F(\mathbf{x}, t)$ is the probability that the walker, starting at the origin, arrives at $\mathbf{x}$ for the first time at t. Then $p(\mathbf{x}, t)$ can be expressed in terms of $F(\mathbf{x}, t)$ by the relation

$$p(\mathbf{x}, t) = \sum_{t'=0}^{t} F(\mathbf{x}, t'). \tag{12.7}$$

This is because if the site $\mathbf{x}$ is visited by time t, the walker must have arrived for the first time at $\mathbf{x}$ at time $t' \leq t$. Now define the generating function

$$\tilde{p}(\mathbf{x}, z) = \sum_{n=1}^{\infty} p(\mathbf{x}, n)\, z^n\,, \quad \text{and} \quad \tilde{F}(\mathbf{x}, z) = \sum_{n=1}^{\infty} F(\mathbf{x}, n)\, z^n\,. \tag{12.8}$$

Taking the generating function of Eq. (12.7), it follows that for $\mathbf{x} \neq \mathbf{0}$

$$\tilde{p}(\mathbf{x}, z) = \frac{\tilde{F}(\mathbf{x}, z)}{1 - z}, \tag{12.9}$$

where we used the fact that the $p(\mathbf{x}, 0) = \delta_{\mathbf{x},\mathbf{0}}$, and hence, for $\mathbf{x} \neq \mathbf{0}$, we have $p(\mathbf{x}, 0) = 0$. The generating function $\tilde{F}(\mathbf{x}, z)$ of the first-passage probability can in turn be expressed in terms of the free propagator as follows. Let $G(\mathbf{x}, t)$ denote the probability of arriving at $\mathbf{x}$ at time t, not necessarily for the first time. Then, for

$\mathbf{x} \neq \mathbf{0}$, its generating function can be written in terms of $\tilde{F}(\mathbf{x}, z)$ by the geometric series

$$\tilde{G}(\mathbf{x}, z) = \tilde{F}(\mathbf{x}, z) + \tilde{F}(\mathbf{x}, z)\tilde{F}(\mathbf{0}, z) + \tilde{F}(\mathbf{x}, z)\tilde{F}(\mathbf{0}, z)^2 + \ldots, \quad (12.10)$$

where $\tilde{F}(\mathbf{0}, z)$ denotes the generating function of the return probability to the origin for the first time. This relation (12.10) is easy to understand. The first term refers to trajectories that arrive at $\mathbf{x}$ for the first time at t. The second term corresponds to trajectories that arrive at $\mathbf{x}$ once before t and then returns from $\mathbf{x}$ to $\mathbf{x}$ once. We also used the fact that this return probability from $\mathbf{x}$ to $\mathbf{x}$ is independent of $\mathbf{x}$ and, hence, is given by $F(\mathbf{0}, t)$. Similarly, the third term refers to trajectories with a first arrival at $\mathbf{x}$ followed by two returns to $\mathbf{x}$. Summing up the geometric series in Eq. (12.10) one gets, for $\mathbf{x} \neq \mathbf{0}$

$$\tilde{G}(\mathbf{x}, z) = \frac{\tilde{F}(\mathbf{x}, z)}{1 - \tilde{F}(\mathbf{0}, z)}. \quad (12.11)$$

On the other hand, for $\mathbf{x} = \mathbf{0}$, the analog of Eq. (12.10) reads

$$\tilde{G}(\mathbf{0}, z) = 1 + \tilde{F}(\mathbf{0}, z) + \tilde{F}(\mathbf{0}, z)^2 + \ldots = \frac{1}{1 - \tilde{F}(\mathbf{0}, z)}. \quad (12.12)$$

Here, the first term 1 comes from the initial condition $p(\mathbf{0}, 0) = 1$. Therefore, putting these results in Eq. (12.9) gives, for $\mathbf{x} \neq \mathbf{0}$, the desired relation

$$\tilde{p}(\mathbf{x}, z) = \frac{\tilde{G}(\mathbf{x}, z)}{(1 - z)\,\tilde{G}(\mathbf{0}, z)}. \quad (12.13)$$

For $\mathbf{x} = \mathbf{0}$, we just have $\tilde{p}(\mathbf{0}, z) = \tilde{F}(\mathbf{0}, z)/(1 - z)$ and, hence, using (12.12) one gets

$$\tilde{p}(\mathbf{0}, z) = \frac{1}{1 - z}\left[1 - \frac{1}{\tilde{G}(\mathbf{0}, z)}\right]. \quad (12.14)$$

Since the generating function of the lattice propagator is well known (see e.g. [11, 12]), one can infer the exact generating function of $p(\mathbf{x}, t)$.

In this chapter, we are interested in the long-time behaviors of $\langle C_N(t)\rangle$ and $\langle D_N(t)\rangle$. It turns out that the large-time behaviors of $\langle D_N(t)\rangle$ and $\langle C_N(t)\rangle$ only require the asymptotic behavior of $p(\mathbf{x}, t)$ for large $x = |\mathbf{x}|$ and large t. This asymptotic behavior of $p(\mathbf{x}, t)$ can in principle be derived from the generating function as discussed above. However, it turns out that this long-time/large-distance behavior of $p(\mathbf{x}, t)$ can be derived directly by replacing the lattice random walker in discrete time by a Brownian motion in continuous space and of duration t in continuous time. This continuum space-time limit can be achieved from the discrete space-time by taking the limit of the lattice constant $a \to 0$ and the time step

$\Delta t \to 0$, while keeping the ratio $a^2/(2\Delta t) = D$ fixed where D refers to the diffusion constant of the Brownian motion. Consequently, the discrete sums over $\mathbf{x}$ in Eqs. (12.5) and (12.6) can be replaced by integrals over continuous space. Thus, for both quantities $\langle D_N(t) \rangle$ and $\langle C_N(t) \rangle$, the central object is to compute $p(\mathbf{x}, t)$ for a single Brownian motion of duration t, which we derive in the next subsection.

In fact, once we know the central quantity $p(\mathbf{x}, t)$, there are other interesting observables, in addition to $\langle D_N(t) \rangle$ and $\langle C_N(t) \rangle$, which can be computed. As an example, we consider $\langle V_{K,N}(t) \rangle$ denoting the mean number of sites that are visited exactly by K walkers ($1 \le K \le N$) up to time t. A site $\mathbf{x}$ is visited before t by a single walker with probability $p(\mathbf{x}, t)$. Using the independence of the walkers and summing over all $\mathbf{x}$, it then follows that

$$\langle V_{K,N}(t) \rangle = \binom{N}{K} \sum_{\mathbf{x}} [p(\mathbf{x}, t)]^K \, [1 - p(\mathbf{x}, t)]^{N-K} . \tag{12.15}$$

For $K = N$, we just have $\langle V_{N,N}(t) \rangle = \langle C_N(t) \rangle$. Thus, the mean number of common sites is just a special case of $\langle V_{K,N}(t) \rangle$ with $K = N$.

12.2.1 The Central Quantity $p(\mathbf{x}, t)$

Consider then a single Brownian motion of length t and diffusion constant D in d-dimensions starting at the origin. We are interested in $p(\mathbf{x}, t)$, the probability that the site $\mathbf{x}$ is visited (at least once) by the walker up to time t. Let τ denote the last time before t that the site $\mathbf{x}$ was visited by the walker. Then, using the Markov property of the walk, it follows that

$$p(\mathbf{x}, t) = \int_0^t G(\mathbf{x}, \tau) q(t - \tau) \, d\tau , \tag{12.16}$$

where $G(\mathbf{x}, \tau)$ is the propagator of the Brownian motion

$$G(\mathbf{x}, \tau) = \frac{1}{(4\pi D \tau)^{d/2}} \, e^{-x^2/(4D\tau)} , \tag{12.17}$$

measuring the probability density of reaching $\mathbf{x}$ at time τ, starting from the origin at time $t = 0$. The quantity $q(\tau)$ in Eq. (12.16) denotes the probability that the walker, starting at $\mathbf{x}$, does not return to $\mathbf{x}$ up to time τ. Due to the translation invariance of the walk, $q(\tau)$ does not depend on $\mathbf{x}$ is thus also the probability that the walker, starting at the origin, does not return to the origin up to time τ.

The no-return probability $q(\tau)$ for a d-dimensional Brownian motion can be computed as follows. It is useful first to relate it to the first-return probability $F(\tau)$ to the origin by the relation $q(\tau) = \int_\tau^\infty F(\tau')\, d\tau'$. Here $F(\tau)\, d\tau$ denotes the probability that the walker, starting at the origin at $\tau = 0$, returns to the origin for the first time between τ and $\tau + d\tau$. In other words,

$$\frac{dq}{d\tau} = -F(\tau). \tag{12.18}$$

We also define their Laplace transforms

$$\tilde{q}(s) = \int_0^\infty q(t)\, e^{-st}\, dt, \quad \tilde{F}(s) = \int_0^\infty F(t)\, e^{-st}\, dt. \tag{12.19}$$

Taking Laplace transform of Eq. (12.18) and using $q(0) = 1$, we get the well-known relation [18, 19]

$$\tilde{q}(s) = \frac{1 - \tilde{F}(s)}{s}. \tag{12.20}$$

The Laplace transform of the first-passage probability $\tilde{F}(s)$ can further be related to the Laplace transform of the propagator $G(\mathbf{0}, t)$ as follows

$$G(\mathbf{0}, t) = \delta(t) + \sum_{n=1}^{\infty} \int \left(\prod_{i=1}^{n} F(t_i) dt_i \right) \delta \left(\sum_{i=1}^{n} t_i - t \right), \tag{12.21}$$

where the m-th term ($m \geq 1$) of the sum corresponds to trajectories with exactly $m - 1$ returns to the origin before time t and the m-th return exactly at t. The first term $\delta(t)$ just reflects the initial condition that the walker starts at the origin at time $t = 0$. Taking Laplace transform on both sides and evaluating the sum as a geometric series gives

$$\tilde{G}(\mathbf{0}, s) = \frac{1}{1 - \tilde{F}(s)}. \tag{12.22}$$

Eliminating $1 - \tilde{F}(s)$ between Eqs. (12.20) and (12.22) gives

$$\tilde{q}(s) = \frac{1}{s\, \tilde{G}(\mathbf{0}, s)}. \tag{12.23}$$

Let us recall that $\tilde{G}(\mathbf{0}, s) = \int_0^\infty e^{-s\,t} G(\mathbf{0}, t)\, dt$ where the propagator $G(\mathbf{x}, t)$ is given in (12.17). If we substitute $\mathbf{x} = \mathbf{0}$ in Eq. (12.17), one would get $G(\mathbf{0}, t) = 1/(4\pi D\, t)^{d/2}$. However, as we will see later, one would need a lattice cutoff $\mathbf{a}$ in order to regularize the integrals over t, in particular for $d > 2$. Hence, we will use the following regularized expression for $G(\mathbf{0}, t)$,

$$G(\mathbf{0}, t) \simeq \frac{1}{(4\pi Dt)^{d/2}}\, e^{-\frac{a^2}{4Dt}}, \tag{12.24}$$

where a denotes the norm of $\mathbf{a}$, i.e., $a = |\mathbf{a}|$.

We will see later that, for $d < 2$, we can take eventually the $a \to 0$ limit and the answer will be finite. In contrast, for $d > 2$, we need to keep a nonzero cutoff a, and the result for $p(\mathbf{x}, t)$ will depend explicitly on the cutoff a.

12.2.2 *The Scaling Analysis of* $p(\mathbf{x}, t)$ *in All Dimensions*

From Eq. (12.16), we see that, to extract the scaling behavior of $p(\mathbf{x}, t)$ for large $|\mathbf{x}|$ and large t, we need to know the behavior of the no-return probability to the origin up to time t, i.e., $q(t)$ for large t. Below, we first extract the large t behavior of $q(t)$ in all dimensions and then substitute this asymptotic behavior in Eq. (12.16) to extract the scaling behavior of $p(\mathbf{x}, t)$. This is the approach that was used in Ref. [15].

12.2.2.1 Asymptotic Behavior of $q(t)$ for Large t

For this asymptotic analysis of $q(t)$, we start from its exact Laplace transform in Eq. (12.23), with $\tilde{G}(\mathbf{0}, s)$ given by

$$\tilde{G}(\mathbf{0}, s) \simeq \int_0^\infty e^{-st} \frac{1}{(4\pi D t)^{d/2}}\, e^{-\frac{a^2}{4Dt}}\, dt = \frac{1}{(4\pi D)^{d/2}} \frac{1}{s^{1-d/2}} \int_0^\infty \frac{e^{-\frac{a^2 s}{4Dy}}\, e^{-y}}{y^{d/2}}\, dy, \tag{12.25}$$

where we used Eq. (12.24) in the first relation and then made a change of variable $y = s\,t$ in the second equality. We would now like to consider the small s behavior of the integral in Eq. (12.25) in three separate cases: (i) $d < 2$, (ii) $d > 2$, and (iii) $d = 2$.

(i) $d < 2$: In this case, we can take the limit $s \to 0$ in the integral, which remains convergent. This then gives, to leading order as $s \to 0$

$$\tilde{G}(\mathbf{0}, s) \simeq \frac{\Gamma\left(1 - \frac{d}{2}\right)}{(4\pi D)^{d/2} s^{1-d/2}}. \tag{12.26}$$

Substituting this in Eq. (12.23) gives, for small s,

$$\tilde{q}(s) \sim \frac{(4\pi D)^{d/2}}{\Gamma(1 - d/2)} \frac{1}{s^{d/2}}. \tag{12.27}$$

To invert this Laplace transform, we use the identity

$$\mathcal{L}^{-1}_{s\to t}\left[\frac{1}{s^\alpha}\right] = \frac{1}{\Gamma(\alpha)\, t^{1-\alpha}}, \quad \text{for} \quad \alpha > 0. \tag{12.28}$$

Using this identity with $\alpha = d/2$ gives the leading large t behavior of $q(t)$ as

$$q(t) \simeq \frac{A_d}{t^{1-d/2}}, \quad A_d = \frac{(4\pi D)^{d/2}}{\pi} \sin\left(\frac{\pi d}{2}\right). \tag{12.29}$$

Note that the leading large t behavior of $q(t)$ is independent of the cutoff a for $d < 2$.

(ii) $d > 2$: In this case, we see from the definition of $\tilde{G}(\mathbf{0}, s)$ in Eq. (12.25) that, when $s \to 0$, we need a finite cutoff $a > 0$ for the integral to be convergent. Thus, setting $a = 0$ gives, for small s,

$$\tilde{G}(\mathbf{0}, s) \underset{s \to 0}{\longrightarrow} \int_0^\infty \frac{1}{(4\pi D t)^{d/2}} e^{-\frac{a^2}{4Dt}} dt \equiv \frac{1}{E_d}, \tag{12.30}$$

where the constant E_d depends explicitly on the cutoff a. Substituting this in Eq. (12.23) gives, for small s,

$$\tilde{q}(s) \simeq \frac{E_d}{s}. \tag{12.31}$$

Inverting trivially the Laplace transform gives, for large t

$$q(t) \simeq E_d, \quad \text{as} \quad t \to \infty. \tag{12.32}$$

Thus, E_d is exactly the no-return or eventual "escape" probability of the Brownian walker in $d > 2$

(iii) $d = 2$: we start from the definition

$$\tilde{G}(\mathbf{0}, s) \simeq \int_0^\infty \frac{1}{4\pi D t} e^{-\frac{a^2}{4Dt}} e^{-st} dt. \tag{12.33}$$

In this case, if we directly set $s = 0$, we see that we need a nonzero cutoff $a > 0$ in order that the integral is convergent. However, the leading small s behavior turns out to be independent of the cutoff a as we show now. Indeed, taking derivative of Eq. (12.33) with respect to s gives

$$\frac{dG(\mathbf{0}, s)}{ds} = -\frac{1}{4\pi D} \int_0^\infty e^{-\frac{a^2}{4Dt}} e^{-st} dt = -\frac{1}{4\pi D s} \int_0^\infty e^{-\frac{a^2 s}{4D}} e^{-y} dy. \tag{12.34}$$

We can now take the limit $s \to 0$ limit inside the integral since it remains convergent, and this gives $\frac{dG(\mathbf{0},s)}{ds} \simeq -\frac{1}{4\pi D}\frac{1}{s}$. Integrating it back, we get, to leading order as $s \to 0$,

$$G(\mathbf{0}, s) \simeq -\frac{1}{4\pi D} \ln s, \tag{12.35}$$

which is clearly independent of the cutoff a. We substitute this back in Eq. (12.23) to obtain the leading small s behavior

$$\tilde{q}(s) \simeq -\frac{4\pi D}{s \ln s}. \tag{12.36}$$

Inverting this Laplace transform, we get for large t

$$q(t) \simeq \frac{4\pi D}{\ln t}. \tag{12.37}$$

Thus, to summarize, the leading large t behavior of $q(t)$ in different dimensions is given by

$$q(t) \simeq \begin{cases} \dfrac{A_d}{t^{1-d/2}}, & d < 2, \\[2ex] \dfrac{4\pi D}{\ln t}, & d = 2, \\[2ex] E_d, & d > 2, \end{cases} \tag{12.38}$$

where the constants A_d and E_d are given respectively in Eqs. (12.29) and (12.30). This shows that, for $d \leq 2$, the no-return probability $q(t) \to 0$ as $t \to \infty$, indicating that the walk is recurrent. In contrast, for $d > 2$, it approaches a nonzero constant, indicating that the walk can escape to infinity with a finite probability E_d. Thus, this result in Eq. (12.38) illustrates the well-known [18, 20] fact that the Brownian walker is recurrent for $d \leq 2$, while it is transient for $d > 2$.

12.2.2.2 Asymptotic Behavior of $p(\mathbf{x}, t)$ for Large $|\mathbf{x}|$ and Large t

We start with Eq. (12.16) and first make a change of variable $\tau = tu$, with $0 < u < 1$. This gives

$$p(\mathbf{x}, t) = \frac{t^{1-d/2}}{(4\pi D)^{d/2}} \int_0^1 \frac{e^{-z^2/u}}{u^{d/2}}\, q\,(t(1-u))\, du, \quad \text{where} \quad z = \frac{|\mathbf{x}|}{\sqrt{4Dt}}. \tag{12.39}$$

Thus, keeping the scaling variable z fixed, if we take the large t limit, the behavior of the integral is controlled by the large argument behavior of the no-return probability $q(\tau)$. Hence, we can directly substitute in the integral the large argument behavior of $q(\tau)$ from Eq. (12.38). Thus, once again, we analyze three different cases: (i) $d < 2$, (ii) $d > 2$, and (iii) $d = 2$.

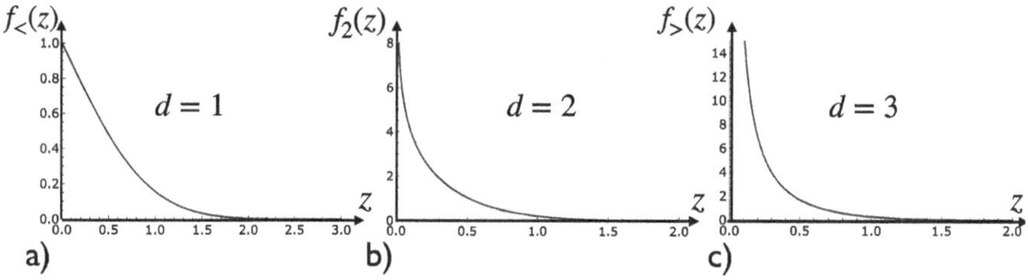

Fig. 12.2 (**a**) Plot of the scaling function $f_<(z)$ vs z for $d = 1$. (**b**) Plot of the scaling function $f_2(z)$ vs z ($d = 2$). (**c**) Plot of the scaling function $f_>(z)$ vs z for $d = 3$

(i) $d < 2$: In this case, substituting the behavior of $q(\tau)$ from the first line of Eq. (12.38) in Eq. (12.39) we get

$$p(\mathbf{x}, t) \simeq f_< \left(z = \frac{|\mathbf{x}|}{\sqrt{4Dt}} \right),$$

$$\text{where} \quad f_<(z) = \frac{\sin\left(\frac{\pi d}{2}\right)}{\pi} \int_0^1 e^{-z^2/u} u^{-d/2} (1-u)^{d/2-1}\, du. \tag{12.40}$$

It is easy to show that the function $f_<(z)$ decreases monotonically with increasing z. In Fig. 12.2a we show a plot of this function for $d = 1$ where $f_<(z) = \text{erfc}(z) = (2/\sqrt{\pi}) \int_z^\infty e^{-u^2} du$. For general $d < 2$, the scaling function has the following asymptotic behaviors

$$f_<(z) \simeq \begin{cases} 1 - \dfrac{1}{\Gamma\left(2 - \frac{d}{2}\right)} z^{2-d}, & z \to 0, \\ \\ \dfrac{\sin\left(\frac{\pi d}{2}\right)}{\pi} \dfrac{e^{-z^2}}{z^d}, & z \to \infty. \end{cases} \tag{12.41}$$

(ii) $d > 2$: Using $q(t) \simeq E_d$ for large t from Eq. (12.38) in Eq. (12.39), we get the following scaling behavior

$$p(\mathbf{x}, t) \simeq \frac{E_d}{(4\pi D)^{d/2}}\, t^{1-d/2}\, f_> \left(z = \frac{|\mathbf{x}|}{\sqrt{4Dt}} \right),$$

$$\text{where} \quad f_>(z) = \int_0^1 \frac{e^{-z^2/u}}{u^{d/2}}\, du. \tag{12.42}$$

A plot of this function for $d = 3$ is shown in Fig. 12.2c. The asymptotic behaviors of this scaling function $f_>(z)$ are given by

$$f_>(z) \simeq \begin{cases} \dfrac{\Gamma(d/2-1)}{z^{d-2}}, & z \to 0, \\ \\ \dfrac{e^{-z^2}}{z^2}, & z \to \infty. \end{cases} \tag{12.43}$$

(iii) $d = 2$: In this case, substituting $q(t) \simeq (4\pi D)/\ln t$ for large t in Eq. (12.39) gives the scaling behavior

$$p(\mathbf{x}, t) \simeq \frac{1}{\ln t} f_2\left(z = \frac{|\mathbf{x}|}{\sqrt{4Dt}}\right), \quad \text{where} \quad f_2(z) = \int_0^1 \frac{e^{-z^2/u}}{u}\, du. \tag{12.44}$$

A plot of this function is shown in Fig. 12.2b. The scaling function $f_2(z)$ has the following asymptotic behaviors

$$f_2(z) \simeq \begin{cases} -2\ln z, & z \to 0\,, \\ \\ \dfrac{e^{-z^2}}{z^2}, & z \to \infty. \end{cases} \tag{12.45}$$

Thus, summarizing the scaling forms for $p(\mathbf{x}, t)$ in different dimensions, we get [15]

$$p(\mathbf{x}, t) \simeq \begin{cases} f_<\left(\frac{|\mathbf{x}|}{\sqrt{4Dt}}\right), & d < 2, \\ \\ \dfrac{1}{\ln t} f_2\left(\frac{|\mathbf{x}|}{\sqrt{4Dt}}\right), & d = 2 \\ \\ \frac{E_d}{(4\pi D)^{d/2}}\, t^{1-d/2}\, f_>\left(\frac{|\mathbf{x}|}{\sqrt{4Dt}}\right), & d > 2, \end{cases} \tag{12.46}$$

where the scaling functions $f_<(z)$, $f_2(z)$ and $f_>(z)$ are given respectively in Eqs. (12.40), (12.45), and (12.41).

12.2.3 The Asymptotic Behavior of $\langle D_N(t)\rangle$ and $\langle C_N(t)\rangle$

In this subsection, we use the scaling behavior of $p(\mathbf{x}, t)$ from Eq. (12.46) in Eqs. (12.5) and (12.6) to compute the mean number of distinct and common sites visited by N Brownian walkers up to time t for large t.

12.2.3.1 $\langle D_N(t)\rangle$ for Large t

The average number of distinct sites visited up to time t clearly depends on the dimension d. We thus consider the three cases separately: (i) $d < 2$, (ii) $d > 2$, and (iii) $d = 2$.

(i) $d < 2$: We substitute the scaling behavior of $p(\mathbf{x}, t)$ from the first line of Eq. (12.46) in Eq. (12.5). Performing the change of variable $z = |\mathbf{x}|/\sqrt{4Dt}$ and using the spherical symmetry, we get

$$\langle D_N(t)\rangle \simeq B_N(d)\, t^{d/2} \quad \text{as} \quad t \to \infty, \tag{12.47}$$

where the prefactor $B_N(d)$ is given by

$$B_N(d) = (4D)^{d/2} S_d \int_0^\infty dz\, z^{d-1} \left[1 - (1 - f_<(z))^N\right]. \tag{12.48}$$

Here $S_d = 2\pi^{d/2}/\,\Gamma(d/2)$ is the surface area of the unit sphere in d dimensions and the scaling function $f_<(z)$ is given in Eq. (12.40). It is hard to compute this prefactor $B_N(d)$ in Eq. (12.48) explicitly for general N. However, one can easily extract the asymptotic large N behavior of $B_N(d)$. For large N, the integral in Eq. (12.48) is dominated by the large z behavior of $f_<(z)$ given in Eq. (12.41). Substituting this asymptotic behavior in Eq. (12.48), one finds

$$(1 - f_<(z))^N \sim \exp\left[-N \frac{\sin\left(\frac{\pi d}{2}\right)}{\pi} \frac{e^{-z^2}}{z^d}\right]. \tag{12.49}$$

As a function of z, the right-hand side of Eq. (12.49) approaches 1 as $z \to \infty$ and vanishes as $z \to 0$. As $N \to \infty$, this jump from 0 to 1 occurs at $z \simeq \sqrt{\ln N}$ where the argument of the exponential vanishes. Thus, for large N, we can approximate the right-hand side of Eq. (12.49) by a step function $\theta\left(\sqrt{\ln N} - z\right)$. Using this large N approximation in Eq. (12.48), we see that we can cutoff the integral at $z = \sqrt{\ln N}$, and this gives, to leading order for large N,

$$B_N(d) \simeq \frac{(4\pi D)^{d/2}}{\Gamma\left(\frac{d}{2}+1\right)} (\ln N)^{d/2} \quad \text{as} \quad N \to \infty. \tag{12.50}$$

(ii) $d > 2$: Here also, we substitute the scaling behavior of $p(\mathbf{x}, t)$ from the third line of Eq. (12.46) in Eq. (12.5). This gives, using the spherical symmetry

$$\langle D_N(t)\rangle \simeq S_d \int_0^\infty dx\, x^{d^1} \left[1 - \left(1 - \frac{E_d}{(4\pi D)^{d/2}} t^{1-d/2} f_>\left(\frac{|\mathbf{x}|}{\sqrt{4Dt}}\right)\right)^N\right]. \tag{12.51}$$

For $d > 2$ and large t, the amplitude of the term containing $f_>(z)$ is small, and hence, expanding in Taylor series and keeping only the leading term gives

$$\begin{aligned}\langle D_N(t)\rangle &\simeq S_d \frac{N\,E_d}{(4\pi D)^{d/2}} \int_0^\infty dx \frac{x^{d-1}}{t^{d/2-1}} f_>\left(\frac{x}{\sqrt{4Dt}}\right) \\ &= E_d\, N\, t \frac{S_d}{\pi^{d/2}} \int_0^\infty dz\, z^{d-1} f_>(z). \end{aligned} \tag{12.52}$$

Performing the integral over z, and using $S_d = 2\pi^{d/2}/\Gamma(d/2)$ gives the very simple answer

$$\langle D_N(t)\rangle \simeq E_d\, N\, t. \tag{12.53}$$

There is an alternative way of arriving at the same result. Starting from the definition in Eq. (12.5), we see that, at late times, $p(\mathbf{x}, t) \to 0$ as $t^{1-d/2}$ for $d > 2$, for fixed $z = |\mathbf{x}|/\sqrt{4Dt}$ (see the third line of Eq. (12.46)). Expanding in Taylor series and keeping only the leading term gives

$$\langle D_N(t)\rangle \simeq N \int d^d x\; p(\mathbf{x}, t). \tag{12.54}$$

Substituting the expression for $p(\mathbf{x}, t)$ from Eq. (12.16) and using $\int d^d x\, G(\mathbf{x}, t) = 1$ gives

$$\langle D_N(t)\rangle \simeq N \int_0^t q(t-\tau)d\tau = N \int_0^t q(\tau)d\tau. \tag{12.55}$$

Using the fact that, for $d > 2$, the no-return probability $q(\tau) \to E_d$ for large τ [see the third line of Eq. (12.38)], immediately reproduces the result in Eq. (12.53). Physically, this result has the following implication. Consider first a single walker at late times t in $d > 2$ dimensions. The total number of sites visited is t, while in this limit the probability that a given site is not revisited approaches E_d at late times. Hence, the mean number of distinct sites visited by a single walker approaches $E_d\, t$ asymptotically. For large N and $d > 2$, the independent walkers hardly overlap and each of them visits on an average $E_d\, t$ distinct sites. This gives the result in Eq. (12.53).

(iii) $d = 2$: In this case, we see from the second line of Eq. (12.46), that for fixed $z = |\mathbf{x}|/\sqrt{4Dt}$, the probability $p(\mathbf{x}, t)$ still decays to zero as $t \to \infty$, albeit very slowly as $1/\ln t$. Hence, as in the $d > 2$ case above, we expand Eq. (12.5) in a Taylor series for small $p(\mathbf{x}, t)$ and keep only the leading term, which again gives, to leading order for large t,

$$\langle D_N(t)\rangle \simeq N \int_0^t q(t-\tau)d\tau = N \int_0^t q(\tau)d\tau. \tag{12.56}$$

Using the asymptoptic behavior of $q(\tau) \simeq 4\pi D/\ln\tau$ from the second line of Eq. (12.38), we get, to leading order for large t

$$\langle D_N(t)\rangle \simeq N\frac{4\pi D\,t}{\ln t}. \tag{12.57}$$

Thus, the asymptotic non-overlapping behavior of the number of distinct sites visited by N independent walkers remains true even for $d = 2$, as in the $d > 2$ case.

Let us finish this subsection by summarizing the leading large t behavior of $\langle D_N(t)\rangle$ in different dimensions. We obtain

$$\langle D_N(t)\rangle \simeq \begin{cases} B_N(d)\, t^{d/2}, & d < 2, \\ N\dfrac{4\pi D\,t}{\ln t}, & d = 2, \\ E_d\, N\, t, & d > 2, \end{cases} \tag{12.58}$$

where $B_N(d)$ and E_d are given respectively in Eqs. (12.48) and (12.30). Thus, the temporal growth of $\langle D_N(t)\rangle$ for large t for N independent walkers is identical to that of a single walker and the N-dependence emerges only in the prefactor of this asymptotic growth. In Ref. [14], the time dependence of $\langle D_N(t)\rangle$ was analyzed in $d = 1, 2$ and 3 using a slightly different approach. The late time results in Eq. (12.58) are consistent with the results of Ref. [14] for $d = 1, 2$ and 3. In addition, the authors of Ref. [14] also found an intermediate regime where the time dependence is different from the asymptotic growth. This intermediate regime can also be recovered from our approach, though we do not provide details here.

12.2.3.2 $\langle C_N(t)\rangle$ for Large t

Here, we derive the asymptotic large t behavior of $\langle C_N(t)\rangle$. As in the case of $\langle D_N(t)\rangle$, the asymptotic behavior of $\langle C_N(t)\rangle$ changes at $d = 2$. But as we will see below, that for $d > 2$, the asymptotic behavior of $\langle C_N(t)\rangle$ is much richer than that of $\langle D_N(t)\rangle$. In fact, it turns out that there is another critical dimension $d_c(N) = 2N/(N-1)$ such that, for $2 < d < d_c(N)$, the mean number of common sites grows as a power law $\langle C_N(t)\rangle \sim t^{\nu}$ for large t where the exponent $\nu = N - d(N-1)/2$ depends on both d and N [15]. In contrast, for $d > d_c(N)$, $\langle C_N(t)\rangle$ approaches a constant at late times. Below we discuss these cases separately.

(i) $d < 2$: We start from the formula in Eq. (12.6) and substitute the scaling form of $p(\mathbf{x}, t)$ from the first line in Eq. (12.46). Using spherical symmetry, we then get

$$\langle C_N(t)\rangle \simeq b_N(d)\, t^{d/2} \quad \text{as} \quad t \to \infty, \tag{12.59}$$

where the amplitude $b_N(d)$ is given by

$$b_N(d) = (4D)^{d/2}\, S_d \int_0^\infty dz\, z^{d-1}\, [f_<(z)]^N\,. \tag{12.60}$$

Once again, it is difficult to compute this integral explicitly for arbitrary $d < 2$ and N. However, one can extract the large N behavior of $b_N(d)$ for fixed d as follows. For large N, the integral is dominated by the small z behavior of $f_<(z)$. To see this, we first substitute the small z behavior of $f_<(z)$ from the first line of Eq. (12.41). This gives

$$b_N(d) \simeq (4D)^{d/2} S_d \int_0^\infty dz\; z^{d-1} \left[1 - \frac{1}{\Gamma\,(2-d/2)}\, z^{d-2}\right]^N . \tag{12.61}$$

In the limit $N \to \infty$, the leading contribution to the integral comes from the region where $z \sim N^{1/(2-d)}$. In this regime, to leading order for large N, one can make the approximation

$$\left[1 - \frac{1}{\Gamma\,(2-d/2)}\, z^{d-2}\right]^N \simeq \exp\left[-\frac{N}{\Gamma\,(2-d/2)}\, z^{d-2}\right]. \tag{12.62}$$

Substituting this behavior back into Eq. (12.61) and performing the integral over z explicitly, we get the leading large N behavior of the amplitude $b_N(d)$ as

$$b_N(d) \simeq \frac{\tilde{b}(d)}{N^{\frac{d}{(2-d)}}} \quad \text{with} \quad \tilde{b}(d) = (4D)^{d/2}\, \frac{S_d}{d}\, \Gamma\left(\frac{2}{2-d}\right) \left[\Gamma\left(\frac{4-d}{2}\right)\right]^{\frac{d}{2-d}}, \tag{12.63}$$

where we recall that $S_d = 2\pi^{d/2}/\,\Gamma(d/2)$ is the surface area of the unit sphere in d dimensions. By comparing with Eq. (12.50), we see that, while both $\langle D_N(t)\rangle$ and $\langle C_N(t)\rangle$ grow as $t^{d/2}$ for large t, the amplitude of this growth has very different dependence on N. In the case of $\langle D_N(t)\rangle$, the amplitude $B_N(d)$ grows as $(\ln N)^{d/2}$ for large N, while for $\langle C_N(t)\rangle$, the amplitude $b_N(d)$ decreases as a power law $\sim N^{-d/(2-d)}$ for large N.

(ii) $d > 2$: We now consider the case $d > 2$ where the scaling behavior of $p(\mathbf{x}, t)$ is given in the third line of Eq. (12.46). We substitute this scaling form in Eq. (12.6), replace the sum over $\mathbf{x}$ by a continuous integral, and perform this integral using the spherical symmetry. This gives

$$\langle C_N(t)\rangle \simeq t^{\nu}\, \frac{E_d^N}{(4\pi D)^{Nd/2}} S_d\, (4D)^{d/2} \int_0^\infty dz\, z^{d-1}\, [f_>(z)]^N, \tag{12.64}$$

where the exponent ν is given by

$$\nu = N - \frac{d(N-1)}{2}, \tag{12.65}$$

while the scaling function $f_>(z)$ is given in Eq. (12.42). This result, of course, makes sense only when the integral on the right-hand side is convergent. To find the condition of this convergence, we only examine the small z behavior of the integrand [for large z, the integral is convergent anyway given the asymptotic behavior given in the second line of Eq. (12.41)]. As $z \to 0$, from the first line of Eq. (12.41), we have $f_>(z) \sim 1/z^{d-2}$. Consequently, the integrand $z^{d-1}\,[f_>(z)]^N \sim z^{2N-d(N-1)-1}$. Thus, the integral is convergent in the lower limit $z \to 0$ provided $d < d_c(N)$ where

$$d_c(N) = \frac{2N}{N-1}. \tag{12.66}$$

Thus, for $d < d_c(N)$, Eq. (12.64) predicts that $\langle C_N(t)\rangle$ grows algebraically for large time as

$$\langle C_N(t)\rangle \simeq \alpha_N(d)\, t^{\nu}, \tag{12.67}$$

with $\nu = (N-1)(d_c - d)/2 > 0$ given in Eq. (12.65). The amplitude $\alpha_N(d)$ is then given by

$$\alpha_N(d) = S_d\,(4D)^{d/2}\,\frac{E_d^N}{(4\pi D)^{dN/2}} \int_0^\infty dz\, z^{d-1}\,[f_<(z)]^N\,. \tag{12.68}$$

A different situation occurs for $d > d_c(N)$. First, we note that the scaling form in Eq. (12.46) holds for $|\mathbf{x}| = O(\sqrt{t})$, for large t. It is clearly not valid when $|\mathbf{x}| = O(1) \ll \sqrt{t}$. In this range, $p(\mathbf{x},t) \sim O(1)$. For example, exactly at $\mathbf{x} = \mathbf{0}$, by definition, $p(\mathbf{0},t) = 1$. Thus, in the sum $\langle C_N(t)\rangle = \sum_{\mathbf{x}} [p(\mathbf{x},t)]^N$, we can separate the non-scaling part and the scaling part. For the scaling part, we can again use the third line of Eq. (12.46), but the integral over z is cut off at $z = a/\sqrt{4\,D\,t}$, where a is a short-distance cutoff. Using the small z behavior of the integrand $z^{d-1}\,[f_>(z)]^N \sim z^{2N-d(N-1)-1}$ (as discussed above), one finds that the contribution from the lower limit scales for large t as $t^{-\nu}$ where ν is given in Eq. (12.67). This exactly cancels the prefactor $\sim t^{\nu}$ in Eq. (12.64). Hence, for large t, we get

$$\langle C_N(t)\rangle \simeq \text{const.} \quad \text{for} \quad d > d_c(N), \tag{12.69}$$

where the constant evidently depends on the cutoff, i.e., on the details of the lattice and hence is non-universal. Physically, this indicates that for $d > d_c(N)$, the common sites visited by all the walkers are typically close to the origin, and they get visited at relatively early times. At late times, the walkers hardly overlap, and hence, $\langle C_N(t)\rangle$ ceases to grow with time.

Exactly at $d = d_c(N)$, the exponent $\nu = 0$ from Eq. (12.65). Using the asymptotic small z behavior of $f_>(z)$ from the first line of Eq. (12.41), one finds that the integrand in Eq. (12.64) behaves as $[\Gamma(1/(N-1))]^N z^{-1}$. Note that in this case also, we need to keep the lower limit as $a/\sqrt{4Dt}$ with a cutoff a. Consequently, at large time, the integral in Eq. (12.64) is dominated by the lower limit $z \to 0$, and we get for large t

$$\langle C_N(t)\rangle \simeq a_c(N), \ln t, \text{ with } \quad a_c(N) = \frac{1}{2}\left[\frac{E_{d_c}\,\Gamma\left(\frac{1}{N-1}\right)}{(4\pi D)^{d_c/2}}\right]^N (4\,D)^{d_c/2}, \tag{12.70}$$

where we used the shorthand notation $d_c \equiv d_c(N)$.

Summarizing the large t behavior of $\langle C_N(t)\rangle$ for $d > 2$, we obtain

$$\langle C_N(t)\rangle \simeq \begin{cases} \alpha_N(d)\, t^{\nu}, & 2 < d < d_c(N), \\ a_c(N)\, \ln t, & d = d_c(N), \\ \text{const.}, & d > d_c(N), \end{cases} \tag{12.71}$$

where the expression for the amplitudes $\alpha_N(d)$ and $a_c(N)$ are given above and $\nu = N - \frac{d(N-1)}{2}$.

(iii) $d = 2$: In this case, we substitute the scaling form of $p(\mathbf{x}, t)$ from the second line of Eq. (12.46) into the expression for $\langle C_N(t)\rangle$ in Eq. (12.6), replace the sum by an integral as usual, and use the spherical symmetry to obtain

$$\langle C_N(t)\rangle \simeq b_2(N)\,\frac{t}{(\ln t)^N}, \qquad b_2(N) = 8\pi D \int_0^{\infty} [f_2(z)]^N\, z\, dz, \tag{12.72}$$

where the function $f_2(z)$ is given in Eq. (12.44).

Let us summarize the main results discussed above, which originally appeared in Ref. [15]. One obtains a host of rich growth behavior of $\langle C_N(t)\rangle$ for large t, depending on the two parameters d and N. These growth laws are summarized in the phase diagram in the (N, d) plane in Fig. 12.3. Even though N and d are integers, and so is $C_N(t)$, once we have formulated the problem in terms of $p(\mathbf{x}, t)$ (the probability to visit a site $\mathbf{x}$ before time t), one can consider the mean $\langle C_N(t)\rangle$ as a continuous variable. Moreover, it can be analytically continued to real d (for example, to fractal lattices) and also to real N. One of the most interesting results of this analysis is the existence of this anomalous regime $2 < d < d_c(N)$ regime in the (N, d) plane (see Fig. 12.3) where $\langle C_N(t)\rangle \sim t^{\nu}$ grows algebraically for large t, but with an exponent $\nu = N - (d/2)(N-1)$ that depends continuously on d and N.

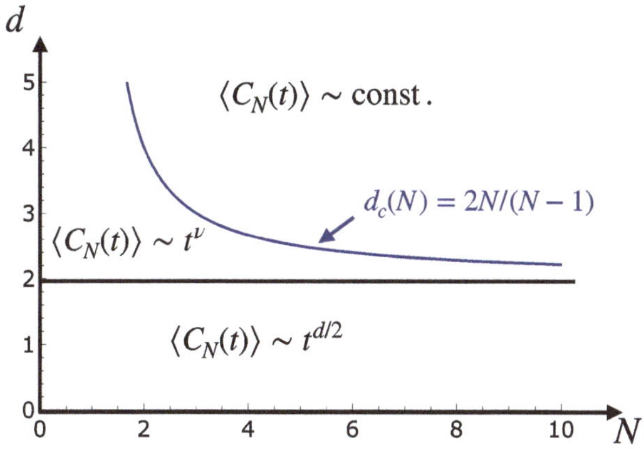

Fig. 12.3 Phase diagram in the (N, d) plane for the asymptotic large t growth of the mean number of common sites $\langle C_N \rangle$ visited by N independent random walkers in d dimensions. There are two critical lines: $d = 2$ (the black horizontal line) and $d_c(N) = 2N/(N-1)$ that separate three regimes of growth. In the regime $2 < d < d_c(N)$, one obtains an algebraic growth $\langle C_N(t) \rangle \sim t^{\nu}$ where the exponent $\nu = N - (d/2)(N-1)$ is anomalous and depends on the parameters d and N

For example, for $N = 2$, the critical dimension $d_c(2) = 4$. Hence, $d = 3$ would fall in this anomalous regime where $\langle C_2(t) \rangle t^{1/2}$. This analysis also shows that exactly at $d = d_c(N)$, the mean number $\langle C_N(t) \rangle \sim \ln t$ grows very slowly with time. For example, for $N = 3$, where $d_c(3) = 3$, if we consider $d = 3$, then our results predict $\langle C_3(t) \rangle \sim \ln t$. Some of these predictions have been verified in numerical simulations in Ref. [15].

12.2.4 Asymptotic Behavior of $\langle V_{K,N}(t) \rangle$

In this section, we derive the asymptotic late time growth of $\langle V_{K,N}(t) \rangle$ denoting the mean number of sites visited by exactly K out of the N walkers up to time t (with $1 \leq K \leq N$). The exact formula for $\langle V_{K,N}(t) \rangle$ in terms of the central quantity $p(\mathbf{x}, t)$ is given in Eq. (12.15), which already appeared in Ref. [15], but it was not analyzed for general K. We recall that for $K = N$, one recovers the mean number of common sites $\langle V_{N,N}(t) \rangle = \langle C_N(t) \rangle$ analyzed in the previous subsection. In this subsection, we extend this analysis for other values of K. Our strategy again is to inject the asymptotic scaling behavior of $p(\mathbf{x}, t)$ from Eq. (12.46) into Eq. (12.15) valid for large t and then analyze the sum in Eq. (12.15) upon replacing it by an integral and computing it using spherical symmetry. The details are exactly similar as in the case $K = N$ case, and therefore, we provide only the main results. As in the $K = N$ case, it turns out that $d = 2$ is a critical line, and we consider the three regimes separately: (i) $d < 2$, (ii) $d > 2$, and (iii) $d = 2$.

(i) $d < 2$: We start with the simplest case $d < 2$. Following the strategy mentioned above, it is easy to verify that, for large t,

$$\langle V_{K,N}(t)\rangle \sim t^{d/2} \quad \text{for} \quad \text{all } 1 \leq K \leq N. \tag{12.73}$$

Thus, the growth exponent $d/2$ is independent of K, while the K and N-dependence appear only in the amplitude, which can in principle be computed.

(ii) $d > 2$: In this case, $p(\mathbf{x}, t) \sim t^{1-d/2} f_>(|\mathbf{x}|/\sqrt{4Dt})$ from the third line of Eq. (12.46). Substituting this behavior in Eq. (12.15), one finds that, due to the decaying prefactor $t^{1-d/2}$ of $f_>(|\mathbf{x}|/\sqrt{4Dt})$, at large times, one can approximate $(1 - p(\mathbf{x}, t))^{N-k} \simeq 1$. Consequently, Eq. (12.15) reduces to

$$\langle V_{K,N}(t)\rangle \simeq \binom{N}{K} \sum_{\mathbf{x}} [p(\mathbf{x}, t)]^K . \tag{12.74}$$

Thus, up to this prefactor $\binom{N}{K}$, this is exactly identical to Eq. (12.6) with N replaced by K. Thus, all the scaling analysis done before for $\langle C_N(t)\rangle$ will go through after the replacement of N by K. In particular, we thus get a critical line $d_c(K) = 2K/(K-1)$. Thus, replacing N by K in Eq. (12.71), we get for $K > 1$, up to unimportant prefactors

$$\langle V_{K,N}(t)\rangle \sim \begin{cases} t^{\nu}, & 2 < d < d_c(K), \\ \ln t, & d = d_c(K), \\ \text{const.}, & d > d_c(K), \end{cases} \tag{12.75}$$

where $\nu = K - (d/2)(K-1)$. The case $K = 1$ is special because for this case $d_c(1) \to \infty$ and $\nu = 1$. Indeed, in this case, from Eq. (12.74), one finds

$$\langle V_{1,N}(t)\rangle \simeq N \sum_{\mathbf{x}} p(\mathbf{x}, t) = N\langle D_1(t)\rangle, \tag{12.76}$$

where the last equality follows from Eq. (12.3). Note that $D_1(t)$ is just the number of distinct sites visited by a single walker. Using Eqs. (12.54) and (12.55) for $N = 1$, one finds that $\langle D_1(t)\rangle \simeq E_d\, t$ for large t where E_d is the escape probability of a single walker. Consequently, one gets, from Eq. (12.76)

$$\langle V_{1,N}(t)\rangle \simeq N\, E_d\, t, \tag{12.77}$$

for large t. This result implies that for $d > 2$, each site is visited on an average by only one walker and only once.

(iii) $d = 2$: In this case, using $p(\mathbf{x}, t) \sim [1/\ln(t)]\, f_2\left(|\mathbf{x}|/\sqrt{4Dt}\right)$ and approximating $(1 - p(\mathbf{x}, t))^{N-k} \simeq 1$, one finds, after replacing N by K in Eq. (12.72), that at late times

$$\langle V_{K,N}(t) \rangle \sim \frac{t}{(\ln t)^K}. \tag{12.78}$$

Thus, for each $1 < K \leq N$, one can draw a phase diagram in the (K, d) plane, similar to the $K = N$ case in Fig. 12.3. Essentially, there are three regimes in the (K, d) plane, where the asymptotic behaviors are given by

$$\langle V_{K,N} \rangle \sim \begin{cases} t^{d/2}, & d < 2, \\ t^{\nu}, & 2 < d < d_c(K), \\ \text{const.}, & d > d_c(K), \end{cases} \tag{12.79}$$

where $\nu = K - (d/2)(K - 1)$ and $d_c(K) = 2K/(K - 1)$. Exactly at $d = d_c(K)$, the mean value grows as $\langle V_{K,N} \rangle \sim \ln t$, while at $d = 2$ it behaves as $\sim t/(\ln t)^K$ for large t.

12.3 The Exact Distributions of $D_N(t)$ and $C_N(t)$ in One Dimension and the Link to Extreme Value Statistics

In the previous sections, we have calculated exactly the mean number of distinct and common sites visited up to time t by N independent random walks in d dimensions. For large time, these random walks converge to Brownian motions, and these two observables become easier to compute in the Brownian limit. However, calculating the higher moments or the full distributions of $D_N(t)$ (the number of distinct sites up to t) and $C_N(t)$ (the number of common sites up to t) is a very difficult problem in arbitrary dimension. However, in $d = 1$, one can compute the full distribution of $D_N(t)$ and $C_N(t)$ [17]. In this section, we briefly outline the salient features leading to these exact results.

We consider N independent Brownian motions each of duration t and each starting at the origin. Let $x_i(\tau)$ denote the position of the i-th walker at time τ (with $0 \leq \tau \leq t$)—see Fig. 12.4 for $N = 2$. For each walk, we first identify the global maximum M_i and the global minimum m_i. Note that the maxima M_i's are necessarily non-negative, i.e., $M_i \geq 0$, while the minima m_i's are necessarily non-positive, i.e., $m_i \leq 0$. Let us now define the pair of variables M_+ and m_-

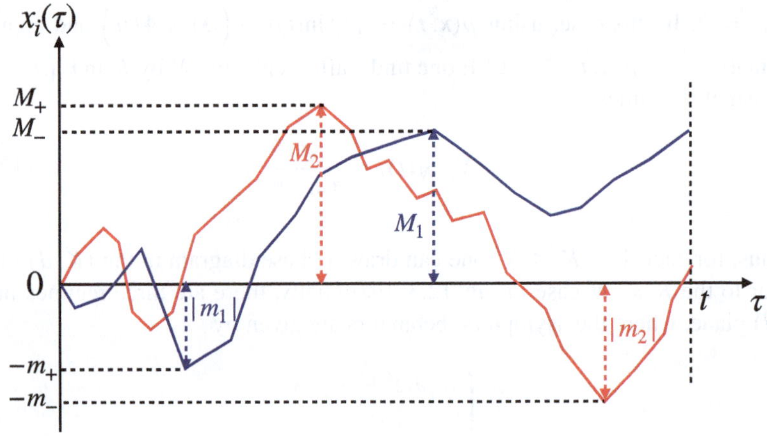

Fig. 12.4 Schematic trajectories of $N = 2$ independent Brownian motions, each of duration t. The variables M_1, M_2 (denoting the global maximum of each walker) and m_1, m_2 (denoting the global minimum of each walker) are indicated. Similarly, $M_+ = \max(M_1, M_2)$, $M_- = \min(M_1, M_2)$, $m_- = \max(|m_1|, |m_2|)$ and $m_+ = \min(|m_1|, |m_2|)$ are also indicated

$$M_+ = \max\{M_1, M_2, \cdots, M_N\} = \max_{1 \le i \le N} M_i, \tag{12.80}$$

$$m_- = -\min\{m_1, m_2, \cdots, m_N\} = \max\{|m_1|, |m_2|, \cdots, |m_N|\} = \max_{1 \le i \le N} |m_i| . \tag{12.81}$$

Here M_+ denotes the global maximum of all the N Brownian motions, while m_- denotes the absolute value of the global minimum of these walks up to time t. For brevity of notations, we do not exhibit the explicit time dependence of these random variables. The number of distinct sites visited up to time t by all the walkers on the positive side of the origin, in this continuous time limit, is just $M_+(t)$, since it represents the span of the walkers on the positive side. Similarly, $m_-(t)$ represents the number of distinct sites visited by the walkers on the negative side up to time t. Hence, the number of distinct sites $D_N(t)$ is just the sum of these two observables

$$D_N(t) = M_+ + m_- . \tag{12.82}$$

Similarly, let us define the pair of variables M_- and m_+

$$M_- = \min\{M_1, M_2, \cdots, M_N\} = \min_{1 \le i \le N} M_i, \tag{12.83}$$

$$m_+ = -\max\{m_1, m_2, \cdots, m_N\} = \min\{|m_1|, |m_2|, \cdots, |m_N|\} = \min_{1 \le i \le N} |m_i|. \tag{12.84}$$

These four observables M_+, M_-, m_+ and m_- are indicated in Fig. 12.4 for $N = 2$ walkers.

The variable M_- denotes the smallest of the N maxima, while m_+ denotes the smallest of the absolute values of the N minima. Therefore, M_- denotes the intersection of all the sites visited by N walkers on the positive side of the origin, while m_+ denotes the intersection on the negative side. Hence, the number of common sites $C_N(t)$ visited by all the N walkers up to time t is given by the sum

$$C_N(t) = M_- + m_+ \,. \tag{12.85}$$

Thus, in order to compute the distribution of $D_N(t)$ using the relation in Eq. (12.82), we need the joint distribution of M_+ and m_-. Similarly, for the distribution of $C_N(t)$ in Eq. (12.85), we need the joint distribution of M_- and m_+. These joint distributions can be computed explicitly using the fact that the walkers are independent. This brings us to the extreme value statistics (EVS) of N independent and identically distributed (IID) random variables [19, 21]. To proceed, we first make a simple observation that, for a single Brownian motion, the joint distribution of M_i and m_i is only a function of the rescaled variables $\widetilde{M}_i = M_i/\sqrt{4\,D\,t}$ and $\widetilde{m}_i = m_i/\sqrt{4\,D\,t}$ [21]. In other words, the dependence on t only appears through these rescaled variables. For simplicity, we set $D = 1/2$ and define the rescaled variables

$$\widetilde{M}_\pm = \frac{M_\pm}{\sqrt{2t}}, \quad \widetilde{m}_\pm = \frac{m_\pm}{\sqrt{2t}} \,. \tag{12.86}$$

In terms of the rescaled variables, one can think of N Brownian motions defined on the unit time interval. In the following two subsections, we compute the distribution of $D_N(t)$ and $C_N(t)$ separately.

12.3.1 Distribution of the Number of Distinct Sites

Let us start with the computation of the PDF of $D_N(t)$ in Eq. (12.82). For this, we need to compute the joint distribution of M_+ and m_-. It turns out to be convenient to consider their cumulative distribution, which, in terms of the rescaled variables, is defined as

$$P_D(\ell_1, \ell_2) = \text{Prob.}\left(\widetilde{M}_+ \le \ell_1, \widetilde{m}_- \le \ell_2\right). \tag{12.87}$$

If we know this joint cumulative distribution, the joint PDF is simply given by

$$p_D(\ell_1, \ell_2) = \frac{\partial^2}{\partial \ell_1 \partial \ell_2} P_D(\ell_1, \ell_2). \tag{12.88}$$

Using the independence of the N Brownian motions, the cumulative distribution is given by

$$P_D(\ell_1, \ell_2) = [g(\ell_1, \ell_2)]^N \,, \tag{12.89}$$

where

$$g(\ell_1, \ell_2) = \text{Prob.}(\widetilde{M} \le \ell_1, \widetilde{m} \ge -\ell_2). \tag{12.90}$$

Here $\widetilde{M}$ and $\widetilde{m}$ denote respectively the maximum and the minimum of a single Brownian motion on the unit time interval. This joint cumulative distribution for a single Brownian motion can be computed explicitly by solving the diffusion equation in a box $[-\ell_2, \ell_1]$ with absorbing boundary conditions at the two edges. This can be done by using a standard eigenfunction expansion of the diffusion equation [18, 19]. It reads

$$g(\ell_1, \ell_2) = \frac{2}{\pi} \sum_{n=0}^{\infty} \frac{1}{n+\frac{1}{2}} \sin\left(\frac{2(n+1)\pi \ell_2}{\ell_1+\ell_2}\right) \exp\left[-\left(\frac{(n+\frac{1}{2})\pi}{\ell_1+\ell_2}\right)^2\right]. \tag{12.91}$$

Therefore, using Eqs. (12.82), (12.88) and (12.89), the PDF of $D_N(t)$ is given by

$$p_N^D(s) = \text{Prob.}(D_N(t) = s) = \int_0^\infty d\ell_1 \int_0^\infty d\ell_2 \, \delta(s-\ell_1-\ell_2) \frac{\partial^2}{\partial \ell_1 \partial \ell_2} [g(\ell_1, \ell_2)]^N \,. \tag{12.92}$$

For small values of N, one can compute this double integral explicitly and numerical simulations confirm this result [17]. For general N, it is hard to compute the distribution explicitly from Eq. (12.92). However, the tails of the distribution for small and large s can be computed for general N, and they are given by (details can be found in Ref. [17, 22]). For $N \ge 2$, one gets

$$p_N^D(s) \simeq \begin{cases} a_N \, s^{-5} \exp\left[-N\pi^2/(4\,s^2)\right], & s \to 0 \\ \\ b_N \exp(-s^2/2), & s \to \infty, \end{cases} \tag{12.93}$$

where the prefactors a_N and b_N are given by

$$a_N = 4\pi^{3/2} N(N-1) \left(\frac{4}{\pi}\right)^{N-2} \frac{\Gamma\left(\frac{N-1}{2}\right)}{\Gamma\left(\frac{N}{2}\right)}, \quad b_N = \frac{2^{3/2}}{\sqrt{\pi}} N(N-1). \tag{12.94}$$

For $N = 1$ the asymptotic behaviors are the same as in Eq. (12.93) with the prefactors $a_1 = 2\pi^2$ and $b_1 = 8/\sqrt{\pi}$.

It turns out that, for large N, an interesting scaling behavior emerges. The typical scale of the fluctuations of $D_N(t)/\sqrt{2t}$ can be estimated from the connection to the EVS of IID variables, using Eqs. (12.82) and (12.83). The rescaled variables $\widetilde{M}_i$'s, which are the maxima of the ith Brownian motion on the unit interval, are IID variables. Their common PDF is a half-Gaussian [19], $p(M) = (2/\sqrt{\pi})e^{-M^2}$, $M > 0$. The same holds for the rescaled variables $-\widetilde{m}_i$'s. Hence, for large N, standard results of EVS [21, 23] state that the typical value of $\widetilde{M}_+ = \max_{1\le i\le N}\widetilde{M}_i$ is $O(\sqrt{\log N})$ while its fluctuations are of order $1/\sqrt{\log N}$ and governed by a Gumbel distribution. This means that, to leading order for large N, the random variable $\widetilde{M}_+$ can be expressed as

$$\widetilde{M}_+ \simeq \sqrt{\ln N} + \frac{1}{2\sqrt{\ln N}} G_1, \tag{12.95}$$

where G_1 is a random variable of $O(1)$, which is distributed via the Gumbel law

$$\text{Prob.}(G_1 \le \ell_1) = e^{-e^{-\ell_1}}. \tag{12.96}$$

Similarly, one can express

$$\widetilde{m}_- \simeq \sqrt{\ln N} + \frac{1}{2\sqrt{\ln N}} G_2, \tag{12.97}$$

where G_2 is again distributed via the Gumbel law. For large N, these two extremes $\widetilde{M}_+$ and $\widetilde{m}_-$ become uncorrelated as the global maximum and global minimum are most likely not reached by the same walker. Hence, to leading order for large N, the variables $\widetilde{M}_+$ and $\widetilde{m}_-$ can be considered as independent. This implies that the associated Gumbel variables G_1 and G_2 are also uncorrelated. Consequently, we can write

$$g^N\left[\mu_N + \frac{\tilde{\ell}_1}{2\mu_N}, \mu_N + \frac{\tilde{\ell}_2}{2\mu_N}\right] \underset{N\to+\infty}{\longrightarrow} e^{-e^{-\tilde{\ell}_1}} e^{-e^{-\tilde{\ell}_2}} \tag{12.98}$$

with $\mu_N = \sqrt{\ln N}$. Inserting (12.98) in (12.92) with $\widetilde{s} = 2\mu_N(s - 2\mu_N)$, one finds

$$p_N^D(s) \sim 2\sqrt{\ln N} \int_{-\infty}^{\infty} d\widetilde{\ell}_2\, e^{-\widetilde{s}} e^{-e^{-\widetilde{\ell}_2}} e^{-e^{-(\widetilde{s}-\widetilde{l}_2)}}, \tag{12.99}$$

which can be evaluated explicitly. This gives the scaling form for the PDF of $D_N(t)$ for large N

$$p_N^D(s) \simeq 2\sqrt{\ln N}\, \mathcal{D}(2\sqrt{\ln N}(s - 2\ln N)), \tag{12.100}$$

where the scaling function $\mathcal{D}(z)$ is given by

$$\mathcal{D}(z) = 2\,e^{-z}\, K_0(2\,e^{-z/2}). \tag{12.101}$$

Here $K_0(z)$ is the modified Bessel function of index 0. This result was verified in numerical simulations for $N = 50$ and $N = 100$ in Ref. [17]. Here we just provide a plot of the scaling function $\mathcal{D}(z)$ in Fig. 12.5. Its asymptotic behaviors are given by

$$\mathcal{D}(z) \simeq \begin{cases} \sqrt{\pi}\, e^{3|z|/4 - 2e^{|z|/2}}, & z \to -\infty \\ \\ z\, e^{-z}, & z \to \infty. \end{cases} \tag{12.102}$$

This fact that the convolution of two independent Gumbel variables is distributed via the modified Bessel function also appeared in other contexts, e.g., for the maximum of a log-correlated gas of particles on a circle [24, 25].

12.3.2 Distribution of the Number of Common Sites

We now turn to the PDF of $C_N(t)$ in Eq. (12.85). This requires the computation of the joint PDF of M_- and m_+. It turns out to be convenient again to consider the cumulative distribution of the rescaled variables

$$P_C(j_1, j_2) = \text{Prob.}\left(\widetilde{M}_- \geq j_1, \widetilde{m}_+ \geq j_2\right). \tag{12.103}$$

If we know this joint cumulative distribution, the joint PDF is simply given by

$$p_C(j_1, j_2) = \frac{\partial^2}{\partial j_1 \partial j_2} P_C(j_1, j_2). \tag{12.104}$$

Using the independence of the N Brownian motions, the cumulative distribution is given by

$$P_C(j_1, j_2) = \left[h(j_1, j_2)\right]^N, \tag{12.105}$$

where

$$h(j_1, j_2) = \text{Prob.}(\widetilde{M} \geq j_1, |\widetilde{m}| \geq j_2) = \text{Prob.}(\widetilde{M} \geq j_1, \widetilde{m} \leq -j_2), \tag{12.106}$$

where $\widetilde{M}$ and $\widetilde{m}$ denote respectively the maximum and the minimum of a single Brownian motion on the unit time interval. In writing the last equality in Eq. (12.106), we used the fact that $\widetilde{m} \leq 0$. In fact, by using inclusion–exclusion principle of probability, it is easy to see that $h(j_1, j_2)$ is related to the function $g(\ell_1, \ell_2)$ in Eq. (12.91) via

$$h(j_1, j_2) = 1 - \text{erf}(j_1) - \text{erf}(j_2) + g(j_1, j_2), \tag{12.107}$$

where $\mathrm{erf}(x) = 2/\sqrt{\pi}\int_0^x e^{-u^2\,du}$. Note that, here, we used the fact that Prob.$(\tilde{M} \le j_1) = \mathrm{erf}(j_1)$ and Prob.$(\tilde{m} \ge -j_2) = \mathrm{erf}(j_2)$ [19, 21]. Therefore using Eqs. (12.85), (12.104) and (12.105), the PDF of $C_N(t)$ is given by

$$\begin{aligned} p_N^C(w) &= \text{Prob.}(C_N(t) = w) \\ &= \int_0^\infty d\ell_1 \int_0^\infty d\ell_2\, \delta(w - j_1 - j_2) \frac{\partial^2}{\partial j_1 \partial j_2} \left[h(j_1, j_2)\right]^N . \end{aligned} \tag{12.108}$$

The asymptotic behavior of $p_N^C(w)$ for $N \ge 2$ are given by [17]

$$p_N^C(w) \simeq \begin{cases} c_N\, w, & w \to 0 \\ \\ d_N\, w^{1-N} \exp\left(-N w^2\right), & w \to \infty, \end{cases} \tag{12.109}$$

where the constants c_N and d_N are given by

$$c_N = \frac{4}{\pi} N(N-1), \quad d_N = \frac{8N}{\pi^{N/2}}. \tag{12.110}$$

For $N = 1$, i.e., for a single Brownian motion, the number of distinct and common sites are identical, i.e., $D_1(t) = C_1(t)$. Hence, the asymptotic behavior of $p_1^C(w)$ can be read off Eq. (12.93) with $N = 1$, for which $a_1 = 2\pi^2$ and $b_1 = 8/\sqrt{\pi}$.

As in the case of $D_N(t)$, the PDF of $C_N(t)$ also exhibits an interesting scaling behavior for large N. The typical scale of the fluctuations of $C_N(t)/\sqrt{2t}$ can be estimated from the connection to the EVS of IID variables, using Eqs. (12.85) and (12.80). The rescaled variables $\widetilde{M}_i$'s, which are the maxima of the ith Brownian motion on the unit interval, are IID variables. Their common PDF is a half-Gaussian [19], $p(M) = (2/\sqrt{\pi})e^{-M^2}$, $M > 0$. The same holds for the rescaled variables $-\widetilde{m}_i$'s. Hence, for large N, standard results of EVS [21, 23] state that the PDF of their minimum scales as $\widetilde{M}_- \simeq X/N$ where the random variable X is distributed via the Weibull law

$$\text{Prob.}(X \ge z) = e^{-z/\sqrt{\pi}}. \tag{12.111}$$

Similarly, for large N, we have $m_+ \simeq Y/N$ where Y is also distributed by the same law as in Eq. (12.111). For large N, the minimum on the positive side and the minimum on the negative side are not achieved by the same walker. Hence, X and Y can be considered as independent random variables, each of which being distributed via (12.111). Consequently, their sum $C_N(t)$ is a convolution of two exponentials, and it is easy to see that the PDF of $C_N(t)$ can be written in the scaling form

$$p_N^C(w) \simeq N\, C(N\, w), \tag{12.112}$$

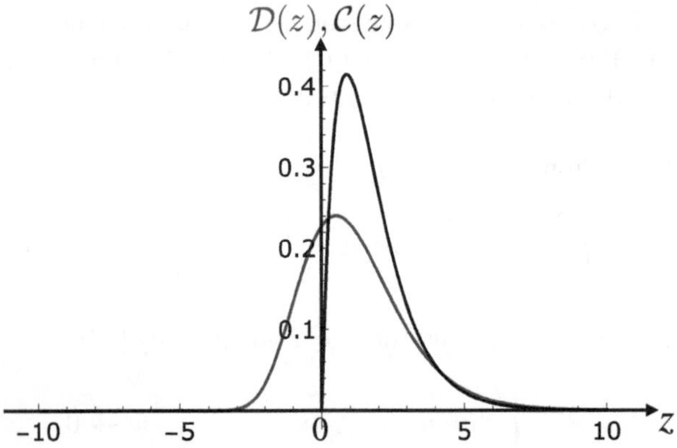

Fig. 12.5 Plot of the scaling function $\mathcal{D}(z)$—blue solid line—given in Eq. (12.101) and of $C(z)$—black solid line—given in Eq. (12.112) as a function of z

where the scaling function $C(z)$ is given by

$$C(z) = \frac{4}{\pi}\, z\, e^{-2z/\sqrt{\pi}}, \quad z \geq 0. \tag{12.113}$$

This scaling function is plotted in Fig. 12.5, along with $\mathcal{D}(z)$.

12.4 Extension to Other Models

The method presented here for computing the statistics of $D_N(t)$ and $C_N(t)$ for N independent Brownian walkers can be easily extended to other models where the walkers remain independent, but their individual stochastic motion need not be Brownian. We briefly discuss two simple examples below.

12.4.1 N Independent Brownian Bridges

A Brownian bridge is a Brownian motion, which is constrained to come back to its starting point (here the origin) after a fixed time t. In models of animals foraging, the Brownian bridge often plays an important role since animals typically come back to their nest at the end of the day, after foraging. It is then natural to ask what is the number of distinct and common sites visited by N Brownian bridges in d dimensions. The method presented in Sect. 12.2 for N Brownian motions involves the central quantity $p(\mathbf{x}, t)$ denoting the probability that the site $\mathbf{x}$ is visited by a single walker before time t. The same method, in terms of $p(\mathbf{x}, t)$, goes through for N independent Brownian bridges except that $p(\mathbf{x}, t)$ for a Brownian bridge of duration t is not exactly identical to that of a Brownian motion. However, one can show that $p(\mathbf{x}, t)$ for large $|\mathbf{x}|$ and large t, for a Brownian bridge, exhibits exactly similar scaling forms as in Eq. (12.46) for the Brownian motion, except that the scaling functions $f_<(z)$ (for $d < 2$), $f_2(z)$ (for $d = 2$), and $f_>(z)$ (for $d > 2$) for the Brownian bridges are different from their Brownian motion counterparts. Therefore, the large-time behavior of $\langle D_N(t)\rangle$ and $\langle C_N(t)\rangle$ for N Brownian bridges will be exactly similar to those of the Brownian motions, respectively in Eqs. (12.58) and (12.71), except that the prefactors will be different. Consequently, the phase diagram shown in Fig. 12.3 will also be similar for the Brownian bridges.

Furthermore, in $d = 1$, one can compute the distribution of $D_N(t)$ and $C_N(t)$ for N independent Brownian bridges, as in the Brownian motion case. For finite N, these distributions for the bridges are different from those of Brownian motions. However, for large N, using the universality of the EVS of IID random variables [19, 21], it is possible to show that the distribution of $D_N(t)$ and $C_N(t)$—up to some trivial scale factors—are exactly the same as those of Brownian motions, given respectively in Eqs. (12.100) and (12.112). Thus, the two scaling functions $\mathcal{D}(z)$ and $C(z)$ in Eqs. (12.101) and (12.113) are universal, i.e., they are independent of the bridge constraint.

12.4.2 N Independent Resetting Brownian Motions

A resetting Brownian motion (RBM) is a Brownian motion that resets to the origin with a constant rate r and has been studied extensively during the last decade (for a review, see [26]). In the context of animal foraging, RBM can be used to model the fact that an animal can return to the nest from time to time. In this context, one can also study the resetting Brownian bridge (RBB) [27]. For a single RBM, the mean number of distinct sites $\langle D_1(t)\rangle$ has been studied in all dimensions and has been shown to grow extremely slowly as $\langle D_1(t)\rangle \sim (\ln t)^d$ for large t [28]. This is considerably slower compared to the Brownian motion, and in addition, the special role played by $d = 2$ disappears in the case of RBM. In addition, the distribution of $D_1(t)$ has also been computed exactly in $d = 1$ [28]. However, the statistics of $D_N(t)$ and $C_N(t)$ for N independent RBM or RBB have not yet been studied and thus remain an interesting open problem. The method used in this chapter may be useful to solve this problem.

12.5 Conclusion

The motion of a foraging animal can often be modeled by a random walker/Brownian motion. In this chapter, we considered a simple model of N independent foraging animals, each of them performing independent Brownian motion. Even though each walker is independent, the statistical properties of the territory covered (for example, the home range [29]) by the animals can have nontrivial statistics. There are several measures for characterizing the area of the territory covered by the N animals. For example, one popular measure is the convex hull of the union of the trajectories of these Brownian motions, and the statistics of this convex hull has been studied extensively [30–35]. Another interesting measure of the home range is the number of distinct sites visited by N walkers up to time t [14]. A related question concerns the statistics of the number of sites visited by all the N walkers [15]. In this chapter, we reviewed the recent results on the last two quantities for N independent Brownian motions and also discussed the extension of these results to two other models, namely the N independent Brownian bridges and N independent resetting Brownian motions.

Let us summarize the main results presented here. We have studied, in all dimensions d, the mean number of distinct ($\langle D_N(t)\rangle$) and common sites ($\langle C_N(t)\rangle$) up to time t by N independent random walks, which, in the large time limit, can be studied in terms of N independent Brownian motions. We have shown that, to compute these two quantities, one needs to study the late time scaling behavior of a central quantity $p(\mathbf{x}, t)$, denoting the probability that the site $\mathbf{x}$ has been visited by a single walker before time t. We have shown that $\langle D_N(t)\rangle \sim t^{d/2}$ for $d < 2$ and $\langle D_N(t)\rangle \sim t$ for $d > 2$, as for a single walker, with N-dependent prefactors. In contrast, $\langle C_N(t)\rangle$ exhibits a much richer behavior. While for $d < 2$ it grows as $\langle C_N(t)\rangle \sim t^{d/2}$ with an N-dependent prefactor, it has very different behaviors for $d > 2$. In particular, there exists a critical dimension $d_c(N) = 2N/(N-1)$ such that, for $2 < d < d_c(N)$, the asymptotic growth is given by $\langle C_N(t)\rangle \sim t^{\nu}$ where the exponent $\nu = N - d(N-1)/2$ depends on d and N. For $d > d_c(N)$, the mean number of common sites approaches a constant at late times, while exactly at $d = d_c(N)$, it grows logarithmically with an N-dependent prefactor. These behaviors in the (N, d) plane are summarized in the phase diagram in Fig. 12.3. In addition, we have also computed the mean number of sites visited by exactly $1 \leq K \leq N$ walkers up to time t. The critical dimension in this case is $d_c(K) = 2K/(K-1)$. Finally, in $d = 1$, we have computed the full distribution of $D_N(t)$ and $C_N(t)$ for any N and showed that they exhibit an interesting scaling behavior for large N. This provides a nice and interesting application of the extreme value statistics.

In principle, it should be possible to extend our method to random walks on lattices with arbitrary jump size distribution with a finite variance. In this case, one would expect the central limit theorem to be valid and the description in terms of Brownian motion (with appropriate renormalized diffusion constant) should hold. Similarly, we would expect our late time asymptotic results to apply to continuous-time random walks on lattices, but with a waiting time distribution that has a finite mean and variance. In principle, the methods presented here can also be extended to the case when the waiting time distribution has a power law tail, such that the mean/variance may be divergent. It would be interesting to derive the statistics of $D_N(t)$ and $C_N(t)$ in such cases.

Finally, the method presented here is general enough to be extended to other types of stochastic processes. For example, we have discussed two applications, namely the case of N independent Brownian bridges and N independent resetting Brownian motions/bridges. In particular, the latter case has been studied only for $N = 1$ and possible extension to general N is an interesting open problem. Furthermore, the full distributions of $D_N(t)$ and $C_N(t)$ are known only in $d = 1$, and computing them for $d > 1$ remains a challenging problem. Finally, we considered here N noninteracting Brownian motions. However, in reality, the animals are always interacting, and it would be interesting to study the statistics of $D_N(t)$ and $C_N(t)$ for interacting random walkers. These interactions can be either attractive or repulsive. For example, ants typically leave pheromones along their trajectories and other ants get attracted to these trails. In contrast, some animals like tigers like to mark their territories by leaving chemical smell, which other animals avoid, leading to repulsive interactions between trajectories. It would be interesting to see how such interactions modify the asymptotic behaviors of $D_N(t)$ and $C_N(t)$.

Acknowledgments We would like to thank M. Biroli, A. Kundu, F. Mori, and M. Tamm for useful discussions and collaborations on this subject. We acknowledge support from ANR, Grant No. ANR- 23- CE30-0020-01 EDIPS.

References

1. Beeler, R.J., Delaney, A.J.: Order-disorder events produced by single vacancy migration. Phys. Rev. A **130**, 962 (1963)
2. Beeler, R.J.: Distribution functions for the number of distinct sites visited in a random walk on cubic lattices: relation to defect annealing. Phys. Rev. A **134**, 1396 (1964)
3. Blumen, A., Klafter, J., Zumofen, G.: In: Zschokke, I. (ed.) Optical Spectroscopy of Glasses, pp. 199–265. Reidel, New York (1986)
4. Czech, R.: Number of distinct sites visited by random walks in lattice gases. J. Chem. Phys. **91**, 2498 (1989)
5. Bordewijk, P.: Defect-diffusion models of dielectric relaxation. Chem. Phys. Lett. **32**, 592 (1975)
6. Condat, C.A.: Target relaxation by walkers departing from a finite domain. Phys. Rev. A **41**, 3365 (1990)
7. Edelstein-Keshet, L.: Mathematical Models in Biology. Random House, New York (1988)
8. Pielou, E.C.: An Introduction to Mathematical Ecology. Wiley-Interscience, New York (1969)
9. Cattuto, C., Barrat, A., Baldassari, A., Schehr, G., Loreto, V.: Collective dynamics of social annotation. Proc. Natl. Acad. Sci. USA **106**, 10511 (2009)
10. Dvoretzky, A., Erdös, P.: In: Proceedings of the Second Berkeley Symposium on Mathematical Statistics and Probability. University of California Press, Berkeley (1951)
11. Vineyard, G.H.: The number of distinct sites visited in a random walk on a lattice. J. Math. Phys. **4**, 1191 (1963)
12. Montroll, E.W., Weiss, G.H.: Random walks on lattices. II. J. Math. Phys. **6**, 167 (1965)
13. van Wijland, F., Hilhorst, H.J.: Universal fluctuations in the support of the random walk. J. Stat. Phys. **89**, 119 (1997)
14. Larralde, H., Trunfio, P., Havlin, S., Stanley, H.E., Weiss, G.H.: Territory covered by N diffusing particles, Nature (London) **355**, 423 (1992); Number of distinct sites visited by N random walkers, Phys. Rev. A **45**, 7128 (1992)
15. Majumdar, S.N., Tamm, M.V.: Number of common sites visited by N random walkers. Phys. Rev. E **86**, 021135 (2012)

16. Turban, L.: On the number of common sites visited by N random walkers. Preprint arXiv: 1209.2527
17. Kundu, A., Majumdar, S.N., Schehr, G.: Exact distributions of the number of distinct and common sites visited by N independent random walkers. Phys. Rev. Lett. **110**, 220602 (2013)
18. Redner, S.: A Guide to First-Passage Processes. Cambridge University Press, Cambridge (2001)
19. Majumdar, S.N., Schehr, G.: Statistics of Extremes and Records in Random Sequences. Oxford University Press, Oxford (2024)
20. Feller, W.: Introduction to Probability Theory and Its Applications, vol. 2. Wiley: New York (1966)
21. Majumdar, S.N., Pal, A., Schehr, G.: Extreme value statistics of correlated random variables: a pedagogical review. Phys. Rep. **840**, 1 (2020)
22. A. Kundu, S. N. Majumdar, and G. Schehr, Erratum: Exact Distributions of the Number of Distinct and Common Sites Visited by N, Phys. Rev. Lett. **111**, 049904 (2013)
23. Gumbel, E.J.: Statistics of Extremes. Dover, New York (1958)
24. Fyodorov, Y.V., Bouchaud, J.-Ph.: Freezing and extreme-value statistics in a random energy model with logarithmically correlated potential. J. Phys. A: Math. Theor. **41**, 372001 (2008)
25. Subag, E., Zeitouni, O.: Freezing and decorated Poisson point processes. Commun. Math. Phys. **337**, 55 (2015)
26. Evans, M.R., Majumdar, S.N., Schehr, G.: Stochastic resetting and applications. J. Phys. A: Math. Theor. **53**, 193001 (2020)
27. De Bruyne, B., Majumdar, S.N., Schehr, G.: Optimal resetting Brownian bridges via enhanced fluctuations. Phys. Rev. Lett. **128**, 200603 (2022)
28. Biroli, M., Mori, F., Majumdar, S.N.: Number of distinct sites visited by a resetting random walker. J. Phys. A: Math. Theor. **55**, 244001 (2022)
29. Giuggioli, L., Kenkre, V.M.: Consequences of animal interactions on their dynamics: emergence of home ranges and territoriality. Movement Ecol. **2**, 1 (2014)
30. Randon-Furling, J., Majumdar, S.N., Comtet, A.: Convex hull of N planar Brownian motions: exact results and an application to ecology. Phys. Rev. Lett. **103**, 140602 (2009)
31. Majumdar, S.N., Comtet, A., Randon-Furling, J.: Random convex hulls and extreme value statistics. J. Stat. Phys. **138**, 955 (2010)
32. Claussen, G., Hartmann, A.K., Majumdar, S.N.: Convex hulls of random walks: large-deviation properties. Phys. Rev. E **91**, 052104 (2015)
33. Hartmann, A.K., Majumdar, S.N., Schawe, H., Schehr, G.: The convex hull of the run-and-tumble particle in a plane. J. Stat. Mech. **2020**(5), 053401 (2020)
34. Majumdar, S.N., Mori, F., Schawe, H., Schehr, G.: Mean perimeter and area of the convex hull of a planar Brownian motion in the presence of resetting. Phys. Rev. E **103**, 022135 (2021)
35. Singh, P., Kundu, A., Majumdar, S.N., Schawe, H.: Mean area of the convex hull of a run and tumble particle in two dimensions. J. Phys. A: Math. Theor. **55**, 225001 (2022)

Chapter 13
Collective Foraging and Behavioural Syndromes in Ants: First-Passage Statistics with Heterogeneous Walkers on a Honeycomb Lattice

Daniel Marris, Pol Fernández-López, Frederic Bartumeus, and Luca Giuggioli

13.1 Introduction

Social groups are rarely uniform or cohesive units; instead, individuals within these groups often exhibit consistent behavioural differences, commonly referred to as "personalities" [1]. Understanding individual heterogeneity is essential for grasping the variability in life histories and population dynamics [2]. Moreover, it significantly influences the functionality of key ecological responses. In eusocial species, such as ants and termites, individual heterogeneity is exploited to associate different roles to different tasks [3], e.g. typically one female and several males are reproductively active, while non-breeding individuals care for the young, protect the colony and provide resources. For the non-breeding individuals a key debate in this regard is whether these roles represent specialised task forces designed to address different needs or if individuals switch between tasks to adapt to inherent environmental uncertainties [4].

Daniel Marris and Pol Fernández-López contributed equally to this work.

D. Marris · L. Giuggioli (✉)
School of Engineering Mathematics and Technology, University of Bristol, Bristol, UK
e-mail: luca.giuggioli@bristol.ac.uk

P. Fernández-López
Theoretical and Computational Ecology, Centre for Advanced Studies of Blanes, Girona, Spain

F. Bartumeus
Theoretical and Computational Ecology, Centre for Advanced Studies of Blanes, Girona, Spain

Centre for Research on Ecology and Forestry Applications (CREAF), Barcelona, Spain

Catalan Institution for Research and Advanced Studies (ICREA), Passeig de Lluís Companys, Barcelona, Spain

L. Giuggioli, P. K. Maini (eds.), *The Mathematics of Movement*,
https://doi.org/10.1007/978-3-032-00873-2_13

In the context of foraging processes, ants are well known to have evolved various strategies for efficient localisation and exploitation of food resources [3] with two common strategies being group recruitment and mass recruitment. In group recruitment, a scout discovers a food source and returns to the nest to recruit a small number of nestmates by leaving a chemical trail that only the scout itself can trace. This strategy, where nestmates directly follow scouts, suits smaller or scattered food sources. On the other hand, mass recruitment, where the scout lays a strong pheromone trail that may attract many nestmates, is faster and involves more ants, ideal for large or abundant food sources. Key determinants to select the appropriate strategy are the environmental conditions and the type of food available. For instance, some species primarily rely on individual foraging but switch to group or mass recruitment as food sources increase in quality and predictability, allowing ant colonies to optimise resource acquisition across different scenarios. Regardless the recruitment strategy, the number of scouts and recruits involved in food collection is also flexible, varying dynamically as the food is being collected and with the amount, quality, and predictability of resources. This adaptability raises intriguing questions about the role of these distinct behaviours, how colonies adjust their numbers and how these adjustments impact collective success.

Experiments that accurately track animal movement are a convenient approach to learn about foraging strategies. Y-maze laboratory arenas, where the ants are confined to the perimeters of tiled hexagons, are particularly suited to understanding ant behaviour as they offer a reliable and practical means of quantifying decision-making [5]. Here we employ the empirical set-up used in previous experiments [6] to observe the foraging behaviour of ants (*Aphaenogaster senilis*) within a large ($2 \times 1\ \text{m}^2$) honeycomb lattice arena.

A. senilis has been selected because of its known group recruitment foraging strategies whereby scouts and recruits have distinct roles [7, 8]. Recent studies on a Y-maze arena have identified significant disparities in the movement patterns of these two types of ants [9], with quantifiable differences in the probabilities of moving left, right or backwards at each step. Scouts exhibit less sinuous movement and are less likely to move backwards compared to recruits. In contrast, recruits display movement patterns more akin to a diffusive walk, with an approximately uniform sampling of the three possible directions at each vertex of the honeycomb lattice. Consequently, scouts travel greater distances from the nest and spend more time exploring the arena, while recruits typically remain within a confined region, usually no farther than 25 cm from the nest. One may thus expect that the differences in numbers and exploratory behaviours of scouts and recruits would lead to markedly different spatial spreading and coverage patterns.

While ants are known to perform central place foraging [10], returning to a central location (like a nest or den) after foraging trips, it is not clear what movement strategies they employ in environments with multiple food patches, as new patches may be discovered while previous ones are still being exploited. In these complex, multi-target scenarios, both the nest and the exploited patches could serve as departure points for further searches by either scouts or recruits. Here we investigate whether exploratory ants follow a strict central place foraging strategy (nest-to-patch search) or use previously discovered patches as starting points for further exploration (patch-to-patch search).

To answer quantitatively these questions, we develop a discrete space-time, non-Markovian, random walk model on a bounded honeycomb lattice that faithfully represents the experimental arena. Since the honeycomb lattice is non-Bravais, to represent the dynamics of search in the experimental arena, we utilise the theory of random walks with internal states to capture the periodic heterogeneity in the domain sites [11, 12] and include a one-step memory in the movement model via the theory of persistent random walks [13, 14]. We use this model first with a single walker to investigate the spreading (mean square displacement and mean number of distinct visited sites) in the infinite lattice, quantifying how quickly scouts and recruits explore the space compared to a diffusive walker. We then study the so-called first first-passage time (fFPT) distribution of a population of independent and heterogeneous walkers. We do so by extending well-known theory [15] to account for a population of distinct walkers, that is, with unequal movement statistics. We further generalise the theory to consider multiple targets and calculate the splitting probability, defined as the time dependent probability that one member of the population reaches specific target(s) before any other individual has reached any other target. Similarly to the case of one single walker [14, 16], with multiple walkers one may expect that a large increase in persistence can hinder search efficiency. Using the splitting probability, we show in fact that a highly persistent population may miss close-by targets. In contrast, when one uses experimentally measured parameters, we show that the local target is found first.

By employing a semi-analytic procedure, we also study the ant search behaviour on an exact replica of the experimental arena with hard reflecting boundaries. We explore the fFPT distribution for the $\mathcal{N}$ ants observed in nine different experimental realisations and demonstrate that in the majority of cases, the empirical measurements coincide much closer to the mode of the fFPT distribution than its mean. Moreover, by identifying the differences in the ordering of events in nest-to-patch and patch-to-patch search, we elucidate that the empirical timescales to find the second patch match closer with the nest-to-patch case.

This chapter is organised as follows. The details of the experimental procedures are described in Sect. 13.2. In Sect. 13.3, we derive the formalism of the honeycomb correlated random walk on the unbounded (infinite) lattice and then use it to explore coverage efficiency in Sect. 13.4. We then turn our attention to the fFPT of $\mathcal{N}$ walkers for which we derive a general formalism for the splitting probability with multiple targets when the population is composed of non-identical walkers. We study some theoretical examples on bounded lattices in Sect. 13.5.1, while the experimentally relevant model on the Y-maze arena is worked out in Sect. 13.5.2. Empirical findings follow in Sect. 13.6, and conclusions form Sect. 13.7.

13.2 Experimental Protocol in the Y-Maze Arena

Foraging experiments were carried out with *Aphaenogaster senilis*, a Mediterranean ant species displaying flexible foraging strategies, ranging from individual to group recruitment [17]. To reproduce the natural spatiotemporal scales at which these ants forage, we recorded 3 hours of video per trial, capturing their movements within a 2 metre by 1 metre honeycomb-like arena (see Figure 1 in ref. [6]). This Y-maze pattern is widely utilised in studies of animal decision-making (e.g. [5]) and simplifies the quantification of animal behaviour and movement since at each vertex, the walker has only three choices: to go left, right or backwards. Additionally, spatial discreteness of the arena lends itself to the use of the present random walk theory and other statistical physics frameworks such as spin glasses [6, 18].

The main purpose of the experiments was to test how ant foraging dynamics adapted to highly unpredictable environments. Accordingly, food sources were randomly distributed throughout the experimental arena, changing food placement at each experimental trial (nine experiments with different food placements). The food resources consisted of two *Tenebrio molitor* larvae cut into six pieces arranged in the vertices of two hexagons of the arena, signifying food patches.

We tracked individual ant trajectories from the videos at a temporal resolution of 2 Hz. We analysed several aspects of their movement and classified them into scouts or recruits according to the tracks' average distance to the nest. After, the tracks were discretised in time at 1-second intervals, each position was assigned to the closest node (Y-shaped intersection) in the lattice. We then estimated the turning probabilities, computing whether the ant remained in the same node (c), or turned to the right (r), left (l) or backwards (b), relative to its last position. This was performed separately for scouts and recruits. The relevant parameters, averaged over all experiments, for the recruits are $r \approx 0.0818$, $b \approx 0.0539$, $l \approx 0.0821$ and $c \approx 0.7822$, while for the scouts we found $r \approx 0.1039$, $b \approx 0.0358$ and $l \approx 0.1008$ and $c \approx 0.7595$, with all movement parameters rounded to four decimal places. In Fig. 13.1, we represent some example tracks for both scouts and recruits, to showcase their different movement patterns.

13.3 Master Equation for a Correlated Walk In a Honeycomb Lattice in Unbounded Space

To model the correlated movement of the ants on the non-Bravais honeycomb lattice, we use the theory of lattice random walk with internal states [12, 19]. The model requires six internal states, representing both the physical site types (spatial heterogeneity) [11, 20] and the direction last travelled (one-step memory) [13, 14]. For the spatial degree of freedom, we follow Montroll's procedure of embedding the honeycomb lattice into the $\mathbb{Z}^2$ square lattice [12], where each unit cell contains two site types ($\triangleright$ and $\triangleleft$), referred to henceforth as positive ($\mathfrak{p}$) and negative ($\mathfrak{n}$)

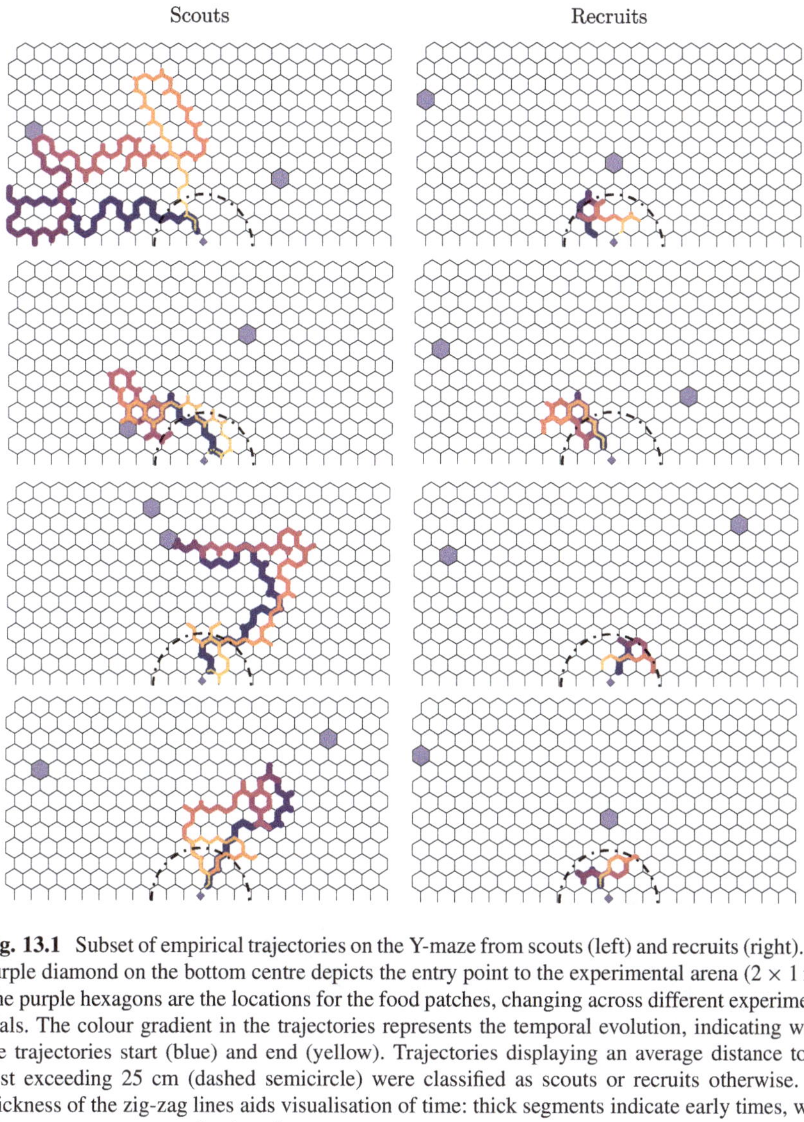

Fig. 13.1 Subset of empirical trajectories on the Y-maze from scouts (left) and recruits (right). The purple diamond on the bottom centre depicts the entry point to the experimental arena ($2 \times 1\ \text{m}^2$). The purple hexagons are the locations for the food patches, changing across different experimental trials. The colour gradient in the trajectories represents the temporal evolution, indicating where the trajectories start (blue) and end (yellow). Trajectories displaying an average distance to the nest exceeding 25 cm (dashed semicircle) were classified as scouts or recruits otherwise. The thickness of the zig-zag lines aids visualisation of time: thick segments indicate early times, while thin segments correspond to late times

sites. As described in Sect. 13.2, the movement parameters of the model represent the probability of making a right turn (r) a left turn (l) or a backtrack (b). Since the direction previously travelled must be known to determine in which direction the walker will turn, each physical site is then decomposed into three internal states representing the movement direction last travelled. We also allow the walker at each time step to remain on a site with probability c, such that $l + r + b + c = 1$. This construction allows to express the non-Markov movement steps on the honeycomb lattice as a higher-order Markov chain [21].

More specifically, we describe the one-step non-Markov dynamics via the coupled master equation

$$
\begin{aligned}
P(\boldsymbol{n},1,t+1) &= bP(n_1,n_2,4,t)+lP(n_1,n_2,5,t)+rP(n_1,n_2,6,t)+cP(\boldsymbol{n},1,t),\\
P(\boldsymbol{n},2,t+1) &= rP(n_1,n_2+1,4,t)+bP(n_1,n_2+1,5,t)\\
&\quad +lP(n_1,n_2+1,6,t)+cP(\boldsymbol{n},2,t),\\
P(\boldsymbol{n},3,t+1) &= lP(n_1-1,n_2,4,t)+rP(n_1-1,n_2,5,t)+bP(n_1-1,n_2,6,t)\\
&\quad +cP(\boldsymbol{n},3,t),\\
P(\boldsymbol{n},4,t+1) &= bP(n_1,n_2,1,t)+lP(n_1,n_2,2,t)+rP(n_1,n_2,3,t)+cP(\boldsymbol{n},4,t),\\
P(\boldsymbol{n},5,t+1) &= rP(n_1,n_2-1,1,t)+bP(n_1,n_2-1,2,t)+lP(n_1,n_2-1,3,t)\\
&\quad +cP(\boldsymbol{n},5,t),\\
P(\boldsymbol{n},6,t+1) &= lP(n_1+1,n_2,1,t)+rP(n_1+1,n_2,2,t)+bP(n_1+1,n_2,3,t)\\
&\quad +cP(\boldsymbol{n},6,t),
\end{aligned}
\tag{13.1}
$$

where the notation $P(\boldsymbol{n},m,t) = P(n_1,n_2,m,t)$ denotes the probability at time t of occupying internal state m at site $\boldsymbol{n} = (n_1,n_2)$, and each timestep represents a 1-second interval. States $m \in \{1,3\}$ represent the movement directions into positive states, while states $m \in \{4,6\}$ represent the movement directions into negative states (see Fig. 13.2 for pictorial representation).

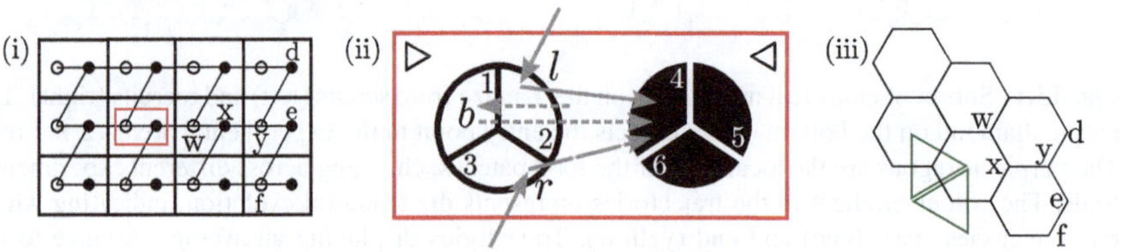

Fig. 13.2 (Colour online). A schematic showing the theoretical domain of interest. (i) The Montroll [12] embedding of the honeycomb lattice into the $\mathbb{Z}^2$ lattice, whereby a visual comparison between (i) and (iii) confirms that one requires one more unit cell along each dimension in the integer lattice (i) than the number of hexagons in that direction in the hexagonal lattice (iii). Panel (ii) details the decomposition of the positive (▷) and negative (◁) sites into their memory components and represents a 'blow-up' of the red box in plot (i). We show, via the arrows, the possibilities to enter state 4 (with sojourns omitted for visual clarity). These are taking a left turn from state 2 (having previously lowered the n_2 coordinate) a right turn from state 3 (having previously raised the n_2 coordinate) or making a backtrack from state 1 (having previously moved from negative to positive site in the unit cell). Subfigures (i) and (iii) are in the style of Fig. 2 of ref. [12]

Equation (13.1) is highly general and models a wide range of movement behaviour. The diffusive limiting case treated in [11, 12, 20] can be recovered by taking $r = l = b = q/3$ $(0 < q \leq 1)$. In contrast, if $r = 1$ or $l = 1$, the motion is deterministic whereby the walker traverses the perimeter of one hexagon in a prescribed direction for all times.

To seek the solution of Eq. (13.1), it is more convenient to re-write the coupled scalar equation in matrix form as

$$\begin{aligned} \boldsymbol{P}(n_1, n_2, t+1) = \mathbb{A} \cdot \boldsymbol{P}(n_1 - 1, n_2, t) + \mathbb{B} \cdot \boldsymbol{P}(n_1 + 1, n_2, t) \\ +\mathbb{C} \cdot \boldsymbol{P}(n_1, n_2 - 1, t) \\ +\mathbb{D} \cdot \boldsymbol{P}(n_1, n_2 + 1, t) + \mathbb{E} \cdot \boldsymbol{P}(n_1, n_2, t), \end{aligned} \tag{13.2}$$

where here, and elsewhere, capitalised bold symbols denote column vectors and where the matrices containing the transition probabilities, viz. $\mathbb{A}, \mathbb{B} \ldots, \mathbb{E}$, are defined in Appendix 1. We introduce the initial condition $\boldsymbol{P}(n_1, n_2, 0) = \delta_{\boldsymbol{n},\boldsymbol{n_0}} \boldsymbol{U}_{m_0}$, which states that the walkers are intitally localised at some site $\boldsymbol{n}$, with the probability of beginning their trajectory in some internal state m distributed by $\boldsymbol{U}_{m_0}$. In our case, the elements of $\boldsymbol{U}_{m_0}$ are dependent on whether the walker starts in a positive or negative state as well as in which direction the walker is assumed to have entered that site from. With little information on the direction travelled at $t = 0$ within the nest, we assume the walker entered via a right, left or back turn with equal probability leading to $\boldsymbol{U}_{\mathrm{p}} = [1/3, 1/3, 1/3, 0, 0, 0]^{\mathsf{T}}$ if the walker started in a positive state and $\boldsymbol{U}_{\mathrm{n}} = [0, 0, 0, 1/3, 1/3, 1/3]^{\mathsf{T}}$ otherwise.

By performing the Fourier transform, $\widehat{f}(\xi) = \sum_{n \in \mathbb{Z}} \mathrm{e}^{i\xi n} f(n)$, and the z-transform, $\widetilde{f}(z) = \sum_{t=0}^{\infty} z^t f(t)$, on Eq. (13.2) we find

$$\widehat{\widetilde{\boldsymbol{P}}}_{\boldsymbol{n_0}, m_0}(\xi_1, \xi_2, z) = \mathrm{e}^{i[\xi_1 n_{01} + \xi_2 n_{02}]} \left[\mathbb{I} - z\boldsymbol{\lambda}(\xi_1, \xi_2)\right]^{-1} \cdot \boldsymbol{U}_{m_0}, \tag{13.3}$$

where $\boldsymbol{\xi} = (\xi_1, \xi_2)$ are the Fourier components along each Cartesian axis and

$$\boldsymbol{\lambda}(\xi_1, \xi_2) = \begin{array}{c} \\ \nwarrow \\ \swarrow \\ \longrightarrow \\ \searrow \\ \nearrow \\ \longleftarrow \end{array} \begin{array}{c} \begin{array}{cccccc} \nwarrow & \swarrow & \longrightarrow & \searrow & \nearrow & \longleftarrow \end{array} \\ \begin{bmatrix} c & 0 & 0 & b & l & r \\ 0 & c & 0 & r\mathrm{e}^{-i\xi_2} & b\mathrm{e}^{-i\xi_2} & l\mathrm{e}^{-i\xi_2} \\ 0 & 0 & c & l\mathrm{e}^{i\xi_1} & r\mathrm{e}^{i\xi_1} & b\mathrm{e}^{i\xi_1} \\ b & l & r & c & 0 & 0 \\ r\mathrm{e}^{i\xi_2} & b\mathrm{e}^{i\xi_2} & l\mathrm{e}^{i\xi_2} & 0 & c & 0 \\ l\mathrm{e}^{-i\xi_1} & r\mathrm{e}^{-i\xi_1} & b\mathrm{e}^{-i\xi_1} & 0 & 0 & c \end{bmatrix} \end{array}. \tag{13.4}$$

The arrows around the matrix $\boldsymbol{\lambda}(\xi_1, \xi_2)$ represent the movement on the honeycomb lattice encoded by the element in the matrix where $\boldsymbol{\lambda}(\xi_1, \xi_2)_{i,j}$ governs the movement from element j to element i. To illustrate, consider the element $\boldsymbol{\lambda}(\xi_1, \xi_2)_{5,2}$. To move from state 2 to state 5, one must hop from a positive state to the negative state in the unit cell above via a backtracking from the previous movement direction. This corresponds to a movement in the South-West direction followed by a movement back in the North-East direction as may be evinced by the arrows surrounding Eq. (13.4) and the schematic in Fig. 13.2. We note that the choice of the 'positive' Fourier transform, $\widehat{f}(\xi) = \sum_{n\in\mathbb{Z}} e^{i\xi n} f(n)$, leads to exponents in the structure function, which reflects the direction of movement. For example, an element $pe^{i\xi_1}$ represents the possibility of a step to the right with probability p [22]. An alternative definition of the Fourier transform with negative exponent may be taken instead [23], which leads to an equivalent representation of the dynamics as Eq. (13.3), whereby $(\xi_1, \xi_2) \to (-\xi_1, -\xi_2)$.

From Eq. (13.3), we take the inverse Fourier and z-transforms to find the unbounded propagator or lattice Green's function

$$\boldsymbol{P}_{\boldsymbol{n}_0,m_0}(n_1, n_2, t) = \frac{1}{4\pi^2} \int_{-\pi}^{\pi} \int_{-\pi}^{\pi} e^{-i[\xi_1(n_1-n_{01})+\xi_2(n_2-n_{02})]} \boldsymbol{\lambda}(\xi_1, \xi_2)^t \cdot \boldsymbol{U}_{m_0} d\xi_1 d\xi_2, \tag{13.5}$$

which represents a 6×1 column vector. The probability of occupying the positive state in (n_1, n_2) is $P^{(\mathfrak{p})}_{\boldsymbol{n}_0,m_0}(n_1, n_2, t) = \sum_{i=1}^{3} \boldsymbol{P}_{\boldsymbol{n}_0,m_0}(n_1, n_2, t)_i$, equivalently $P^{(\mathfrak{p})}_{\boldsymbol{n}_0,m_0}(n_1, n_2, t) = \sum_{i=1}^{3} P_{\boldsymbol{n}_0,m_0}(n_1, n_2, i, t)$, while for the corresponding negative state, one takes $P^{(\mathfrak{n})}_{\boldsymbol{n}_0,m_0}(n_1, n_2, t) = \sum_{i=4}^{6} \boldsymbol{P}_{\boldsymbol{n}_0,m_0}(n_1, n_2, t)_i$.

13.4 Coverage Metrics

There are two fundamental quantities to elucidate how efficiently a random walker explores the space over time: the mean-squared displacement (MSD), $\Delta\boldsymbol{n}(t)$ and the mean number of distinct visited sites (MNDVS) up to time t, $S(t)$. We study here both quantities for scouts and recruits on the unbounded lattice.

13.4.1 Mean-Squared Displacement

We begin by defining displacement on the honeycomb lattice. Although other definitions have been proposed [24], the most natural definition of displacement on networks is simply the shortest path [25]. Here, as we embed the honeycomb lattice into the square lattice, we define displacement as the shortest path on the square lattice. While this definition provides an underestimation of the shortest path on the honeycomb lattice, it allows for exact calculations.

To calculate the MSD, we begin by assuming without loss of generality that the walker starts in the positive site, $P(\boldsymbol{n}, 0) = \delta_{n_1,0}\delta_{n_2,0}\delta_{m_0,\mathfrak{p}}$. Using this initial condition in Eq. (13.3) leads to

$$\widehat{\widetilde{\boldsymbol{P}}}_{\boldsymbol{n}_0,\mathfrak{p}}(\xi_1, \xi_2, z) = [\mathbb{I} - z\boldsymbol{\lambda}(\xi_1, \xi_2)]^{-1} \cdot \boldsymbol{U}_{\mathfrak{p}}. \tag{13.6}$$

From Eq. (13.6), we find the MSD via the second derivative of the propagator in Fourier space as

$$\Delta \boldsymbol{n}(t) = -\mathcal{Z}^{-1}\left\{\sum_{i=1}^{2}\frac{\partial^2}{\partial \xi_i^2}\widehat{\widetilde{P}}^{(\mathrm{p})}_{\boldsymbol{n}_0,\mathrm{p}}(\xi_1,\xi_2,z)+\sum_{i=1}^{2}\frac{\partial^2}{\partial \xi_i^2}\widehat{\widetilde{P}}^{(\mathrm{n})}_{\boldsymbol{n}_0,\mathrm{p}}(\xi_1,\xi_2,z)\right\}\Bigg|_{(\xi_1,\xi_2)=(0,0)}, \tag{13.7}$$

where $\mathcal{Z}^{-1}$ denotes the inverse z-transform. After some algebra, we find

$$\widetilde{\Delta \boldsymbol{n}}(z) = \frac{2(c-1)z + 2(c-1)(b-3c)z^2 + \delta z^3 + \gamma z^4}{3(z-1)^2\,(z(2c-1)-1)\left(1+\alpha z+\beta z^2\right)}, \tag{13.8}$$

where

$$\begin{aligned}
\alpha &= 2-4c-3(l+r), \quad \beta = 1+4c^2-c\,[4-6(l+r)]+3\left[l^2+(r-1)(l+r)\right],\\
\delta &= 2(c-1)\left[4c^2-(l-1)^2-(l-2)r-r^2\right],\\
\gamma &= 2(c-1)\left[3r-1-4c^2(c+b)+4c\left[(l-1)^2+(l-2)r+r^2\right]-3(r^2-l(c+b))\right].
\end{aligned} \tag{13.9}$$

The complex poles of Eq. (13.8) can be found exactly, and we are thus able to extract the time-dependent MSD. However the expression is very cumbersome and we refrain from presenting it. Instead, we show analytically some limiting cases and display the experimentally relevant cases in Fig. 13.3a.

We begin with the diffusive limit $r = l = b = (1-c)/3$, whose time-dependent MSD is (see Appendix 2)

$$\Delta \boldsymbol{n}(t) = 9^{-1}\left[4(1-c)t + 1 - (2c-1)^t\right]. \tag{13.10}$$

The first term of Eq. (13.10) displays the linear MSD with the dependence on the sojourn probability c that one expects from diffusive motion. The second time-dependent term, which might be surprising on first inspection, arises because at each time step, one of the three movement options keeps the walker in the same unit cell. Therefore, after one step, the MSD over the embedded lattice (with $c = 0$) is $\Delta \boldsymbol{n}(t) = 2/3$ compared to $\Delta \boldsymbol{n}(t) = 1$ over the real lattice.

Some other limiting cases include the backtracking limit ($b = 1$) and the circular limit ($r = 1$ or $l = 1$). In the former case, we find $\Delta \boldsymbol{n}(t) = 3^{-1}\left[1+(-1)^t\right]$, which represents a walker which hops between two sites for all time. In the latter case, the hexagonal structure of the honeycomb lattice leads to circular-like motion, whereby the walker traverses the perimeter of one hexagon for all time. Hence, the MSD is a piecewise linear function, $\Delta \boldsymbol{n}(t) = 9^{-1}\left[9-(-1)^t-8\cos\left(\frac{\pi t}{3}\right)\right]$ which periodically returns to zero with period 6. In contrast, in the so-called directed $r = l = 1/2$ case, we find $\Delta \boldsymbol{n}(t) = 9^{-1}\left[2^{(4-t)}-(-1)^t+3(4t-5)\right]$, with super-linear dependence at short times before diffusive spread quickly takes over.

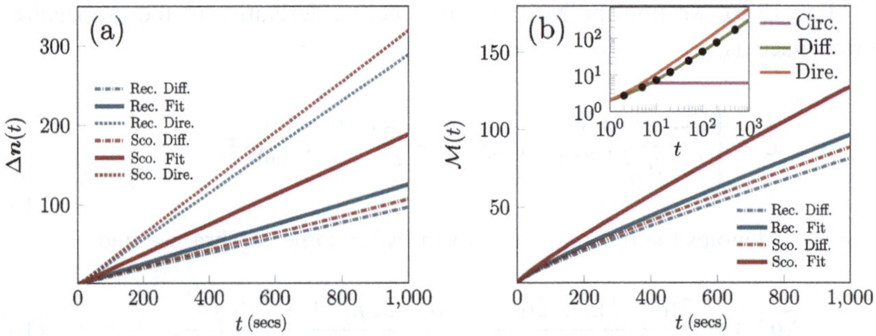

Fig. 13.3 A comparison between the scout and recruit spreading statistics over the infinite honeycomb lattice with the mean-squared displacement in (**a**) and the mean number of visited sites in (**b**). The estimated model parameters from data (i.e. r, l, b and c) for both scouts and recruiters are given explicitly in Sect. 13.2. We display the curves with these parameters with the label 'Fit' in the legend and compare them against the diffusive and directed limits. We take the sojourn probability c to be $c \approx 0.7594$ (scouts) and $c \approx 0.7821$ (recruits), leading to $b = 0$, and $r = l = (1 - c)/2$ and $r = b = l = (1 - c)/3$ for the directed and diffusive cases, respectively. In the inset of panel (**b**), we plot three limiting cases (with dimensionless time), each with $c = 0$. The black dots show excellent agreement with the results obtained iteratively in ref. [20] for the diffusive walk, while the zero gradient past $t = 6$ in the circular case is to be expected since the circular walker will repeat its journey around one hexagon for all time

13.4.2 Mean Number of Distinct Visited Sites

An alternate measure of exploration efficiency is the MNDVS, which has attracted lots of attention in the diffusive limit and has been studied on lattices with different coordination numbers [12, 20]. Somewhat surprisingly, this quantity has been studied to a much lesser extent when the movement is persistent with some results for the square lattice appearing only recently [13].

Currently, we define a site as visited if any of its internal states have been occupied. More precisely, if for example, state 1 of site $\boldsymbol{n}$ is visited, then the positive site is counted, and any visits to states 2 or 3 will not increase the count. On the other hand, if state 4 is visited, then the negative site in $\boldsymbol{n}$ is counted. As such, unlike the MSD, the MNDVS is unaffected by the inclusion of the spatial internal degrees of freedom.

It is well known that the MNDVS is related to the first-passage probability via a summation of the first-passage over all states up to some time t [26, 27], which in our case is written as

$$\mathcal{M}(t) = \sum_{\boldsymbol{n}} \sum_{j \in \{\mathrm{p},\mathrm{n}\}} \sum_{t'=1}^{t} F^{(j)}_{\boldsymbol{n_0},m_0}(\boldsymbol{n}, t'). \tag{13.11}$$

Via the celebrated renewal equation, the first-passage probability is known to be related to a simpler quantity, the occupation probability [27, 28]. In the case of the correlated random walk, if the condition $\widetilde{P}^{(\mathfrak{n})}_{\boldsymbol{n},\mathfrak{n}}(\boldsymbol{n}, z) = \widetilde{P}^{(\mathfrak{p})}_{\boldsymbol{0},\mathfrak{p}}(\boldsymbol{0}, z)$ is met, then the renewal equation reduces to the Markovian case [13, 14]. Here this condition is satisfied allowing us to express the first-passage to a positive site as $\widetilde{F}^{(\mathfrak{p})}_{\boldsymbol{n_0},m_0}(\boldsymbol{n}, z) = \widetilde{P}^{(\mathfrak{p})}_{\boldsymbol{n_0},m_0}(\boldsymbol{n}, z)/\widetilde{P}^{(\mathfrak{p})}_{\boldsymbol{n},\mathfrak{p}}(\boldsymbol{n}, z)$. Passing this relation into the generating function of Eq. (13.11) and again making the harmless assumption that the walker started in a positive state, we find

$$\widetilde{\mathcal{M}}(z) = \frac{1}{1-z}\sum_{\boldsymbol{n}}\left[\frac{\widetilde{P}^{(\mathfrak{p})}_{\boldsymbol{n_0},\mathfrak{p}}(\boldsymbol{n}, z)}{\widetilde{P}^{(\mathfrak{p})}_{\boldsymbol{n},\mathfrak{p}}(\boldsymbol{n}, z)} + \frac{\widetilde{P}^{(\mathfrak{n})}_{\boldsymbol{n_0},\mathfrak{p}}(\boldsymbol{n}, z)}{\widetilde{P}^{(\mathfrak{n})}_{\boldsymbol{n},\mathfrak{n}}(\boldsymbol{n}, z)}\right], \tag{13.12}$$

where if a negative site was chosen, one would replace the $\boldsymbol{n_0}, \mathfrak{p}$ in each numerator with $\boldsymbol{n_0}, \mathfrak{n}$. The equivalent expression for the cases without internal states has been presented in ref. [26].

Equation (13.12) represents the general formulation of the MNDVS for multi-state Markov walks, valid for both unbounded and bounded domains. If the aforementioned condition is met ($\widetilde{P}^{(\mathfrak{n})}_{\boldsymbol{n},\mathfrak{n}}(\boldsymbol{n}, z) = \widetilde{P}^{(\mathfrak{p})}_{\boldsymbol{0},\mathfrak{p}}(\boldsymbol{0}, z)$), i.e. the probability of a walker occupying its initial site is invariant to the location and site type of the initial condition, we may simplify Eq. (13.12) further to

$$\begin{aligned}\widetilde{\mathcal{M}}(z) &= \frac{1}{(1-z)\widetilde{P}^{(\mathfrak{p})}_{\boldsymbol{0},\mathfrak{p}}(\boldsymbol{0}, z)}\sum_{\boldsymbol{n}}\left[\widetilde{P}^{(\mathfrak{p})}_{\boldsymbol{n_0},\mathfrak{p}}(\boldsymbol{n}, z) + \widetilde{P}^{(\mathfrak{n})}_{\boldsymbol{n_0},\mathfrak{p}}(\boldsymbol{n}, z)\right]\\ &= \frac{1}{(1-z)^2\widetilde{P}^{(\mathfrak{p})}_{\boldsymbol{0},\mathfrak{p}}(\boldsymbol{0}, z)},\end{aligned} \tag{13.13}$$

where we have used, in the second line, the property that the sum of the occupation probability over all the sites and states equals unity to recover the well-known form of the MNDVS [27, 29]. For non-isotropic lattices, e.g. in the presence of reflecting domains, one needs to employ Eq. (13.12) contrary to what is erroneously implied in ref. [23].

Upon substitution of the generating function of Eq. (13.5) into Eq. (13.13) and performing some algebra, one arrives at

$$\mathcal{M}(t) = \frac{2\pi}{i}\oint_{|z|<1}\frac{1}{(1-z)^2}\left[\sum_{j=1}^{3}\left(\int_{-\pi}^{\pi}\int_{-\pi}^{\pi}[\mathbb{I} - z\boldsymbol{\lambda}(\xi_1, \xi_2)]^{-1}\cdot \boldsymbol{U}_{\mathfrak{p}}\mathrm{d}\xi_1\mathrm{d}\xi_2\right)_j\right]^{-1}\mathrm{d}z. \tag{13.14}$$

While evaluating Eq. (13.14) analytically is a formidable task, it may be calculated straightforwardly via numerical procedures. To do so, the inner integral is computed via a two-dimensional cubature scheme [30], the output of which is then fed into a numerical contour scheme, utilising standard fast-Fourier techniques to compute the contour integral [31] .

In Fig. 13.3, we plot the MSD and MNDVS for recruits and scouts, with comparison to diffusive and directed spread. It is seen that the scouts explore the space more efficiently than the recruits, and both behavioural types explore quicker than a diffusive walker. These results align with expectations, as animals with frontal and bilateral sensory systems are known to exhibit more persistent directional movement compared to pure random walkers [32, 33]. Similarly, scouts are expected to be more effective explorers than recruits, as the latter play a more critical role during food exploitation after scouts make initial patch discoveries.

13.5 Search of $\mathcal{N}$ Distinct Walkers

We first present general theoretical findings in Sects. 13.5.1 and 13.5.1.1 pertaining to search with $\mathcal{N}$ independent random walkers, while in Sect. 13.5.2, we consider model predictions pertaining to the ant experiment. Specifically, we consider the fFPT, which is the probability that the first walker in the population reaches any target in the set $\mathcal{T}$ for the first time at time t, in a heterogeneous population with individuals displaying different movement characteristics. This theoretical development for the fFPT is particularly relevant now given the recent surge in closed-form expressions for the first-passage and splitting probability generating function in a range of Markovian [23, 35, 36] and non-Markovian [14] walks. From the biological perspective, while in general, ants share information in the food gathering process [37], the application of the fFPT is pertinent to our experiments as recent work on *A. senilis* has elucidated that during the exploration phase, i.e. before any food is found, individual ants move largely independent of each other [6], justifying the assumption of independent walkers.

13.5.1 Theory of First First-Passage Processes in a Heterogeneous Population with Multiple Targets

The probability that none of $\mathcal{N}$ independent random walkers have reached any site in the set of targets $\mathcal{T} = \{s_1, \ldots, s_{\mathcal{M}}\}$ is expressed via the survival probability of the entire population $\mathcal{S}_{\boldsymbol{n}_0}(t; \mathcal{N}) = \prod_{j=1}^{\mathcal{N}} \mathcal{S}_{\boldsymbol{n}_{0_j}}^{(j)}(t)$ (see e.g. refs. [38, 39] and references therein), where $\mathcal{S}_{\boldsymbol{n}_{0_j}}^{(j)}(t)$ is the survival probability of walker j at time t given it was located at $\boldsymbol{n}_{0_j}$ at $t = 0$. When the movement statistics of each walker are identical and the walkers start from the same initial position, which represents the nest in our case, one has $\mathcal{S}_{\boldsymbol{n}_0}(t; \mathcal{N}) = \mathcal{S}_{\boldsymbol{n}_0}(t)^{\mathcal{N}}$. In continuous time, the direct relation of the survival probability and the first-passage probability to any of the targets in the set $\mathcal{T}$, namely $F_{\boldsymbol{n}_0}(\mathcal{T}, t) = -\frac{\mathrm{d}}{\mathrm{d}t}\mathcal{S}_{\boldsymbol{n}_0}(t)$, leads to the well-known expression [15]

$$F_{\boldsymbol{n}_0}(\mathcal{T}, t; \{1, \mathcal{N}\}) = \mathcal{N} F_{\boldsymbol{n}_0}(\mathcal{T}, t)\mathcal{S}_{\boldsymbol{n}_0}(t)^{\mathcal{N}-1}, \tag{13.15}$$

which represents the fFPT time of $\mathcal{N}$ independent and identical random walkers, all starting from $\boldsymbol{n}_0$. Note that in continuous time, Eq. (13.15) implies that the probability of more than one walker reaching the target for the first time at identical times is exactly zero. In discrete time, this is no longer the case since events of simultaneous spatial coincidence at the target of more than one walker (for the first time) are possible.

To allow for such a situation, we modify Eq. (13.15) to

$$F_{\boldsymbol{n}_0}(\mathcal{T}, t; \{1, \mathcal{N}\}) = \sum_{k=1}^{\mathcal{N}} \binom{\mathcal{N}}{k} F_{\boldsymbol{n}_0}(\mathcal{T}, t)^k \mathcal{S}_{\boldsymbol{n}_0}(t)^{\mathcal{N}-k}, \tag{13.16}$$

where $\binom{n}{r}$ is the binomial coefficient. Equation (13.16) accounts for two distinct first-passage events: (i) the target has been reached for the first time by exactly one walker and none of the others, and (ii) more than one walker reaches the target for the first time. The event (i) is accounted for by the $k = 1$ term of the summation, which represents the case in Eq. (13.15), while the contribution of the $k = 2, 3, \ldots, \mathcal{N}$ terms quantifies the probability of event (ii). For a given $k \geq 2$, there are k walkers among the $\mathcal{N}$ that may reach the target at the same time. As we do not know which member(s) of the population reached the target, the binomial coefficient, multiplying each element of the series, takes into account all the possible ways to pick k walkers among the $\mathcal{N}$. The independence of the events then demands to take the product of the probability for k first-passage events with the probability that the remaining $\mathcal{N} - k$ walkers have not reached the target. Note that while outcome (ii) encompasses a larger range of statistically realisable events, it is generally much rarer than (i). However, under certain conditions, event (ii) may become more frequent. For instance, with a large number of persistent walkers, the probability that more than one walker occupies lattice sites at equivalent distance from the initial condition at the same time is non-negligible.

In the case where the search involves more than one target ($\mathcal{n} > 1$), we extend the theory to account for the splitting probability $F_{\boldsymbol{n}_0}(\boldsymbol{s}_i | \boldsymbol{s}_{\{1,\ldots,\mathcal{n}\}-\{i\}}, t)$, that is, the probability of a first first-passage event at target $\boldsymbol{s}_i$ conditioned on no member of the population reaching any other target in the set $\boldsymbol{s}_{\{1,\ldots,\mathcal{n}\}-\{i\}}$ in the times up to and including t. In such a scenario, one decomposes the event (ii) into two mutually exclusive first-passage events: (iia) more than one walker simultaneously reach target $\boldsymbol{s}_i$ and (iib) one or more walkers reach target $\boldsymbol{s}_i$ while one or more walkers reach one of the remaining sites in the target set $\boldsymbol{s}_{\{1,\ldots,\mathcal{n}\}-\{i\}}$. For mathematical convenience, we take the event (iia) as a success and (iib) as a failure, and we leave the derivation of a general formalism that accounts also for events (iib) to a future publication.

To derive the expression for the splitting probability, we observe that by construction, the survival probability is dependent on a walker having not reached any target in the set. Since each successful walker must be at $\boldsymbol{s}_i$, a replacement of the first passage in Eq. (13.16) with the single walker splitting probability to $\boldsymbol{s}_i$ provides the required extension:

$$F_{\boldsymbol{n}_0}(\boldsymbol{s}_i|\boldsymbol{s}_{\{1,\ldots,\mathcal{n}\}-\{i\}}, t; \{1, \mathcal{N}\}) = \sum_{k=1}^{\mathcal{N}} \binom{\mathcal{N}}{k} F_{\boldsymbol{n}_0}(\boldsymbol{s}_i|\boldsymbol{s}_{\{1,\ldots,\mathcal{n}\}-\{i\}}, t)^k \mathcal{S}_{\boldsymbol{n}_0}(t)^{\mathcal{N}-k}. \tag{13.17}$$

Equation (13.17) represents the time-dependent first-splitting probability for one or more walkers among $\mathcal{N}$, starting at $\boldsymbol{n}_0$, to reach target $\boldsymbol{s}_i$ before any other in the set of $\mathcal{n}$ targets. As the event (iib) is excluded from all possible outcomes making up $F_{\boldsymbol{n}_0}(\boldsymbol{s}_i|\boldsymbol{s}_{\{1,\ldots,\mathcal{n}\}-\{i\}}, t; \{1, \mathcal{N}\})$, one needs to keep in mind that for $\mathcal{N} \neq 1$, $\sum_{i=1}^{\mathcal{n}} F_{\boldsymbol{n}_0}(\boldsymbol{s}_i|\boldsymbol{s}_{\{1,\ldots,\mathcal{n}\}-\{i\}}, t; \{1, \mathcal{N}\}) \neq F_{\boldsymbol{n}_0}(\mathcal{T}, t; \{1, \mathcal{N}\})$.

We now further extend the notion of search for a heterogeneous population to allow for the different movement patterns and/or different initial conditions of the individuals. We do so by partitioning the population into $h \leq \mathcal{N}$ subsets, where each subset is comprised of $\mathcal{N}_h$ walkers with the same movement statistics such that $\sum_h \mathcal{N}_h = \mathcal{N}$ and $\mathcal{N}_1 \geq \mathcal{N}_2 \geq \ldots \geq \mathcal{N}_h$. By considering all permutations of possible successful search outcomes, the generalisation of Eq. (13.16), to a heterogeneous population, denoted by the superscript $\mathfrak{h}$, is given by

$$\begin{aligned} F_{\boldsymbol{n}_0}^{(\mathfrak{h})}(\mathcal{T}, t; \{1, \mathcal{N}\}) &= \sum_{i=1}^{h} \sum_{k_i=1}^{\mathcal{N}_i} \binom{\mathcal{N}_i}{k_i} F_{\boldsymbol{n}_{0_i}}^{(i)}(\mathcal{T}, t)^{k_i} \prod_{j=1}^{h} \mathcal{S}_{\boldsymbol{n}_{0_j}}^{(j)}(t)^{\mathcal{N}_j - \delta_{i,j} k_j} \\ &+ \sum_{k_1=1}^{\mathcal{N}_1} \cdots \sum_{k_h=1}^{\mathcal{N}_h} \prod_{i=1}^{h} \binom{\mathcal{N}_i}{k_i} F_{\boldsymbol{n}_{0_i}}^{(i)}(\mathcal{T}, t)^{k_i} \mathcal{S}_{\boldsymbol{n}_{0_i}}^{(i)}(t)^{\mathcal{N}_i - k_i}, \end{aligned} \tag{13.18}$$

which reduces to Eq. (13.16) when $\mathcal{N}_1 = \mathcal{N}$ ($\mathcal{N}_i = 0, \forall i > 1$) as we define the nested sum in the second term to be zero in this case. This makes sense physically as the first term represents the case of one or more walkers from the same subset reaching the target at the same time (summed over each subset), while the second term represents the cases where a mixed population of walkers reaches the target for the first time. The splitting probability expression with a heterogeneous population may be derived via an analogous procedure to the homogeneous case, i.e. by replacing the first-passage probabilities in Eq. (13.18) with the splitting probabilities for the type of walkers belonging to the subset population.

The general expression in Eq. (13.18) is important both from a theoretical and an applied perspective. The theoretical advance lies in the ability to study first-passage processes to an arbitrary number of targets for multiple independent walkers, with each individual following their own movement statistics. The applied convenience stems from the computational cost in evaluating $F_{\boldsymbol{n}_0}^{(\mathfrak{h})}(\mathcal{T}, t; \{1, \mathcal{N}\})$: despite its appearance, Eq. (13.18) has time complexity on the same order as the one walker case since one determines the survival probability for each subset h prior to evaluating the full summations.

13.5.1.1 First First-Passage Probability in Bounded Domains

Most studies of the fFPT with $\mathcal{N}$ walkers have focused on large homogeneous populations of walkers searching in unbounded space [15, 39–41]. Here instead, guided by the observations of *A. senilis*, we consider a small population of $\mathcal{N}$ ($\mathcal{N} \leq 10$) individuals with unequal movement statistics. As an illustrative example, we consider a search process in bounded periodic space and study two cases: a population comprising of both strongly persistent and diffusive walkers on the square lattice, and a population of correlated walkers, representing scouts and recruits, on a honeycomb lattice.

For the square lattice case, we take the well-known diffusive periodic propagator (see e.g. refs. [23, 42, 43]) and the corresponding correlated propagator given by Eq. (14) of ref. [14]. For correlated motion on the bounded honeycomb lattice, we introduce a periodic boundary on Eq. (13.5) via standard procedures (see e.g., [42]) to find

$$\widetilde{\boldsymbol{P}}^{(p)}_{\boldsymbol{n}_0,m_0}(\boldsymbol{n},z) = \frac{1}{N^2}\sum_{k_1=0}^{N-1}\sum_{k_2=0}^{N-1} \mathrm{e}^{-\frac{2\pi i}{N}\boldsymbol{k}\cdot[\boldsymbol{n}-\boldsymbol{n}_0]^{\mathsf{T}}}\left[\mathbb{I} - z\boldsymbol{\lambda}\left(\frac{2\pi k_1}{N}, \frac{2\pi k_2}{N}\right)\right]^{-1}\cdot U_{m_0}, \tag{13.19}$$

where the inversion to the time-domain is trivial, and the superscript (p) denotes that the domain is bounded periodically.

From ref. [14], the generating function of the single walker first-passage distribution for the correlated RW is found via

$$\widetilde{F}_{\boldsymbol{n}_0}(\mathcal{T},z) = \sum_{j=1}^{M}\sum_{i=1}^{S}\alpha_{m_{s_i}}\frac{\det[\mathbb{H}^{(i)}(\boldsymbol{n}_0,m_{0_j},z)]}{\det[\mathbb{H}(z)]}, \tag{13.20}$$

and the splitting probability is given by

$$\widetilde{F}_{\boldsymbol{n}_0}(\boldsymbol{s}_{\{i\}}|\boldsymbol{s}_{\{1,\ldots,\mathcal{n}\}-\{i\}},z) = \sum_{j=1}^{M}\alpha_{m_{s_i}}\frac{\det[\mathbb{H}^{(i)}(\boldsymbol{n}_0,m_{0_j},z)]}{\det[\mathbb{H}(z)]}, \tag{13.21}$$

where $\mathbb{H}(z)_{l,k} = \alpha_{m_{s_k}}\widetilde{P}_{s_k,m_{s_k}}(s_l,m_{s_l},z)$, $\mathbb{H}(z)_{k,k} = \alpha_{m_{s_k}}\widetilde{P}_{s_k,m_{s_k}}(s_k,m_{s_k},z)$ and $\mathbb{H}^{(i)}(\boldsymbol{n}_0,m_0,z)$ is the same, but with the ith column replaced with $\mathcal{F}(n_0,m_{0_j},z) = \alpha_{m_{0_j}}\left[\widetilde{P}_{\boldsymbol{n}_0,m_{0_j}}(\boldsymbol{s}_1,m_{s_1},z),\ldots,\widetilde{P}_{\boldsymbol{n}_0,m_{0_j}}(s_{\mathcal{n}},m_{s_{\mathcal{n}}},z)\right]^{\mathsf{T}}$. The notation $\alpha_{m_{0_j}}P_{\boldsymbol{n}_0,m_{0_j}}(\boldsymbol{n},m,t) = \alpha_{m_{0_j}}\boldsymbol{e}_m^{\mathsf{T}}\cdot\boldsymbol{P}_{\boldsymbol{n}_0}(\boldsymbol{n},t)\cdot\boldsymbol{e}_{m_{0_j}}$, is the occupation probability that the walker occupies site $\boldsymbol{n}$ and state m given the initial condition $(\boldsymbol{n}_0,m_{0_j})$, which is weighted by $\alpha_{m_{0_j}}$. Currently, due to the assumption of each direction being equi-probable at time zero $\alpha_i = 1/3$, $\forall i$. Moreover, since each site can be accessed via three distinct internal states, to represent a fully absorbing trap, we set

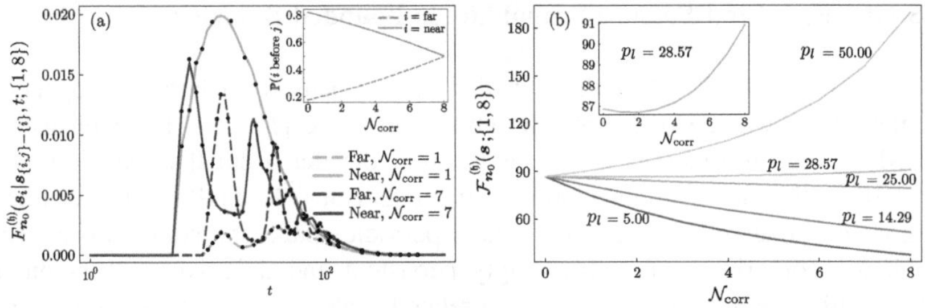

Fig. 13.4 The fFPT for different heterogeneous populations of random walkers on the periodic square lattice. The population comprises $\mathcal{N} = \mathcal{N}_{\text{corr}} + \mathcal{N}_{\text{diff}} = 8$ random walkers. In panel (a), we evaluate Eq. (13.18) with the first-passage terms replaced with the splitting probability for the single persistent walker given by Eq. (13.21). The population starts at $\boldsymbol{n}_0 = (10, 10)$ and is searching for a nearby target at $\boldsymbol{s} = (7, 7)$ and a far away target at $\boldsymbol{s} = (1, 9)$ in a domain size $N_1 \times N_2 = 20 \times 20$. Hence, with the square lattice requiring four internal states to represent the possible previous movement directions, and the system contains two physical target sites, we have $\mathcal{n} = 8$ target states. For the diffusive population, we take $q = 0.95$, while the persistent population has a persistence length $p_l = 20$. The inset shows the overall probability that each target is found before the other (the summation of the splitting probability over time) as a function of $\mathcal{N}_{\text{corr}}$ for the same parameter regime. Panel (b) shows the MfFPT, $\mathcal{F}_{\boldsymbol{n}_0}^{(\mathfrak{h})}(\mathcal{T}; \{1, \mathcal{N}\} = \sum_t t F_{\boldsymbol{n}_0}^{(\mathfrak{h})}(\mathcal{T}, t; \{1, \mathcal{N}\})$, as a function of $\mathcal{N}_{\text{corr}}$ with varying persistent lengths, all chosen past the optimum for this system $p_l \approx 4.76$. Note that the diffusive persistence length is $p_l = 4(3q)^{-1}$ and for the diffusive members of the population, we take $q = 0.95$ ($p_l \approx 1.40$). In this case, we choose a domain size of $N_1 \times N_2 = 14 \times 14$, $\boldsymbol{n}_0 = (7, 7)$ and $\boldsymbol{s} = (14, 14)$ meaning $\mathcal{n} = 4$. Dots represent the result of 10^6 stochastic simulations

the probability of leaving each internal state in the trap site to zero. For example, to place a target in a positive site, absorbing traps in the states 1, 2 and 3 are required. Hence, for all cases presented here, and elsewhere in this chapter, the target set $\mathcal{T}$ comprises sets of three target states for each physical site. Henceforth, for ease of notation, when discussing specific cases, we denote the set of targets with only the physical location(s) at site(s) $\boldsymbol{s}_i$. The splitting probability for the Markovian case for the diffusive subset can be found in Eq. (11) of ref. [34].

We plot in Fig. 13.4 the fFPT for different heterogeneous populations on the square lattice. To present results on the Bravais lattices, it is convenient to introduce the so-called persistence length [16], that is, the average number of steps taken in the same direction between turns. On the square lattice, it is defined as $p_l = [b+2\tau]^{-1}$, where b is the probability of backtracking and 2τ the probability of turning left or right (probability τ in each direction). In Fig. 13.4a we consider a heterogeneous population of eight walkers searching on a domain of 20×20 lattice sites. The domain contains two targets, one close-by to the initial position and the other far away and just off the persistence path, (see caption of Fig. 13.4 for the exact locations). By increasing the proportion of persistent walkers in the population (inset Fig. 13.4a), the overall probability $\mathbb{P}$ of finding the far target increases until it is almost equi-probable as many trajectories head in a straight line out of the initial condition bypassing the closer target.

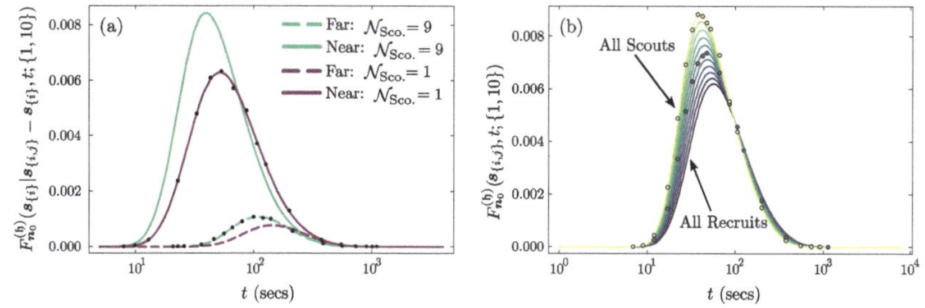

Fig. 13.5 The splitting probability to either site (**a**) and first first-passage probability to both sites (**b**) of an $\mathcal{N} = \mathcal{N}_{\mathrm{Sco.}} + \mathcal{N}_{\mathrm{Rec.}} = 10$ heterogeneous population of scouts and recruits (see Sect. 13.2 for parameter values) in an $N \times N = 12 \times 12$ (288 sites) periodic honeycomb lattice. For all curves, the walkers start in the positive site of $\boldsymbol{n}_0 = (6, 6)$. The target set $\mathcal{T}$ comprises the three internal states in a nearby target in the negative site of $\boldsymbol{s}_1 = (9, 9)$ and the three internal states in the far target in the positive site of $\boldsymbol{s}_2 = (11, 12)$ (see caption of Fig. 13.2 for definition of positive and negative sites). In both panels, lines represent theoretical results (Eq. (13.18) evaluated via Eqs. (13.20) and (13.21), respectively) while dots are the result of 10^6 stochastic simulations, which we omit from some of the curves for visual clarity

In panel (a), one notices clear qualitative differences between the splitting probabilities of the search with different populations. Persistence may lead to oscillatory behaviour in the first-passage distribution of one walker [14], something that is greatly suppressed as the movement parameters approach the diffusive regime. These oscillations, which are not present with diffusive walkers, have low amplitude when one persistent walker is introduced, while they increase in regularity and amplitude as the number of persistent walkers increase.

To gain further insights on the role of persistence in the first-passage processes, we interpret our findings also in terms of the mean fFPT (MfFPT), for which it is known that there is an optimal persistence length p_l, whose value depends on the search parameters, which minimises the global search time on the square lattice [16]. If the persistence parameter greatly surpasses p_l, one may find very high mean first-passage time if the target is off the persistence path. Here in Fig. 13.4b we show persistence lengths starting from just past the optimum for the chosen system ($p_l \approx 4.76$). In such a parameter regime, an increase in the persistence length hinders the search and weakens the effect of adding more persistent walkers until a population of diffusive walkers outperforms a population of correlated walkers. Around the level of persistence which tips the walker beyond the diffusive efficacy (shown presently as $p_l = 28.57$), the MfFPT becomes non-monotonic as a function of the number of correlated walkers in the population as shown in the inset of panel (b).

For the experimentally relevant case, we consider scouts and recruits in a honeycomb lattice, and we plot the first-passage dynamics in Fig. 13.5. In panel (a), we show the splitting probability to a nearby and far target. In such a case, the closest target is always more likely to be found regardless of the proportion of scouts to recruits, which is not only expected to be more energetically efficient, but it also matches the exploratory patterns observed in different ant species [8, 9]. Panel (b) shows the entire fFPT distribution for each ratio of scouts to recruits. Since an increase in the proportion of scouts allows the space to be explored more efficiently, the so-called direct trajectories become more effective causing the mode of the fFPT distribution to shift upwards and towards the left as it becomes more probable to find the target at shorter times.

13.5.2 Experimentally Relevant Search on the Y-Maze Arena

We now consider a model which faithfully represents the domain size and boundaries; food and nest locations; and the number of ants in the system for each experimental realisation. Since the ant population in the arena was dynamic, with ants constantly entering and leaving the nest, we estimated the number of ants by considering how many were exploring the arena at the time when any food target was found (end of exploration phase) as the static number $\mathcal{N}$ for the model. The experimental set-up is modelled as a honeycomb lattice confined in a rectangular space, with hard, 'bouncing' reflecting walls, which causes persistent walkers to change direction at the boundary and continue its motion in the opposite direction [14].

To implement a reflecting boundary condition on the honeycomb lattice, note that although there are positive and negative states occupying the unit cells that lie at the edge of the domain, if the site is not at the corner, only one of the two states will interact with the boundary creating the so-called zig-zag boundary condition [11, 44]. To elucidate, consider Fig. 13.2 (iii) where, although the positive site labelled y, lies in unit cell at the edge of the square lattice, all three movement options are available to the walker leaving the site. In contrast, travelling horizontally East from the negative site labelled e would cause a reflection to occur. In general, for the $\boldsymbol{n} = (N_1, n_2)$ boundary, the walker may only hit the boundary if it leaves a negative site, while at the $\boldsymbol{n} = (n_1, N_2)$ boundary, it is the positive site that interacts with the boundary, meaning at $\boldsymbol{n} = (N_1, N_2)$, both sites are part of the reflecting boundary.

To implement this zig-zag reflecting boundary condition via a fully analytical procedure is a formidable task when the movement is not diffusive. To proceed we use a semi-analytical iterative procedure [14, 45], which has been used to study persistent motion with complicated boundary conditions where the analytical representation of the occupation probability is not available. This procedure requires the decomposition of the movement dynamics into matrices that govern only the jumps from each state i to every other state j, denoted by $\mathbb{M}(i, j)$. In our case, this leads to 36 sparse transition matrices, of size $N_1 N_2 \times N_1 N_2$. More concretely, the matrices that govern movement in the bulk have $1/N_1 N_2$ ($\approx$0.31%) non-zero elements, while those governing movement on the boundary have $1/N_1^2 N_2$ ($\approx$0.022%) or $1/N_1 N_2^2$ ($\approx$0.0135%) non-zero elements. Hence, all matrices benefit from the use of sparse methods leading to efficient computation. See Appendix 3 for a detailed discussion of the transition matrices.

Encoded into $\mathbb{M}(i, j)$ is also the probability of absorption, i.e. the outgoing probabilities of the trap site, i.e. the site in which the first food source is found, sum up to $1 - \rho$, where ρ is the probability of absorption (finding the food) at the target. Experimentally, we estimated ρ by measuring the result of each ant-food interaction. Namely, given an interaction event (i.e. an ant finds a food item), we consider a success (absorption) the ant collecting the food item, or a failure otherwise. Subsequent interactions were considered independent when the ant had walked away from the food for two or more nodes. The sum of successes divided by the number of interactions results in an empirical estimate of $\rho \approx 0.82$.

Once the elements of transition matrices are generated, the master equation for the bounded process can be written as a set of six coupled equations as

$$\boldsymbol{P}(j, t+1) = \sum_{i=1}^{6} \mathbb{M}(i, j) \cdot \boldsymbol{P}(i, t), \quad (\forall j \in \{1, 6\}), \tag{13.22}$$

where $\mathbb{M}(i, j)$ is the probability of jumping from state i to state j. Then via $\mathcal{S}_{\boldsymbol{n}_0}(t) = \sum_{i=1}^{6} \sum_{n=1}^{N_1 N_2} \boldsymbol{P}(i, t)_n$, and $F_{\boldsymbol{n}_0}(\mathcal{T}, t) = \mathcal{S}_{\boldsymbol{n}_0}(t-1) - \mathcal{S}_{\boldsymbol{n}_0}(t)$, where the inclusion of $\rho \neq 0$ ensures $0 \leq \mathcal{S}_{\boldsymbol{n}_0}(t) \leq 1$, one has all the information required to evaluate the fFPT via Eq. (13.18). Once the entire fFPT distribution is known, the MfFPT and mode fFPT time are extracted trivially via $\mathcal{F}_{\boldsymbol{n}_0}(\mathcal{T}; \{1, \mathcal{N}\}) = \sum_t t F_{\boldsymbol{n}_0}(\mathcal{T}, t; \{1, \mathcal{N}\})$ and $\mathfrak{F}_{\boldsymbol{n}_0}(\mathcal{T}; \{1, \mathcal{N}\}) = \arg\max \left[F_{\boldsymbol{n}_0}(\mathcal{T}, t; \{1, \mathcal{N}\})\right]$, respectively.

13.6 Results

We have analysed the full fFPT distribution for homogeneous populations of either scouts or recruits, using the same food placements (see Table 13.1 in Appendix 4 for specific sites) and the number of agents as in the empirical data (see Fig. 13.6). For all distributions, the mean and modal times were lower for the scouts than the recruits. With the exception of experiments 2 and 6, the empirical time of first food item capture (black crosses) falls in a timescale, which may be characterised by the mode of the theoretical distribution, showing that generally, the empirical time of food discovery corresponds to the direct regime of the theoretical distributions. This region, comprising the timescale from the first deviation from zero up to and just past the mode [46], is characterised by the set of trajectories that deviate minimally from paths that connect directly the initial condition to the target. Figure 13.6 also shows that in general, including more walkers to the population leads to a narrower fFPT distribution with sharper tails and shapes that are more symmetric around the mode. In other words, the larger the $\mathcal{N}$, the faster the tails decay, and the smaller the difference between the mean and the mode.

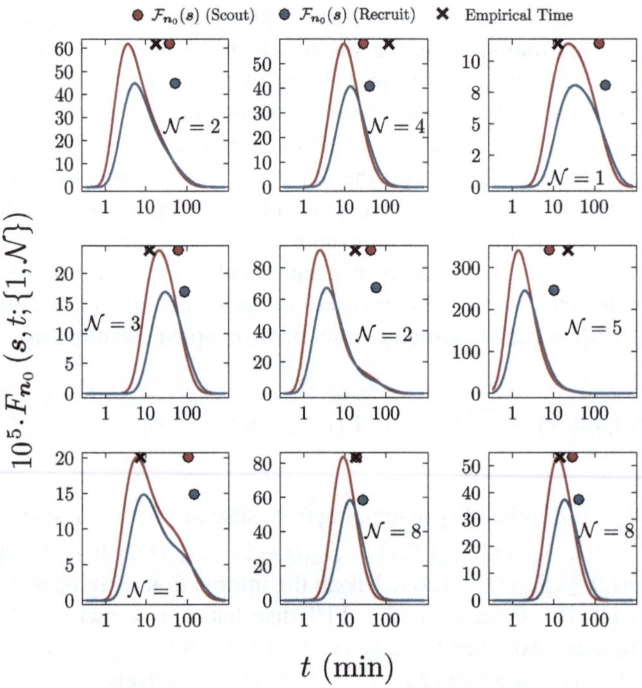

Fig. 13.6 The first first-passage time distributions in Eq. (13.16), evaluated with the output of Eq. (13.22), for a population of scouts (red) and recruits (blue) modelling experiments 1 to 9 in the honeycomb arena with hard reflecting walls. Physical target locations, s, are given in Appendix 4, and the set of targets $\mathcal{T}$ comprises the three internal states within. Results for each experiment are shown in order (from 1 to 9) starting from the top left corner panel and ending at the bottom right corner panel, moving from one panel to the next one by rows. Crosses denote the empirical times for the food item found while red (scouts) and blue (recruits) dots indicate the mean value of the fFPT distributions. The x-axis has been scaled by a division of 60 to give results in minutes. $\mathcal{N}$ shows the number of ants observed in the arena up to the first food item encountered, in each of the 9 experiments

With ant colonies facing scenarios with multiple food sources in nature, we have explored how switching behaviour, e.g. from scout to recruit, may affect search efficiency in a landscape with multiple targets (Fig. 13.7). For each experiment, we assume a homogeneous population comprising $\mathcal{N}$ ants observed experimentally in the laboratory set-up (written explicitly in Fig. 13.6), starting as either scouts or recruits. By assuming that the entire population may modify searching patterns upon finding the first target or when returning back to the nest, we address the question of behavioural switching in two possible foraging strategies; (i) nest-to-patch foraging, i.e. strict central place foraging, in which ants go back to the nest before searching for the next target; and (ii) patch-to-patch foraging, where the ants use the previously discovered patch as a starting point to search for new food targets. The theoretical time for scenario (i) is defined as $\mathfrak{F}_{n_0}(s_2; \{1, \mathcal{N}\}) = \mathfrak{F}_{n_0}(s_1; \{1, \mathcal{N}\}) + \mathfrak{F}_{s_1}(n_0; \{1, \mathcal{N}\}) + \mathfrak{F}_{n_0}(s_2; \{1, \mathcal{N}\})$, with s_1 being the first found

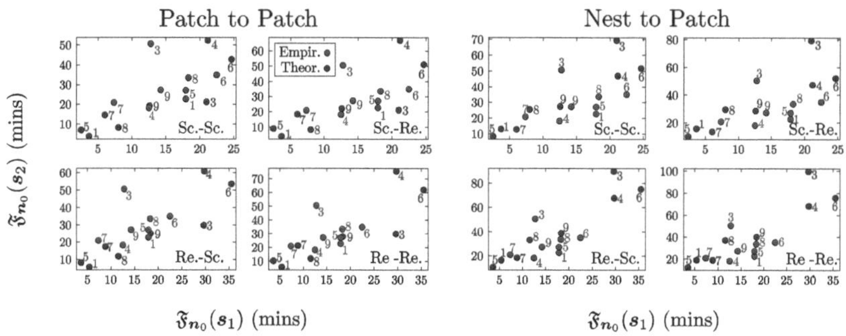

Fig. 13.7 Relationship between the fFPT of the first patch (x-axis) and the second patch (y-axis). We considered two different foraging strategies: using previously found resources as starting points to search for new food targets (Patch to Patch, left panels) and strict central place foraging, wherein ants go back to the nest before searching for subsequent food targets (Nest to Patch, right panels). We allow the entire population to switch between scout (Sc.) and recruit (Re.) upon reaching the first target in the patch-to-patch case or when returning from the nest in the nest-to-patch scenario. As the theoretical values, we show the mode of the fFPT, $\mathfrak{F}_{n_0}(s; \{1, \mathcal{N}\})$ in red, and the first food capture time in blue. The numbers relate to the experimental setup, i.e. target placements and number of ants, given explicitly in Appendix 4 and Fig. 13.6, respectively. Due to the large emprical value in experiment 2, we do not plot this case here

patch, and s_2 the second one, while we define the theoretical time for scenario (ii) as $\mathfrak{F}_{n_0}(s_2; \{1, \mathcal{N}\}) = \mathfrak{F}_{n_0}(s_1; \{1, \mathcal{N}\}) + \mathfrak{F}_{s_1}(s_2; \{1, \mathcal{N}\})$. Note that since the process is bounded by reflecting walls, the first first-passage probability is not symmetric with respect to the initial condition and target, i.e. $F_{n_0}(s, t; \{1, \mathcal{N}\}) \neq F_s(n_0, t; \{1, \mathcal{N}\})$.

To interpret all experiments together, we show the overall trend in Fig. 13.7 where we see that scouts searching for the first patch reduced the time required for finding the targets, i.e. smaller modal fFPT, regardless of the foraging strategy (top panels), while switching from scout to recruit (Sc.-Re.) did not improve search efficiency in any of the set-ups. In agreement with previous results (Figs. 13.4 and 13.6), recruits starting the search resulted in larger modal fFPT irrespective of the context or the behavioural switching (Fig. 13.7, bottom panels). This was particularly evident in some spatial configurations, i.e. set-ups 3, 4 and 6, wherein the fFPT departed significantly from the empirical distribution.

We calculated the Euclidean distance between the mode of the theoretical distribution and the corresponding first capture time in each empirical set-up. The results shown in Fig. 13.8 indicate that *A. senilis* likely uses a central-place foraging strategy (i.e., Nest to Patch), with scouts initially searching for the first patch. The results are similar regardless of whether a behavioural switch from scout to recruit (Sc.-Re.) occurs after a return to the nest following the discovery of the first patch or whether scouts continue the search to the second patch (Sc.-Sc.). Nonetheless, the trend observed in Fig. 13.8 is not statistically supported by pairwise Wilcoxon tests, and the figure highlights a heterogeneous distribution of data points, that is, certain set-ups reveal close alignment between theoretical and experimental behaviours, while others do not. This is consistent with the analyses carried out in Figs. 13.6 and 13.7.

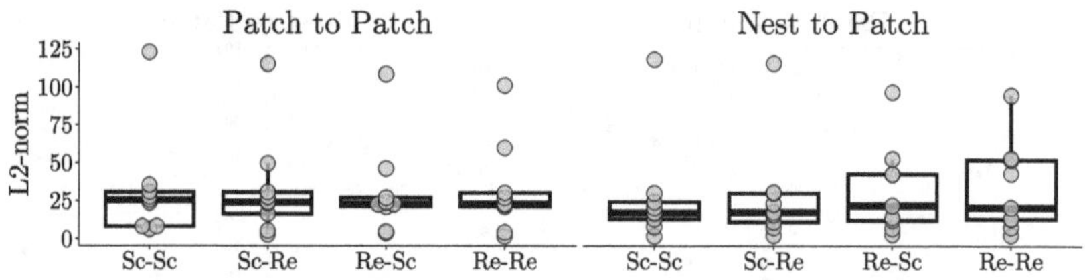

Fig. 13.8 We illustrate the difference between empirical and theoretical data for each combination of foraging strategy and behavioural switch (Fig. 13.7). We measured the Euclidean distance (L2 norm, y-axis) between the mode of the theoretical distribution and the corresponding empirical data point. The results are shown as box-and-whisker plots. The box delimits the interquartile range (IQR), compressed between the first (q1) and the third quartile (q3), with the 50% percentile depicted as a thick line inside the box. The upper whisker is calculated as $\min(\max(\text{data}), q3 + 1.5 \cdot \text{IQR})$, and the lower whisker is calculated as $\max(\min(\text{data}), q1 - 1.5 \cdot \text{IQR})$

13.7 Discussion

When collectively foraging, scouts and recruits are often identified in many ant species. While they are qualitatively acknowledged by their role during foraging, recent studies of the ant *A. senilis* have shown that scouts and recruits are not only located in different positions relative to the nest but also move [9] and behave [8] differently. The dynamics of these two types of ants in group or mass recruitment strategies are often complex, and it is not easy to identify whether these are fixed or switching roles. In this chapter, we studied the macroscopic consequences of locally different movement of both scouts and recruits in terms of their exploratory and search efficiency. In particular, we studied the mean number of sites visited and the mean squared displacement for both ant roles and also explored different proportions of both types of ants in central place and non-central place foraging scenarios. Differences in foraging efficiency allowed us to assess what foraging strategies and combinations and roles of individuals adjusted better to empirical observations.

Our theoretical predictions using the nest-to-patch model aligns more closely with the empirical times than the patch-to-patch model, (Figs. 13.7 and 13.8) providing some evidence that a central place foraging strategy is undertaken by the ants. These results suggest a temporal order in foraging, wherein ants first need to find a patch, convey the information to the rest of the colony on their way back to the nest, e.g. by depositing scent, to start patch exploitation and then restart the search. In this manner, the solution of the patch exploration–exploitation trade-off is a sequential process in time, where the exploration effort and the patch distance to the nest are the main drivers to forage efficiency. However, when there are multiple food patches, ant colonies may parallelise the exploitation as a function of the exploration effort of the colony, i.e. the number of scouts, and the initial positions from where scouting behaviour starts [9].

Moreover, Fig. 13.7 shows that scouts switching to recruit behaviour did not improve the colony performance when searching for multiple patches. In fact, in all exploration (Fig. 13.3) and search (Figs. 13.5 and 13.6) scenarios studied, the scouts always outperformed the recruits. This finding is supported by past theoretical investigations on square lattices showing that introducing some persistence into the motion of a searcher can lower searching times in comparison to a diffusing motion [16]. Since our results show that recruits are statistically close to being diffusive while scouts show persistent motion, one expects the scouts to be more efficient in discovering food patches. Possible future work in this context includes using the results on the honeycomb persistent random walk presented here to determine the optimum level of persistence on the honeycomb lattice to assess the overall efficiency of the scouts' search.

Since all our results point to recruits being less efficient explorers than scouts, we question whether recruits have any role in exploration. If diffusive or sub-diffusive searching by recruits around an area is important in detecting nearby food around already-discovered patches, recruits may have less of a role in explorative tasks associated with food patch discovery. Because of their motion, recruits often end up aggregated, maintaining a minimum group cohesion in certain areas. We hypothesize that this pattern, might be crucial for rapid information transfer and exploitation of known resources. To provide conclusive answers to these questions, further research is needed to explore the roles that different movement patterns and switching behaviours could play in exploration–exploitation trade-offs while foraging.

From a theoretical standpoint, there has been a wealth of studies on the scaling of the MfFPT when the number of walkers is large and homogeneous [15, 39–41]. Here, we have focused on experimentally relevant $\mathcal{N}$ and extended the theory to account for discrete time, the calculation of the splitting probability and also the possibility of a heterogeneous population comprising of walkers with different movement characteristics. As well as addressing the biologically motivated examples, we have utilised the theory to study some theoretical scenarios on the square lattice. We have uncovered (Fig. 13.4) oscillatory splitting probabilities when persistence is introduced into the population and also shown that the number of persistent walkers in the population impacts the likelihood that each target in a multi-target search is found before the other. We have confirmed, in Fig. 13.4b, that similarly to the one-walker case [14, 16], setting the persistence length too high in a multi-walker search may cause an increase in the MfFPT (Fig. 13.4b).

To explain this aspect, consider that the MFPT may be thought of as a measure of the effectiveness of the direct trajectories [47]. With more effective walkers in the population, the chance of a direct trajectory being successful increases, which causes the MFPT to fall at a lower time as the mean becomes less skewed by long meandering trajectories. As a result, the fFPT distribution narrows and becomes more symmetric around the mode. If, on the other hand, a large subset of the population is ineffective, such as those that are highly persistent and do not explore the space efficiently, the tails of the distribution will be very long and the MfFPT will increase. In such extreme cases, where the persistence length is much larger than the domain size, we show that replacing diffusive walkers with persistent walkers in the population actually decreases search efficiency.

We note that while the mean (and global mean) first-passage times [16, 48–50] can be useful in quantifying the overall efficiency of searches, more recent studies have suggested that the MFPT can often over-estimate the most-probable value for the first-passage time [51] and often falls in the region associated with the indirect trajectories [47]. Here, the results of Fig. 13.6 show clearly that the experimental times in which the food is found are associated with the direct trajectories and are more reminiscent of the modal timescale than the mean of the fFPT distribution. This corroborates with what we see experimentally where trajectories in the exploration phase tend to be direct (Fig. 13.1). As exemplified by these trajectories, the ants perform foraging bouts, with looping trajectories of different lengths, which do not meander and explore the entire space. Instead, these trajectories extend from the nest outwards, and then, after an unsuccessful forage, return to the nest.

These spatio-temporal patterns may explain why two of the nine theoretical predictions (exps. 2 and 6) in Fig. 13.6 underestimate the empirical findings and show the empirical time falling in a region of near-zero probability of a first first-passage event. After an unsuccessful foraging bout, the ant may return to, and remain in, the nest before performing another foraging excursion. Experimentally, the clock is not stopped if this occurs. This behavioural trait is not captured by the model, which, by construction, assumes that once a trajectory is started, the walker will search indefinitely in the arena until they find the target site. Therefore, a natural extension to the model would be to include a mechanism for preferential returns to the initial condition, which may be included via a discrete space equivalent to the Smoluchowski-type diffusion equation [52] or via the inclusion of a process of resetting, whose discrete formulation has been recently derived [53, 54].

It is known that after the first food source is found, not only do interactions between ants increase, but the number of ants in the arena also increases [6]. More specifically, it was observed that during the period between the first and second food source being found, the number of ants in the system increased monotonically [6]. Recognising a need to incorporate a varying number of ants into the fFPT calculations when studying foraging strategies, future work may also include the introduction of a dynamic population size into the fFPT, which is currently known only for continuous one-dimensional diffusion [55]. Furthermore, the calculation of the fFPT to the second food item assumes that the excursion from the nest to the first found patch is independent of the excursion from the first patch to the second (or back to the nest), an assumption which is clearly violated in practice. This highlights the need to develop a theory on *sequential* first-passage times, that is, the first-passage probability to a target s_2 conditioned on the trajectory first travelling through a previous target s_1.

In summary, we have presented a theoretical model to quantify the foraging behaviour of the ant species *Aphaenogaster senilis* in laboratory experiments. The model extends known theory on persistent random walks to account for the underlying Y-maze (honeycomb) lattice and on the fFPT of $\mathcal{N}$ independent random walkers, allowing the population to be heterogeneous in behaviour. By identifying microscopic differences between the ant types, we have shown that scouts outperform recruits in both search and exploration. We have also highlighted that the characteristic fFPT time of the foraging ants corresponds to the timescale related to the mode of the fFPT distribution rather than its mean. Real ant foraging is a much more complex process with varying number of ants and sequential first-passage times, which challenges the current model and opens new research pathways and model extensions to be seeked in the future.

Acknowledgments DM acknowledges useful conversations with Márton Balázs and funding from an EPSRC DTP studentship, grant number EP/T517872/1. LG, DM and FB thank the Isaac Newton Institute for Mathematical Sciences, Cambridge, for support and hospitality during the programmes Mathematics of Movement: an interdisciplinary approach to mutual challenges in animal ecology and Cell Biology and Stochastic Systems for Anomalous Diffusion, where work was, respectively, initiated and concluded supported by EPSRC grants EP/R014604/1 and EP/Z000580/1. LG also acknowlegdes funding from NERC grant NE/W00545X/1 and support by a grant from the Simons Foundation. FB and PF also acknowledge the funding from the Ministry of Science, Innovation, and Universities in Spain through grant no. PID2021-122893NB-C21.

Appendix 1: Definition of Matrices in Original Master Equation

The matrices in Eq. (13.2) are defined as follows:

$$\begin{aligned}
&\mathbb{A} = \{a_{i,j}\} \text{ where } a_{3,4} = l,\ a_{3,5} = r,\ a_{3,6} = b,\\
&\mathbb{B} = \{\mathscr{b}_{i,j}\} \text{ where } \mathscr{b}_{6,1} = l,\ \mathscr{b}_{6,2} = r,\ \mathscr{b}_{6,3} = b,\\
&\mathbb{C} = \{c_{i,j}\} \text{ where } c_{5,1} = r,\ c_{5,2} = b,\ c_{5,3} = l,\\
&\mathbb{D} = \{\mathscr{d}_{i,j}\} \text{ where } \mathscr{d}_{2,4} = r,\ \mathscr{d}_{2,5} = b,\ \mathscr{d}_{2,6} = l,\\
&\mathbb{E} = \{e_{i,j}\} + c\mathbb{I} \text{ where } e_{4,1} = e_{1,4} = b,\ e_{4,2} = e_{1,5} = l,\ e_{4,3} = e_{1,6} = r,
\end{aligned} \tag{13.23}$$

where $\mathbb{I}$ is the 6×6 identity matrix and where all other indices are zero.

Appendix 2: The z-Inversion of the Mean-Squared Displacement

In the diffusive limit Eq. (13.8) of the main text reduces to

$$\widetilde{\Delta \boldsymbol{n}}(z) = \frac{2z(c-1)\,[z(4c-1)-3]}{9(z-1)^2[(2c-1)z-1]}, \tag{13.24}$$

which we invert to the time domain by evaluating the counterclockwise contour integral $\Delta \boldsymbol{n}(t) = 1/(2\pi i) \oint_{|z|<1} z^{-t-1} \widetilde{\Delta \boldsymbol{n}}(z) \mathrm{d}z$ via Cauchy's residue theorem.

One can easily see the poles of $\widetilde{\Delta \boldsymbol{n}}(z)$ lie at $z = 1$ (second order) and at $z = 1/(1-2c)$ (first order). Since both these poles are outside the contour region, we make the change of variable to $s = 1/z$, such that the transform becomes $\Delta \boldsymbol{n}(t) = 1/(2\pi i) \oint_{|s|\le 1} s^{t-1} \widetilde{\Delta \boldsymbol{n}}(s) \mathrm{d}s$ with the contour running counterclockwise. We also treat the $c = 1/2$ separately.

Starting with the $c = 1/2$ case, $\Delta \boldsymbol{n}(t) = \frac{1}{18\pi i} \oint_{|s|\le 1} s^{t-1}(3s-1)(s-1)^{-2} \mathrm{d}s$, which is trivially evaluated to $\Delta \boldsymbol{n}(t) = 9^{-1}(2t+1)$. With $c \neq 1/2$, via Cauchy's residue theorem we find $\Delta \boldsymbol{n}(t) = \left[4(1-c)t + 1 - (2c-1)^t\right]$. This procedure is equivalent for all cases presented in the main text.

Appendix 3: Construction of the Transition Matrices

When constructing the transition matrices, before resizing them to size $N_1 N_2 \times N_1 N_2$, it is instructive to consider them as $N_1 \times N_2 \times N_1 \times N_2$ sized tensors where the probability at index $(\alpha, \beta, \gamma, \delta)$ governs transitions from site $\boldsymbol{n} = (\gamma, \delta)$ to $\boldsymbol{n} = (\alpha, \beta)$. Therefore, the probability of each movement, which raises (lowers) the n_1 coordinate is located on the $(\gamma+1, \beta, \gamma, \delta)$ $((\gamma-1, \beta, \gamma, \delta))$ diagonal, while those that raise (lower) the n_2 coordinate may be found on the $(\alpha, \delta+1, \gamma, \delta)$ $((\alpha, \delta-1, \gamma, \delta))$ diagonal. Sojourn probabilities are placed on the $(\alpha, \alpha, \alpha, \alpha)$ elements.

The matrices are sparse as there is one unique way to move between each pair of states. For example, as seen from Eqs. (13.1) and (13.4) of the main text, the walker may never leave a positive (negative) site and find itself in a positive (negative) site in the subsequent time step when it is situated in the bulk of the domain. Hence, the 12 matrices that govern transitions from positive (negative) to positive (negative) sites contain only boundary interactions. Moreover, the six matrices governing the movement from states i to i ($\forall i \in \{1, 6\}$) are identity matrices, with the sojourn probability c on the main diagonal, with perturbations at the boundary elements. Finally, we have 18 matrices governing bulk movement, which each dictate one movement option.

Reflections are included by perturbing the transition matrices at the boundary elements and we model 'bouncing off the wall' as the walker changing its internal state at the boundary site. More specifically, the walker moves to the internal state that corresponds to the direction that the walker would have been travelling if it had physically hit the wall and turned around.

We elucidate this using the upper $\boldsymbol{n} = (n_1, N_2)$ edge as an example. On this boundary, it is only the positive site that interacts with the boundary. If a bounce occurs, on the $\mathbb{Z}^2$ embedding this is akin to the walker heading North hitting the wall, turning and heading South. This results in a re-wiring of the left and right turns, which take the walker North out of states 1 and 3, back into state 2 of the same site. Backtracks out of states 1 and 3 are considered as sojourns and re-wired into themselves. The implementation of this boundary condition leads to a uniform steady state at long times [14].

Finally, we include absorbing targets by multiplying the outgoing probability of the trap site by $(1-\rho)$, where ρ is the probability of absorption. More concretely, a trap in a positive state in site $\boldsymbol{n} = (n_1, n_2)$ would be included by a multiplication of $(1-\rho)$ to the $(\gamma, \delta) = (n_1, n_2)$ element of each $\mathbb{M}(i, j)$ $(3 \geq i \geq 6)$.

Appendix 4: Experimentally Relevant Parameters

Table 13.1 We use the symbol '–' to denote that the entry is equivalent to the entry above. The two coordinates in the value of the target coordinates refer to the first-found and second-found locations, respectively. These locations are found by matching the distance from the initial condition used in the experimental set-up

Parameter	Meaning	Value
N_1	Number of squares in the x-direction	23
N_2	Number of squares in the y-direction	14
(n_{01}, n_{02}, m_0)	Initial Condition (Location of Nest)	$(12, 1, \mathfrak{p})$
(s_1, s_2, m_s)	Locations of targets in experiment one.	$(15, 8, \mathfrak{p})$, $(16, 10, \mathfrak{n})$
–	Locations of targets in experiment two	$(2, 10, \mathfrak{n})$, $(19, 12, \mathfrak{n})$
–	Locations of targets in experiment three	$(9, 11, \mathfrak{n})$, $(10, 12, \mathfrak{p})$
–	Locations of targets in experiment four	$(3, 8, \mathfrak{p})$, $(16, 4, \mathfrak{p})$
–	Locations of targets in experiment five	$(8, 3, \mathfrak{p})$, $(15, 8, \mathfrak{p})$
–	Locations of targets in experiment six	$(1, 10, \mathfrak{n})$, $(12, 5, \mathfrak{p})$
–	Locations of targets in experiment seven	$(7, 5, \mathfrak{p})$, $(15, 5, \mathfrak{p})$
–	Locations of targets in experiment eight	$(18, 12, \mathfrak{p})$, $(17, 13, \mathfrak{p})$
–	Locations of targets in experiment nine	$(13, 14, \mathfrak{p})$, $(17, 8, \mathfrak{n})$

References

1. Jolles, J.W., King, A.J., Killen, S.S.: The role of individual heterogeneity in collective animal behaviour. Trends Ecol. Evol. **35**(3), 278–291 (2020)
2. Chesson, P., Rosenzweig, M.: Behavior, heterogeneity, and the dynamics of interacting species. Ecology **72**(4), 1187–1195 (1991)
3. Lach, L., Parr, C.L., Abbott, K.L.: Ant Ecology Oxford University Press, Oxford (2010)
4. Gordon, D.M.: From division of labor to the collective behavior of social insects. Behav. Ecol. Sociobiol. **70**(7), 1101–1108 (2016)
5. Czaczkes, T.J.: Using T- and Y-mazes in myrmecology and elsewhere: a practical guide. Insectes Soc. **65**(2), 213–224 (2018)

6. Cristín, J., Fernández-López, P., Lloret-Cabot, R., Genovart, M., Méndez, V., Bartumeus, F., Campos, D.: Spatiotemporal organization of ant foraging from a complex systems perspective. Sci. Rep. **14**(1), 12801 (2024)
7. Campos, D., Bartumeus, F., Méndez, V., Espadaler, X.: Reorientation patterns in central-place foraging: internal clocks and klinokinesis. J. R. Soc. Interface **11**(90), 20130859 (2014)
8. Martin, R., Leroy, C., Maák, I., D'Ettorre, P.: Group phenotypic composition drives task performances in ants. Biol. Lett. **20**(1), 20230463 (2024)
9. Fernández-López, P., Oro, D., Lloret-Cabot, R., Genovart, M., Garriga, J., Bartumeus, F.: Foraging ants as liquid brains: Movement heterogeneity shapes collective efficiency. Proc. Natl. Acad. Sci. U.S.A. 122 (31) e2506930122 (2025)
10. Oriens, G.H., Pearson, N.E.: On the theory of central place foraging. In: Analysis of Ecological Systems, pp. 154–177. Ohio State University Press, Columbus (1979)
11. Marris, D., Sarvaharman, S., Giuggioli, L.: Exact spatiotemporal dynamics of lattice random walks in hexagonal and honeycomb domains. Phys. Rev. E **107**(5), 054139 (2023)
12. Montroll, E.W.: Random walks on lattices III. Calculation of first-passage times with application to exciton trapping on photosynthetic units. J. Math. Phys. **10**(4), 753–765 (1969)
13. Larralde, H.: First-passage probabilities and mean number of sites visited by a persistent random walker in one-and two-dimensional lattices. Phys. Rev. E **102**(6), 062129 (2020)
14. Marris, D., Giuggioli, L.: Persistent and anti-persistent motion in bounded and unbounded space: resolution of the first-passage problem. New J. Phys. **26**(7), 073020 (2024)
15. Weiss, G.H., Shuler, K.E., Lindenberg, K.: Order statistics for first passage times in diffusion processes. J. Stat. Phys. **31**, 255–278 (1983)
16. Tejedor, V., Voituriez, R., Bénichou, O.: Optimizing persistent random searches. Phys. Rev. Lett. **108**(8), 088103 (2012)
17. Cerdá, X., Angulo, E., Boulay, R., Lenoir, A.: Individual and collective foraging decisions: a field study of worker recruitment in the gypsy ant Aphaenogaster senilis. Behav. Ecol. Sociobiol. **63**(4), 551–562 (2009)
18. Cristín, J., Bartumeus, F., Méndez, V., Campos, D.: Occupancy patterns in superorganisms: a spin-glass approach to ant exploration. R. Soc. Open Sci. **7**(12), 201250 (2020)
19. Weiss, G.H.: Aspects and Applications of the Random Walk. North-Holland, Amsterdam (1994)
20. Zumofen, G., Blumen, A.: Energy transfer as a random walk. II. Two-dimensional regular lattices. J. Chem. Phys. **76**(7), 3713–3731 (1982)
21. van Kampen, N.G. Remarks on non-Markov processes. Braz. J. Phys. **28**, 90–96 (1998)
22. Klafter, J., Sokolov, I.M.: First Steps in Random Walks: From Tools to Applications Oxford University Press, Oxford (2011)
23. Giuggioli, L.: Exact spatiotemporal dynamics of confined lattice random walks in arbitrary dimensions: a century after Smoluchowski and Pólya. Phys. Rev. X **10**, 021045 (2020)
24. Wilson, D.B., Baker, R.E., Woodhouse, F.G.: Displacement of transport processes on networked topologies. SIAM J. Appl. Math. **79**(5), 1892–1915 (2019)
25. Gallos, L.K.: Random walk and trapping processes on scale-free networks. Phys. Rev. E **70**(4), 046116 (2004)
26. Biroli, M., Mori, F., Majumdar, S.N.: Number of distinct sites visited by a resetting random walker. J. Phys. A-Math. Theor. **55**(24), 244001 (2022)
27. Montroll, E.W., Weiss, G.H.: Random walks on lattices. II. J. Math. Phys. **6**(2), 167–181 (1965)
28. Redner, S.: A Guide to First-Passage Processes. Cambridge University Press, Cambridge (2001)
29. Blumen, A., Zumofen, G.: Energy transfer as a random walk on regular lattices. J. Chem. Phys. **75**(2), 892–907 (1981)
30. Genz, A.C., Malik, A.A.: Remarks on algorithm 006: an adaptive algorithm for numerical integration over an N-dimensional rectangular region. J. Comput. Appl. Math. **6**(4), 295–302 (1980)
31. Abate, J., Whitt, W.: The Fourier-series method for inverting transforms of probability distributions. Queueing Syst. **10**, 5–87 (1992)

32. Stevens, M.: Sensory Ecology, Behaviour, and Evolution. Oxford University Press, Oxford (2013)
33. Turchin, P.: Quantitative Analysis of Movement: Measuring and Modeling Population Redistribution in Animals and Plants. Sinauer Associates, Sunderland (1998)
34. Giuggioli, L., Sarvaharman, S.: Spatio-temporal dynamics of random transmission events: from information sharing to epidemic spread. J. Phys. A: Math. Theor. **55**(37), 375005 (2022)
35. Sarvaharman, S., Giuggioli, L.: Particle-environment interactions in arbitrary dimensions: a unifying analytic framework to model diffusion with inert spatial heterogeneities. Phys. Rev. Res. **5**(4), 043281 (2023)
36. Giuggioli, L., Sarvaharman, S., Das, D., Marris, D., Kay, T.: Multi-target search in bounded and heterogeneous environments: a lattice random walk perspective. In: Target Search Problems, pp. 107–133. Springer, Cham (2024)
37. Dornhaus, A., Powell, S.: Foraging and defence strategies. In: Ant Ecology, pp. 210–230. Oxford University Press, Oxford (2010)
38. Grebenkov, D., Metzler, R., Oshanin, G.: From single-particle stochastic kinetics to macroscopic reaction rates: fastest first-passage time of n random walkers. New J. Phys. **22**(10), 103004 (2020)
39. Lawley, S.D.: Universal formula for extreme first passage statistics of diffusion. Phys. Rev. E **101**(1), 012413 (2020)
40. Lawley, S.D., Madrid, J.B.: A probabilistic approach to extreme statistics of Brownian escape times in dimensions 1, 2, and 3. J. Nonlinear Sci. **30**(3), 1207–1227 (2020)
41. Meerson, B., Redner, S.: Mortality, redundancy, and diversity in stochastic search. Phys. Rev. Lett. **114**(19), 198101 (2015)
42. Hughes, B.D.: Random Walks and Random Environments Volume 1: Random Walks. Clarendon Press, Oxford (1995)
43. Montroll, E.W.: Random walks in multidimensional spaces, especially on periodic lattices. J. Soc. Ind. Appl. Math. **4**(4), 241–260 (1956)
44. Henry, B.I., Batchelor, M.T.: Random walks on finite lattice tubes. Phys. Rev. E **68**(1), 016112 (2003)
45. Guéneau, M., Touzo, L.: Siegmund duality for physicists: a bridge between spatial and first-passage properties of continuous-and discrete-time stochastic processes. J. Stat. Mech.: Theory Exp. **2024**(8), 083208 (2024)
46. Godec, A., Metzler, R.: First passage time distribution in heterogeneity controlled kinetics: going beyond the mean first passage time. Sci. Rep. **6**(1), 20349 (2016)
47. Godec, A., Metzler, R.: Optimization and universality of Brownian search in a basic model of quenched heterogeneous media. Phys. Rev. E **91**(5), 052134 (2015)
48. Bénichou, O., Voituriez, R.: From first-passage times of random walks in confinement to geometry-controlled kinetics. Phys. Rep. **539**(4), 225–284 (2014)
49. Condamin, S., Bénichou, O., Moreau, M.: First-passage times for random walks in bounded domains. Phys. Rev. Lett. **95**(26), 260601 (2005)
50. Condamin, S., Bénichou, O., Tejedor, V., Voituriez, R., Klafter, J.: First-passage times in complex scale-invariant media. Nature **450**(7166), 77–80 (2007)
51. Mattos, T.G., Mejía-Monasterio, C., Metzler, R., Oshanin, G.: First passages in bounded domains: when is the mean first passage time meaningful? Phys. Rev. E **86**(3), 031143 (2012)
52. Kac, M.: Random walk and the theory of Brownian motion. Am. Math. Mon. **54**(7), 369–391 (1947)
53. Das, D., Giuggioli, L.: Discrete space-time resetting model: Application to first-passage and transmission statistics. J. Phys. A: Math. Theor. **55**(42), 424004 (2022)
54. Barbini, A., Giuggioli, L.: Lattice random walk dynamics with stochastic resetting in heterogeneous space. J. Phys. A: Math. Theor. **57**(42), 425001 (2024)
55. Campos, D., Méndez, V.: Dynamic redundancy as a mechanism to optimize collective random searches. Phys. Rev. E **109**, 064109 (2024)

Chapter 14
Prospection and Dispersal in Metapopulations: A Perspective from Opinion Dynamics Models

Daniela Molas and Daniel Campos

14.1 Introduction

The scientific study of decision-making in living beings poses a significant challenge across multiple disciplines. Traditionally associated with psychology and sociology, this field has recently garnered growing interdisciplinary interest, driven by advancements in experimental techniques and theoretical insights from ecology, mathematics, and physics. These developments have enabled more integrative approaches to understanding decision-making processes. Notable examples include the study of collective movement in animal groups through the frameworks of complexity science [1] and fluid dynamics [2–4], as well as the analysis of cooperative behavior among rational agents in evolutionary game theory scenarios, such as the prisoner's dilemma [5, 6], or via decentralized communication systems, as observed in eusocial species [7, 8].

A particularly important aspect within this context concerns how groups of individuals prospect for information from peers to guide their decisions during extended decision-making processes. Information sharing and prospection serve as essential mechanisms for fostering behavioral and/or opinion consensus. This is of intrinsic interest for understanding collective phenomena in ecological and social systems, including positive outcomes (e.g., cooperation) and negative consequences (e.g., the tragedy of the commons).

The Voter Model [9] is one of the most prominent mathematical frameworks for studying opinion consensus in such contexts. In its simplest form, the Voter Model assumes that individuals (voters) make sequential choices from a discrete set of options (e.g., political parties) by imitating the decisions of their peers. Despite

---

D. Molas · D. Campos (✉)
Physics Department, Universitat Autonoma de Barcelona, Barcelona, Spain
e-mail: daniela.molas@uab.cat; daniel.campos@uab.cat

L. Giuggioli, P. K. Maini (eds.), *The Mathematics of Movement*,
https://doi.org/10.1007/978-3-032-00873-2_14

its simplicity, the Voter Model and its numerous extensions [10–20] have attracted sustained theoretical interest over the past decades. This model plays a central role in the statistical physics of social systems [10, 11], providing a robust framework for exploring the dynamics of consensus formation in both natural and artificial populations.

14.1.1 The Voter Model in Biology

While copying and imitation are common mechanisms in living systems, most studies on the Voter Model have been restricted to theoretical contexts, with real-life applications remaining relatively unexplored [21]. This is somewhat surprising, as natural selection itself can be viewed as a copying mechanism, where different options (genotypes) spread and compete, leading to either consensus (survival of the fittest) or polarization (coexistence through specialization or resource partitioning).

Several factors may explain the limited application of Voter-like models to experimental biology:

(i) Most versions of the Voter Model are restricted to populations with a fixed number of individuals, which is often unrealistic in biological systems.
(ii) Opinion consensus in biological populations typically arises through local and dynamic interactions. However, spatially explicit versions of the Voter Model remain scarce. The Sznajd model provides some notable exceptions; see [22, 23] for details.
(iii) As highlighted in [21], consensus among sepparate individuals is not always meaningful in biological systems. Instead, population-level or site-level descriptions are more relevant, yet extending Voter Models to these scales is nontrivial.

Despite these challenges, Voter-like models have been successfully applied to genetic drift and evolutionary dynamics, starting with Moran's pioneering work [24] and extending to more recent studies [25–30]. Applications to fields like behavioral and movement ecology, however, remain rare (but see [31–34]).

Imitation and majority rules, central to many Voter Models, are recognized as powerful drivers of both individual and social decision-making in animal groups [35–37]. A particularly relevant case involves seasonal species that annually select breeding sites. In such cases, individual experience is often complemented by social information gathered from the decisions of conspecifics, a process known as "prospection" [38]. The effect of prospection on fitness and reproductive success has only recently been explored, as it requires tracking animal movements over large temporal and spatial scales. Some datasets, particularly for seabirds [39], have become available recently through advances in GPS technology and other tracking methods, allowing a notable expansion of the field.

In this work, we explore how Voter-like models can be adapted to study the role of prospection in metapopulation dynamics [40], where different sites or patches

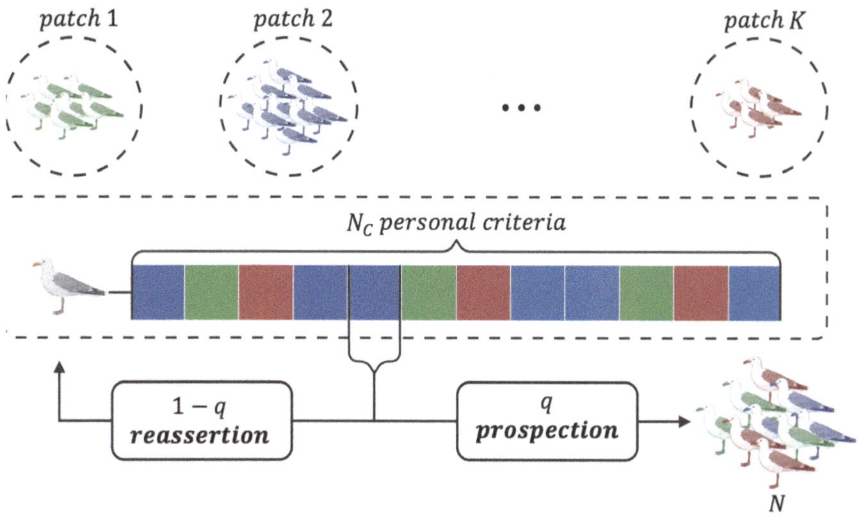

Fig. 14.1 Model schematics. Each individual in the group has personal information based on N_C criteria, each supporting a specific patch or breeding site (indicated by colors). Individual choices at each time step are influenced by these criteria, which are updated either by prospection (probability q) or by reinforcing the individual's current option (reassertion, probability $1 - q$). See the main text for further details

represent the options available to "voters." Breeding decisions based solely on prior experience correspond to "stubborn voters," who repeatedly choose the same option, while prospection introduces an imitation dynamic within the model.

Our primary objective is to investigate how prospection influences opinion consensus, a typical outcome in classical Voter Models. Specifically, we examine: (i) how the time to achieve consensus varies with prospection rates or efficiency, and (ii) the emergence of alternative outcomes, such as polarization or stalemates. In the case of polarization, individuals coexist within distinct subpopulations, each maintaining differentiated opinions or preferences. Consequently, the system ceases to evolve, ultimately leading to an absorbing state. On the other hand, stalemates refer to stationary scenarios where these subpopulations coexist and interact by exchanging individuals but without reaching a unified consensus so, unlike polarization, system fluctuations hinder the attainment of an absorbing state.

To this end, we propose a version of the Voter Model where individuals accumulate personal information through prospection (Fig. 14.1). Similar approaches incorporating personal information have recently been explored, revealing mechanisms that can promote polarized states [12, 16]. In our adaptation, individuals base their decisions on personal/internal criteria, which, in a biological context, could represent attributes such as proximity to food or predation pressure. These criteria are updated through prospection visits to neighboring patches. Spatially explicit effects are expected to play a significant role, and we analyze these under two idealized scenarios: global prospection (where individuals can prospect for any patch) and local prospection (restricted to neighboring patches).

The structure of this work is as follows. In Sect. 14.2, we introduce the Voter-like model with personal criteria and describe its parameters. Section 14.3 examines the effects of personal information in the mean-field case, where all patches are equally accessible (global prospection). Section 14.4 addresses spatially explicit effects by considering local prospection. Finally, in Sect. 14.5, we summarize the main findings and propose ideas for future research.

14.2 Model

We consider a system of N individuals that at each time step have to choose one from a set of K different options available (with $K \geq 2$). In the classical Voter Model, such decisions are simply taken by imitating what the other individuals do, though several variants exist, which take into account different aspects of animal or human decision-making. One such aspect, not much studied previously, is the idea that observing others' decisions might not necessarily lead to instantaneous imitation, but it just provides an external source of information thats adds to the background knowledge of the individual. To take into account this aspect, we assume that decisions are based on a set of N_C personal criteria, which represent relevant features of the options. So that, the decision is expected to be the result of combining or prioritizing such features/criteria.

Let us illustrate this with the specific example of birds selecting a suitable patch annually for nesting (Fig. 14.1). To make their decision, the birds consider their individual preferences based on a set of N_C criteria, such as proximity to food sources, predation risk, competition with conspecifics, and availability of resting areas.

Each individual has its own preferences (or personal information) regarding these criteria. If we denote individuals by the index $i = 1, 2, \ldots, N$, where N is the total number of individuals, and criteria by the index $j = 1, 2, \ldots, N_C$, where N_C is the total number of criteria considered, then the j-th criterion of the i-th individual, defined as $c_{i,j}$, can take possible values from $\{1, 2, \ldots, K\}$, where K is the number of available options.

The values of these criteria are updated based on either prior experience or information acquired by the individual through exploring other patches. At each time step, (i) one criterion from each individual (say i_1) is chosen at random, (ii) a neighboring individual i_2 is selected randomly, and (iii) the chosen criterion $c_{i_1,j}$ is updated such that it adopts the option (patch) held by the neighboring individual i_2 with probability q, or with probability $1 - q$, it retains the option chosen by the individual i_1 itself. Hence, q represents the probability that individual i_1 acquires new information from its neighbors and updates its preferences accordingly; otherwise, the individual reaffirms its current preferences.

Next, following these personal preferences, the corresponding decisions are made as follows. At each time step, (i) we choose one individual at random, (ii) we choose one of its personal criteria at random, and (iii) the choice made by the individual is updated to the option supported by the selected criterion. So, the agent is assumed to be prioritizing such particular criterion/feature at that moment.

By introducing this dual dynamics (a dynamics for taking decisions, and another one for updating criteria), we dissociate what individuals "do" from what individuals "think." Of course, if criteria are updated at a rate fast enough, then they will reflect the real options chosen by the population, and then individuals' decisions and criteria will be aligned. But there are also mechanisms (e.g., noisy or uncertain prospection) that could lead the individuals to "behave" and "think" differently. While this is a very attractive possibility to explore, in the present contribution, we will rather focus for simplicity on simple situations where individuals are expected on average to "think" and "behave" similarly.

We define $n_\alpha(t)$ as the number of individuals that are holding option α at time step t, and $n_{\alpha,\beta}(t)$ as the total number of criteria for these individuals that give support to option β. As a result, $N = \sum_\alpha n_\alpha$ and $N_C = \sum_\beta n_{\alpha,\beta}$ hold, and the case $n_{\alpha,\alpha} = N_C$ would represent a situation in which all the criteria for the individuals holding option α give support to that option.

From the definitions above, we can define a set of *magnetizations* as

$$m_\alpha \equiv n_\alpha - \sum_{\beta \neq \alpha} n_{\alpha,\beta} = 2n_\alpha - N, \tag{14.1}$$

for $\alpha = 1, 2, \ldots, K$. This magnetization explicitly depends on the fraction of individuals holding option α and reflects the level of prevalence of this option over the others. Defined in this way, the consensus of the population, where all individuals support the same option α, satisfies $m_\alpha = N$, and $m_{\alpha'} = -N$ for $\alpha' \neq \alpha$.

Following the usual guidelines from the classical Voter Model with multiple options (see, e.g., [14, 16]), a master equation for the evolution of these magnetizations can be written in the mean-field approximation. By introducing $P(m_\alpha, t)$ as the probability that the magnetization m_α takes a particular value at time step t, the master equation reads

$$P(m_\alpha, t+1) = \sum_{\beta \neq \alpha}^{K} \left[\frac{n_\beta}{N} \frac{n_{\alpha,\beta}}{N_C} P(m_\alpha - 2, t) + \frac{n_\alpha}{N} \frac{n_{\beta,\alpha}}{N_C} P(m_\alpha + 2, t) \right] + \\ + \left(\frac{n_\alpha}{N} \frac{n_{\alpha,\alpha}}{N_C} + \sum_{\beta \neq \alpha} \frac{n_\beta}{N} \frac{\sum_{\gamma \neq \alpha} n_{\gamma,\beta}}{N_C} \right) P(m_\alpha, t). \tag{14.2}$$

Let us analyze each term one by one:

1. The first term represents the probability that an individual holding option β is chosen at random at time t (n_β/N), and that the criterion assigned randomly

supports option α ($n_{\alpha,\beta}/N_C$), causing the individual to switch from β to α. As a result, n_α increases by 1, which causes m_α to increase by 2.
2. The second term represents the possibility that the individual chosen at random holds option α, but the assigned criterion supports option β, causing the individual's option to change from $\alpha \to \beta$. As a result, n_α decreases by 1, which causes m_α to decrease by 2.
3. The third term stands for the situation where individuals do not switch their option. The expression $\frac{n_\alpha}{N}\frac{n_{\alpha,\alpha}}{N_C}$ represents the case where the option held by the individual and that supported by the criteria are both α, so nothing changes. Similarly, $\sum_{\beta\neq\alpha}\frac{n_\beta}{N}\frac{\sum_{\gamma\neq\alpha} n_{\gamma,\beta}}{N_C}$ stands for those cases where both the option of the individual and the option supported by the criterion are different from α, for which the value of m_α remains the same.

In our particular model, the imitation rule is not explicitly included in the dynamics above of how the individuals take options, but on how the criteria are updated. For this reason, it is more convenient and instructive to work with *magnetizations* defined at the level of the criteria. These can be defined as

$$\sigma_\alpha \equiv n_{\alpha,\alpha} - \sum_{\beta\neq\alpha} n_{\alpha,\beta}, \tag{14.3}$$

since the criteria are part of a set of options. It follows that $\sigma_\alpha = N_C$ whenever $n_{\alpha,\alpha} = N_C$, which implies that the set of criteria is aligned with the same option α.

The master equation associated with $\rho(\sigma_\alpha, t)$, which represents the probability density of σ_α at time t, takes a form similar to that above for $P(m_\alpha, t)$. However, this equation consists of two distinct terms: the first represents the reassertion mechanism, with probability $(1-q)$, and the second describes the dynamics induced by imitation rules, with complementary probability q:

$$\begin{aligned}\rho(\sigma_\alpha, t+1) = (1-q)\sum_{\beta\neq\alpha}^{K}\left[\frac{n_\beta}{N}\frac{n_{\alpha,\beta}}{N_C}\rho(\sigma_\alpha - 2, t) + \frac{n_\alpha}{N}\frac{n_{\beta,\alpha}}{N_C}\rho(\sigma_\alpha + 2, t)\right] +\\ +q\sum_{\beta\neq\alpha}\left[\frac{n_\beta}{N}\frac{\sum_{\gamma=1}^{K} n_{\alpha,\gamma}}{KN_C}\rho(\sigma_\alpha - 2, t) + \frac{n_\alpha}{N}\frac{\sum_{\gamma=1}^{K} n_{\beta,\gamma}}{KN_C}\rho(\sigma_\alpha + 2, t)\right] +\\ +\left(\frac{n_\alpha}{N}\frac{n_{\alpha,\alpha}}{N_C} + \sum_{\beta\neq\alpha}\frac{n_\beta}{N}\frac{\sum_{\gamma\neq\alpha} n_{\gamma,\beta}}{KN_C}\right)\rho(\sigma_\alpha, t),\end{aligned} \tag{14.4}$$

where the quotient $\frac{\sum_{\gamma=1}^{K} n_{\alpha,\gamma}}{KN_C}$ represents the fraction of criteria supporting option α for the whole population, and similarly, $\frac{\sum_{\gamma=1}^{K} n_{\beta,\gamma}}{KN_C}$ represents the fraction of criteria supporting option β. Note that in the particular case $K = 2$ (only two options available), the previous expression simplifies to

$$\rho(\sigma, t+1) = (1-q)\left[\frac{n_2}{N}\frac{n_{1,2}}{N_C}\rho(\sigma-2,t) + \frac{n_1}{N}\frac{n_{2,1}}{N_C}\rho(\sigma+2,t)\right] +$$
$$+ q\left[\frac{n_2}{N}\frac{n_{1,1}+n_{1,2}}{2N_C}\rho(\sigma-2,t) + \frac{n_1}{N}\frac{n_{2,1}+n_{2,2}}{2N_C}\rho(\sigma+2,t)\right] +$$
$$+ \left(\frac{n_1}{N}\frac{n_{1,1}}{N_C} + \sum_{\beta\neq\alpha}\frac{n_\beta}{N}\frac{\sum_{\gamma\neq\alpha} n_{\gamma,\beta}}{2N_C}\right)\rho(\sigma,t), \qquad (14.5)$$

with $\sigma \equiv (n_{2,1}+n_{2,2}) - (n_{1,1}+n_{1,2})$. While this equation does not admit a simple solution, this simplified case allows us to gain some understanding about the effect that each mechanism plays on the dynamics of the system. For this, in the following, we consider separately the effect of the prospection and the reassertion mechanisms.

14.2.1 *Adiabatic Approximation*

Before interpreting the system dynamics, let us introduce a key concept to understand the arguments and conclusions that we present below. As mentioned, our model distinguishes between what individuals "do" (the options they take) and what they "think" (the options supported by their criteria). When imitation is highly effective and criteria are updated rapidly, these criteria faithfully reflect the options chosen by individuals in the group. In this situation, individuals' decisions mimic what others are doing, as in the classical Voter Model. We refer to this situation as an *adiabatic approximation*, as it allows us to separate the timescale of criteria updates (fast scale) from that of individual decisions (slow scale). From this perspective, the classical Voter Model can be seen as an adiabatic approximation of our model with personal criteria. Mathematically, this approximation is expressed as:

$$\frac{n_\alpha}{N} = \frac{n_{\alpha,\beta}}{N_C}, \qquad (14.6)$$

for any α and β.

In this context, the statistics of our model can be directly derived from the classical model whenever the adiabatic approximation holds. In particular, the times required to reach consensus or for an option to disappear in the classical model with multiple options are defined as follows:

- The consensus time, represented by T_c, is defined as the time it takes for all individuals within the system to reach the same option ($n_\alpha = N$ for some α).
- Extinction time, represented by T_e, is defined as the time required for one of the options to disappear completely from the system ($n_\alpha = 0$ and $n_{i,\alpha} = 0$ for any i, for option α).

Under these definitions, the average (computed over the different stochastic realizations of the system) of the consensus and extinction times in the classical model is known to satisfy [15]:

$$\langle T_c \rangle = \frac{K-1}{K} N, \qquad \langle T_e \rangle = \frac{N}{K(K-1)}, \tag{14.7}$$

where K is the number of available options and N the total number of individuals. Under simplified conditions, we will verify in the following sections how these exact results also apply to our model with personal criteria.

14.2.2 Prospection Mechanism (Case $q = 1$)

We consider the case $q = 1$ in Eq. (14.5). Using a Kramers–Moyal expansion in time and space, one obtains the following Fokker–Planck equation:

$$\frac{\partial \rho}{\partial t} = \left[\frac{n_{2,1} + n_{2,2} - n_{1,1} - n_{1,2}}{N N_C}\right] \frac{\partial \rho}{\partial \sigma} + \left[\frac{(n_{2,1} + n_{2,2})^2 + (n_{1,1} + n_{1,2})^2}{2N N_C}\right] \frac{\partial^2 \rho}{\partial \sigma^2}, \tag{14.8}$$

where the explicit dependences in $\rho(\sigma, t)$ have been omitted for ease of notation.

After some algebraic manipulations, it is possible to rewrite this equation in terms of σ and $m \equiv n_2 - n_1$, leading to:

$$\frac{\partial \rho}{\partial t} = \left[\frac{m - \sigma}{N N_C}\right] \frac{\partial \rho}{\partial \sigma} + \left[1 - \frac{m\sigma}{N N_C}\right] \frac{\partial^2 \rho}{\partial \sigma^2}. \tag{14.9}$$

From this, the effects of the drift and diffusion terms in (14.9) on the system can be analyzed. The drift term drives the system toward the condition $m = \sigma$, aligning the options with the criteria. This effectively brings the system into a regime where the adiabatic approximation (14.6) holds.

The diffusion term determines the magnitude of the fluctuations around the condition $m = \sigma$, as promoted by the drift term. Similarly to the classical Voter Model, cases with vanishing magnetization ($m \approx \sigma \approx 0$) produce larger fluctuations. Conversely, as consensus is approached, with either $m \approx N, \sigma \approx N_C$ or $m \approx -N, \sigma \approx -N_C$, fluctuations tend to diminish. Consequently, system fluctuations favor departures from the homogeneous state $m = 0$, where both options 1 and 2 coexist, and promote the consensus state where only one option prevails, resulting in extreme magnetization values.

14.2.3 Reassertion Mechanism (Case $q = 0$)

By carrying out the same analysis as in the previous section (using the Kramers–Moyal expansion), the resulting Fokker–Planck equation for the dynamics of m is given by:

$$\frac{\partial \rho}{\partial t} = \left[\frac{n_{2,1} - n_{1,2}}{N N_C}\right] \frac{\partial \rho}{\partial \sigma} + \left[\frac{(n_{2,1} + n_{2,2})^2 + (n_{1,1} + n_{1,2})^2}{2 N N_C}\right] \frac{\partial^2 \rho}{\partial \sigma^2}. \tag{14.10}$$

In this case, it is not possible to express the drift and diffusion coefficients in terms of m and σ, as the update of an individual's option depends solely on their personal criteria, which introduces local effects into the model.

From (14.10), we observe that the drift term has a rather neutral effect, promoting the symmetry condition $n_{2,1} = n_{1,2}$ between the two options, which aligns with the adiabatic approximation. Meanwhile, the diffusion term remains the same as in Eq. (14.9), ensuring that consensus is eventually reached through random changes in individuals' decisions driven by imitation (alternatively, consensus can also be influenced by an imbalance in the initial conditions or in the update rules, favoring one of the two options).

In conclusion, the dynamics of the system is largely governed by the imitation dynamics in the criteria update, which promote the emergence of the adiabatic approximation conditions if such imitation effects are sufficiently fast or strong. In the next section, we explore these ideas numerically.

14.3 Global Prospection (Mean-Field Case)

In previous studies [16], two possible outcomes were identified in the voter model with personal information: consensus and population polarization. Our main objective is to identify possible mechanisms by which these two outcomes can emerge or be promoted and to analyze their characteristic timescales. Exact analytical solutions are generally unattainable for our model, so we focus on numerical results. However, in cases where the adiabatic approximation mentioned above is fulfilled, we find that most of our results can be explained using the classical Voter Model.

If the imitation rule allows all criteria to be updated by imitating any individual in the population, the only possible equilibrium state of the system is *consensus*, defined by

$$n_\beta = \delta_{\beta\alpha} N, \qquad n_{\beta,\gamma} = \delta_{\beta\alpha}\delta_{\gamma\alpha} N_C, \tag{14.11}$$

where $\delta_{\alpha,\beta}$ represents the Kronecker delta function, and α is the option reached after consensus. Essentially, condition (14.11) implies that (i) all individuals in the system share the same option, and (ii) all their criteria support that option.

One of the most interesting questions concerns how the different parameters in the system affect the consensus time, T_c, required to reach this unique equilibrium state. For multi-option imitation models, consensus is reached through a cascade of events [14], in which one of the K available options first disappears (as soon as there are no criteria supporting it), leaving the system in a steady situation. Subsequently, a second option disappears, and so on, until only one option remains. The *first extinction time*, T_e, refers to the time required for the first of the K initial options to disappear completely from the system, thereby reducing it to a $(K - 1)$-option model. In other words, it marks the first event in which an option becomes entirely absent from the system. This quantity is of particular interest as it represents the onset of the cascade of option eliminations that eventually leads to consensus.

In Fig. 14.2, we present numerical results for the mean-field case, showing both the average first extinction times (left column) and consensus times (right column) for different values of K (the number of available options) and imitation probabilities q, as a function of the number of personal criteria N_C. In this case, instead of considering local interactions between neighbors, it is assumed that each individual is equally influenced by the global average of opinions in the population.

The results reveal that in the regime $N_C \ll N$, extinction and consensus dynamics are largely governed by prospection. This is evident as the average times for both processes are inversely proportional to q in this regime. When imitation operates at a fast rate, it accelerates the consensus dynamics and promotes convergence to the equilibrium state at a rate proportional to the imitation rate.

For $N_C \gg N$, a more complex effect emerges. In this case, extinction and consensus times converge slowly for all values of q (complete convergence is not observed numerically in Fig. 14.2, as computation times scale with N_C^2, limiting simulations for large N_C). Here, imitation dynamics alone is insufficient to achieve consensus; it also requires alignment between individual options and the criteria. While this alignment is nearly instantaneous for small N_C, it requires significantly long times for large N_C, introducing a new timescale into the model. Specifically, when q is large, imitation is faster, but this causes criteria to change frequently, slowing the alignment process. Conversely, small q values slow down imitation but accelerate alignment.

We find that first extinction dynamics occur more rapidly in systems with a larger number of options, as expected from (14.7). This confirms that extinction of each particular option occurs with independent probability, so the greater the number of options, the shorter the time required for at least one to become extinct.

Finally, we observe two distinct regimes influenced by the number of criteria, N_C, corresponding to the conditions $N_C \ll N$ and $N_C \gg N$. In the former, consensus for all criteria of a single individual is achieved much faster than consensus among individuals. The bottleneck for reaching the equilibrium condition (14.11) is therefore achieving consensus among individuals. This implies that while criteria for a single individual can align quickly, the overall alignment of all individuals becomes the limiting factor. In other words, the time required for the entire population to agree on a single option dictates the pace of the system's approach to equilibrium. As a result, the classical prediction (14.7) applies in this

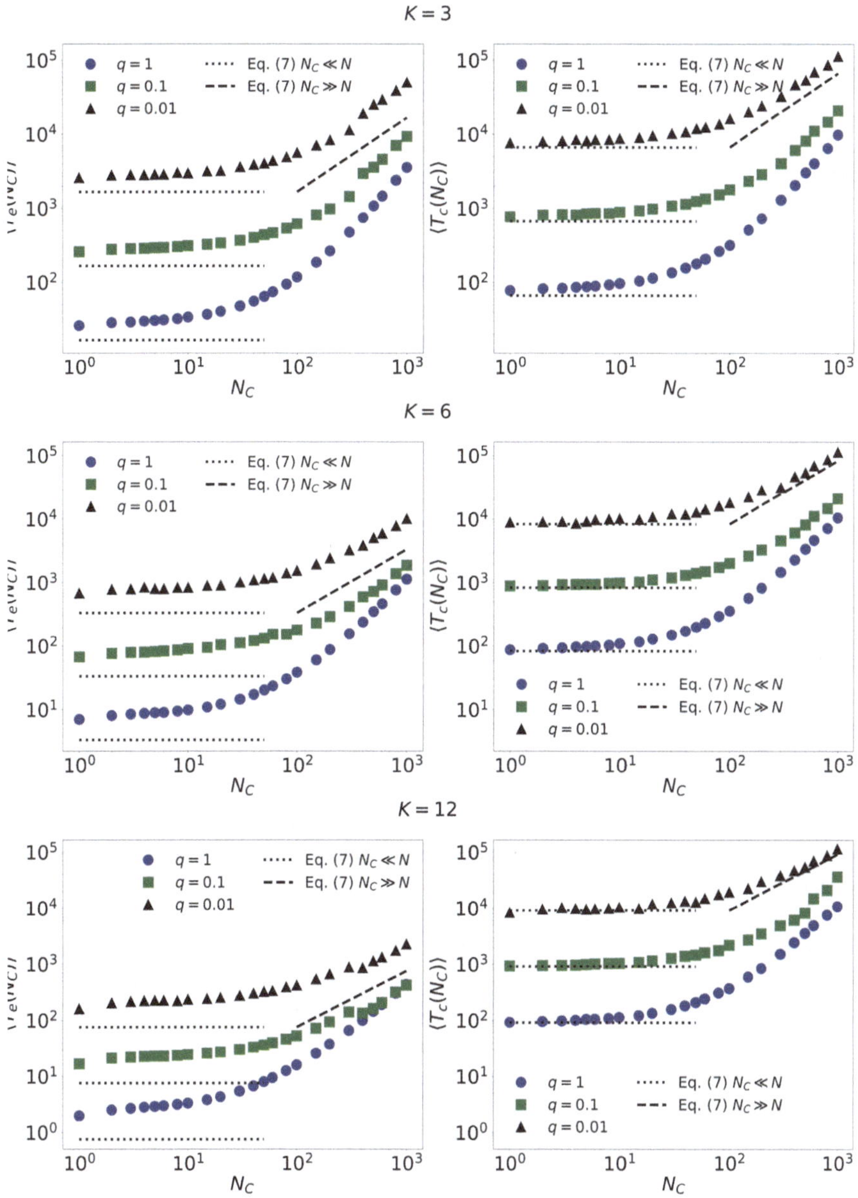

Fig. 14.2 Average extinction (left) and consensus (right) times as a function of the number of criteria N_C. Dashed and horizontal dotted lines correspond to predictions from the classical model in Eq. (14.7) after parameter reinterpretation (see main text for details)

regime (see horizontal dotted lines). Furthermore, extinction and consensus times remain nearly constant in this regime, though numerical simulation data admit power-law fits for $\langle T_e \rangle \sim N_C^{h_1}$ and $\langle T_c \rangle \sim N_C^{h_2}$, with exponents h_1 and h_2 taking values between 0 and 0.1 for the range of parameters explored here.

For the regime $N_C \gg N$, the bottleneck for reaching consensus is determined by the second condition in (14.11). In this case, the classical result (14.7) still applies, but N should be replaced by $N \times N_C$ (the total number of criteria in the system), so both extinction and consensus times satisfy the scaling $\langle T \rangle \sim N \cdot N_C$.

Interestingly, modifying the rate at which criteria are updated does not significantly affect first extinction and consensus times, provided the imitation probability q remains constant. To test this, we modified the model algorithm so that at each time step, only a fraction f of the personal criteria of each individual were updated on average. The initial model corresponds to $f = 1/N_C$ (i.e., only one criterion is updated per time step). The value of f has a minor impact on extinction and consensus dynamics for both $N_C \ll N$ and $N_C \gg N$ (not shown). This confirms that equilibrium is governed primarily by the imitation rate q. Rapid updates of personal criteria are ineffective unless imitation is sufficiently effective.

14.4 Local Prospection

In this section, we address the question: "How can prospection dynamics be modified to allow for the coexistence of multiple options?" Note that complete consensus is the only possible state under global prospection. However, from an ecological perspective, consensus will lead to an inefficient resource exploitation, due to the entire population converging on a single option. In contrast, the stable coexistence of options or species, referred to as polarization in the context of the Voter Model, requires alternative mechanisms to counteract imitation and prospection dynamics, as explored in [16].

As a potential mechanism to promote polarization, we consider here that in real systems, prospection predominantly occurs within a local context. For example, birds or other animals primarily prospect neighboring patches to extract information, while visits to distant patches are infrequent or nonexistent. To simplify, we consider a scenario where options (or breeding places) are arranged in a one-dimensional lattice (labeled again from 1 to K), and prospection is restricted to consecutive or neighboring places.

Restricting prospection to local vicinity causes the imitation dynamics to lead to random extinction of different options at different regions. Consequently, the system can fragment into small subgroups of patches isolated from each other, with consensus dynamics operating independently within each subgroup. Over time, only a discrete set of isolated options survive; this corresponds to the *polarized* state.

One key question is how the number of available options, K, influences polarization dynamics. When fewer options are available, isolated populations are less likely to form because the available options are strongly connected, allowing a single option to dominate and absorb the others, resulting in consensus. Conversely, as K

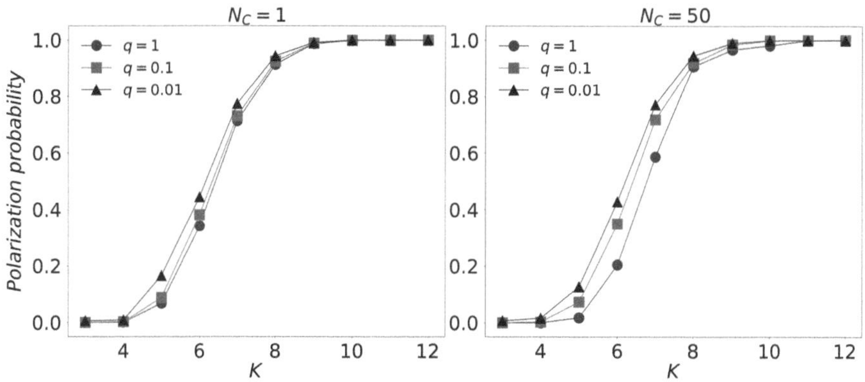

Fig. 14.3 Polarization probability as a function of the number of options K. All data were obtained by considering $N = 100$ individuals and averaging over 10,000 realizations of the process

increases, the probability of isolated populations emerging becomes significantly higher.

Figure 14.3 illustrates the numerical results, showing how the probability of reaching polarization (i.e., stable coexistence of isolated subpopulations) exhibits a phase transition for K between approximately 6 and 8. Remarkably, this behavior is largely independent of the model parameters (q, N_C). It is primarily driven by the number of available options and the likelihood of isolation during the cascade of successive extinctions. This confirms that local prospection alone can drive the system toward polarized states opening avenues for exploring more realistic dynamics in future work, such as prospection governed by specific dispersal kernels, to better understand metapopulation dynamics and coexistence in real habitats or environments.

For completeness, Fig. 14.4 shows the distribution of subpopulation sizes after the system reaches its final state. In cases of consensus, the size always corresponds to N, whereas in polarized states, different sizes are observed. The size distribution exhibits a peak at N that decreases as K increases, indicating that polarization becomes more likely. For sufficiently large K, the population fragments into smaller subpopulations, and the distribution skews toward smaller sizes.

Finally, in Fig. 14.5, we examine how extinction times decay with the number of options K, confirming that the extinction of any option occurs with constant and independent probability. This dependence is related to the number of connections between patches. In the one-dimensional lattice considered, the number of initial connections is $K - 1$. Thus, the extinction time scales approximately as $\langle T_e \rangle \sim \langle T_e \rangle^* / (K - 1)$, where $\langle T_e \rangle^*$ represents the extinction time for $K = 2$. This theoretical prediction, represented by dashed lines in Fig. 14.5, aligns well with numerical results, particularly in the thermodynamic limit where N, N_C, and K are very large, ensuring that the adiabatic approximation holds.

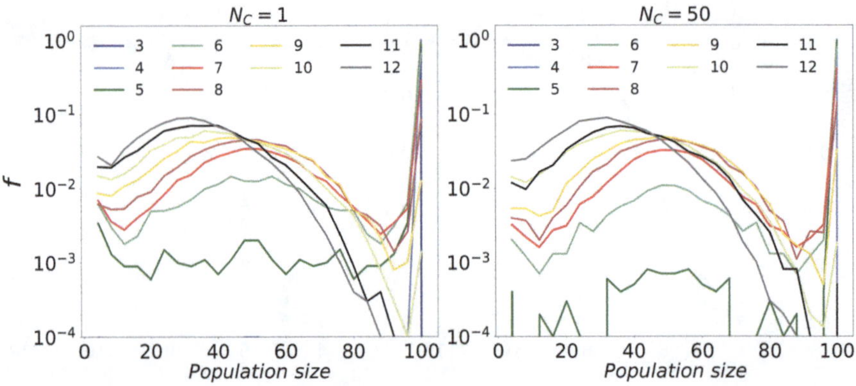

Fig. 14.4 Population size distribution. Data were obtained by considering $N = 100$, $q = 1$, and averaging over 10,000 realizations of the process

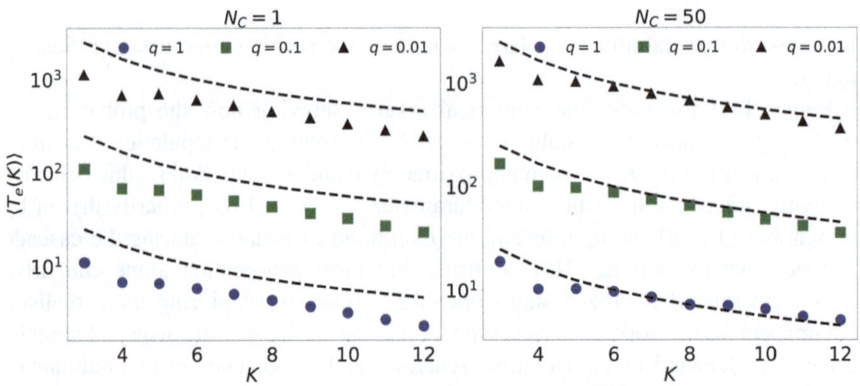

Fig. 14.5 Average extinction time as a function of the number of options K. Dashed lines represent the theoretical approximation $\langle T_e \rangle \sim \langle T_e \rangle^* / (K-1)$ (see main text for details). The colors correspond to different values of K, while the parameter f represents the fraction of personal criteria updated at each time step, expressed as a value between 0 and 1. All data were obtained by considering $N = 100$ individuals and averaging over 10,000 realizations of the process

The colors in the figure correspond to different values of q, while the parameter f represents the fraction of personal criteria updated at each time step, expressed as a value between 0 and 1.

14.5 Conclusions

In this work, we have introduced an extension of the classical Voter Model by incorporating the concept that prospecting information from others does not result in immediate imitation but instead contributes to an information accumulation process that eventually leads to such behavior. Specifically, we considered a dual dynamics framework in which individuals make decisions based on internal personal criteria, while these criteria are updated through an imitation rule akin to that in the classical model. The dynamics of these criteria act as an intermediate process that delays the system's convergence to consensus, requiring alignment between individuals' actual choices and their criteria.

We have demonstrated that our model with personal criteria can be effectively understood in terms of the classical Voter Model, provided the alignment process occurs rapidly and does not become a significant bottleneck for consensus. This scenario corresponds to the *adiabatic approximation*, which we have used as a simplified regime to illustrate the model's applicability. Extending the model to the multi-option case ($K > 2$), we have also shown how a local prospection mechanism can naturally lead to polarized states, characterized by the coexistence of isolated populations reaching each a particular consensus locally. The existence of this polarized regime is seen to be essentially driven by the number of options available, K, while the other parameters in the model (q, N_C) simply rescale the time required to reach that state.

Future work should explore scenarios where the duality between (i) decision-making and (ii) criteria updating becomes more complex and less straightforward, potentially breaking the adiabatic approximation, as a consequence of memory effects, biased/persistent dynamics in the choice of the criteria, etc. This could lead to intriguing phenomena, such as populations that "act" and "think" differently, or scenarios where consensus is unattainable due to noisy or uncertain prospection. Such cases represent promising avenues for further research.

Additionally, experimental testing and calibration of the model appear feasible using large-scale tracking datasets (e.g., GPS) for seasonal birds, as in [38, 39]. In this context, the criteria in the model could be linked to relevant fitness metrics, and individuals' visits to neighboring patches could serve as a measurable proxy for prospection. These extensions illustrate the broader applicability and potential interest of the approach presented here.

Acknowledgments DC acknowledges the organizers of the Isaac Newton Institute program on *Mathematics of movement: an interdisciplinary approach to mutual challenges in animal ecology and cell biology* for their kind invitation (EPSRC grant EP/R014604/1) and for discussions, which have substantially contributed to the present work. DM acknowledges financial support from the Coordination of the National Scholarship Program Carlos Antonio López (BECAL). The authors acknowledge the financial support of the Spanish government under grant PID2021-122893NB-C22.

References

1. Ioannou, C.C., Laskowski, K.L.: A multi-scale review of the dynamics of collective behaviour: from rapid responses to ontogeny and evolution. Philos. Trans. Roy. Soc. B **378**, 20220059 (2023)

2. Vicsek, T., Czirók, A, Farkas, I.J, Helbing, D.: Application of statistical mechanics to collective motion in biology. Physica A **274**, 182–189 (1999)
3. Fodor, E., Marchetti, M.C.: The statistical physics of active matter: from self-catalytic colloids to living cells. Physica A **504**, 106–120 (2018)
4. Ouellette, N.T.: The most active matter of all. Matter **2**, 297–299 (2019)
5. Roca, C.P., Cuesta, J.A., Sánchez, A.: Evolutionary game theory: temporal and spatial effects beyond replicator dynamics. Phys. Life Rev. **6**(4), 208–249 (2009)
6. McAvoy, A., Allen, B., Nowak, M.A.: Social goods dilemmas in heterogeneous societies. Nat. Hum. Behav. **4**(8), 819–831 (2020)
7. Gordon, D.M.: Ant encounters: Interaction Networks and Colony Behavior. Cambridge University Press, Cambridge (2010)
8. Piñero, J., Solé, R.: Statistical physics of liquid brains. Philos. Trans. R. Soc. B **374**, 20180376 (2019)
9. Holley, R.A., Liggett, T.M.: Ergodic theorems for weakly interacting infinite systems and the voter model. Ann. Probab. **3**(4), 643–663 (1975)
10. Redner, S.: Reality-inspired voter models: a mini-review. Comptes Rendus Physique **20**(4), 275–292 (2019)
11. Jedrzejewski, A., Sznajd-Weron, K.: Statistical physics of opinion formation: is it a SPOOF? Comptes Rendus Physique **20**(4), 244–261 (2019)
12. De Marzo, G., Zaccaria, A., Castellano, C.: Emergence of polarization in a voter model with personalized information. Phys. Rev. Res. **2**, 043117 (2020)
13. Liggett, T.M.: Interacting Particle Systems, vol. 2. Springer, New York (1985)
14. Ramirez, L., San Miguel, M., Galla, T.: Local and global ordering dynamics in multistate voter models. Phys. Rev. E **106**(5), 054307 (2022)
15. Starnini, M., Baronchelli, A., Pastor-Satorras, R.: Ordering dynamics of the multi-state voter model. J. Stat. Mech.: Theor. Exp. **2012**(10), P10027 (2012)
16. Iannelli, G., De Marzo, G., Castellano, C.: Filter bubble effect in the multistate voter model. Chaos **32**(4), 043103 (2022)
17. Littler, R.A.: Loss of variability at one locus in a finite population. Math. Biosci. **25**(1–2), 151–163 (1975)
18. Kimura, M.: Random genetic drift in multi-allelic locus. Evolution **9**(4), 419–435 (1955)
19. Mobilia, M.: Does a single zealot affect an infinite group of voters? Phys. Rev. Lett. **91**(2), 028701 (2003)
20. Herrerías-Azcué, F., Galla, T.: Consensus and diversity in multistate noisy voter models. Phys. Rev. E **100**(2), 022304 (2019)
21. Oborny, B.: Lost in translation? Caveat to the application of the voter model in ecology and evolutionary biology. Sci. Prog. **106**(2), 368504231175324 (2023)
22. Sznajd-Weron, K., Sznajd, J.: Opinion evolution in closed community. Int. J. Mod. Phys. C **11**(6), 1157–1165 (2000)
23. Sznajd-Weron, K., Sznadj, J., Weron, T.: A review on the Sznajd model—20 years after. Physica A **565**, 125537 (2021)
24. Moran, P.A.P.: Random processes in genetics. Math. Proc. Camb. Philos. Soc. **54**, 60–71 (1958)
25. Constable, G.W.A., McKane, A.J.: Stationary solutions for metapopulation Moran models with mutation and selection. Phys. Rev. E **91**(3), 032711 (2015)
26. Constable, G.W.A., McKane, A.J.: Models of genetic drift as limiting forms of the Lotka-Volterra competition model. Phys. Rev. Lett. **114**(3), 038101 (2015)
27. Schneider, D.M., Martins, A.B., de Aguiar, M.A.M.: The mutation–drift balance in spatially structured populations. J. Theor. Biol. **402**, 9–17 (2016)
28. Molina, C., Earn, D.J.D.: The mutation-drift balance in spatially structured populations. J. Math. Biol. **76**(3), 645–678 (2018)
29. Heinrich, L., Müller, J., Teller, A., Zivkovic, D.: Effects of population- and seed bank size fluctuations on neutral evolution and efficacy of natural selection. Theor. Pop. Biol. **123**, 45–69 (2018)

30. Svoboda, J., Joshi, S., Chatterjee, K.: Amplifiers of selection for the Moran process with both Birth-death and death-Birth updating. PLoS Comput. Biol. **20**(3), e1012008 (2024)
31. Chave, J.: Spatial patterns and persistence of woody plant species in ecological communities. Am. Nat. **157**(1), 51–65 (2001)
32. Hidalgo, J., Suweis, S., Maritan, A.: Species coexistence in a neutral dynamics with environmental noise. J. Theor. Biol. **413**, 1–10 (2017)
33. Tu, C., Suweis, S., Grilli, J., Formentin, M., Maritan, A.: Reconciling cooperation, biodiversity and stability in complex ecological communities. Sci. Rep. **9**, 5580 (2019)
34. Martínez-García, R., López, C., Vázquez, F.: Species exclusion and coexistence in a noisy voter model with a competition-colonization tradeoff. Phys. Rev. E **103**, 032406 (2021)
35. Wang, C., et al.: Decision-making process during collective movement initiation in golden snub-nosed monkeys (Rhinopithecus roxellana). Sci. Rep. **10**, 480 (2020)
36. Winklmayr, C., Kao, A.B., Bak-Coleman, J.B., Romanczuk, P.: The wisdom of stalemates: consensus and clustering as filtering mechanisms for improving collective accuracy. Proc. R. Soc. B **287**, 20201802 (2020)
37. Rajendran, H., Haluts, A., Gov, N.S., Feinerman, O.: Ants resort to majority concession to reach democratic consensus in the presence of a persistent minority. Curr. Biol. **32**(3), 645–653 (2022)
38. Oro, D., Becares, J., Bartumeus, F., Arcos, J.M.: High frequency of prospecting for informed dispersal and colonisation in a social species at large spatial scale. Oecologia **197**, 395–409 (2021)
39. Kralj, J., et. al.: Active breeding seabirds prospect alternative breeding colonies. Oecologia **201**, 341–354 (2023)
40. Hanski, I.: Metapopulation dynamics. Nature **396**, 41–49 (1998)

Chapter 15
Starving Random Walks

L. Régnier, M. Dolgushev, and O. Bénichou

15.1 Introduction

When it comes to the exploration of a territory by an animal or biological tracers, a fundamental property in quantifying this exploration is the number $\mathcal{N}(t)$ of sites visited by time t [1, 2]. This quantity has been the focus of many mathematical [3–7] and physical [8–13] studies, in which the average, variance, and sometimes the full distribution have been explicitly obtained for a wide range of stochastic processes. These processes include nearest-neighbor jumps on a $d-$dimensional lattice, on fractal media, or RWs with long-range jumps, which model a large variety of biological motions. However, since $\mathcal{N}(t)$ is a cumulative quantity, it does not provide significant information about the past exploration dynamics of the walk: were all the visitations occurring regularly, or were there large time gaps where the RW did not encounter any food units? This differentiation becomes crucial when considering living organisms that rely on finding resources to propel their motion.

Recently, a new quantity has been introduced to enable this differentiation [8, 14–17]: the time τ_k. This represents the time it takes for a RW to find a site it has never visited before, given that k sites have already been visited. These random variables encapsulate the full dynamics of RW exploration and have been shown to be particularly useful in characterizing the main properties of the starving RW problem [14, 18], a minimal model to describe depletion-controlled dynamics of

L. Régnier · M. Dolgushev
Sorbonne Université, Laboratoire de Physique Théorique de la Matière Condensée (LPTMC), Paris, France
e-mail: leo.regnier.pro@outlook.fr; maxim.dolgushev@sorbonne-universite.fr

O. Bénichou (✉)
Sorbonne Université, CNRS, Laboratoire de Physique Théorique de la Matière Condensée (LPTMC), Paris, France
e-mail: benichou@lptmc.jussieu.fr

L. Giuggioli, P. K. Maini (eds.), *The Mathematics of Movement*,
https://doi.org/10.1007/978-3-032-00873-2_15

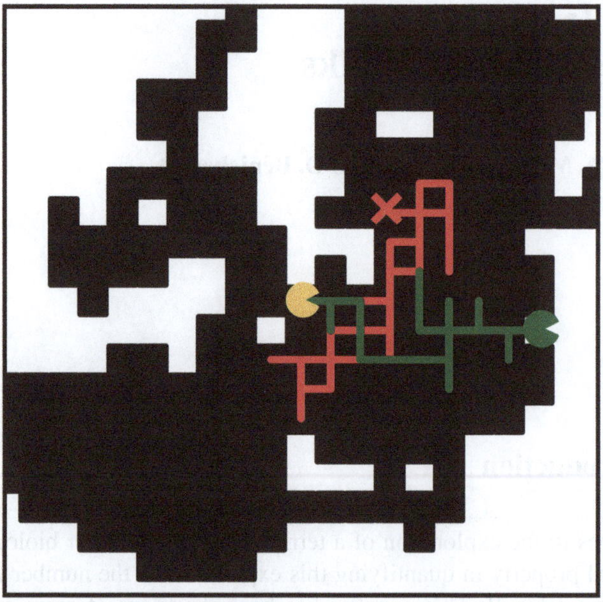

Fig. 15.1 The starving RW model. In the lattice, each site initially holds one food unit, which is consumed upon visit by the RW. The RW can make $\mathcal{S}$ steps (here $\mathcal{S} = 50$) without encountering food before starving. Two trajectories (green and red) are depicted. A forager (yellow) has consumed $\mathcal{N}(t) = 1000$ food units. Areas devoid of food are depicted in black. Along the green trajectory, the forager (green) promptly discovers a new food unit, so that $\tau_{1000} < \mathcal{S}$. Conversely, following the red trajectory, it fails to locate food before succumbing to starvation (depicted by the red cross), $\tau_{1000} \geq \mathcal{S}$

an RW [19, 20] as encountered in bacteria [21], animals [22], or robots [23]. In the original formulation of the problem, food units are placed on every site of the lattice, and the RW moves through the medium, consuming each food unit upon encounter. The RW starves if $\mathcal{S}$ time steps (the metabolic time) pass without a food encounter (see Fig. 15.1 for an illustration). This definition establishes a profound connection between τ_k and starvation: starvation occurs the first time we observe a time between visits to new sites, and thus, the finding of a new food unit, larger than the metabolic time. Consequently, understanding the long inter-visit times is fundamental for the search for food, necessitating the study of the extreme-value statistics [24–28]. We will explore this connection further in the following sections.

In this chapter, we examine the visitation properties of general classes of RWs, including anomalous diffusion. We investigate symmetric Markovian RWs that:

- Move in a medium of fractal dimension d_f, such that the number of sites within a ball of radius r grows as r^{d_f}.
- Have a distance to the origin r_t after t jumps that grows as t^{1/d_w}, where d_w is the walk dimension [29]. More precisely, the distribution of $r_t/t^{1/d_\text{w}}$ is independent of t for large t.

We note that diffusive RWs in hypercubic lattices of dimension d ($d_{\mathrm{w}} = 2$ and $d_{\mathrm{f}} = d$), as well as Lévy flights (jumps whose length distribution $p(\ell)$ is distributed algebraically with exponent $1+\alpha$, $p(\ell) \propto 1/|\ell|^{1+\alpha}$, such that $d_{\mathrm{w}} = 1/\alpha$ for $\alpha < 2$) are included in our description.

We begin with a brief summary of results on the visitation process of RWs, focusing particularly on the inter-visit time statistics τ_k of the general RWs under consideration, whose properties depend mainly on the ratio $\mu \equiv d_{\mathrm{f}}/d_{\mathrm{w}}$. Then, we delve into the two original cases in which the starving RW problem was examined [14, 18]: the unidimensional and infinite-dimensional cases, which can be viewed as the two limiting scenarios. Finally, we present the general framework for obtaining the properties of both the number N_S of sites visited at starvation and the lifetime T_S of the starving RW.

15.2 Visitation Properties

In this section, we review the results concerning the exploration dynamics of general Markovian processes, which will be useful throughout this chapter.

15.2.1 Number of Sites Visited

In the context of symmetric Markovian RWs characterized by a fractal dimension d_{f} and a walk dimension d_{w}, it was shown in Refs. [9, 30–32] that there exist two main different classes of exploration: either $\mu \equiv d_{\mathrm{f}}/d_{\mathrm{w}} \leq 1$, and the RW visits every site infinitely often, or $\mu > 1$, and the RW will never visit some sites. The first type of RWs is said to be recurrent, while the second type is said to be transient. Of note, the case $\mu = 1$ is differentiated and called marginally recurrent, which we will abbreviate to marginal in the following.

This distinction is particularly striking when one considers the average number of sites visited by time t (see Refs. [5, 6, 33]),

$$\langle N(t) \rangle \propto \begin{cases} t^{\mu}, & [\mu < 1], \\ \frac{t}{\ln t}, & [\mu = 1], \\ t, & [\mu > 1]. \end{cases} \tag{15.1}$$

The difference between the behavior at $\mu < 1$ and $\mu > 1$ can be understood as follows: while recurrent RWs typically visit the entirety of the ball of volume $r_t^{d_{\mathrm{f}}} \propto t^{\mu}$ (the number t of steps they perform is much larger than the number of accessible sites $r_t^{d_{\mathrm{f}}} = t^{\mu} \ll t$), transient RWs rarely return to previous sites and thus visit a new site at almost every time step. Such strong differences also appear in the exploration dynamics for both types of processes, as we proceed to show in the following section.

15.2.2 Inter-Visit Times Statistics

Below, we summarize the inter-visit time statistics, represented through their probability density $F_k(\tau) \equiv \mathbb{P}(\tau_k = \tau)$, obtained for general Markovian RWs in [15].

The results are as follows:

- For recurrent RWs, $\mu < 1$: there are two distinct time regimes. The first one occurs at times $\tau \ll t_k = k^{1/\mu}$, for which $F_k(\tau)$ decays algebraically with exponent $1 + \mu$, independently of k. Then, for $\tau \gg t_k$, $F_k(\tau)$ decays exponentially, at a rate of $1/k^{1/\mu}$. Overall, $F_k(\tau)$ admits a scaling form (for $\tau, k \gg 1$, $\tau/k^{1/\mu}$ fixed):

$$F_k(\tau) \sim \frac{1}{k^{1+\mu}} \psi(\tau/k^{1/\mu}), \tag{15.2}$$

where ψ is a scaling function that depends on the process.
- For transient RWs, $\mu > 1$: there are two distinct time regimes once again, but the early time regime $\tau \ll T_k = k^{1+1/\mu}$ is now characterized by an exponential decay in the variable $\tau^{\mu/(1+\mu)}$ (stretched exponential in τ), and the late time regime is an exponential decay in the variable $\tau/k^{1/\mu}$, as for the recurrent case. There is no algebraic decay.
- For marginal RWs, $\mu = 1$: there are now three distinct regimes. Firstly, the distribution $F_k(\tau)$ decays algebraically for $\tau \ll \sqrt{k}$, with an exponent of 2. This is followed by an exponential decay in the variable $\sqrt{\tau/\sqrt{k}}$ for $\sqrt{k} \ll \tau \ll k^{3/2}$, and finally, an exponential decay for $\tau \gg k^{3/2}$ with a rate of $1/k$. More precisely, in the cross-over regime $\tau \sim \sqrt{k}$, which will be particularly useful in the following, it was shown [34] that $F_k(\tau)$ admits a scaling form (for $\tau, k \gg 1$, $\tau/\sqrt{k}$):

$$F_k(\tau) \sim \frac{1}{k^{3/2}} \phi(\tau/\sqrt{k}), \tag{15.3}$$

where ϕ decays algebraically with an exponent of 3 at small arguments and exponentially with $\sqrt{\tau/\sqrt{k}}$ at large arguments. With this we get in the cross-over an algebraic decay τ^{-3} followed by the stretched exponential regime indicated in Table 15.1. This rich cross-over behavior comes from the analysis of the exit times from a cascade of the largest visited regions, see Ref. [34] for details.

Table 15.1 Summary of the time dependence of $F_k(\tau)$ for the three classes of RWs—recurrent, marginal, and transient. The constants are independent of k and τ. The crossover times t_k and T_k are given up to logarithmic prefactors

$F_k(\tau)$	t_k	T_k	$1 \ll \tau \ll t_k$	$t_k \ll \tau \ll T_k$	$T_k \ll \tau$
$\mu < 1$ [recurrent]	$k^{1/\mu}$	$k^{1/\mu}$			
$\mu = 1$ [marginal]	$\sqrt{k}$	$k^{3/2}$	$\tau^{-(1+\mu)}$	$\exp\left[-\text{const}\,(\tau/t_k)^{\mu/(1+\mu)}\right]$	$\exp\left[-\text{const}\,\tau/k^{1/\mu}\right]$
$\mu > 1$ [transient]	1	$k^{(\mu+1)/\mu}$			

Before applying these fundamental results describing the general τ_k properties to the starving RW problem, let us start with the two simplest cases were the inter-visit time statistics can be derived exactly.

15.3 Low- and High-Dimensional Cases

In this section, we review the two limit cases of unidimensional and infinite dimensional starving RWs, which were solved in [14, 18].

15.3.1 The 1d Case

We start with the case of a symmetric nearest-neighbor RW on a $1d$ line. In this model, the RW jumps with equal probability to one of its two neighboring lattice sites, on the left or on its right. In this situation, the domain visited by the RW is simple, as it is simply an interval, and the RW starts its new exploration of a new site on the border of the interval (new visits occur only on the left or right border). Thus, the time τ_k to visit a new site when k sites have already been visited is simply given by the time to exit this interval of length k. Consequently, the precise distribution $F_k(\tau)$ of τ_k is known, and given by the first exit time distribution of the interval starting on one of its border [35] in the diffusive scaling limit ($\tau, k \gg 1$, τ/k^2 fixed):

$$F_k(\tau) \sim \frac{2\pi}{k^2}\sum_{j=0}^{\infty}(2j+1)\sin\left(\frac{2j+1}{k}\pi\right)\exp\left[-(2j+1)^2\pi^2\tau/2k^2\right]. \qquad (15.4)$$

One can check that indeed Eq. (15.4) is contained in Table 15.1 for the value $\mu = 1/2$, as $d_{\mathrm{f}} = 1$ and $d_{\mathrm{w}} = 2$ (the RW is diffusive). Additionally, we note that the τ_k are independent, as the distribution of τ_k only depends on the size k of the visited interval. In the following, we see how the knowledge of the τ_k provides the distribution of the number $N_{\mathcal{S}}$ of sites visited at starvation and then of the lifetime $T_{\mathcal{S}}$.

15.3.1.1 Number of Sites Visited at Starvation

First, we connect the inter-visit time statistics to the starving RW problem using the following remark: to have visited at least n sites before starvation, it means that all the n first inter-visit times $\tau_0, \tau_1, \ldots, \tau_{n-1}$ are smaller than the metabolic time $\mathcal{S}$, or in other words, their maximum is smaller than $\mathcal{S}$. It states that there was no time interval longer than $\mathcal{S}$ between two visits. From this remark, we deduce that:

$$\mathbb{P}\left(N_{\mathcal{S}} \geq n\right) = \mathbb{P}\left(\max_{k=0,\ldots,n-1} \tau_k < \mathcal{S}\right) = \mathbb{P}\left(\tau_k < \mathcal{S},\ k = 0, \ldots, n-1\right). \tag{15.5}$$

We observe that, using the inter-visit times τ_k, determining the distribution of the number of sites visited at starvation can be seen as an extreme-value problem involving independent but non-identically distributed random variables. This type of problems has attracted increasing attention in both the mathematics and physics communities [36–39] in recent years. From Eq. (15.5) and the independence of the τ_k, we set that (in the limit of large n and $\mathcal{S}$):

$$\mathbb{P}\left(N_{\mathcal{S}} \geq n\right) = \prod_{k=0}^{n-1}\left(1 - \sum_{\tau=\mathcal{S}}^{\infty} F_k(\tau)\right) \sim \exp\left[-\int_0^n \mathrm{d}k \int_{\mathcal{S}}^{\infty} \mathrm{d}\tau\, F_k(\tau)\right]. \tag{15.6}$$

Finally, inserting the exact expression of $F_k(\tau)$ given by Eq. (15.4) in Eq. (15.6), we obtain:

$$-\ln \mathbb{P}\left(N_{\mathcal{S}} \geq n\right) \sim \sum_{j=0}^{\infty} \int_0^n \frac{4\mathrm{d}k}{k} \mathbb{E}^{-\frac{(2j+1)^2\pi^2\mathcal{S}}{2k^2}} = \sum_{j=0}^{\infty} \int_1^{\infty} \frac{2\mathrm{d}x}{x} \mathbb{E}^{-\frac{x(2j+1)^2\pi^2\mathcal{S}}{2n^2}}, \tag{15.7}$$

where we used that $\sin\left(\frac{2j+1}{k}\pi\right) \sim \frac{2j+1}{k}\pi$ at large k and performed the change of variable $x = n^2/k^2$. We note that Eq. (15.7) has a scaling form in the variable $\mathcal{S}/n^2$,

$$\mathbb{P}\left(N_{\mathcal{S}} \geq n\right) \sim \Xi_{1/2}\left(\frac{\mathcal{S}}{n^2}\right), \tag{15.8}$$

where $\Xi_{1/2}$ is the scaling function, independent of n and $\mathcal{S}$. From Eq. (15.8), we deduce the scaling with $\mathcal{S}$ of the number of sites visited at starvation:

$$\begin{aligned} \langle N_{\mathcal{S}} \rangle &\sim \sum_{n=0}^{\infty} \mathbb{P}\left(N_{\mathcal{S}} \geq n\right) \\ &\sim \int_0^{\infty} \mathrm{d}n\ \Xi_{1/2}\left(\frac{\mathcal{S}}{n^2}\right) = \sqrt{\mathcal{S}} \int_0^{\infty} \mathrm{d}u\ \Xi_{1/2}\left(\frac{1}{u^2}\right) \approx 2.9\sqrt{\mathcal{S}}. \end{aligned} \tag{15.9}$$

Additionally, one can derive the asymptotic properties of $\Xi_{1/2}(\mathcal{S}/n^2)$ at large and small arguments. For small number of sites visited at starvation $n/\sqrt{\mathcal{S}} \ll 1$, the sum in Eq. (15.7) is dominated by the $j = 0$ term, such that:

$$-\ln \mathbb{P}(N_{\mathcal{S}} \geq n) \sim \int_1^\infty \frac{2\mathrm{d}x}{x} \exp\left[-x\frac{\pi^2 \mathcal{S}}{2n^2}\right] \sim \frac{4n^2}{\pi^2 \mathcal{S}} \exp\left[-\frac{\pi^2 \mathcal{S}}{2n^2}\right]. \tag{15.10}$$

In the opposite limit of large number of sites visited at starvation $n/\mathcal{S}^2 \gg 1$, we use that $F_k(\tau)$ converges to the first passage time distribution $F_\infty(\tau)$ to the site at distance one as the RW does not see the other border of the visited interval. The expression of $F_\infty(\tau)$ is well-known and given by [35]:

$$F_\infty(\tau) = \frac{1}{\sqrt{2\pi\tau^3}} \mathbb{E}^{-1/2\tau}. \tag{15.11}$$

Once we insert this expression in (15.6), we get:

$$-\ln \mathbb{P}(N_{\mathcal{S}} \geq n) = \sum_{k=0}^{n-1} \int_{\mathcal{S}}^\infty F_\infty(\tau)\mathrm{d}\tau = n \int_{\mathcal{S}}^\infty \frac{\mathrm{d}\tau}{\sqrt{2\pi\tau^3}} \mathbb{E}^{-1/2\tau} \sim \frac{n}{\sqrt{8\pi\mathcal{S}}}. \tag{15.12}$$

15.3.1.2 Lifetime

We now proceed to determine the lifetime $T_{\mathcal{S}}$. Initially, we calculate the average lifetime of the RW, which is the sum of inter-visit times before starvation plus the metabolic time, representing the final search for a food unit. This can be expressed as:

$$\langle T_{\mathcal{S}} \rangle = \left\langle \sum_{k=1}^{N_{\mathcal{S}}} \tau_k \right\rangle + \mathcal{S} = \sum_{n=1}^{\infty} \left(\sum_{k=1}^{n} \langle \tilde{\tau}_k \rangle \right) \mathbb{P}(N_{\mathcal{S}} = n) + \mathcal{S}. \tag{15.13}$$

Here, $\tilde{\tau}_k$ denotes the kth inter-visit time conditioned to be less than the metabolic time $\mathcal{S}$. The average of the conditional inter-visit time is given by:

$$\langle \tilde{\tau}_k \rangle = \frac{\int_0^{\mathcal{S}} \tau F_k(\tau)\mathrm{d}\tau}{\int_0^{\mathcal{S}} F_k(\tau)\mathrm{d}\tau}. \tag{15.14}$$

Considering the limit $k \to \infty$ for large visited intervals, τ_k follows an asymptotic distribution noted F_∞, representing the time to exit a semi-infinite interval starting one unit length from the border as given in Eq. (15.11). Consequently, the limit as $k \to \infty$ of Eq. (15.14) is asymptotically:

$$\langle \tilde{\tau}_\infty \rangle = \frac{\int_0^{\mathcal{S}} \tau F_\infty(\tau) \mathrm{d}\tau}{\int_0^{\mathcal{S}} F_\infty(\tau) \mathrm{d}\tau} \sim \sqrt{\frac{2}{\pi}} \sqrt{\mathcal{S}}. \tag{15.15}$$

Utilizing Eq. (15.13), we obtain:

$$\begin{aligned} \langle T_{\mathcal{S}} \rangle - \mathcal{S} &\sim \sum_{n=1}^{\infty} \left(\sum_{j=1}^{n} \langle \tilde{\tau}_\infty \rangle \right) \mathbb{P}\left(N_{\mathcal{S}} = n \right) = \langle \tilde{\tau}_\infty \rangle \sum_{n=1}^{\infty} n \mathbb{P}\left(N_{\mathcal{S}} = n \right) \\ &= \langle \tilde{\tau}_\infty \rangle \langle N_{\mathcal{S}} \rangle \approx 2.3\mathcal{S}. \end{aligned} \tag{15.16}$$

We can demonstrate more generally that the nth cumulant of the lifetime $T_{\mathcal{S}}$ scales as $\mathcal{S}^n$, implying that the random variable $T_{\mathcal{S}}/\mathcal{S}$ is independent of $\mathcal{S}$ for large $\mathcal{S}$. Denoting by $\kappa_n(\ldots)$ the nth cumulant function ($n > 1$), we find that for the nth cumulant of the sum of $N_{\mathcal{S}}$ random variables [40], we have:

$$\kappa_n\left(T_{\mathcal{S}}\right) \sim \sum_{k_1,\ldots,k_n \in S_n} \frac{n! \kappa_{\sum_{i=1}^n k_i}\left(N_{\mathcal{S}}\right)}{k_1! \ldots k_n! (1!)^{k_1} \ldots (n!)^{k_n}} \kappa_1(\tilde{\tau}_\infty)^{k_1} \ldots \kappa_n(\tilde{\tau}_\infty)^{k_n}. \tag{15.17}$$

where S_n is the set of all vectors $(k_1, \ldots, k_n) \in \mathbb{N}^n$ such that $\sum_{i=1}^n i k_i = n$. Then, from Eq. (15.11), we find for any n,

$$\left\langle \tilde{\tau}_\infty^n \right\rangle = \frac{\int_0^{\mathcal{S}} \tau^n F_\infty(\tau) \mathrm{d}t}{\int_0^{\mathcal{S}} F_\infty(\tau) \mathrm{d}\tau} \sim \sqrt{\frac{2}{\pi(2n-1)^2}} \mathcal{S}^{n-1/2}. \tag{15.18}$$

which implies that $\kappa_n\left(\tilde{\tau}_\infty\right) \sim \left\langle \tilde{\tau}_\infty^n \right\rangle$. Finally, we obtain that for any vector $(k_1, \ldots, k_n)$ of S_n, using that $N_{\mathcal{S}}$ is scale-invariant from Eq. (15.8) such that, for any m, $\kappa_m(N_{\mathcal{S}}) \propto \mathcal{S}^m$,

$$\begin{aligned} \kappa_{\sum_{i=1}^n k_i}(N_{\mathcal{S}}) \kappa_1(\tilde{\tau}_\infty)^{k_1} \ldots \kappa_n(\tilde{\tau}_\infty)^{k_n} &\propto \left(\sqrt{\mathcal{S}}\right)^{\sum_{i=1}^n k_i} \mathcal{S}^{k_1(1-1/2)} \ldots \mathcal{S}^{k_n(n-1/2)} \\ &\propto \mathcal{S}^{\sum_{i=1}^n i k_i} = \mathcal{S}^n. \end{aligned} \tag{15.19}$$

All the terms in the sum of Eq. (15.17) scale as $\mathcal{S}^n$. This leads to the following scaling with $\mathcal{S}$ for all the cumulants of the lifetime,

$$\kappa_n\left(T_{\mathcal{S}}\right) \propto \mathcal{S}^n. \tag{15.20}$$

In particular, we deduce that the lifetime is scale-invariant and scales linearly with the metabolic time.

Several important properties stem from the analysis of the $1d$ starving RW:

1. The number of sites visited at starvation grows diffusively with the metabolic time $\mathcal{S}$, $N_{\mathcal{S}} \propto \sqrt{\mathcal{S}}$.
2. The average lifetime grows linearly with the metabolic time $\mathcal{S}$, $T_{\mathcal{S}} \propto \mathcal{S}$.
3. The distribution of the number of sites visited at starvation and the lifetime are scale-invariant, in the sense that $N_{\mathcal{S}}/\sqrt{\mathcal{S}}$ and $T_{\mathcal{S}}/\mathcal{S}$ are non-deterministic and independent of $\mathcal{S}$ for large $\mathcal{S}$. Additionally, the distribution of $N_{\mathcal{S}}$ decays exponentially at large number of distinct sites visited.

These properties are in agreement with what could be expected from the recurrence of the 1D RW (see Sect. 15.2.1): because the RW is recurrent, the starving process stops rapidly, as the RW often comes back to previously visited sites. Besides, we expect the RW not to find a lot of food units before starving. These two observations are in agreement with the linearity of the lifetime with the metabolic time, which is the minimal dependence with $\mathcal{S}$ (as the RW performs at least $\mathcal{S}$ steps before starving). The number of sites visited at starvation corresponds to the domain visited during this short lifetime (see Eq. (15.1) for $\mu = 1/2$), $N_{\mathcal{S}} \propto \sqrt{T_{\mathcal{S}}} \propto \sqrt{\mathcal{S}}$.

One might wonder whether these results hold when increasing the dimension. As a first example, the other limit case of infinite dimension was considered in [14, 18].

15.3.2 The Infinite Dimensional Case

In this section, we consider the starving RW in a medium of large dimension d. A first remark is that in the case of a nearest-neighbor RW in dimension d, the constant prefactor in (15.1) in the transient case can be shown to go to 1 as the dimension increases (see Refs. [9, 41]). This implies that the RW visits a new site at almost every time step, and thus, the effect of depletion is small. Consequently, for the RW not visiting a new site, the most likely scenario is to repeat the previous step, returning to its previous position, which occurs with probability $1/(2d)$, a small value. This justifies our assumption that at every time step, the RW visits a new site with probability λ, independent of time, a value that diminishes in larger dimensions (where λ can be interpreted as a generalization of $1/(2d)$ to an RW in a large-dimensional space). Additionally, visitation events are assumed to be independent: they occur rarely, and the visited domain contains almost no loops, as it seldom retraces its past trajectory.

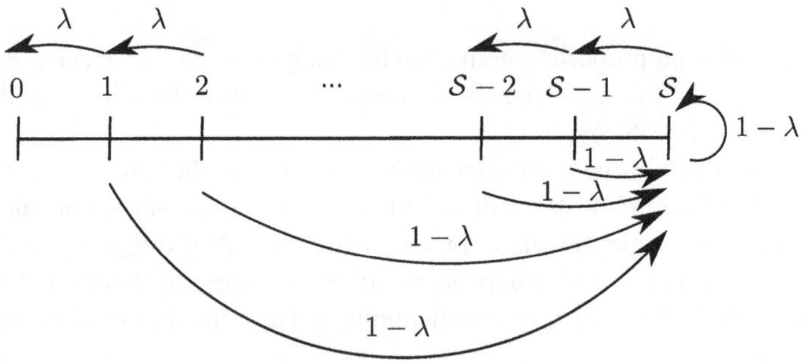

Fig. 15.2 Effective walk representing the starving RW in infinite dimension. The effective RW starts at position $\mathcal{S}$, takes one step closer to 0 with probability λ

These approximations lead to the effective starving RW model represented in Fig. 15.2 (see Ref. [18]): the effective RW starts at position $\mathcal{S}$, takes one step closer to 0 with probability λ, and returns to $\mathcal{S}$ with probability $1-\lambda$ (signifying the finding of a food unit). Similar to the $1d$ case, the derivation begins with determining the number of sites visited at starvation, followed by determining the lifetime.

15.3.2.1 Number of Sites Visited at Starvation

As a first step, we note that Eq. (15.5) holds, as one can reexpress the probability of visiting at least n sites before starving via the statistics of the τ_k, specifically the probability that each of the τ_k is smaller than $\mathcal{S}$ for $k = 0, \ldots, n-1$. The probability that $\tau_k \geq \mathcal{S}$ is exactly given by the probability that the effective RW, represented in Fig. 15.2, arrives at 0 before returning to $\mathcal{S}$, starting from $\mathcal{S}$ (since the RW just visited a new site). This probability, denoted as $1 - P_{\mathcal{S}} \equiv \mathbb{P}(\tau_k \geq \mathcal{S})$, is independent of k and is given by $1 - P_{\mathcal{S}} = \lambda^{\mathcal{S}}$, the probability that none of the $\mathcal{S}$ steps lead back to $\mathcal{S}$ (this scaling is in agreement with the $\mu \to \infty$ case of Table 15.1). Particularly, we note that the τ_k are again independent, as the state at which any visit starts is the same (the effective RW starts at $\mathcal{S}$).

Using Eq. (15.5), we obtain that:

$$\mathbb{P}\left(N_{\mathcal{S}} \geq n\right) = \prod_{k=0}^{n-1} \mathbb{P}(\tau_k < \mathcal{S}) = P_{\mathcal{S}}^n = (1 - \lambda^{\mathcal{S}})^n \sim \mathbb{E}^{-n\lambda^{\mathcal{S}}} \qquad (15.21)$$

in the limit of large n and small λ. In particular, we note that the distribution of $N_{\mathcal{S}}$ is exponential with a rate of $\lambda^{\mathcal{S}}$, such that:

$$\langle N_{\mathcal{S}} \rangle \sim 1/\lambda^{\mathcal{S}}. \qquad (15.22)$$

Finally, we notice that the distribution of $N_{\mathcal{S}}/\langle N_{\mathcal{S}}\rangle = N_{\mathcal{S}}\lambda^{\mathcal{S}}$ is again independent of $\mathcal{S}$, as in the $1d$ case, see Eq. (15.8).

15.3.2.2 Lifetime

Next, we determine the distribution of the lifetime $T_{\mathcal{S}}$. To do so, we first compute the probability $F(n,t)$ for the effective RW to arrive at 0 for the first time at step t starting from position n. This distribution obeys a recurrence relation,

$$F(0,t) = \delta_t,\ \ F(n,t) = \lambda F(n-1,t-1) + (1-\lambda)F(\mathcal{S},t-1), \tag{15.23}$$

where δ_t is the Kronecker symbol (1 if $t = 0$, 0 otherwise). To solve this recurrence equation, we Laplace transform Eq. (15.23), defined for any n and s by $\hat{F}(n,s) = \mathcal{L}(F(n,t)) \equiv \sum_{t=0}^{\infty} F(n,t)\mathbb{E}^{-st}$. This leads to the following equation for $\hat{F}(n,s)$:

$$\hat{F}(0,s) = 1,\ \ \mathbb{E}^{-s}\hat{F}(n,s) = \lambda\hat{F}(n-1,s) + (1-\lambda)\hat{F}(\mathcal{S},s). \tag{15.24}$$

The solution is given by:

$$\hat{F}(n,s) = \left(\lambda\mathbb{E}^s\right)^n\left(1 - \frac{\mathbb{E}^s(1-\lambda)}{1-\lambda\mathbb{E}^s}\hat{F}(\mathcal{S},s)\right) + \frac{\mathbb{E}^s(1-\lambda)}{1-\lambda\mathbb{E}^s}\hat{F}(\mathcal{S},s). \tag{15.25}$$

In particular, Eq. (15.25) gives the average lifetime expression, by noting that $\mathbb{P}(T_{\mathcal{S}} = t) = F(\mathcal{S},t)$:

$$\langle T_{\mathcal{S}}\rangle = \sum_{t=0}^{\infty} tF(\mathcal{S},t) = -\partial_s\hat{F}(\mathcal{S},s)_{s=0} = \lambda^{-\mathcal{S}}\left(\frac{1-\lambda^{\mathcal{S}}}{1-\lambda}\right) \sim 1/\lambda^{\mathcal{S}}, \tag{15.26}$$

which shows that the average lifetime grows exponentially with the metabolic time $\mathcal{S}$. Additionally, we note that in the limit of small λ and large metabolic time $\mathcal{S}$, keeping $s/\lambda^{\mathcal{S}}$ fixed,

$$\mathcal{L}(\mathbb{P}(T_{\mathcal{S}} = t)) \sim \frac{\lambda^{\mathcal{S}}}{s+\lambda^{\mathcal{S}}}, \tag{15.27}$$

so that:

$$\mathbb{P}(T_{\mathcal{S}} = t) \sim \lambda^{\mathcal{S}}\mathbb{E}^{-t\lambda^{\mathcal{S}}} = \frac{1}{\langle T_{\mathcal{S}}\rangle}\mathbb{E}^{-t/\langle T_{\mathcal{S}}\rangle}, \tag{15.28}$$

which is exponential and scale-invariant in the sense that $T_{\mathcal{S}}/\langle T_{\mathcal{S}}\rangle$ is a non-deterministic random variable independent of $\mathcal{S}$.

Several important properties stem from the analysis of the infinite-dimensional starving RW:

1. The number of sites visited at starvation grows exponentially with the metabolic time $\mathcal{S}$.
2. The average lifetime grows exponentially with the metabolic time $\mathcal{S}$.
3. The distribution of both the number of sites visited at starvation and the lifetime are scale-invariant, in the sense that $N_{\mathcal{S}}\lambda^{\mathcal{S}}$ and $T_{\mathcal{S}}\lambda^{\mathcal{S}}$ are independent of $\mathcal{S}$ for large $\mathcal{S}$. Additionally, both asymptotic distributions of the rescaled variables decay exponentially at large argument.

These observations are in agreement with what could be expected for a transient RW: it rarely comes back to previously visited sites, such that the starving process lasts for a long time and finds a lot of sites before starving. This is in agreement with the exponential growth with the metabolic time of both the lifetime and the number of sites visited.

Let us now bridge the gap between the one- and infinite-dimensional cases.

15.4 General d−Dimensional Case: The Maximum of Inter-Visit Times

As highlighted in the calculations of the unidimensional and infinite-dimensional cases, the derivation of the starving RWs' main observables starts with the determination of the probability that the n first intervisit times are smaller than the metabolic time $\mathcal{S}$, as in Eqs. (15.5) and (15.21). The probability of this event can be obtained from the statistics of the maximum $M_n \equiv \max\limits_{k=0,\ldots,n-1} \tau_k$:

$$\{\tau_k < \mathcal{S},\ k = 0,\ \ldots,\ n-1\} = \left\{ \max_{k=0,\ldots,n-1} \tau_k < \mathcal{S} \right\} = \{M_n < \mathcal{S}\}. \tag{15.29}$$

However, the τ_k are a priori correlated in the general case: indeed, even though the RW is Markovian, values of τ_k can be representative of different geometries of the visited domain. In the following, we neglect these correlations, because we expect them to be too weak to affect the statistics of the maximum (see [34, 42] for a quantitative argument about situations where the correlations can be neglected). This is the reason why we make use of the following approximate expression for the maximum's cumulative distribution function:

$$\mathbb{P}(M_n \leq T) \approx \prod_{k=0}^{n-1} \mathbb{P}(\tau_k \leq T) \sim \exp\left[-\sum_{k=0}^{n-1} \int_T^{\infty} F_k(\tau)\mathrm{d}\tau \right]. \tag{15.30}$$

15.4.1 Recurrent Walks

For recurrent walks ($\mu < 1$), the probability distribution function of τ_k presents a scaling form (see Eq. (15.2)), where the scaling function ψ is algebraic at small argument and exponential at large argument (see Table 15.1). Pluging this information into Eq. (15.30) implies that the cumulative distribution function of M_n also has a scaling form:

$$-\ln \mathbb{P}(M_n \leq T) \approx \sum_{k=0}^{n-1} \int_T^{\infty} k^{-1-1/\mu} \psi(\tau/k^{1/\mu}) \mathrm{d}\tau$$
$$\sim \int_0^{n/T^\mu} \frac{\mathrm{d}v}{v} \int_{v^{-1/\mu}}^{\infty} \psi(u) \mathrm{d}u = -\ln \Xi_\mu(T/n^{1/\mu}). \quad (15.31)$$

Let us now consider the limit behaviors of $\Xi_\mu(m)$ when $m \equiv T/n^{1/\mu} \ll 1$ and $m \gg 1$.

First, in the limit $m \gg 1$, using Table 15.1,

$$-\ln \Xi_\mu(m) \propto \int_0^{1/m^\mu} \frac{\mathrm{d}x'}{x'} \mathbb{E}^{-A/x'^{1/\mu}} \propto \int_m^{\infty} \frac{\mathrm{d}x'}{x'} \mathbb{E}^{-Ax'} \propto \mathbb{E}^{-Am}/m. \quad (15.32)$$

Then, in the limit $m \ll 1$

$$-\ln \Xi_\mu(m) \propto \text{const.} + \int_1^{1/m^\mu} \frac{\mathrm{d}x'}{x'} x' \propto m^{-\mu}. \quad (15.33)$$

We will see the implications of these scalings on the starving RW problem in Sect. 15.5.

15.4.2 Marginal Walks

Based on the functional form (15.3) for $F_k(\tau)$, we expect $M_n/\sqrt{n}$ to converge to a nontrivial distribution at large n. Indeed, for $m > 0$ fixed and $n \to \infty$, we have that:

$$-\ln \mathbb{P}(M_n \leq m\sqrt{n}) \approx \int_0^n \mathrm{d}k \int_{m\sqrt{n}}^{\infty} \mathrm{d}\tau F_k(\tau)$$
$$= \int_0^{m^{2/3}n^{1/3}} \mathrm{d}k \int_{m\sqrt{n}}^{\infty} \mathrm{d}\tau \exp\left[-\text{const.}\ \tau/k\right] + \int_{m^{2/3}n^{1/3}}^{n} \frac{\mathrm{d}k}{k} \int_{m\sqrt{n}}^{\infty} \frac{\mathrm{d}\tau}{\sqrt{k}} \phi\left(\frac{\tau}{\sqrt{k}}\right)$$
$$\sim \int_0^{1/m^2} \frac{\mathrm{d}u}{u} \int_{1/\sqrt{u}}^{\infty} \mathrm{d}t \phi(t) = -\ln \Xi_1(m). \quad (15.34)$$

From this, it is apparent that $M_n/\sqrt{n}$ has an asymptotic limit distribution, denoted by Ξ_1. The properties of $\Xi_1(m)$ at small and large m are directly derived from the properties of ϕ introduced in Eq. (15.3) and following similar steps as Eqs. (15.32) and (15.33):

$$-\ln \Xi_1(m) \propto \begin{cases} m^{-2}, & \text{for } m \ll 1, \\ \mathbb{E}^{-A\sqrt{m}}, & \text{for } m \gg 1, \end{cases} \tag{15.35}$$

where A is a process-dependent constant and all algebaric prefactors have been neglected in the $m \gg 1$ regime.

15.4.3 Transient Walks

For transient walks, $\mu > 1$, according to Table 15.1, the decrease of the distribution of τ_k is first exponential in $\tau^{\mu/(1+\mu)}$ (stretched exponential in τ of exponent $\delta = \mu/(1+\mu)$), a regime that lasts up to time $T_k = k^{1+1/\mu}$, and next exponential in $\tau/k^{1/\mu}$. On the other hand, the maximum M_n of n random variables with a stretched exponential tail of exponent δ has an average, which grows as $\langle M_n \rangle \sim a\,(\ln n)^{1/\delta} = a\,(\ln n)^{1+1/\mu}$ (a a model-dependent constant) and a standard deviation, which grows as $\sqrt{\mathrm{Var}\,(M_n)} = \sqrt{\langle M_n^2 \rangle - \langle M_n \rangle^2} \propto (\ln n)^{1/\delta - 1} = (\ln n)^{1/\mu}$, as shown in Ref. [24]. In particular, this means that the maximum is peaked around its mean (standard deviation much smaller than its mean), which is itself small compared to the time $n^{1+1/\mu}$ (typical value of T_k for k between 0 and $n - 1$) at which the stretched exponential regime would stop. We conclude that the asymptotic distribution of the maximum M_n of the τ_k is indeed not influenced by the exponential decay at large times. Thus, the limit law (centered and normalized) of M_n corresponds to the celebrated Gumbel distribution [43] of zero mean and unit variance of cumulative distribution function we note Ξ_∞. For any value of μ, the cumulative distribution function of M_n centered and normalized is given by:

$$-\ln \Xi_\mu(m) = -\ln \Xi_\infty(m) = \mathbb{E}^{-m\mathbb{E}^{-\gamma_E}}, \tag{15.36}$$

where γ_E is the Euler constant and the rescaled random variable m is defined in Eq. (15.37).

Important

By considering the rescaled random variable (a a model-dependent constant)

$$m \equiv \begin{cases} M_n/n^{1/\mu}, & [\mu < 1], \\ M_n/\sqrt{n}, & [\mu = 1], \\ \frac{M_n - a(\ln n)^{1/\mu+1}}{(\ln n)^{1/\mu}}, & [\mu > 1], \end{cases} \tag{15.37}$$

Table 15.2 Asymptotic properties of the cumulative distribution Ξ_μ of the properly rescaled maximum for different types of processes. All constants have been put to one for simplicity

$-\ln \Xi_\mu(m)$	$m \ll 1$	$m \gg 1$
$\mu < 1$ [recurrent]	$m^{-\mu}$	$\mathbb{E}^{-m}/m$
$\mu = 1$ [marginal]	m^{-2}	$\mathbb{E}^{-\sqrt{m}}$
$\mu > 1$ [transient]	$\mathbb{E}^{-m}$	$\mathbb{E}^{-m}$

we obtain its asymptotic cumulative distribution Ξ_μ, whose behavior at large and small arguments is summarized in Table 15.2.

15.5 Starving Random Walks

We now show that the knowledge of the cumulative distribution function of M_n allows one to determine the statistics of the number of sites visited at starvation and the lifetime of a starving RW.

15.5.1 *Number of Sites Visited at Starvation*

We start from the remark that having M_n smaller than the metabolic time $\mathcal{S}$ is the same as having visited at least n sites before starvation, thus

$$\mathbb{P}(N_\mathcal{S} \geq n) = \mathbb{P}(M_n < \mathcal{S}) . \tag{15.38}$$

This means that one can deduce the distribution of $N_\mathcal{S}$ directly from that of M_n.

- For recurrent RWs: The distribution of $N_\mathcal{S}$ is scale-invariant (so that $N_\mathcal{S}/\mathcal{S}^\mu$ is independent of $\mathcal{S}$) and its asymptotics is given by:

$$-\ln \mathbb{P}(N_\mathcal{S}/\mathcal{S}^\mu \geq x) = -\ln \Xi_\mu(x^{-1/\mu}) \propto \begin{cases} \mathbb{E}^{-A/x^{1/\mu}} x^{1/\mu}, & \text{for } x \ll 1, \\ x, & \text{for } x \gg 1. \end{cases} \tag{15.39}$$

  Of note, this result is consistent with what we found in Sect. 15.3.1, Eqs. (15.7), (15.12) corresponding to $\mu = 1/2$.
- For marginal RWs: As $\mathbb{P}(N_{\mathcal{S}=\sqrt{n}x} \geq n) = \mathbb{P}(M_n \leq \sqrt{n}x) \to \Xi_1(x)$, with $\Xi_1(x)$ of Eq. (15.35), we get that $\mathbb{P}(N_\mathcal{S}/\mathcal{S}^2 \geq x) \to \Xi_1(1/\sqrt{x})$. In particular, the distribution has a single-parameter scaling at large starvation index $\mathcal{S}$ and number of visited sites at starvation $N_\mathcal{S}$, and its average and standard deviation scale both as $\mathcal{S}^2$. Based on Eq. (15.35), up to logarithmic corrections,

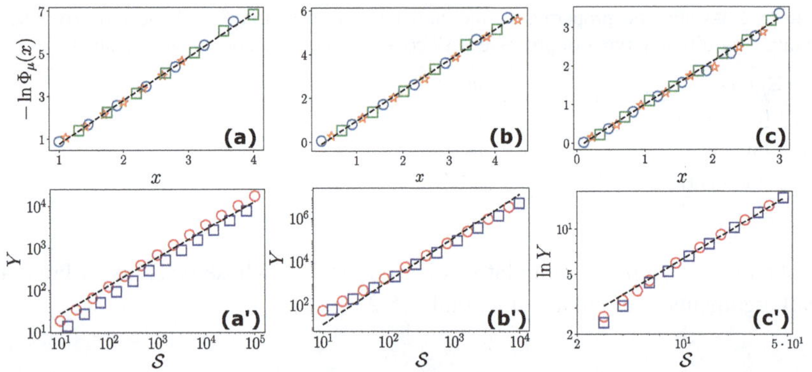

Fig. 15.3 **Number of sites visited at starvation** $N_{\mathcal{S}}$**.** **(a)–(c)** Distributions of the rescaled variable $x \equiv N_{\mathcal{S}}/\langle N_{\mathcal{S}}\rangle$ (insets illustrate the small x–values) and **(a′)–(c′)** the corresponding averages (red circles) and standard deviations (blue squares) of $N_{\mathcal{S}}$. The black dashed lines correspond to the best fit of the theory (see Eqs. (15.39), (15.40), (15.41), and (15.42)). Different universality classes are represented by **(a)** $1d$ Lévy flights of parameter $\alpha = 1.5$ (recurrent), $\mathcal{S} = 14667$, 31622 and 68129; **(b)** nearest neighbor $2d$ RWs, $\mu = 1$ (marginal), $\mathcal{S} = 2335$, 4832 and 10^4; **(c)** nearest neighbor $3d$ RWs, $\mu = 3/2$ (transient), $\mathcal{S} = 12$, 17 and 22. Increasing values of $\mathcal{S}$ are represented by blue circles, orange stars, and green squares

$$-\ln \mathbb{P}(N_{\mathcal{S}}/\mathcal{S}^2 \geq x) \propto \begin{cases} \mathbb{E}^{-A/x^{1/4}}, & \text{for } x \ll 1, \\ x, & \text{for } x \gg 1. \end{cases} \tag{15.40}$$

- For transient RWs, considering $\langle M_n\rangle \sim a(\ln n)^{\frac{1+\mu}{\mu}}$ and $\sqrt{\mathrm{Var}(M_n)} \sim b(\ln n)^{\frac{1}{\mu}} \equiv \frac{a(1+\mu)\sqrt{6}}{\pi\mu}(\ln n)^{\frac{1}{\mu}}$ [24], the probability $\mathbb{P}\left(\frac{M_n - a(\ln n)^{1/\mu+1}}{b(\ln n)^{1/\mu}} \leq x\right)$ converges to a Gumbel distribution $\Xi_\infty(x)$ with zero mean and unit variance. Consequently, $\mathbb{P}(N_{\mathcal{S}=xb(\ln n)^{1/\mu}+a(\ln n)^{1/\mu+1}} \geq n)$ also converges to $\Xi_\infty(x)$ for large n. By inverting $\mathcal{S} = xb(\ln n)^{1/\mu} + a(\ln n)^{1/\mu+1}$ in the limit of large n, we obtain:

$$\mathbb{P}(N_{\mathcal{S}}/\exp\left[(\mathcal{S}/a)^{\mu/(1+\mu)}\right] \geq x) \sim \Xi_\infty\left(-\frac{\sqrt{6}}{\pi}\ln x\right) = \exp\left[-x\mathbb{E}^{-\gamma_E}\right]. \tag{15.41}$$

  This implies that $N_{\mathcal{S}}$ asymptotically follows a single-parameter scaling: both its average and standard deviation grow exponentially with $\mathcal{S}^{\mu/(1+\mu)}$. Moreover, the asymptotic distribution of $N_{\mathcal{S}}$ is exponential, akin to the infinite-dimensional case (refer to Eq. (15.21)). Finally, by taking the limit $\mu \to \infty$ in the expression for the average of $N_{\mathcal{S}}$, we recover the exponential growth of the average number of visited sites at starvation with the metabolic time, as obtained for the infinite-dimensional case in Eq. (15.22).

The average, standard deviation, and asymptotic distribution of $N_{\mathcal{S}}$ in the recurrent, marginal, and transient cases are checked numerically in Fig. 15.3 for representative RWs of each class.

Table 15.3 Asymptotic properties of the minus log tail distribution Φ_μ of the rescaled number of sites visited at starvation for different type of processes. All constants have been put to one for simplicity

$-\ln \Phi_\mu(x)$	$x \ll 1$	$x \gg 1$
$\mu < 1$ [recurrent]	$\mathbb{E}^{-1/x^{1/\mu}}/x^{1/\mu}$	x
$\mu = 1$ [marginal]	$\mathbb{E}^{-1/x^{1/4}}$	x
$\mu > 1$ [transient]	x	x

> **Important**
>
> By considering the rescaled number of sites visited at starvation $\frac{N_S}{\langle N_S \rangle}$, where the asymptotics of $\langle N_S \rangle$ is given by
>
> $$\langle N_S \rangle \sim \begin{cases} S^\mu, & [\mu < 1], \\ S^2, & [\mu = 1], \\ \exp\left[S^{\mu/(1+\mu)}\right], & [\mu > 1], \end{cases} \tag{15.42}$$
>
> we obtain its asymptotic tail distribution $\mathbb{P}\left(\frac{N_S}{\langle N_S \rangle} \geq x\right) = \Phi_\mu(x)$, whose behavior at large and small arguments is summarized in Table 15.3.

15.5.2 *Lifetime*

We proceed by demonstrating that the distribution of N_S gives access to the distribution of the lifetime T_S.

The lifetime T_S is determined by the accumulation of inter-visit times $\tilde{\tau}_k$ ($k < N_S$), each corresponding to τ_k conditioned on being less than the metabolic time S. Expressing the distribution of T_S yields:

$$\begin{aligned} &\mathbb{P}(T_S = t) \\ &= \int_0^\infty dn \mathbb{P}\left(\sum_{k=0}^{N_S - 1} \tau_k + S = t | N_S = n\right) \mathbb{P}(N_S = n) \\ &\approx \int_0^\infty dn \delta\left(n \langle \tilde{\tau}_\infty \rangle + S - t\right) \mathbb{P}(N_S = n). \end{aligned} \tag{15.43}$$

In this derivation, we exploit the fact that for large k, the distribution of $\tilde{\tau}_k$ becomes independent of k, and the sum follows the law of large numbers, $\sum_{k=0}^{n-1} \langle \tilde{\tau}_k \rangle \sim n \lim_{k \to \infty} \langle \tilde{\tau}_k \rangle = n \langle \tilde{\tau}_\infty \rangle$. This leads to the tail distribution:

$$\mathbb{P}(T_{\mathcal{S}} \geq t) \approx \mathbb{P}\left(N_{\mathcal{S}} \geq \frac{t-\mathcal{S}}{\langle \tilde{\tau}_\infty \rangle}\right). \tag{15.44}$$

Then, we proceed with the characterization of $\tilde{\tau}_\infty$, which varies for each type of RW.

- For recurrent RWs: At early times, the distribution $F_k(\tau)$ of inter-visit time τ_k behaves as $F_k(\tau) \propto \tau^{-1-\mu}$ (see Table 15.1). Consequently, the average value of $\tilde{\tau}_\infty$ scales as:

$$\langle \tilde{\tau}_\infty \rangle \propto \frac{\sum_{\tau=1}^{\mathcal{S}} \tau^{-\mu}}{\sum_{\tau=1}^{\mathcal{S}} \tau^{-1-\mu}} \sim c\mathcal{S}^{1-\mu}, \tag{15.45}$$

  where c is a model-dependent constant. Since $0 < \mu < 1$, the scaling $\mathcal{S}^{1-\mu}$ arises from the numerator. By rescaling $T_{\mathcal{S}}$ by $\mathcal{S}$ and using Eq. (15.39), we find:

$$\mathbb{P}(T_{\mathcal{S}}/\mathcal{S} \geq t) = \mathbb{P}\left(N_{\mathcal{S}} \geq c(t-1)\mathcal{S}^{\mu}\right) = \Xi_\mu((c(t-1))^{-1/\mu}), \tag{15.46}$$

  where Ξ_μ is defined in Eq. (15.31). This reveals that $T_{\mathcal{S}}$ scales with $\mathcal{S}$ irrespective of the value of the exponent $\mu < 1$, generalizing the $1d$ scenario (see Eq. (15.13)).
- For marginal RWs: Analogously to recurrent RWs, the average value of $\tilde{\tau}_\infty$ is given by:

$$\langle \tilde{\tau}_\infty \rangle \propto \frac{\sum_{\tau=1}^{\mathcal{S}} \tau^{-1}}{\sum_{\tau=1}^{\mathcal{S}} \tau^{-2}} \sim c \ln \mathcal{S}. \tag{15.47}$$

  Hence, by scaling $T_{\mathcal{S}}$ by $\mathcal{S}^2 \ln \mathcal{S}$, we obtain:

$$\mathbb{P}(T_{\mathcal{S}}/(\mathcal{S}^2 \ln \mathcal{S}) \geq t) = \mathbb{P}\left(N_{\mathcal{S}} \geq t\mathcal{S}^2/c\right) = \Xi_1(\sqrt{c/t}), \tag{15.48}$$

  resulting in a scaling form for the distribution of $T_{\mathcal{S}}$.
- For transient RWs: In this case, the inter-visit times' distribution has finite moments, leading to $\langle \tilde{\tau}_\infty \rangle \to \langle \tau_\infty \rangle = c$. By rescaling $T_{\mathcal{S}}$ by $\exp\left[(\mathcal{S}/a)^{\mu/(1+\mu)}\right]$, we derive:

$$\mathbb{P}\left(T_{\mathcal{S}}/\exp\left[(\mathcal{S}/a)^{\mu/(1+\mu)}\right] \geq t\right) \sim \mathbb{P}\left(N_{\mathcal{S}} \geq t\exp\left[(\mathcal{S}/a)^{\mu/(1+\mu)}\right]/c\right)$$
$$\sim \Xi_\infty\left(-\frac{\sqrt{6}}{\pi}\ln(t/c)\right) = \exp\left[-\frac{t}{c}\mathbb{E}^{-\gamma_E}\right], \tag{15.49}$$

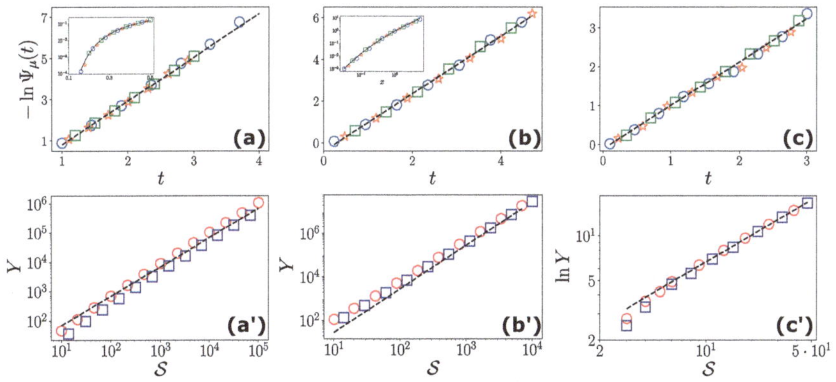

Fig. 15.4 Lifetime in Starving Random Walks. **(a)–(c)** Distributions of the rescaled variable $t \equiv T_{\mathcal{S}}/\langle T_{\mathcal{S}}\rangle$ (insets illustrate behavior at small t-values), and **(a′)–(c′)** corresponding averages (red circles) and standard deviations (blue squares) of $T_{\mathcal{S}}$. The black dashed lines denote the best fit from theory (see Eqs. (15.46), (15.48), (15.49), and (15.50)). Various universality classes are depicted: **(a)** $1d$ Lévy flights with parameter $\alpha = 1.5$ (recurrent), $\mathcal{S} = 14667$, 31622, and 68129; **(b)** nearest-neighbor $2d$ random walks, $\mu = 1$ (marginal), $\mathcal{S} = 2335$, 4832, and 10^4; **(c)** nearest-neighbor $3d$ random walks, $\mu = 3/2$ (transient), $\mathcal{S} = 12$, 17, and 22. Increasing values of $\mathcal{S}$ are denoted by blue circles, orange stars, and green squares

where a is the same constant as in Eqs. (15.37) and (15.41). The lifetime $T_{\mathcal{S}}$ follows an exponential distribution, akin to the number $N_{\mathcal{S}}$ of sites visited at starvation, as observed in the infinite-dimensional scenario (see Eq. (15.26)).

The average, standard deviation, and asymptotic distribution of $T_{\mathcal{S}}$ in the recurrent, marginal, and transient cases are checked numerically in Fig. 15.4 for representative RWs of each class.

> Important

By considering the rescaled lifetime $\frac{T_{\mathcal{S}}}{\langle T_{\mathcal{S}}\rangle}$, where the asymptotic of $\langle T_{\mathcal{S}}\rangle$ is given by

$$\langle T_{\mathcal{S}}\rangle \sim \begin{cases} \mathcal{S}, & [\mu < 1], \\ \mathcal{S}^2, & [\mu = 1], \\ \exp\left[\mathcal{S}^{\mu/(1+\mu)}\right], & [\mu > 1], \end{cases} \tag{15.50}$$

we obtain the asymptotic tail distribution $\mathbb{P}\left(\frac{T_{\mathcal{S}}}{\langle T_{\mathcal{S}}\rangle} \geq t\right) = \Psi_\mu(t)$, which has the properties at large and small arguments summarized in Table 15.4.

Table 15.4 Asymptotic properties of the minus log tail distribution Ψ_μ of the rescaled lifetime for different type of processes. All constants have been put to one for simplicity

$-\ln \Psi_\mu(t)$	$x \ll 1$	$x \gg 1$
$\mu < 1$ [recurrent]	$\mathbb{E}^{-1/t^{1/\mu}}/t^{1/\mu}$	t
$\mu = 1$ [marginal]	$\mathbb{E}^{-1/t^{1/4}}$	t
$\mu > 1$ [transient]	t	t

15.5.3 Discussion

In this section, we finally bridged the gap between the unidimensional and infinite-dimensional starving RWs studied in Sect. 15.3.

Several important properties stem from the analysis of the general Markovian starving RW:

1. The number of sites visited at starvation grows algebraically with the metabolic time $\mathcal{S}$ for recurrent ($N_\mathcal{S} \propto \mathcal{S}^\mu$) and marginal ($\propto \mathcal{S}^2$) RWs, while it grows significantly faster for transient RWs ($\propto \exp\left[\mathcal{S}^{\frac{\mu}{1+\mu}}\right]$).
2. The average lifetime grows linearly with $\mathcal{S}$ for recurrent RWs ($T_\mathcal{S} \propto \mathcal{S}$), while it grows quadratically for marginal RWs ($\propto \mathcal{S}^2$) and stretched exponentially for transient RWs ($\propto \exp\left[\mathcal{S}^{\frac{\mu}{1+\mu}}\right]$).
3. The distribution of both the number of sites visited at starvation and the lifetime are scale-invariant, in the sense that $N_\mathcal{S}/\langle N_\mathcal{S}\rangle$ and $T_\mathcal{S}/\langle T_\mathcal{S}\rangle$ are independent of $\mathcal{S}$ for large $\mathcal{S}$. Additionally, both asymptotic distributions of the rescaled variables decay exponentially at large argument.

15.6 Conclusion

To conclude, we have shown how to derive key properties of the starving random walk (RW) problem, such as the number of sites visited at starvation ($N_\mathcal{S}$) and the lifetime of the RW ($T_\mathcal{S}$). We have demonstrated that understanding the universal properties of the time between visits (τ_k) leads to universal characteristics for these quantities for general symmetric Markovian RWs. We have highlighted important differences based on a single parameter μ, which characterizes recurrence or transience of RWs. For example, exploration in one or two dimensions (i.e., by recurrent walks) differs significantly from that in three dimensions (i.e., by transient walks), where the lifetime of a tracer with a specific metabolism time changes from algebraic (for recurrent walks) to stretched exponential (for transient walks).

This chapter's methods and findings serve two purposes:

First, they provide a basic framework for studying systems with limited resources, useful in fields like animal ecology [22] or the study of biological species in crowded environments [21]. In these situations, significant memory effects arise because the RW is influenced by its past path, which is affected by resource depletion.

Second, our results furnish a null model against which the exploration dynamics of biological tracers can be assessed. Departures from this model indicate the potential influence of intricate and realistic strategies not encompassed in this study, including the tracer's memory of its trajectory (using biological cues, such as ants or MDCK cells do [1, 2]), renewal of the resources with time [44], or the option for the tracer to abstain from consuming food when unnecessary [45]. Other extensions include the food detection at long distances [46, 47], penalized long moves [48], the intermittent motion [49]. The starving RW problem provides a relatively straightforward model yielding numerous insights and results, with the methods presented in this chapter laying the foundation for a comprehensive understanding of resource-limited motion in a broad range of systems.

References

1. d'Alessandro, J., Barbier-Chebbah, A., Cellerin, V., Bénichou, O., Mège, R.M., Voituriez, R., Ladoux, B.: Nat. Commun. **12**(1), 4118 (2021)
2. Gordon, D.M.: Anim. Behav. **49**(3), 649 (1995)
3. Dvoretzky, A., Erdös, P. In: Proceeding of the Second Berkeley Symposium, pp. 353–367 (1951)
4. Jain, N.C., Pruitt, W.E. In: *Proceeding of the Sixth Berkeley Symposium*, vol. 3, pp. 31–50 (1972)
5. Gillis, J.E., Weiss, G.H.: J. Math. Phys. **11**(4), 1307 (1970)
6. Le Gall, J.F., Rosen, J.: Ann. Probab. **19**(2), 650–705 (1991)
7. Vineyard, G.H.: J. Math. Phys. **4**(9), 1191 (1963)
8. Régnier, L., Dolgushev, M., Redner, S., Bénichou, O.: Phys. Rev. E **105**, 064104 (2022)
9. Hughes, B.D.: Random Walks and Random Environments. Oxford University Press, New York (1995)
10. Turban, L.: On the Number of Common Sites Visited by *n* Random Walkers (2016)
11. Kundu, A., Majumdar, S.N., Schehr, G.: Phys. Rev. Lett. **110**, 220602 (2013)
12. Majumdar, S.N., Schehr, G.: Number of Distinct and Common Sites Visited by *n* Independent Random Walkers (2024)
13. Biroli, M., Mori, F., Majumdar, S.N.: J. Phys. A: Math. Theor. **55**(24), 244001 (2022)
14. Bénichou, O., Redner, S.: Phys. Rev. Lett. **113**, 238101 (2014)
15. Régnier, L., Dolgushev, M., Redner, S., Bénichou, O.: Nat. Commun. **14**(1), 618 (2023)
16. Régnier, L., Bénichou, O., Krapivsky, P.L.: Phys. Rev. Lett. **130**, 227101 (2023)
17. Régnier, L., Dolgushev, M., Bénichou, O.: Phys. Rev. E **109**, 064101 (2024)
18. Bénichou, O., Chupeau, M., Redner, S.: J. Phys. A: Math. Theor. **49**(39), 394003 (2016)
19. Sowinski, D.R., Carroll-Nellenback, J., Markwick, R.N., Piñero, J., Gleiser, M., Kolchinsky, A., Ghoshal, G., Frank, A.: PRX Life **1**, 023003 (2023)
20. Klages, R. In: Diffusive Spreading in Nature, Technology and Society, pp. 53–74. Springer, Berlin (2023)

21. Passino, K.M. In: Innovations and Developments of Swarm Intelligence Applications, pp. 219–234. IGI Global, New York (2012)
22. Orlando, C.G., Tews, A., Banks, P., McArthur, C.: Biol. Lett. **16**(7), 20200329 (2020)
23. Winfield, A.F.: Distributed Autonomous Robotic Systems **8**, 185–192 (2009)
24. Majumdar, S.N., Pal, A., Schehr, G.: Phys. Rep. **840**, 1 (2020)
25. Majumdar, S.N., Schehr, G.: J. Stat. Mech. **2014**(1), P01012 (2014)
26. Majumdar, S.N., Krapivsky, P.L.: Phys. Rev. E **62**(6), 7735 (2000)
27. Ben-Naim, E., Krapivsky, P.L., Majumdar, S.N.: Phys. Rev. E **64**(3), 035101 (2001)
28. Kay, T., Giuggioli, L.: J. Phys. A: Math. Theor. **56**(34), 345002 (2023)
29. Ben-Avraham, D., Havlin, S.: Diffusion and Reactions in Fractals and Disordered Systems. Cambridge University, Cambridge (2000)
30. Pólya, G.: Mathematische Annalen **84**(1–2), 149 (1921)
31. Shepp, L.A.: Bull. Am. Math. Soc. **70**(6), 540 (1964)
32. Bäumler, J.: Electron. J. Probab. **28**, 1 (2023)
33. Mariz, A., Van Wijland, F., Hilhorst, H., Gomes Júnior, S., Tsallis, C.: J. Stat. Phys. **102**, 259 (2001)
34. Régnier, L., Dolgushev, M., Bénichou, O.: Phys. Rev. Lett. **132**, 127101 (2024)
35. Redner, S.: A Guide to First-Passage Processes. Cambridge University, Cambridge (2001)
36. Rider, B.: J. Phys. A Math. Gen. **36**(12), 3401 (2003)
37. Chafaï, D., Péché, S.: J. Stat. Phys. **156**, 368 (2014)
38. Krug, J.: J. Stat. Mech. Theor. Exp. **2007**(07), P07001 (2007)
39. Lacroix-A-Chez-Toine, B., Grabsch, A., Majumdar, S.N., Schehr, G.: J. Stat. Mech. Theor. Exp. **2018**(1), 013203 (2018)
40. Niki, N., Shigekazu, N., Inoue, H.: Commun. Stat. Theory Methods **19**(5), 1857 (1990)
41. Watson, G.N.: Quart. J. Math. Oxford **10**, 266 (1939)
42. Carpentier, D., Le Doussal, P.: Phys. Rev. E **63**(2), 026110 (2001)
43. Gumbel, E.J.: Statistics of Extremes. Columbia University, Columbia (1958)
44. Chupeau, M., Bénichou, O., Redner, S.: Phys. Rev. E **93**(3), 032403 (2016)
45. Bénichou, O., Bhat, U., Krapivsky, P.L., Redner, S.: Phys. Rev. E **97**, 022110 (2018)
46. Sanhedrai, H., Maayan, Y., Shekhtman, L.M.: Europhys. Lett. **128**(6), 60003 (2020)
47. Sanhedrai, H., Maayan, Y.: Phys. Rev. E **103**, 012114 (2021)
48. Krishnan, N., Kilpatrick, Z.P.: Phys. Rev. E **98**, 052406 (2018)
49. Campos, D., Cristín, J., Méndez, V.: Phys. Rev. E **103**(5), 052109 (2021)

Chapter 16
Methods for Quantifying Self-Organization in Biology: A Forward-Looking Survey and Tutorial

Alexandria Volkening

16.1 Introduction

Pattern formation driven by the interactions of agents is found across the natural and social world, spanning the population scale to the intracellular scale. Large-scale examples include pedestrian movement in crowds [1–4], honeybee aggregation [5], schooling fish [6, 7], fireflies flashing synchronously [8], and marching locusts [9]. On the scale of cells and tissues [10], neural-crest cell migration [11], color pattern formation in fish skin [12, 13], synchrony in neural firing [14], and cell dynamics during wound healing are examples of self-organization. At still smaller scales, proteins and filaments regulate transport and shape within cells [16]. As these examples illustrate, patterns that arise due to self-organization may be temporal or spatial. In the case of spatial self-organization, studying such complex systems—whether empirically or through mathematical modeling—often leads to qualitative pattern data in the form of images. Being able to quantify these spatial data opens the door to a broader perspective on self-organization and makes complex systems more amenable to interdisciplinary investigation. With this motivation, in this tutorial chapter written for an interdisciplinary audience, I discuss a selection of methods for making qualitative data quantitative that apply to disparate applications.

Across applications in self-organization, a general goal is to describe agent interactions and predict how altered behaviors affect collective dynamics. In a developmental-biology setting, doing so could mean understanding how genes regulate cell behavior and organism phenotype. In a pedestrian-movement example, shedding light on crowd dynamics could help engineers design buildings that allow for quicker evacuation during emergencies. Studying such systems often involves developing predictive models and characterizing the group dynamics and patterns

---

A. Volkening (✉)
Purdue University, Department of Mathematics, West Lafayette, IN, USA
e-mail: avolkening@purdue.edu

L. Giuggioli, P. K. Maini (eds.), *The Mathematics of Movement*,
https://doi.org/10.1007/978-3-032-00873-2_16

that arise from these models under different parameter values and model rules. As I show in Fig. 16.1, which focuses specifically on spatial patterns, the task of quantifying qualitative pattern data is largely the same regardless of whether the data are collected empirically or generated by a spatial model. The main difference is that working with empirical images involves the additional step of first identifying the agents of interest (i.e., pedestrians in Fig. 16.1j) in the data, whereas models provide full information on agent positions or density; also see Fig. 16.2.

Quantifying pattern data serves many goals in interdisciplinary studies. For example, when data and a model are available, one may be interested in estimating model parameters [30, 31]. Or the goal may be to use data-driven approaches (i.e., [32–35]) to select model terms from a large library of possible terms, based on some ground-truth data. There are also cases where researchers seek to develop algorithms for producing realistic synthetic data in order to train segmentation algorithms. Across these examples, there is a need to measure the distance between empirical data and model-generated output, necessitating some quantitative description of the empirical and in silico patterns [36]. In other cases, the data may be fully synthetic or fully empirical. In medical settings, distinguishing between images of healthy and unhealthy tissue is a natural goal. Or, in the case of modeling data, one may want to comprehensively characterize the patterns that can emerge from a model under changes to its parameters. In still other settings, the goal may be to understand the relationship between different modeling approaches to the same problem in order to determine how implementation choices affect predictions [37–40].

All of these goals boil down to a need for transforming data from qualitative to quantitative and quantifying the difference between patterns, whether synthetic and empirical, both empirical, or both synthetic. In the case of spatial patterns arising due to self-organization, this a fundamental challenge that arises across disciplines, including biology, ecology, mathematics, engineering, and social science. There are thus many, many approaches to quantifying spatial patterns, and specific methods are more or less common in each research community. This motivates my chapter. By discussing shared features in pattern data and introducing a range of quantification methods, my goal is to support cross-fertilization between applications and perspectives. First, in Sect. 16.2, I discuss how to express pattern data from various sources—whether empirical images, agent-based models, or continuum models—as point clouds. I follow this in Sect. 16.3 by highlighting some quantifiable patterns features that are present across applications. I next overview a selection of quantification methods, covering order parameters (Sect. 16.4.1), pattern-simplicity scores (Sect. 16.4.2), pair-correlation functions (Sect. 16.4.3), and techniques from topological data analysis (Sect. 16.4.4). Throughout this tutorial, my main focus is on spatial patterns, but I also briefly note how select methods (specifically order parameters and topological techniques) can be adapted to temporal patterns. I conclude in Sect. 16.5 with some comments on data-driven modeling.

Before going further, I discuss my perspective on the relationship between patterns, self-organization, and movement in animal ecology and cell biology. I think of "self-organization" as the emergence of macroscopic properties and patterns from the interactions of smaller-scale agents or components without global instructions,

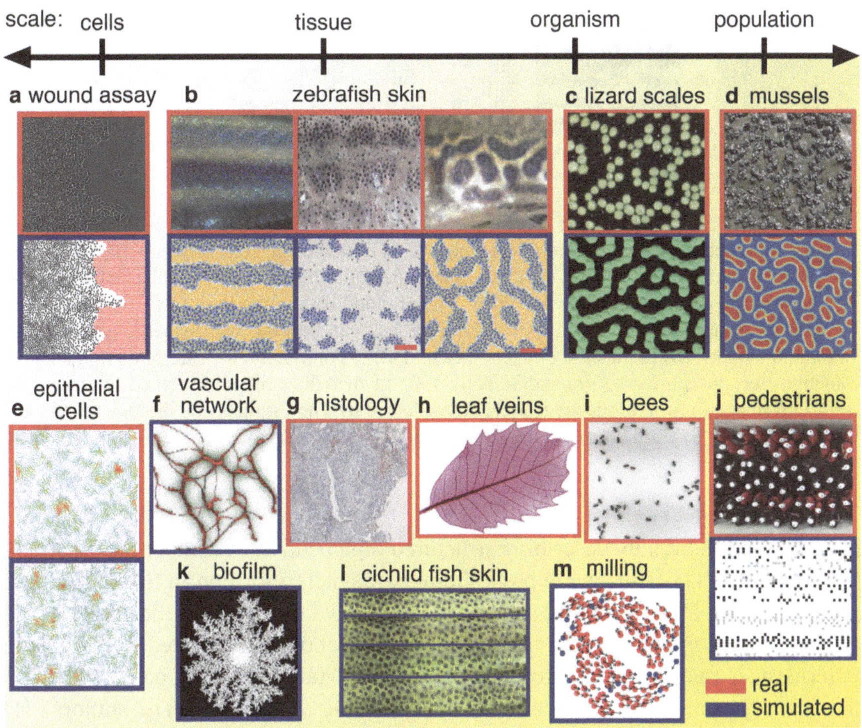

Fig. 16.1 Empirical and model-generated images of spatial self-organization across scales. (**a**) In a scratch-assay experiment, cells spread into empty space [15]; (**b**) zebrafish feature a wide range of skin patterns, all made up of brightly colored pigment cells [17, 18]; (**c**) lizard scales display labyrinth-style patterns [19]; and (**d**) mussels organize into groups [20]. (**e**) Epithelial cells interact and align [15], and (**f**) developing blood vessels exhibit branching [21]. (**g**) Histology images indicate cells in tissue [22]. (**h**) Veins form branching patterns [23], and (**i**) honeybees interact and exchange food [5]. (**j**) Lanes may emerge when pedestrians move in opposite directions [1, 2]. (**k**) Biofilms exhibit branching [24], and (**l**) cichlid fish feature messy stripes, which form during development [25]. (**m**) Flocking, swarming, and milling occur in many biological systems [7, 26–28]. Images in (**a**), (**e**) reproduced from [15], Copyright Sepúlveda et al., and licensed under CC BY; top row of (**b**) reproduced from [17], published by The Company of Biologists Ltd.; and licensed under CC-BY 3.0 (https://creativecommons.org/licenses/by/3.0/); bottom row of (**b**) reproduced from [18] and published by Springer Nature; images in (**c**) reproduced from [29]; images in (**d**) reproduced from [20] with permission from the Proceedings of the National Academy of Sciences, U.S.A.; image in (**f**) reproduced from [21] with permission of Elsevier, Copyright (2006) Elsevier Inc.; image in (**g**) reproduced from [22]; image in (**h**) reproduced from [23]; image (**i**) reproduced from [5]; top panel of (**j**) reproduced from [2]; bottom panel of (**j**) reproduced from [4] with permission from Elsevier, Copyright (2001) Elsevier Inc.; image in (**k**) reproduced from [24] with permission of Elsevier, Copyright (2023) Elsevier Inc.; images in (**l**) reproduced (cropped) from [25], Copyright Liang et al.; image in (**m**) reproduced from [26], Copyright Bhaskar et al. Images in (**c**, **g**, **h**, **i**, **l**, **m**), the bottom row of (**b**), and the top panel of (**j**) licensed under CC-BY 4.0 (https://creativecommons.org/licenses/by/4.0/)

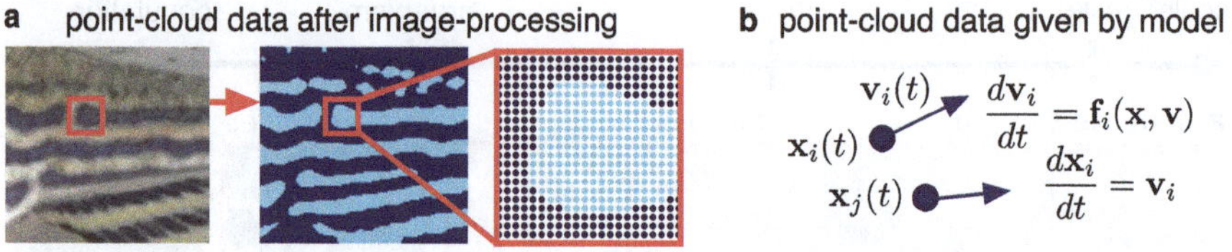

Fig. 16.2 Point-cloud data in self-organization. (**a**) One can use image-processing or machine-learning approaches to process empirical image data. (I used the free software Ilastik [47] to identify dark and light regions of this fish skin pattern.) After taking the centers of the pixels of interest in the processed image, the result is point-cloud data in the form of (x, y)-coordinates for each light-blue and navy pixel. A similar approach can be used with patterns generated by continuum models. (**b**) In the case of agent-based models, point-cloud data are often immediately available. Here the position $\mathbf{x}_i(t)$ and velocity $\mathbf{v}_i(t)$ of the ith agent are specified by the model itself. Left image in (**a**) reproduced from [17], published by The Company of Biologists Ltd.; and licensed under CC-BY 3.0 (https://creativecommons.org/licenses/by/3.0/); I added the red box and cropped the image

and "pattern" as broadly describing structured organization of components spatially or temporally [41]. This encompasses temporal synchronization in firefly flashing, spots or stripes in animal skin, and an aligned crowd of moving pedestrians. In all of these cases, movement (of fireflies, cells, or people) is involved. All of these patterns also may arise due to self-organization, but not every pattern-forming process is driven by the interactions of smaller-scale components. The authors [41] give the example of a marching band to illustrate this point: a marching band can move to produce elaborate patterns, but these patterns might essentially be predetermined by choreography and coach instructions, rather than be the emergent result of interactions between band members. Because I am interested in modeling self-organization in space, I frame this tutorial around patterns arising due to self-organization. However, many of the techniques that I discuss apply to spatial patterns broadly, regardless of whether or not they are the emergent result of agent interactions.

In terms of the themes of this book, which includes the mathematics of animal ecology and cell biology, I come from the cell biology and spatial pattern formation perspective. I stress that this chapter is a selection of some of the quantification methods that I have encountered as a mathematical modeler, mainly of patterning in developmental biology. Importantly, there are a wealth of methods for quantifying spatial data that I do not discuss, such as many in spatial statistics, network-based analysis, Fourier analysis, and morphometric tools. For additional references, I highlight [7, 22, 42–46]. For the approaches that I discuss, I seek to provide a starting point, suggest references to learn more, and encourage work that brings together multiple methods to quantify the same data and actively considers the choices involved. In writing this chapter, I have reflected on what the various approaches in Sect. 16.4 could say about my pattern data. I hope this chapter and the references herein encourage readers to think about their data from a new perspective as well.

16.2 Obtaining Point-Cloud Data on Self-Organization

Throughout this chapter, I mainly discuss quantifying pattern data that are in the form of point clouds, meaning collections of points, typically in two dimensions. As I show in Fig. 16.2a, one can arrive at point-cloud data from an empirical image by thresholding, applying other image-processing tools, or using machine-learning methods; in this example, I used the free software Ilastik [47] to hand-select a few regions of the empirical image in Fig. 16.2a, left panel, and then the software classified the pixels in the rest of the image automatically. After limiting my focus to the pixels with a given color value, the result is point-cloud data on a square grid. Specifically, in Fig. 16.2a, right panel, I plot light-blue and navy points indicating the (x, y)-coordinates of the centers of each light-blue or navy pixel.

For spatial patterns generated by simulating mathematical models, the approach to arriving at point-cloud data depends on the framework. There are two particularly common frameworks for modeling self-organization spatially, namely agent-based (or individual-based) and continuum models [13]. Continuum models track agent density; in spatial settings, these take the form of partial differential equations or integro-differential equations. Reaction-diffusion equations (i.e., [48–51]), Cahn–Hilliard equations (i.e., [20, 52]), and non-local partial differential or integro-differential equations (i.e., [53–56]) have all been used to describe the evolution of agent density. Partial differential equations are often studied through bifurcation analysis, and researchers can quantitatively describe the most unstable wavelengths. In terms of the data that they produce, continuum models of patterning generally provide the density $\rho(\mathbf{x}, t)$ (i.e., of agents) in space $\mathbf{x}$ and time t; I show an example of these data in the bottom panel of Fig. 16.1d.

In order to produce point-cloud data from simulations of continuum models in two dimensions, the approach is broadly similar to working with empirical images—thresholding a continuous agent density leads to point-cloud data. This can be done in various ways, and it depends some on how one solves their continuum model. If the model is solved using finite-difference methods, for example, then the numerical solution is already in the form of values $\rho(x_i, y_j, t)$ at a grid of discrete points $\{(x_i, y_j)\}$ in the domain. This can be presented as a surface plot or as a heatmap with color intensity indicating agent density. One option is to select some cutoff ρ^* and then define a matrix $[M_{ij}]$ by $M_{ij} = 1$ if $\rho(x_i, y_j, t) \geq \rho^*$ and $M_{ij} = 0$ otherwise. This essentially leads to a black and white image. If we consider only the points (x_i, y_j) such that $M_{ij} = 1$ (i.e., white pixels), we arrive at point-cloud data that maintain some grid structure. Alternatively, if one prefers to consider points off lattice, we could follow this step by uniformly at random sampling N query points $\mathbf{z}$ in the domain and keeping only those points $\mathbf{z}$ that fall in white regions of our binarized image $[M_{ij}]$. To implement this using built-in functions in MATLAB, one approach would be to apply the function *bwconncomp* to identify and label groups of white pixels in the binarized image $[M_{ij}]$, extract the bounding area for each group using the function *alphaShape*, and then apply the function *inShape* to each query point $\mathbf{z}$ to determine if it is contained in that group of white pixels.

Off-lattice agent-based (i.e., [18, 57]) and on-lattice agent-based models (i.e., [58, 59]) take a more microscopic perspective than continuum models of agent density, describing the interactions of individual agents during self-organization and offering detailed predictions; see the bottom panels of Fig. 16.1a and b respectively, for examples of patterns generated by on- and off-lattice agent-based models. Agent-based models are common in animal-ecology studies of flocking, swarming, and schooling, as well in research on the behavior of cells in developing and regenerating tissues. Stochastic agent-based models tend to have many parameters, and it is very difficult to broadly characterize the full range of patterns that these models can produce. As I show in Fig. 16.2b, the data generated by simulating off-lattice agent-based models (sometimes called "interacting particle systems") are typically already in point-cloud form as $\{\mathbf{x}_i(t)\}_{i=1,2,\ldots,N}$, where $\mathbf{x}_i(t)$ is the position (e.g., (x, y)-coordinates) of the ith agent at time t. In other cases, we also have the velocity $\mathbf{v}_i(t)$ of each agent. For on-lattice agent-based models, space is discretized, and the pattern data may be lists of grid positions occupied by specific agents at each time; one can use the centers of each grid position of interest to build a point cloud. Other types of microscopic models, such as vertex-based models and cellular Potts models [60, 61], can also give rise to point-cloud data.

It is also useful to briefly note how the discussion above relates to self-organization that results in temporal patterns. In the case of coupled oscillators that exhibit synchronization, the data are often in network form. Quantifying data in network form is outside of the scope of this tutorial. However, there are some settings in which temporal patterning due to self-organization can be adjusted to create a point cloud. As an example, Zabaleta-Ortega et al. [62] extracted point-cloud data from simulations of Kuramoto oscillators [63, 64] and from a network of electronic oscillators. To do so, they considered the time series $y(t)$ of one oscillator in relation to a reference oscillator with time series $x(t)$. Specifically, the authors [62] created their point-cloud data in two dimensions by introducing a delay τ and setting the coordinates of their nth point to be $(x(t_n), y(t_n + \tau))$. They then applied techniques from topological data analysis (see Sect. 16.4.4) to these coordinates.

16.3 Shared Features in Biological Patterns

Before discussing similarities in spatial patterns across scales in biology, I find it useful to highlight three broad perspectives on quantifying patterns, though the distinctions are not completely clearcut. First, in some cases, scientists study patterns without obtaining interpretable quantitative descriptors. For example, machine-learning methods can be used to identify features of clinical images that are related to health or disease [66]. Second, there are patterns in which the specific positions of points in space matter, and then the difference in density in space may be a suitable measurement to compare patterns. For example, in some scratch-assay

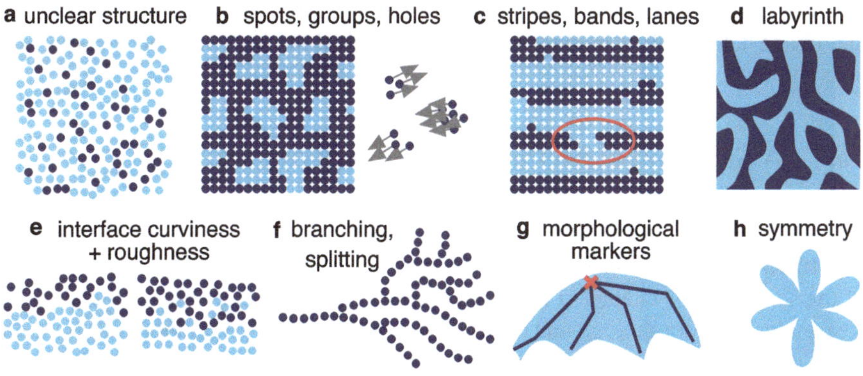

Fig. 16.3 Some example features in qualitative data that researchers often seek to quantify. (**a**) In some cases, the structure in pattern data is difficult to describe or qualitatively judge. Here it is not clear whether the navy points are distributed uniformly at random or if additional structure is present. (**b**) Spots, clusters, and groups appear in many of the patterns in Fig. 16.1, as do (**c**) stripes, bands, and lanes. In (**c**), I highlight that identifying interruptions in stripes can be important in some contexts. (**d**) Labyrinth patterns may feature a characteristic wavelength without directionality. (**e**) Whether working with the wound assay in Fig. 16.1a or stripes in fish skin as in Fig. 16.1l, researchers may quantify curviness or roughness of stripes, interfaces, and boundaries. (**f**) The vascular network and leaf veins in Fig. 16.1 are examples of branching, bifurcating, or splitting patterns. One may also view the labyrinth pattern in Fig. 16.1b as branching, and, in some sense, the navy points in panel (**c**) of this figure are a branching pattern. (**d**) In a different vein (no pun intended), the position of morphological markers is important to describe in some biological images [45]. (**e**) Symmetry (of various kinds) is the feature of interest in other examples of self-organization [65]

experiments, it is sufficient to consider a cross-section (or mean cross-section) of the density of cells spreading into empty space [67]. Third, in messy biological data and stochastic simulations, often it is pattern features—and not precise positions—that are important. For example, in the fish pattern in Fig. 16.1b, one may want to capture that blue spots are present and roughly aligned in stripes, but the individual position of any given spot could be shifted [68]. In this chapter, I focus mainly on the third perspective.

In order to apply methods from disparate applications to one's data, it is useful to look for common features in Fig. 16.1. I summarize some of these features in Fig. 16.3, and note that their names differ depending on the application and field. For example, clusters of points like those in Fig. 16.1b, center, d, or e may be called connected components, spots, or groups. Stripes, lanes, trails, or bands are present in many of the patterns in Fig. 16.1, sometimes with sharp interfaces, and in other cases (i.e., Fig. 16.1l) with rougher, less clear boundaries [25]. Branching, bifurcating, and splitting are other common features of biological patterns, including leaf venation [23, 69], blood-vessel networks [21], and yeast colonies [70]. From a different perspective, networks of blood vessels can also be described as labyrinthine, as can the fish-skin patterns in the right panel of Fig. 16.1b. Various types of symmetry [65] are widespread in natural patterns as well. Finally, quantifying biological patterns and image data involves landmarks or morphological markers in some cases.

16.4 A Selection of Methods for Quantifying Patterns

I now turn to a selection of methods for quantifying pattern data like those in Fig. 16.3a–f. Quantifying morphology and symmetry as in Fig. 16.3g–h is outside of the scope of this chapter; I highlight reference [71] as an example of landmark-free analysis of morphology and the work of Gandhi et al. [65] for more information about symmetry in pattern data. I introduce order parameters in Sect. 16.4.1, pattern simplicity scores in Sect. 16.4.2, pair-correlation functions in Sect. 16.4.3, and topological techniques in Sect. 16.4.4. Except for some examples in Sect. 16.4.2 and some brief comments in Sect. 16.4.4, I consider pattern data in the form of point clouds, typically in two dimensions. As I discuss in Sect. 16.1, the approaches that I illustrate next are a subset of the many quantification methods that researchers use across disciplines.

16.4.1 Order Parameters for Point-Cloud Data

Order parameters have been used to quantify point-cloud data in studies of aggregation [7, 72] across the spectrum of biological scales, including intracellular dynamics [16], cell interactions [73], and animal or insect populations [9, 74]. Broadly, order parameters provide a summarized statement of the degree of order in a system and its overall dynamics; they are particularly useful in identifying transitions from one state or phase to another [7, 75]. What order parameter is most appropriate to describe any given pattern data is application-dependent. As Topaz et al. [72] note, this means order parameters are often defined on a case-by-case basis, using information about the global dynamics that one seeks to identify in data. (Other quantification approaches, like those I discuss in Sects. 16.4.3 and 16.4.4, apply more broadly, though there are also choices involved in these methods.)

In many cases, the point-cloud data that researchers are interested in quantifying using order parameters are in the form of agent positions and/or velocities, as in Figs. 16.1m and 16.2b.) One common order parameter for studies of aggregation introduced by Vicsek et al. [76] is the normalized average velocity across the members of the group, as below:

$$\phi(t) = \text{normalized average velocity} = \frac{1}{N v_{\text{avg}}} \left| \sum_{i=1}^{N} \mathbf{v}_i(t) \right| , \tag{16.1}$$

where N is the number of group members, v_{avg} is their average speed, and $\mathbf{v}_i(t)$ is the velocity of the ith agent at time t [7, 72, 76]. When all members of the group are moving in nearly the same direction, the value of $\phi(t)$ is near 1, and, when all of the agents are moving randomly, $\phi(t)$ is close to 0 [7, 72]. Importantly, if two groups of agents are moving in opposite directions, it is also possible for the normalized average velocity to be zero [72], indicating how this order parameter is able to capture meaningful information in some cases but not in others.

Using multiple-order parameters together can provide a means around this difficulty and increase the number of patterns that one can distinguish. Other order parameters include the polarity and normalized angular momentum [74, 77], and Chuang et al. [77] developed a modified normalized angular momentum order parameter that is able to capture double mills. Together all of these order parameters provide information about coherently moving flocks, single mills, and double mills [77]. As in the example in Eq. (16.1) for pattern data in an ordered phase, the value of its associated order parameter is typically non-zero (and often bounded by one), whereas order parameters are generally defined so that they are zero when disorder is present [7]. Order parameters are not always bounded between 0 and 1, however, and quantities such as the mean distance between nearest neighbors in aggregation data can be considered an order parameter [78]. It also worth noting that order parameters apply to applications beyond self-organization in space; for example, it is common to use order parameters to summarize synchrony in networks of coupled oscillators [63, 64].

16.4.2 Pattern Simplicity Scores and Measurements for Segmented Data

For qualitative data in which the pattern features have been segmented and identified, there are a wealth of summary statistics that one can compute. In the case of Fig. 16.3b, left panel, for example, manual labeling, machine-learning approaches, image-processing tools, clustering methods, or topological techniques [68] (see Sect. 16.4.4) may be used to identify 12 spots, each made up of a set of light-blue pixels. One can then count the number of points per spot and calculate the average distance between spot centers. Similarly, in Fig. 16.3c, once the navy pixels are each assigned to a stripe feature, methods from computational geometry [79] (as in [39]) may be used to quantify stripe curviness. For studies of branching blood-vessel networks as in Fig. 16.3f, researchers have arrived at many summary statistics, including inter-vessel spacing, the number of branches or vessel endpoints, and the length of the longest vessel [80, 81]. When working directly with image data, the built-in methods in the powerful open-source software ImageJ [82] and Fiji [83] offer a wide range of quantitative descriptors for segmented image features.

Because they lend themselves to distinguishing between spot and labyrinth patterns, I highlight pattern simplicity scores [84] as a useful summary statistic that can be computed given labeled data on pattern features. Specifically, in Fig. 16.3d, there are five clusters of light-blue pixels, viewed as five continuous regions of light blue. Depending on the community, these five clusters may be called "connected components" or "contours," and I refer to them broadly as features. Let A_i be the area of the ith feature, and P_i be its perimeter, where $i = \{1, 2, 3, 4, 5\}$ in our example since there are five light-blue features in Fig. 16.3d. Then the pattern simplicity score (PSS) is the average circularity of the features, weighted by their area, as below:

$$\text{PSS} = \sum_{i=1}^{\text{number of pattern features}} \frac{A_i}{\sum_{j=1}^{\text{number of pattern features}} A_j} \times \frac{4\pi A_i}{P_i^2}.$$

Associating higher "pattern simplicity" with rounder features does not make sense for every application, but PSS can be useful when considering a dataset that contains patterns ranging from spotted to labyrinthine (e.g., features with varying circularity). This is natural in the setting of fish-skin patterns, where hybridizing light-spot and dark-spot fish can result in patterns that have maze-like features. Miyazawa et al. [84] used the PSS to study this "pattern blending" in different types of fish, and Djurdjevic et al. [85] focused on pattern simplicity scores and overall color tone in images of trout. The researchers [86] considered the pattern complexity score—which is 1 minus the PSS—to quantify skin-pattern diversity across a very wide range of fish.

16.4.3 Pair-Correlation Functions

Pair-correlation functions (or radial distribution functions) [22, 66, 87–91] belong to the field of spatial statistics and can provide quantitative summaries of a wide range of patterns, including those with spots, stripes, labyrinth features, branching, and unclear structure in Fig. 16.3. Broadly, pair-correlation functions (PCFs) help researchers distinguish between aggregation, segregation, and spatial randomness in point-cloud data [92], and their output can also be further interpreted to describe more pattern features, as I show in Fig. 16.4 and discuss in this subsection. Notably, PCFs had an important role in the history of statistical mechanics, as, for example, in the work of Corner and Lennard–Jones [93] on the distribution of atoms in liquids and gases. More broadly, there is a large literature on PCFs and their application to pattern data, and my discussion below highlights a subset of these directions.

Pair-correlation functions work by describing the density of points at different scales, relative to complete spatial randomness (that is, points appearing completely at random in the domain of interest) or some other null distribution [22, 66]. As I show in Fig. 16.4, the PCF approach involves counting the number of points of a

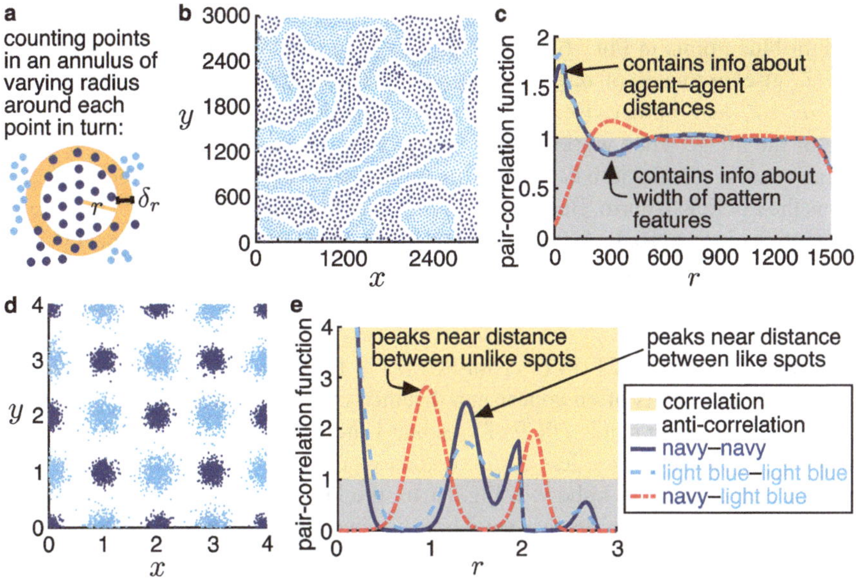

Fig. 16.4 Overview of pair-correlation functions. (**a**) Applying pair-correlation functions to point-cloud data that are continuous in space broadly involves counting the number of points of different types that appear in an annulus centered around each point [22]. (**b**) I show an example labyrinthine pattern in a domain that is periodic in x and y, and (**c**) the corresponding PCFs and cross-PCF according to Eqs. (16.2)–(16.3). (**d**) As another example, I show a spotted pattern in a periodic domain, as well as (**e**) the associated PCF summaries

given type in an annulus of inner radius r and width δ_r centered at each point of that type, and the result of this process is information about clustering and characteristic length scales [94]. Similarly, cross-PCFs quantitatively summarize the relationship between points of two types (i.e., light blue and navy in Fig. 16.4) in data. Ripley's K index [95] is a related measurement in spatial statistics that counts points in circles of increasing size, rather than in annuli.

The specific definitions of the PCF and cross-PCF depend on the choice of null distribution and approach to boundary conditions [22]. In Fig. 16.4, I apply the PCF and cross-PCF definitions of Bull et al. [22] to particles of two types, as below:

$$\text{PCF}(r) = \frac{A}{N_\alpha^2} \sum_{i=1}^{N_\alpha} \frac{1}{A_r(\mathbf{x}_i^\alpha)} \sum_{j=1}^{N_\alpha} \mathbb{1}_{\delta_r}(\|\mathbf{x}_i^\alpha - \mathbf{x}_j^\alpha\| - r), \text{ and} \tag{16.2}$$

$$\text{cross-PCF}(r) = \frac{A}{N_\alpha} \sum_{i=1}^{N_\alpha} \frac{1}{N_\beta} \frac{1}{A_r(\mathbf{x}_i^\alpha)} \sum_{j=1}^{N_\beta} \mathbb{1}_{\delta_r}(\|\mathbf{x}_i^\alpha - \mathbf{x}_j^\beta\| - r), \tag{16.3}$$

where $\{\mathbf{x}_i^\alpha\}_{i=1,2,\ldots,N_\alpha}$ are the points of type α in our data (i.e., the (x, y)-coordinates of light-blue points in Fig. 16.4b) and $\{\mathbf{x}_i^\beta\}_{i=1,2,\ldots,N_\beta}$ are the points of type β (i.e., the (x, y)-coordinates of navy points in Fig. 16.4b); the function $\mathbb{1}_{\delta_r}(d) = 1$ if $0 < d < \delta_r$ and 0 otherwise; A is the area of the domain of interest, which is the area of the image when working with image data; $A_r(\mathbf{x})$ is the area of the annulus of inner radius r and width δ_r centered at point $\mathbf{x}$ and contained in our domain; and $\|\cdot\|$ is the Euclidean norm. Because I use periodic boundary conditions in the x- and y-directions in Fig. 16.4, $A_r(\mathbf{x})$ simplifies to $\pi(r+\delta_r)^2 - \pi r^2$; however, in cases without periodic boundary conditions, more care is needed to compute the PCF [22, 96]. (I have rewritten the equations in [22] in my own notation in Eqs. (16.2)–(16.3).)

When the value of PCF(r) or cross-PCF(r) in Eqs. (16.2)–(16.3) is greater than 1, it means that points often appear at a distance of r from one another in our data [22, 97]. On the other hand, a PCF of less than 1 indicates that points at this distance away from one another are less common than would be the case if the points were randomly dispersed. As I show in Fig. 16.4c, the PCF for like points in Fig. 16.4b peaks first near the mean nearest-neighbor distance and later dips around $r = 300$. The cross-PCF, on the other hand, is less than 1 for low values of r, suggesting anti-correlation between unlike points locally, and reaches a maximum near $r = 300$, which we can interpret as the characteristic width of the labyrinthine stripes in Fig. 16.4b. Pair-correlation functions are very flexible, and I show in Fig. 16.4d–e how they can be used to capture information about spot separation and size in patterns. Averaging PCFs is common when analyzing point-clouds generated from stochastic models or multiple experimental replicates (i.e., [91, 94]).

The examples in Fig. 16.4 illustrate how pair-correlation functions can be applied to off-lattice data, but the PCF approach is also a natural means to quantify lattice point-cloud data, such as the positions of pixels in a segmented image like Fig. 16.2a or grid positions in an on-lattice agent-based model. For example, Binder and Simpson [87] applied PCFs to non-overlapping, discrete data from an agent-based model and to pixel positions in binarized empirical images, and Dini et al. [98] introduced a cross-PCF function for describing correlation and anti-correlation in the positions of agents of different types on a square lattice. Gavagnin et al. [92] developed PCFs for many tessellations, including hexagonal, triangular, and cuboidal lattices.

In some settings (i.e., [70, 90, 92]), whether the data are on or off lattice, the Euclidean distance is not the most suitable for quantifying features of interest. With this motivation, a range of PCFs have been developed that consider distances between points in only one direction, such as in x, in y, or angularly [70, 90]. For example, angular distances are appropriate for quantifying spreading and branching yeast patterns [70]. In other settings, the locations of points are constrained by external forces and obstacles; Johnston and Crampin [99] thus developed a PCF for environments that have obstacles, such as in the case of pedestrians moving in rooms.

16.4.4 Techniques from Topological Data Analysis

Persistent homology is a flexible tool in topological data analysis (TDA) that relies on ideas from algebraic topology to characterize shape [72, 100–102]. Persistent homology describes connected components (i.e., clusters), loops (i.e., holes), and higher-dimensional features in point-cloud data across scales [103]. Notably, persistent homology can be applied to all of the images in Fig. 16.3 and provide meaningful quantitative information. It has been used to shed light on many biological patterns, including empirical and simulated data on fish-skin patterns [39, 68], flocking and milling [26, 72], vascular networks [81], intracellular dynamics [16], honeybee aggregation [5], coral reefs [104], cancer tissue [105], and brain arteries [106]. In this subsection, I discuss persistent homology informally and give examples of how it can be applied to complex systems. See [72, 100, 101, 103, 107] for technical details. I also recommend the recent review [102] for an overview of TDA and its interactions with biology, and the article [103] for guidance on computing persistent homology (along with a discussion of various open-source software libraries involved).

In order to compute persistent homology, one needs to filter through data using some scaling parameter or threshold. I illustrate a common approach to doing this that involves building Vietoris–Rips complexes in Fig. 16.5a. Suppose I have 15 points in my pattern data, and these points, which I show in the left panel of Fig. 16.5a, are the (x, y)-coordinates of 15 people in a room. How many groups are in the room? Perhaps you see a ring of people talking, one cluster, and a lone individual, or you may judge this pattern to contain two groups only. Persistent homology provides a means of quantifying the shape of the positions of people in the room without the need for this judgment call. In order to build a Vietoris–Rips filtration, we place a ball centered at each point and slowly increase the ball diameter d. As the scaling parameter d grows, we keep track of the appearance and disappearance of connected components (dimension-0 features) and holes (dimension-1 features). The result is often visualized using barcodes and persistence diagrams; see Fig. 16.5b and c, respectively. For example, in order to plot a barcode describing the 1-dimensional topological features in data, one draws a horizontal bar for each of these features, where the bar starts at the value d_{birth} of the scaling parameter at which the feature appears (the feature's "birth time") and ends at the value d_{death} of the scaling parameter at which the feature disappears (the feature's "death time") [72].

Once data have been summarized as in Fig. 16.5b, the next step is to interpret and use the resulting barcode [103]. For example, McGuirl et al. [68] interpreted the number of highly persistent dimension-0 and dimension-1 features (i.e., long bars in barcodes) as the number of spots and stripes, respectively, in simulated zebrafish-skin patterns. I highlight how persistent homology can be used to count the number of horizontal stripes on a domain with periodic boundary conditions in x in Fig. 16.6a–c. (Importantly, this process involves the choice of hyper-parameters [39].) In other cases, the shape of data is not as clear as in Fig. 16.6b, and persistence

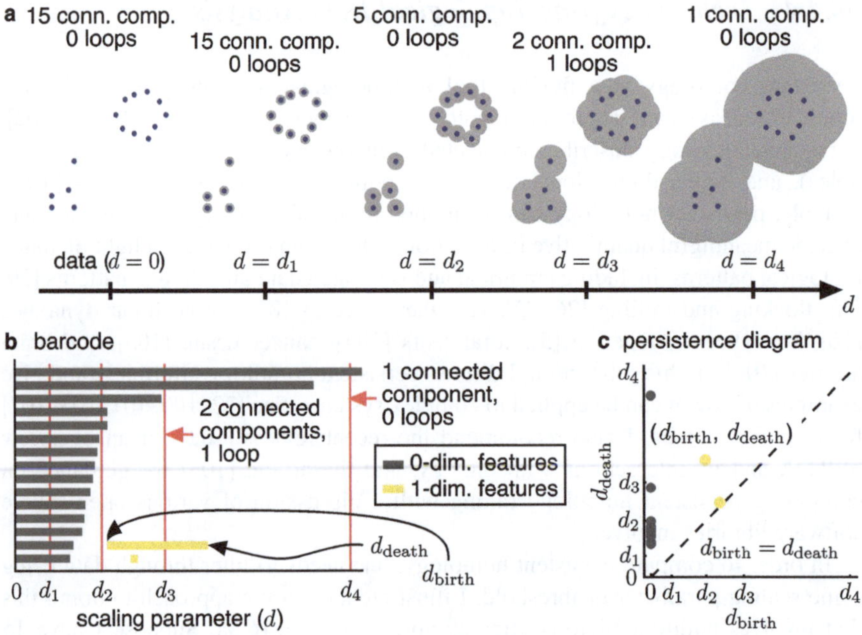

Fig. 16.5 Introduction to persistent homology, a widely used tool in topological data analysis [72, 100–103]. (**a**) To understand the shape of the data in the left-most image, one can place balls around each point and increase the ball diameter d. The birth time of a topological feature is the value of the filtration parameter (d here) at which the feature appears, and the death time is the first filtration-parameter value at which the feature no longer appears. (**b**) Barcodes and (**c**) persistence diagrams summarize how the number of connected components (dimension-0 topological features) and loops (dimension-1 topological features) vary as the scaling parameter increases

landscapes can be useful in this setting [109, 110]. Briefly, a persistence landscape is a decreasing sequence of functions that is defined using the information in persistence diagrams, and persistence landscapes lend themselves to averaging results across patterns and work well with machine learning [109, 110]. Persistence landscapes are not the only means of making the output of persistent-homology computations appropriate for machine learning, and I suggest the survey [111] for more information.

Combining persistent homology with machine learning is a very active area, and references [26, 73, 81, 110, 112, 113] are some examples of this process across various biological systems. For instance, Bonilla et al. [73] applied persistent homology to simulated and real data on collective cell motion for moving interfaces, and they used unsupervised learning to cluster different types of patterns. Working with simulated and empirical patterns that exhibit branching and clustering, Bhaskar et al. [112] partitioned their data by applying hierarchical clustering based on the pairwise Wasserstein distances between persistence diagrams. As another example,

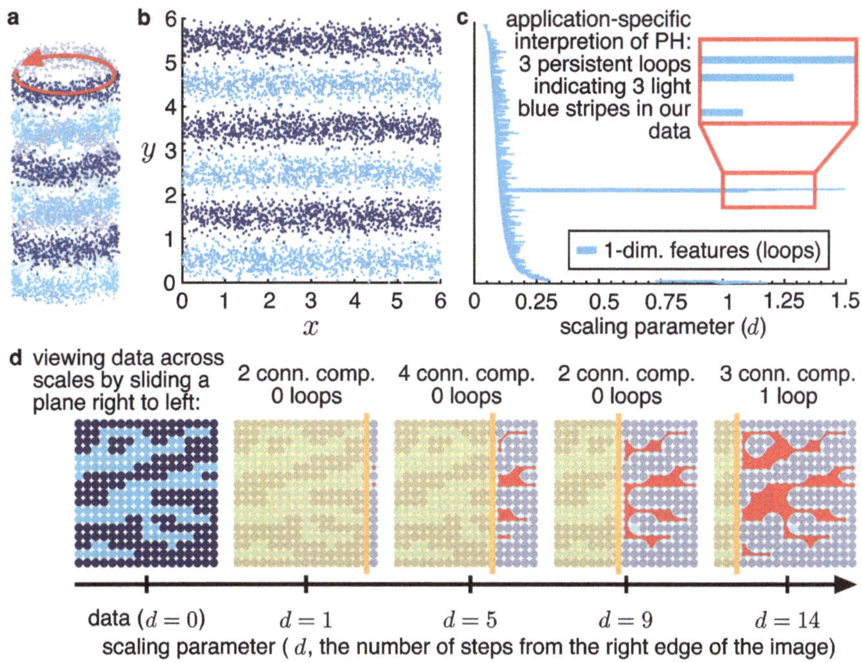

Fig. 16.6 Examples of how persistent homology can be used to quantify stripe and branching patterns. (**a**) After computing persistent homology, the next step is interpreting barcodes and persistence diagrams. As an example, McGuirl et al. [68] showed how to apply persistent homology to stripe-pattern data with periodic boundary conditions in one direction. In this case, the stripes wrap around a cylinder. (**b**) I unwrap the cylinder, indicating three light-blue and three navy stripes, and (**c**) show the associated barcode for dimension-1 topological features, computed using the positions of light-blue points. The three long bars—i.e., three highly persistent features—correspond to three light-blue stripes [39, 68]. (**d**) There are many other choices of filtration in persistent homology, and this choice is application-dependent [102]. In (**c**) and the example in Fig. 16.5, for example, I compute persistent homology by placing balls around each point in the point cloud and increasing the ball size. In the work [81, 108] of Nardini et al. on branching patterns, a sweeping-plane filtration is more natural. This can be thought of as slowly uncovering an image by sweeping a plane across it and counting the number of connected components and holes that emerge during this process

Nardini et al. [108] used persistent homology together with machine learning to characterize and classify segmented images of retinal vasculature.

Just like specific order parameters (see Sect. 16.4.1) and pair-correlation functions (see Sect. 16.4.3) are more or less appropriate for various patterns, the method for filtering through data in Fig. 16.5a is not the most meaningful approach to persistent homology for all patterns. In fact, there are many ways of filtering through data across scales [102]. As an example that is useful for quantifying branching, I highlight the sweeping-plane filtration in Fig. 16.6d [81]. Nardini et al. [81] applied this filtration, among others, to branching patterns generated by a model [114] of angiogenesis. In this case, rather than varying the size of a ball around points, one sweeps through an image, uncovering more of the picture and tracking the topological features that emerge during this process. Persistent homology can also be applied to grayscale images, even without thresholding as in Fig. 16.2a, by filtering through pixels by intensity [102]. Finally, as an example in the setting of temporal self-organization, the weight rank clique filtration provides a means of filtering on edge weights to understand the emergence of synchronization in networks of coupled oscillators [115].

16.5 Discussion and Outlook

As a whole, I see the value in quantifying self-organization as opening the door to a wider perspective, in terms of both biology and mathematics. This encompasses a broader view on details in individual patterns, on variability across patterns, and on alternative models of self-organization capable of producing the same patterns. To conclude, I thus discuss some directions and challenges related to quantifying patterns that I consider particularly exciting for current and future research.

Moving beyond qualitative evaluation of patterns allows us to identify new levels of meaningful detail in biological systems and build models that capture this detail. For example, if our system of interest is the middle picture of zebrafish-skin patterns in Fig. 16.1b [17], our goal may be to adjust parameters in a model of self-organization to achieve "spots." Alternatively, a finer-scale goal would be to use our model to produce "spots roughly aligned in stripes, with some distribution of stray cells outside of spots" [18, 68]. More broadly, when can considering noise and messiness in biological patterns tell us something about the underlying self-organization process? Being able to quantitatively capture the meaningful "messiness" and "imperfections," as well as the dominant pattern features (i.e., spots in Fig. 16.1b, middle panel), makes it possible to begin addressing this question.

However, there are several challenges associated with working with messy patterns. Often the first step in a quantification pipeline is cleaning the data [39], as some methods (i.e., persistent homology) are sensitive to outliers. On top of this, empirical images can contain imperfections that are not meaningful, such as dust or flecks due to lighting conditions. The presence of these types of imperfections can make it very challenging to extract point-cloud data from empirical images. Working with messy patterns thus requires choices about what features should be ignored and which are meaningful, and this introduces additional hyper-parameters into the quantification pipeline. One means of (partly) addressing this issue is to combine multiple approaches: for example, if working with images of empirical patterns, it is helpful to apply several methods for extracting point-cloud data before quantification. This might involve manual labeling by multiple independent scientists, batch-processing many images in the software Ilastik [47] under different labeled training images, or comparing various image-processing techniques.

More broadly, investigating the role of choices and hyper-parameters in quantifying data is an important direction for future work, as all of the approaches in Sect. 16.4 involve choices. For example, in addition to choosing what pair correlation function to apply to data, one must choose the annulus width δ_r. Binder and Simpson [88] discuss the impact of selecting a bandwidth when working with pair-correlation functions and on-lattice point clouds, for instance. In the case of persistent homology in Fig. 16.5c, interpreting barcodes often requires the choice of several hyper-parameters, as Cleveland et al. [39] note. Related to investigating hyper-parameters, many recent studies [22, 26, 44, 72, 81] have brought multiple quantitative perspectives to bear on their data. For example, Nardini et al. [81] quantified angiogenesis used two different persistent-homology filtrations, as well as direct measurements. In a series of papers [22, 44], Bull and collaborators combined a wide range of methods—including pair-correlation functions, other approaches from spatial statistics, and techniques from topological data analysis—to better understand cell organization and cancer data. And Topaz and collaborators [26, 72, 116] investigated persistent homology and order parameters for aggregation.

Studies that bring multiple quantification methods together allow researchers to take a broader perspective on their data, identify the relationship between approaches, and determine how the choice of method may impact conclusions. In my opinion, the more perspectives, the better, as this provides more transparency about choices, and one method may miss an important feature that another captures. However, bringing together many summary statistics can also create a challenge when combining pattern data with machine learning or mechanistic modeling. For example, if the goal is to estimate parameters in a model of self-organization given many quantitative summaries of some empirical patterns, how do we weight the various pieces of information and account for differences in magnitude between them? This likely has to be done in an application-specific way, and it may involve trial and error to identify which summary statistics are most meaningful for our patterns.

Alongside looking more closely at messiness and variability within an individual pattern, another way that quantitative methods can widen perspective is by allowing us to consider questions of plasticity and variability across patterns resulting from the same self-organization dynamics. For example, Fig. 16.1 provides many examples of empirical patterns in biology in comparison to model-generated patterns. However, biology is messy, and many models are stochastic. How do we use the distribution of possible patterns that a specific self-organization process can generate to give insight into the interactions at work? Looking across patterns of the same "type" (i.e., vein patterns in the type of leaf shown in Fig. 16.1h), how do we identify which features are robust and which are not? Quantitative methods open the door to such large-scale perspectives, but one challenge is that this also necessitates having many empirical replicates or sample simulations.

Lastly, being able to quantify complex patterns creates the opportunity for more fully characterizing the behavior of traditionally computational models. There are many exciting directions at the intersection of data-driven modeling and pattern quantification. For example, Johnston et al. [117] estimated parameters in an agent-based model of collective cell spreading, combining pair-correlation functions with cell counts to quantify their data. In the case of branching patterns in kidney tissue, Lambert et. al. [118] used direct measurements of area and branch points in time as their summary statistic before performing Bayesian-style inference in an agent-based model. From the topological-data-analysis perspective, Thorne et al. [119] combined persistent homology with approximate Bayesian methods to infer parameters in a model of angiogenesis. Recently, equation-learning and model-selection approaches (e.g., [32–35, 67, 120, 121]) have emerged as active research directions. In this vein, Lagergren et al. [67] used neural-network approaches to learn partial differential equations describing cell migration in scratch-assay experiments, evaluating the match via one-dimensional cell profiles. And Liu et al. [122] highlighted how choices in quantifying noisy data affect parameter inference and identifiability in the case of continuum models of cell spreading.

The large amount of empirical data available, coupled with methods for quantifying patterns that capture fine details, opens up the door to building models that make use of biological variability and tackle questions about plasticity. To achieve this, it will be important to quantify not only "fully formed" patterns at some final time point of interest, but also the (often noisier and messier) pattern-formation process. As a whole, making messy qualitative data quantitative creates the opportunity for mechanistic modelers to combine their intuition and biological expertise with the power of machine-learning and data-driven approaches. In my opinion, this is a place where the biology is calling for new mathematical methods, so that we can begin to approach richer patterns and make more detailed models tractable in the future. Building inference, equation-learning, and data-driven methods for complex modeling of messy pattern dynamics in time is an exciting direction, and quantifying qualitative data is a key first step to making this process possible.

Acknowledgments A.V. was a participant in the "Mathematics of movement: an interdisciplinary approach to mutual challenges in animal ecology and cell biology" program at the Isaac Newton Institute for Mathematical Sciences, Cambridge, during fall 2023. She gratefully acknowledges the support, hospitality, and excellent environment of the Isaac Newton Institute, as well as a travel grant from the Association for Women in Mathematics. A.V. also thanks the Mathematics of Movement organizers for organizing the Isaac Newton Institute program where work on this chapter was undertaken. This work was supported by EPSRC grant EP/R014604/1.

Competing Interests A.V. has no conflicts of interest to declare that are relevant to this chapter.

Ethics Approval No ethical approval is necessary for the subject of this chapter.

References

1. Zhang, J., Klingsch, W., Schadschneider, A., Seyfried, A.: Ordering in bidirectional pedestrian flows and its influence on the fundamental diagram. J. Stat. Mech. **2012**(02), P02002 (2012)
2. Sieben, A., Schumann, J., Seyfried, A.: Collective phenomena in crowds–Where pedestrian dynamics need social psychology. PLOS One **12**(6), e0177328 (2017)

3. Helbing, D., Molnár, P.: Social force model for pedestrian dynamics. Phys. Rev. E **51**, 4282–4286 (1995)
4. Burstedde, C., Klauck, K., Schadschneider, A., Zittartz, J.: Simulation of pedestrian dynamics using a two-dimensional cellular automaton. Physica A **295**(3), 507–525 (2001)
5. Gharooni-Fard, G., Byers, M., Deshmukh, V., Bradley, E., Mayo, C., Topaz, C.M., Peleg, O.: A computational topology-based spatiotemporal analysis technique for honeybee aggregation. NPJ Complex. **1**(3) (2024)
6. Katz, Y., Tunstrom, K., Ioannou, C.C., Huepe, C., Couzin, I.D.: Inferring the structure and dynamics of interactions in schooling fish. Proc. Natl. Acad. Sci. USA **108**(46), 18720–18725 (2011)
7. Vicsek, T., Zafeiris, A.: Collective motion. Phys. Rep. **517**(3), 71–140 (2012)
8. Sarfati, R., Peleg, O.: Chimera states among synchronous fireflies. Sci. Adv. **8**, eadd6690 (2022)
9. Buhl, J., Sumpter, D., J.T., Couzin, I.D., Hale, J.J., Despland, E., Miller, E.R., Simpson, S.J.: From disorder to order in marching locusts. Science **312**(5778), 1402–1406 (2006)
10. Buttenschön, A., Edelstein-Keshet, L.: Bridging from single to collective cell migration: A review of models and links to experiments. PLoS Comput. Biol. **16**(12), e1008411 (2020)
11. Giniūnaitė, R., Baker, R.E., Kulesa, P.M., Maini, P.K.: Modelling collective cell migration: neural crest as a model paradigm. J. Math. Biol. **80**, 481–504 (2020)
12. Nakamasu, A., Takahashi, G., Kanbe, A., Kondo, S.: Interactions between zebrafish pigment cells responsible for the generation of Turing patterns. Proc. Natl. Acad. Sci. USA **106**(21), 8429–8434 (2009)
13. Volkening, A.: Linking genotype, cell behavior, and phenotype: multidisciplinary perspectives with a basis in zebrafish patterns. Curr. Opin. Genet. Dev. **63**, 78–85 (2020)
14. Salinas, E., Sejnowski, T.: Correlated neuronal activity and the flow of neural information. Nat. Rev. Neurosci. **2**, 539–550 (2001)
15. Sepúlveda, N., Petitjean, L., Cochet, O., Grasland-Mongrain, E., Silberzan, P., Hakim, V.: Collective cell motion in an epithelial sheet can be quantitatively described by a stochastic interacting particle model. PLoS Comp. Biol. **9**(3), e1002944 (2013)
16. Ciocanel, M.V., Juenemann, R., Dawes, A.T., McKinley, S.A.: Topological data analysis approaches to uncovering the timing of ring structure onset in filamentous networks. Bull. Math. Biol. **83**(21) (2021)
17. Frohnhöfer, H.G., Krauss, J., Maischein, H.M., Nüsslein-Volhard, C.: Iridophores and their interactions with other chromatophores are required for stripe formation in zebrafish. Development **140**(14), 2997–3007 (2013)
18. Volkening, A., Sandstede, B.: Iridophores as a source of robustness in zebrafish stripes and variability in *Danio* patterns. Nat. Commun. **9**(3231) (2018)
19. Manukyan, L., Montandon, S.A., Fofonjka, A., Smirnov, S., Milinkovitch, M.C.: A living mesoscopic cellular automaton made of skin scales. Nature **544**(7649), 173 (2017)
20. Liu, Q.X., Doelman, A., Rottschafer, V., de Jager, M., Herman, P.M.J., Rietkerk, M., van de Koppel, J.: Phase separation explains a new class of self-organized spatial patterns in ecological systems. Proc. Natl. Acad. Sci. USA **110**(29), 11905–11910 (2013)
21. Merks, R.M., Brodsky, S.V., Goligorksy, M.S., Newman, S.A., Glazier, J.A.: Cell elongation is key to in silico replication of in vitro vasculogenesis and subsequent remodeling. Dev. Biol. **289**(1), 44–54 (2006)
22. Bull, J.A., Macklin, P.S., Quaiser, T., Braun, F., Waters, S.L., Pugh, C.W., Byrne, H.M.: Combining multiple spatial statistics enhances the description of immune cell localisation within tumours. Sci. Rep. **10**(1), 18624 (2020)
23. Iwamasa, K., Noshita, K.: Network feature-based phenotyping of leaf venation robustly reconstructs the latent space. PLoS Comp. Biol. **19**(7), e1010581 (2023)
24. Wang, J., Wu, J., Li, J., Kong, R., Li, X., Wang, X.: Simulation of various biofilm fractal morphologies by agent-based model. Colloids Surf. B. Biointerfaces **227**, 113352 (2023)
25. Liang, Y., Gerwin, J., Meyer, A., Kratochwil, C.F.: Developmental and Cellular Basis of Vertical Bar Color Patterns in the East African Cichlid Fish *Haplochromis latifasciatus*. Front. Cell Dev. Biol. **8**, 62–62 (2020)

26. Bhaskar, D., Manhart, A., Milzman, J., Nardini, J.T., Storey, K.M., Topaz, C.M., Ziegelmeier, L.: Analyzing collective motion with machine learning and topology. Chaos **29**(12), 123125 (2019)
27. D'Orsogna, M.R., Chuang, Y.L., Bertozzi, A.L., Chayes, L.S.: Self-propelled particles with soft-core interactions: patterns, stability, and collapse. Phys. Rev. Lett. **96**(10), 104302 (2006)
28. Couzin, I.D., Franks, N.: Self-organized lane formation and optimized traffic flow in army ants. Proc. R. Soc. B **270**(1511), 139–146 (2003)
29. Fofonjka, A., Milinkovitch, M.C.: Reaction-diffusion in a growing 3d domain of skin scales generates a discrete cellular automaton. Nat. Commun. **12**(1), 2433–2433 (2021)
30. Toni, T., Welch, D., Strelkowa, N., Ipsen, A., Stumpf, M.P.: Approximate Bayesian computation scheme for parameter inference and model selection in dynamical systems. J. R. Soc. Interface **6**(31), 187–202 (2009)
31. Bode, N.W.F., Ronchi, E.: Statistical model fitting and model selection in pedestrian dynamics research. Collective Dynamics **4**, 1–32 (2019)
32. Mangan, N.M., Kutz, J.N., Brunton, S.L., Proctor, J.L.: Model selection for dynamical systems via sparse regression and information criteria. Proc. R. Soc. A **473**, 20170009 (2017)
33. Brunton, S.L., Proctor, J.L., Kutz, J.N.: Discovering governing equations from data by sparse identification of nonlinear dynamical systems. Proc. Natl. Acad. Sci. USA **113**(15), 3932–3937 (2016)
34. Nardini, J.T., Baker, R.E., Simpson, M.J., Flores, K.B.: Learning differential equation models from stochastic agent-based model simulations. J. R. Soc. Interface **18**(176) (2021)
35. Kemeth, F.P., Bertalan, T., Thiem, T., Dietrich, F., Moon, S.J., Laing, C.R., Kevrekidis, I.G.: Learning emergent partial differential equations in a learned emergent space. Nat. Commun. **13**(1), 3318 (2022)
36. Pargett, M., Umulis, D.M.: Quantitative model analysis with diverse biological data: applications in developmental pattern formation. Methods **62**(1), 56–67 (2013)
37. Plank, M.J., Simpson, M.J.: Models of collective cell behaviour with crowding effects: comparing lattice-based and lattice-free approaches. J. R. Soc. Interface **9**, 2983–2996 (2012)
38. Osborne, J.M., Fletcher, A.G., Pitt-Francis, J.M., Maini, P.K., Gavaghan, D.J.: Comparing individual-based approaches to modelling the self-organization of multicellular tissues. PLoS Comput. Biol. **13**(2), 1–34 (2017)
39. Cleveland, E., Zhu, A., Sandstede, B., Volkening, A.: Quantifying different modeling frameworks using topological data analysis: a case study with zebrafish patterns. SIAM J. Appl. Dyn. Sys. **22**(4), 2649–3430 (2023)
40. Kursawe, J., Baker, R.E., Fletcher, A.G.: Impact of implementation choices on quantitative predictions of cell-based computational models. J. Comp. Phys. **345**, 752–767 (2017)
41. Camazine, S., Deneubourg, J.-L., Franks, N.R., Sneyd, J., Theraulaz, G., Bonabeau, E.: Self-organization in biological systems. In: Princeton Studies in Complexity. Princeton University Press, Princeton (2003)
42. Feng, Y., Yang, T., Zhu, J., Li, M., Doyle, M., Ozcoban, V., Bass, G.T., Pizzolla, A., Cain, L., Weng, S., Pasam, A., Kocovski, N., Huang, Y.K., Keam, S.P., Speed, T.P., Neeson, P.J., Pearson, R.B., Sandhu, S., Goode, D.L., Trigos, A.S.: Spatial analysis with SPIAT and spaSim to characterize and simulate tissue microenvironments. Nat. Commun. **14**(1), 2697 (2023)
43. Law, R., llian, J., Burslem, D.F.R.P., Gratzer, G., Gunatilleke, C.V.S., Gunatilleke, I.A.U.N.: Ecological information from spatial patterns of plants: insights from point process theory. J. Ecol. **97**(4), 616–628 (2009)
44. Bull, J.A., Mulholland, E.J., Moore, J.W., Bosque, J.J., Stolz, B.J., Boen, J., Eggington, H.R., Belnoue-Davis, H.L., Jones, H., Gatenbee, C.D., Anderson, A.R.A., Easton, A., Todd, P., Cunningham, C., Taylor, S., Byrne, H.M., Leedham, S.: Integrating diverse statistical methods to analyse stage-discriminatory cell interactions in colorectal neoplasia. bioRxiv (2024). https://doi.org/10.1101/2024.06.02.597010
45. Bookstein, F.L.: Morphometric Tools for Landmark Data: Geometry and Biology. Cambridge University Press, Cambridge (1997)

46. Summers, H.D., Wills, J.W., Rees, P.: Spatial statistics is a comprehensive tool for quantifying cell neighbor relationships and biological processes via tissue image analysis. Cell Rep. Meth. **2**(11), 100348 (2022)
47. Berg, S., Kutra, D., Kroeger, T., Straehle, C.N., Kausler, B.X., Haubold, C., Schiegg, M., Ales, J., Beier, T., Rudy, M., Eren, K., Cervantes, J.I., Xu, B., Beuttenmueller, F., Wolny, A., Zhang, C., Koethe, U., Hamprecht, F.A., Kreshuk, A.: ilastik: interactive machine learning for (bio)image analysis. Nat. Methods **16**(12), 1226–1232 (2019)
48. Kozák, M., Gaffney, E.A., Klika, V.: Pattern formation in reaction-diffusion systems with piecewise kinetic modulation: an example study of heterogeneous kinetics. Phys. Rev. E **100**(4), 042220 (2019)
49. Turing, A.M.: The chemical basis of morphogenesis. Philos. Trans. R. Soc. Lond. B Biol. Sci. **237**(641), 37–72 (1952)
50. Gierer, A., Meinhardt, H.: A theory of biological pattern formation. Kybernetik **12**(1), 30–39 (1972)
51. Menshykau, D., Michos, O., Lang, C., Conrad, L., McMahon, A.P., Iber, D.: Image-based modeling of kidney branching morphogenesis reveals GDNF-RET based Turing-type mechanism and pattern-modulating WNT11 feedback. Nat. Commun. **10**(239) (2019)
52. Cahn, J.W., Hilliard, J.E.: Free energy of a nonuniform system. I. interfacial free energy. Chem. Phys. **28**(2), 258–267 (1958)
53. Carrillo, J.A., Huang, Y., Schmidtchen, M.: Zoology of a nonlocal cross-diffusion model for two species. SIAM J. Appl. Math. **78**(2), 1078–1104 (2018)
54. Painter, K.J., Bloomfield, J.M., Sherratt, J.A., Gerisch, A.: A nonlocal model for contact attraction and repulsion in heterogeneous cell populations. Bull. Math. Biol. **77**(6), 1132–1165 (2015)
55. Bernoff, A.J., Topaz, C.M.: A primer of swarm equilibria. SIAM J. Appl. Dyn. Sys. **10**(1), 212–250 (2011)
56. Mogilner, A., Edelstein-Keshet, L.: A non-local model for a swarm. J. Math. Biol. **38**, 534–570 (1999)
57. Volkening, A., Sandstede, B.: Modelling stripe formation in zebrafish: an agent-based approach. J. R. Soc. Interface **12**(112), 20150812 (2015)
58. Bullara, D., De Decker, Y.: Pigment cell movement is not required for generation of Turing patterns in zebrafish skin. Nat. Commun. **6**, 6971 (2015)
59. Owen, J.P., Kelsh, R.N., Yates, C.A.: A quantitative modelling approach to zebrafish pigment pattern formation. eLife **9**, e52998 (2020)
60. Glazier, J.A., Graner, F.: Simulation of the differential adhesion driven rearrangement of biological cells. Phys. Rev. E **47**(3), 2128 (1993)
61. Hirashima, T., Rens, E.G., Merks, R.M.H.: Cellular Potts modeling of complex multicellular behaviors in tissue morphogenesis. Dev. Growth. Differ. **59**(5), 329–339 (2017)
62. Zabaleta-Ortega, A., Masoller, C., Guzmán-Vargas, L.: Topological data analysis of the synchronization of a network of Rössler chaotic electronic oscillators. Chaos **33**, 113110 (2023)
63. Kuramoto, Y.: Chemical oscillations, waves, and turbulence. In: Springer Series in Synergetics. Springer, Berlin (1984)
64. Strogatz, S.H.: From Kuramoto to Crawford: exploring the onset of synchronization in populations of coupled oscillators. Physica D **143**, 1–20 (2000)
65. Gandhi, P., Ciocanel, M.V., Niklas, K., Dawes, A.T.: Identification of approximate symmetries in biological development. Philos. Trans. R. Soc. A **379**(2213), 20200273 (2021)
66. Bull, J.A., Mulholland, E.J., Leedham, S.J., Byrne, H.M.: Extended correlation functions for spatial analysis of multiplex imaging data. Biol. Imaging **4**, e2 (2024)
67. Lagergren, J.H., Nardini, J.T., Baker, R.E., Simpson, M.J., Flores, K.B.: Biologically-informed neural networks guide mechanistic modeling from sparse experimental data. PLoS Comp. Biol. **16**(12), e1008462 (2020)
68. McGuirl, M.R., Volkening, A., Sandstede, B.: Topological data analysis of zebrafish patterns. Proc. Natl. Acad. Sci. USA **117**(10), 5113–5124 (2020)

69. Rolland-Lagan, A.G., Amin, M., Pakulska, M.: Quantifying leaf venation patterns: two-dimensional maps. Plant J. **57**(1), 195–205 (2009)
70. Binder, B.J., Sundstrom, J.F., Gardner, J.M., Jiranek, V., Oliver, S.G.: Quantifying two-dimensional filamentous and invasive growth spatial patterns in yeast colonies. PLoS Comp. Biol. **11**(2), e1004070 (2015)
71. Alba, V., Carthew, J.E., Carthew, R.W., Mani, M.: Global constraints within the developmental program of the *Drosophila* wing. eLife **10**, e66750 (2021)
72. Topaz, C.M., Ziegelmeier, L., Halverson, T.: Topological data analysis of biological aggregation models. PLoS One **10**(5), e0126383 (2015)
73. Bonilla, L.L., Carpio, A., Trenado, C.: Tracking collective cell motion by topological data analysis. PLoS Comp. Biol. **16**(12), e1008407 (2020)
74. Couzin, I.D., Krause, J., James, R., Ruxton, G.D., Franks, N.R.: Collective memory and spatial sorting in animal groups. J. Theor. Biol. **218**(1), 1–11 (2002)
75. Clarke, J.B., Hastie, J.W., Kihlborg, L.H.E., Metselaar, R., Thackeray, M.M.: Definitions of terms relating to phase transitions of the solid state (IUPAC recommendations 1994). Pure Appl. Chem. **66**(3), 577–594 (1994)
76. Vicsek, T., Czirók, A., Ben-Jacob, E., Cohen, I.I., Shochet, O.: Novel type of phase transition in a system of self-driven particles. Phys. Rev. Lett. **75**(6), 1226–1229 (1995)
77. Chuang, Y.I., D'Orsogna, M.R., Marthaler, D., Bertozzi, A.L., Chayes, L.S.: State transitions and the continuum limit for a 2D interacting, self-propelled particle system. Physica D **232**, 33–47 (2007)
78. Huepe, C., Aldana, M.: New tools for characterizing swarming systems: a comparison of minimal models. Physica A **387**(12), 2809–2822 (2008)
79. Edelsbrunner, H.: Alpha shapes—a survey. In: Tessellations in the Sciences (2010)
80. Byrne, H.M., Harrington, H.A., Muschel, R., Reinert, G., Stolz, B.J., Tillmann, U.: Topological methods for characterising spatial networks: A case study in tumour vasculature arXiv (2019). https://doi.org/10.48550/arXiv.1907.08711
81. Nardini, J.T., Stolz, B.J., Flores, K.B., Harrington, H.A., Byrne, H.M.: Topological data analysis distinguishes parameter regimes in the Anderson-Chaplain model of angiogenesis. PLoS Comput. Biol. **17**, e1009094 (2021)
82. Schneider, C.A., Rasband, W.S., Eliceiri, K.W.: NIH Image to ImageJ: 25 years of image analysis. Nat. Methods **9**(7), 671–675 (2012)
83. Schindelin, J., Arganda-Carreras, I., Frise, E., Kaynig, V., Longair, M., Pietzsch, T., Preibisch, S., Rueden, C., Saalfeld, S., Schmid, B., Tinevez, J.Y., White, D.J., Hartenstein, V., Eliceiri, K., Tomancak, P., Cardona, A.: Fiji: an open-source platform for biological-image analysis. Nat. Methods **9**(7), 676–682 (2012)
84. Miyazawa, S., Okamoto, M., and Kondo, S.: Blending of animal colour patterns by hybridization. Nat. Commun. **1**, 66 (2010)
85. Djurdjevič, I., Furmanek, T., Miyazawa, S., Bajec, S.S.: Comparative transcriptome analysis of trout skin pigment cells. BMC Genom. **20**(1), 359 (2019)
86. Miyazawa, S.: Pattern blending enriches the diversity of animal colorations. Sci. Adv. **6**(49), eabb9107 (2020)
87. Binder, B.J., Simpson, M.J.: Quantifying spatial structure in experimental observations and agent-based simulations using pair-correlation functions. Phys. Rev. E **88**(2), 022705 (2013)
88. Binder, B.J., Simpson, M.J.: Spectral analysis of pair-correlation bandwidth: application to cell biology images. R. Soc. Open Sci. **2**, 140494 (2015)
89. Møller, J., Waagepetersen, R.P.: Statistical inference and simulation for spatial point processes. In: Monographs on Statistics and Applied Probability; 100. Chapman & Hall/CRC, Boca Raton (2004)
90. Dini, S., Binder, B.J., Fischer, S.C., Mattheyer, C., Schmitz, A., Stelzer, E., H.K., Bean, N.G., Green, J.E.F.: Identifying the necrotic zone boundary in tumour spheroids with pair-correlation functions. J. R. Soc. Interface **13**(123), 20160649 (2016)
91. Treloar, K.K., Simpson, M.J., Binder, B.J., McElwain, D.L.S., Baker, R.E.: Assessing the role of spatial correlations during collective cell spreading. Sci. Rep. **4**(1), 5713 (2014)

92. Gavagnin, E., Owen, J.P., Yates, C.A.: Pair correlation functions for identifying spatial correlation in discrete domains. Phys. Rev. E **97**(6-1), 062104 (2018)
93. Corner, J., Lennard-Jones, D.L.S.: VI. The neighbour distribution function in monatomic liquids and dense gases. Proc. R. Soc. A **178**(975), 401–414 (1941)
94. Agnew, D.L.S., Green, J.E.F., Brown, T.M., Simpson, M.J., Binder, B.J.: Distinguishing between mechanisms of cell aggregation using pair-correlation functions. J. Theor. Biol. **352**, 16–23 (2014)
95. Ripley, B.D.: Modelling spatial patterns. J. R. Stat. Soc. B **39**(2), 172–192 (1977)
96. Baddeley, A., Rubak, E., Turner, R.: Spatial Point Patterns: Methodology and Applications with R. Champan & Hall/CRC Interdisciplinary Statistics Series. CRC Press, Boca Raton (2015)
97. Bull, J.A., Byrne, H.M.: Quantification of spatial and phenotypic heterogeneity in an agent-based model of tumour-macrophage interactions. PLoS Comp. Biol. **19**(3), e1010994 (2023)
98. Dini, S., Binder, B.J., Green, J.E.F.: Understanding interactions between populations: individual based modelling and quantification using pair correlation functions. J. Theor. Biol. **439**, 50–64 (2018)
99. Johnston, S.T., Crampin, E.J.: Corrected pair correlation functions for environments with obstacles. Phys. Rev. E **99**, 032124 (2019)
100. Carlsson, G.: Topology and data. Bull. Am. Math. Soc. **46**(2), 255–308 (2009)
101. Edelsbrunner, H., Harer, J.: Persistent homology—a survey. Contemp. Math. **453**, 257–282 (2008)
102. Amézquita, E.J., Quigley, M.Y., Ophelders, T., Munch, E., Chitwood, D.H.: The shape of things to come: topological Data Analysis and biology, from molecules to organisms. Dev. Dyn. **249**(7), 816–833 (2020)
103. Otter, N., Porter, M.A., Tillmann, U., Grindrod, P., Harrington, H.A.: A roadmap for the computation of persistent homology. EPJ Data Sci. **6**(1), 1–38 (2017)
104. McDonald, R.A., Neuhausler, R., Robinson, M., Larsen, L.G., Harrington, H.A., Bruna, M.: Zigzag persistence for coral reef resilience using a stochastic spatial model. J. R. Soc. Interface **20**(205), 20230280 (2023)
105. Lawson, P., Sholl, A.B., Brown, J.Q., Fasy, B.T., Wenk, C.: Persistent homology for the quantitative evaluation of architectural features in prostate cancer histology. Sci. Rep. **9**(1), 1139 (2019)
106. Bendich, P., Marron, J.S., Miller, E., Pieloch, A., Skwerer, S.: Persistent homology analysis of brain artery trees. Ann. Appl. Stat. **10**(1), 198–218 (2016)
107. Ghrist, R.: Elementary Applied Topology, 1.0 edn. Createspace, California (2014)
108. Nardini, J.T., Pugh, C.W.J., Byrne, H.M.: Statistical and topological summaries aid disease detection for segmented retinal vascular images. Microcirculation **30**(4), e12799 (2023)
109. Bubenik, P.: Statistical topological data analysis using persistence landscapes. J. Mach. Learn. Res. **16**(3), 77–102 (2015)
110. Hartsock, I., Park, E., Toppen, J., Bubenik, P., Dimitrova, E.S., Kemp, M.L., Cruz, D.A.: Topological data analysis of pattern formation of human induced pluripotent stem cell colonies. Sci. Rep. **15**(1), 11544–18 (2025)
111. Ali, D., Asaad, A., Jimenez, M., Nanda, V., Paluzo-Hidalgo, E., Soriano-Trigueros, M.: A survey of vectorization methods in topological data analysis. IEEE Trans. Pattern Anal. Mach. Intell. **45**(12), 14069–14080 (2023)
112. Bhaskar, D., Zhang, W.Y., Wong, I.Y.: Topological data analysis of collective and individual epithelial cells using persistent homology of loops. Soft Matter **17**(17), 4653–4664 (2021)
113. Bhaskar, D., Zhang, W.Y., Volkening, A., Sandstede, B., Wong, I.Y.: Topological data analysis of spatial patterning in heterogeneous cell populations: I. Clustering and sorting with varying cell-cell adhesion. NPJ Sys. Biol. Appl. **9**, 43 (2023)
114. Anderson, A.R.A., Chaplain, M.A.J.: Continuous and discrete mathematical models of tumor-induced angiogenesis. Bull. Math. Biol. **60**(5), 857–899 (1998)
115. Stolz, B.J., Harrington, H.A., Porter, M.A.: Persistent homology of time-dependent functional networks constructed from coupled time series. Chaos **27**(4), 047410 (2017)

116. Ulmer, M., Ziegelmeier, L., Topaz, C.M.: A topological approach to selecting models of biological experiments. PLoS One **14**(3), e0213679 (2019)
117. Johnston, S.T., Simpson, M.J., McElwain, D.L.S., Binder, B.J., Ross, J.V.: Interpreting scratch assays using pair density dynamics and approximate Bayesian computation. Open Biol. **4**, 140097 (2014)
118. Lambert, B., MacLean, A.L., Fletcher, A.G., Combes, A.N., Little, M.H., Byrne, H.M.: Bayesian inference of agent-based models: a tool for studying kidney branching morphogenesis. J. Math. Biol **76**(7), 1673–1697 (2018)
119. Thorne, T., Kirk, P.D.W., Harrington, H.A.: Topological approximate Bayesian computation for parameter inference of an angiogenesis model. Bioinformatics **38**(9), 2529–2535 (2022)
120. González-García, R., Rico-Martínez, R., Kevrekidis, I.G.: Identification of distributed parameter systems: A neural net based approach. Comput. Chem. Eng. **22**, S965–S968 (1998)
121. Raissi, M., Perdikaris, P., Karniadakis, G.E.: Physics-informed neural networks: a deep learning framework for solving forward and inverse problems involving nonlinear partial differential equations. J. Comput. Phys. **378**, 686–707 (2019)
122. Liu, Y., Suh, K., Maini, P.K., Cohen, D.J., Baker, R.E.: Parameter identifiability and model selection for partial differential equation models of cell invasion. J. R. Soc. Interface **21**(212), 20230607 (2024)